PHYSICS OF MAGMATIC PROCESSES

PHYSICS OF
MAGMATIC PROCESSES

R. B. HARGRAVES, EDITOR

1980

PRINCETON UNIVERSITY PRESS

PRINCETON, NEW JERSEY

CONTENTS

PREFACE

In 1928 N. L. Bowen's classic volume, *The Evolution of the Igneous Rocks*, was published. The book was based on a series of lectures that Bowen, of the Carnegie Geophysical Laboratory, delivered at Princeton in 1927. To commemorate its appearance, in 1977 the Department of Geological and Geophysical Sciences at Princeton, together with the Carnegie Geophysical Laboratory, arranged a conference on "The Physics of Magmatic Processes." The conference brought together a distinguished group of scientists whose current research embraces various aspects of both physical and chemical magmatic processes. Whereas the chemical side of igneous petrology has developed prodigiously since the publication of Bowen's book—in fact, has dominated research for many years—the nature of silicate melts and the physical processes associated with generation, transport, and crystallization of magma have been somewhat neglected. However, current research into these more physical aspects is already leading to major advances in understanding of igneous processes, a development clearly anticipated by Bowen more than fifty years ago. It is hoped that the proceedings of the conference, compiled in this volume, will provide a foundation and stimulus for future research.

Financial support was received from the National Science Foundation (Grant EAR 7724116) and endowed funds from the Department of Geological and Geophysical Sciences, Princeton University. The assistance of the Princeton University Conference Office under the direction of William H. O'Brien, Jr., in organizing the Conference from which this volume resulted, is much appreciated.

The editor would like to acknowledge with thanks the secretarial assistance of J. Olsen, and the exceptional contribution to this volume by J. M. Ajdukiewicz, of Princeton University Press.

PHYSICS OF MAGMATIC PROCESSES

Chapter 1

POLYMERIZATION MODEL FOR SILICATE MELTS

PAUL C. HESS

Department of Geological Sciences, Brown University

INTRODUCTION

Attempts to quantify the physical and thermodynamic properties of silicate melts that influence such important igneous processes as diffusion, nucleation, crystal growth, major and trace-element crystal-liquid partitioning, or the formation of immiscible liquids, require at least an elementary understanding of the structure of these melts. It is clear, however, that the word structure has different implications for silicate minerals on the one hand and for melts on the other. Silicate minerals, in common with other crystalline solids, are characterized by systematic arrangement of atoms in a three-dimensional array. The structures of the most common silicate minerals reveal that the Si^{+4} ion occurs in tetrahedral coordination with oxygen over a wide range of temperatures and pressures. Each oxygen surrounding a central silicon ion has the potential of bonding to another silicate ion, thus forming chain, ring, sheet, or framework silicate structures. The linking by bridging oxygens of silicate tetrahedra is often referred to as polymerization; differing degrees of polymerization give rise to the great diversity of silicate structures. In general, a silicate mineral is monodisperse, in that it typically contains only one type of silicate group. For example, olivines contain only isolated SiO_4 tetrahedra and pyroxenes only SiO_3 chains. There are some exceptions, notably the fascinating

© 1980 by Princeton University Press
Physics of Magmatic Processes
0-691-08259-6/80/0003-46$02.30/0 (cloth)
0-691-08261-8/80/0003-46$02.30/0 (paper)
For copying information, see copyright page

biopyribole minerals that contain elements of both sheet and chain structures (Veblen et al., 1977).

In contrast, there are no fixed lattice sites in a silicate melt, and the positions of the ionic groups change instantaneously and continuously in response to the random thermal motions of the atoms. However, the arrangement of atoms in the melt is not totally random. Radial distribution functions of silicate glasses obtained by the techniques of X-ray diffraction prove that the SiO_4 tetrahedra persist in the melt (Warren and Biscoe, 1938; Mozzi and Warren, 1969). Interatomic distances out to 8 Å correspond closely in length to the Si–O, Si–Si and O–O distances around a given silicon ion in silicate minerals. Recent X-ray and neutron-diffraction studies of SiO_2 glass indicate that crystal-like "long-range" ordering may persist even to a distance of 20 Å (Konnert and Karle, 1972).

These data demonstrate that the local structure around a central silicon ion is similar to that obtained in the crystalline state. However, any conclusions pertaining to the structure of the silicate melt on a larger scale remain speculative. The successful application of polymer theory to silicate melts of low SiO_2 composition (Masson et al., 1970; Hess, 1971; among others) and the direct determination of the mass distributions of silicate ionic species by various chromatographic techniques (Baltã et al., 1976) reveal that silicate melts are polydisperse solutions containing a wide distribution of silicate ionic groups of various sizes and shapes. The structure of a given silicate anion is not necessarily fixed but may fluctuate between several related forms. However, the average structure of a melt at a given composition, pressure and temperature is a constant and reproducible function. The melt is therefore in a state of homogeneous equilibrium. As the silica content of a metal-rich liquid increases, the SiO_4 units polymerize and progressively increase in size and complexity. At some composition, infinitely branched chains develop that eventually evolve into the continuous framework structure of pure silica.

Although this model of the silicate melt is widely accepted and is supported by measurements of such physical properties as viscosity, density, expansivity and ionic conductivity (Bockris et al., 1948, 1954, 1955, 1956), the polymerization path by which a silica-poor melt evolves into a silica-rich melt is not understood. Methods of classical polymer theory (Flory, 1953) can be adapted to calculate the size distributions of polymeric ions in a silicate melt (Hess, 1971). However, these methods have not been applied successfully for melts of SiO_2-rich compositions, where silicate anionic polymers of high complexity and infinite size exist. A more general and less specific description of the polymerization reaction is required. This approach is considered in the following section.

POLYMERIZATION MODEL

The formation of a silicate melt is envisaged as a chemical reaction between liquid metal oxide and liquid silica. The reaction is described by the homogeneous equilibrium

Liquid Silica + Liquid Metal Oxide = Silicate Melt

$$Si\text{–}O\text{–}Si + M\text{–}O\text{–}M = 2Si\text{–}O\text{–}M, \tag{1}$$

where M represents a cation other than Si^{+4}. This equilibrium is typically written in a shorthand notation (Toop and Samis, 1962a,b)

$$O^0 + O^{2-} = 2O^-, \tag{2}$$

where O^0, O^{2-} and O^- are called bridging, free, and non-bridging oxygens, respectively. Although this shorthand terminology has the advantage of brevity, it is important to remember that these terms refer to certain bonding configurations and not to discrete oxygen ions in the melt. The bridging oxygen refers to the Si–O–Si complex, the free oxygen to the M–O–M complex and the non-bridging oxygen to the Si–O–M complex. In fact, Hess (1977) showed that in a binary system the chemical potentials of these oxygen species are directly related to the chemical potentials of the oxide and *not* to ionic components. Thus

$$\mu_{O^0} = \tfrac{1}{2}\mu_{SiO_2} \tag{3}$$

$$\mu_{O^{2-}} = \mu_{MO} \tag{4}$$

$$\mu_{O^-} = \tfrac{1}{4}\mu_{M_2SiO_4}. \tag{5}$$

As in all chemical reactions, the homogeneous equilibrium (1) is associated with an equilibrium constant, K, where

$$K = \frac{(O^-)^2}{(O^0)(O^{2-})}, \tag{6}$$

and the bracketed terms refer to the activities of the oxygen species. If molar concentrations are substituted for the activities, then the concentrations of all oxygen species in a binary silicate can be calculated for a given equilibrium constant, K, and composition of the system (Toop and Samis, 1962a,b). The results of three sample calculations are illustrated in Figure 1. A melt for which K = 0 is one in which there is no reaction between liquid silica and liquid metal oxide to form non-bridging oxygen. The silicate melt is a solution of "molecules" of SiO_2 and MO. A melt for which K = ∞ is one in which the reaction to form non-bridging oxygen is complete. Only free and non-bridging oxygen exist between

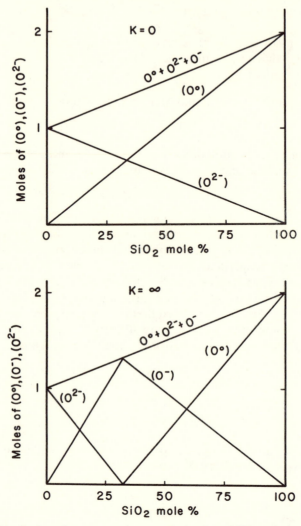

Figure 1. Distribution of oxygen species vs. mole fraction of SiO_2 calculated for $K = \infty$, $K = 17$ and $K = 0$.

0 and $33\frac{1}{3}$ mole % SiO_2 and only non-bridging and bridging oxygen exist between $33\frac{1}{3}$ and 100 mole % SiO_2. Non-bridging oxygen are the sole oxygen species at $33\frac{1}{3}$ mole % SiO_2. Melts characterized by $K = 0$ or $K = \infty$ are hypothetical and do not exist. Finally, a melt for which K is finite but non-zero is characterized by the occurrence of all three oxygen species throughout the compositional range from 0 to 100 mole % SiO_2. As the SiO_2 content of the silicate melt increases, the concen-

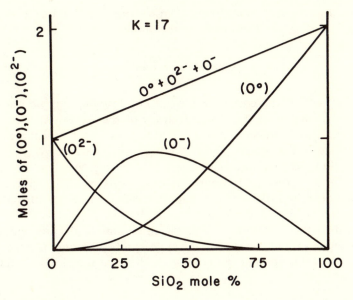

Figure 1. (*continued*)

trations of free oxygen rapidly decrease to low values, the bridging oxygen gradually increase, and the non-bridging oxygen define a maximum at $33\frac{1}{3}$ mole % SiO_2. The maximum is greater for melts characterized by larger K's. The trends reflect the polymerization reactions occurring in the melt as more and more SiO_4 units share common oxygens to form larger silicate anionic complexes. For a given composition, a melt characterized by a smaller K contains more bridging oxygen and by definition is more polymerized than a melt characterized by a larger K.

The ability to calculate concentrations of oxygen species in a given system rests with the assumption that molar concentrations may be substituted for activities in Equation (6). This is possible only if the activity coefficients of the oxygen species are unity or if, by chance, they effectively cancel. Neither case is likely. Solution non-ideality in simple silicate melts is manifested by the occurrence of stable and/or metastable silicate liquid immiscibility in most SiO_2-rich melts. For example, silicate liquid immiscibility occurs in FeO–, MgO– and CaO–SiO_2 melts containing as little as 2 mole % metal oxide at temperatures as high as 1,700°C (Greig, 1927). In this region at least, it is not correct to substitute concentrations for activities. Nevertheless, it is instructive to describe the state of polymerization of a melt by the value of its "equilibrium constant," even though the true value of K, as expressed in terms of the activities of the oxygen species, is not known. The problem is to obtain an estimate of K for a given melt.

Toop and Samis (1962a,b) have approximated the free energy of mixing of a binary silicate melt by the equation

$$\Delta \bar{G} = -\left(\frac{O^-}{2} RT \ln K\right), \tag{7}$$

where O^- equals the concentrations of non-bridging oxygen and K is expressed in terms of the concentrations of the oxygen species. With temperature and pressure fixed, $RT \ln K$ is a constant characteristic of the system. The shape of the $\Delta\bar{G}$-N_{SiO_2} curve, where N_{SiO_2} = mole fraction SiO_2, is therefore similar (but with a negative sign) to that of the O^-–N_{SiO_2} curve. Two important conclusions are deduced from this equation of state. First, the minimum in the free energy of mixing curve should correspond to the maximum in the O^-–N_{SiO_2} curve. Thus, as discussed in the previous section, the minimum in the free energy of mixing curve should occur near $33\frac{1}{3}$ mole % SiO_2. Second, for two melts of the same SiO_2 content but containing different metal oxides, the free

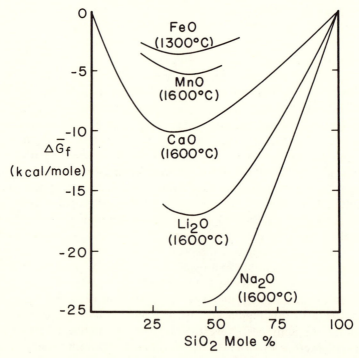

Figure 2. Free energies of mixing relative to the pure liquids. Data are from calculations by Charles (1967; Li_2O, Na_2O), Tewhey and Hess (ms prep, CaO) and calculated from activity data in Masson (1972).

energy of mixing is most negative for the melt with the greatest concentration of nonbridging oxygen. Such melts are characterized by a large value for the equilibrium constant, K.

Although this equation of state is not applicable to regions where silicate liquid immiscibility is encountered, it does mimic the basic features of the free energy of mixing curves for binary silicate melts. Some free energy of mixing curves for various binary systems are shown in Figure 2. All the curves have minima near $N_{SiO_2} = .40$, in good agreement with theory. The free-energy curves for the systems FeO–, MnO–, CaO–, Li_2O–, and Na_2O–SiO_2 are ordered from least to most negative. According to the equation of state (7), these systems should also be ordered from those characterized by small K to those characterized by large K, and from most polymerized to least polymerized melt. Thus, for a given N_{SiO_2}, a FeO–SiO_2 melt contains more bridging—and less non-bridging—oxygen than a Na_2O–SiO_2 melt. By definition, it is the more polymerized melt.

ACID-BASE THEORY

The equilibrium structure of a silicate melt is obtained at constant P, T, and composition when the distribution of oxygen species minimizes the free energy of the solution. If the standard states are the liquids MO and SiO_2 at the same P and T, then the free energy of the solution is

$$\Delta \bar{G} = \Delta \bar{H} - T \Delta \bar{S} = -\left(\frac{O^-}{2}\right) RT \ln K, \qquad (8)$$

where $\Delta \bar{H}$ and $\Delta \bar{S}$ are the molar enthalpy and molar entropy respectively. In order to predict the state of polymerization of a silicate melt, one must be able to correlate the magnitude of this free energy, and therefore the value of K, with some measurable property of the metal cation. Concepts borrowed from acid-base theory have proven very useful. The definition of basicity when applied to molten silicates is open to a number of interpretations and is subject to some controversy. In this paper, basicity refers to the state of the oxygen atom and its ability to donate some of its negative charge to a metal ion (Duffy and Ingram, 1976). The oxygens are then bases, and the cations acids, in the Lewis sense. The ability of oxygen to donate its negative charge to a given cation depends upon the influence of the surrounding cations. The basicity is at a maximum when the oxygen exists as a free oxygen ion removed from the influence of neighboring cations. Such a state is not stable (Cotton and Wilkinson, 1966), but is approached when the co-ordinating cations are of a low electronegativity, i.e., they have only a weak affinity for electrons in the particular bond. When oxygen is co-ordinated to cations of low electronegativity, such as Rb^{+1}, K^{+1} or Ca^{+2}, the basicity of the oxygen and therefore that of the molten oxide is high. When the oxygen is coordinated with Si, a cation of high electronegativity, its basicity is much less because more of the charge is drawn towards the silicon (i.e. covalent bond formation). In this sense, a bridging oxygen is less basic than a non-bridging oxygen, provided that the latter is associated with a cation of low electronegativity.

The heat effect $\Delta \bar{H}$, associated with the formation of non-bridging oxygen, is believed to result from the closer association of the non-bridging oxygen to silicon (Charles, 1969). Non-bridging Si–O (nbr) bonds in crystalline silicates are, in fact, typically shorter than the Si–O (br) bridging bond (Gibbs et al., 1972). According to the acid-base theory, the shortening of the Si–O non-bridging bond is due to the polarization of the electron cloud of oxygen towards, and the oxygen nucleus away from, the silicon (Weyl and Marboe, 1962). The Si–O bond is thus strengthened and the M–O (nbr) is weakened relative to the Si–O (br) and M–O (fr) bonds respectively. In the case for which the total enthalpy change is exothermic, the increased strength of the Si–O (nbr) bond more than compensates for the weak M–O (nbr) bond. The non-bridging oxygens should be most stable in a bond where the coordinating cation has a low electronegativity. Therefore, the addition of oxides of "weak" cations to a SiO_2 melt introduces polarizable oxygens that by reaction (2) form non-bridging oxygens, and thus stabilize the melt phase.

This hypothesis is tested by comparing the heats of formation of crystalline metasilicates ($MSiO_3$) and orthosilicates (M_2SiO_4), for which the crystalline oxides MO and SiO_2 are the standard states. The heat of formation of a metasilicate, $\Delta \bar{H}^f$, for example, is

$$\bar{H}_{MSiO_3} - \bar{H}_{MO} - \bar{H}_{SiO_2} = \Delta \bar{H}^f_{MSiO_3}, \qquad (9)$$

and is the enthalpy change of converting one bridging and one free oxygen to two non-bridging oxygen, i.e.,

$$O^0 + O^{2-} \rightarrow 2O^-. \qquad (10)$$

The heat of formation of the orthosilicate is that obtained by converting two O^0 and two O^{2-} to four O^-. According to the model, the heats of formation of each isostructural series should be related to the ability of the cation to draw electrons from the oxygen, that is, its electronegativity.

Several electronegativity scales have been proposed (Cotton and Wilkinson, 1966). Unfortunately, the correlation of electronegativities with the enthalpies of formation of the silicates is generally poor. One reason for this is that the electronegativity scales are not very sensitive, that is, the electronegativities of Fe^{+2}, Ni^{+2}, Si^{+4} and Pb^{+2} are all 1.8 (Pauling, 1960). A direct measure of the ability of a cation to attract electrons is its ionization potential. This is the minimum energy required to detach one or more electrons from an atom or ion (Cotton and Wilkinson, 1966). Unfortunately, the ionization energies are obtained for atoms in the gaseous state and are not equal to the electron-drawing power of the cation in a bond. However, Mulligan (Pauling, 1960) proposed that the average of the ionization potential and the electron affinity (the energy of the reaction $X + e^- \rightarrow X^-$ where e^- is an electron) was a property closely related to the electronegativity. Thus, although the ionization potential is not a true measure of the electronegativity, it probably is related to it. Consequently, it will be assumed that an ordering of cations according to their ionization potentials should be the same as an ordering obtained from their true electronegativities.

The heats of formation of the metasilicates and orthosilicates are plotted against the sum of the first and second ionization potentials of the divalent cation (Figure 3). In agreement with the proposed model,

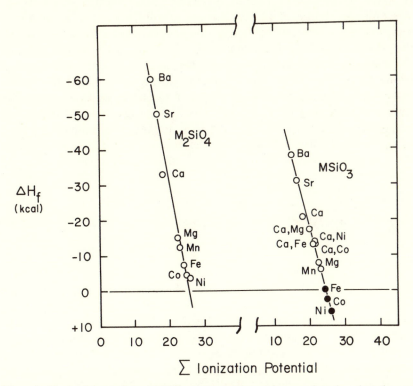

Figure 3. Heats of formation at 986°K relative to the oxides are plotted against the sum of the first and second ionization potentials of the divalent cations. Data are from Navrotsky and Coons (1976) and Navrotsky (1976).

a trend is obtained for each mineral group wherein the heat of formation associated with the formation of non-bridging oxygen decreases with the increasing value of the ionization potential of the cations. Such agreement does not prove the validity of this model, which is certainly much too simple to account for energy changes involving complex ionic and covalent interactions. Nevertheless, it is interesting that the trend for the metasilicate minerals intersects the abscissa near an ionization potential of 24 eV. This predicts that metasilicates formed from divalent cations of ionization potential greater than 24 eV would have a positive $\Delta \bar{H}^f$, and, depending on the magnitude of the entropy of formation, might not form stably at 1 atm. The ionization potentials of the divalent cations of Fe (24 eV), Co (25 eV), Ni (26 eV) and Zn (27 eV) indicate that these metasilicates would have a limited stability. (The stable occurrence of any of these metasilicates, of course, also depends on the free energy change of the reaction $SiO_2 + M_2SiO_4 = 2MSiO_3$.)

The ionization potential of the cation is an important, but certainly not the only, property that must be considered in the analysis of melt structure. For example, Pauling (1960) has demonstrated that both the size and the charge of the cation strongly influence the structures of ionic compounds. Many of the crystal-chemical concepts used to explain the properties of the crystalline phases will be used and referred to extensively throughout this paper. However, the value of the ionization potential of the cation should be a good measure of the state of polymerization of a melt. In general, melts containing cations of high ionization potential should be more polymerized than melts containing cations of low ionization potential. The ordering of the free energies of mixing in Figure 2 from least to most negative follows the ordering of ionization potentials from large to small. This was also interpreted previously as an order of decreasing states of polymerization.

STRUCTURE OF SILICATE MELTS

LIQUID SILICA

The structure of silica glass has been the subject of numerous X-ray diffraction studies, the earliest ones being limited by experimental and theoretical difficulties (Wright and Leadbetter, 1976). For these reasons, the structure of silica glass has attracted considerable controversy. A recent X-ray diffraction study by Mozzi and Warren (1969) has resolved some of these difficulties but has not stilled the controversy. Mozzi and Warren concluded that practically all silicon atoms are tetrahedrally bonded to four oxygens, with an average Si–O distance of 1.62 Å. The latter is typical of Si–O distance in crystalline silica. Crystal-like ordering, however, was recognized out to interatomic distances of only 7 Å. The structure of silica glass was interpreted by Mozzi and Warren in terms of the random network model (Zachariason, 1932). In this model, SiO_2 glass is regarded as a random, continuous, three-dimensional network of linked SiO_4 tetrahedra. The network is characterized by a mean Si–O–Si bond angle and a mean Si–O bond length. Random variations of these parameters contribute to the structural disorder observed beyond 7 Å. A recent reanalysis of Mozzi and Warren's data (Wright and Leadbetter, 1976) indicates that the most probable Si–O–Si bond angle is 152°, a value which is in agreement with the work of Bell and Dean (1972), who found that in order to reproduce the radial distribution function of these glasses it was necessary to build a model with a mean bond angle of 152°. This model includes rings containing four, five, and six SiO_4 tetrahedra, in contrast to cristobalite and tridymite, which contain only six-membered rings. The randomness of the network structure and the existence of ring structures containing other than six tetrahedra distinguishes the glass phase from that of cristobalite or tridymite.

However, recent X-ray and neutron diffraction studies by Konnert and Karle (1972) claim to have discovered significant ordering out to interatomic distances (r) of 20 Å. Their data are consistent with a "structure in which nearly all of the atoms comprise parts of a tridymite-like region." They allow for the possibility of some cristobalite-like ordering at larger interatomic distances. These authors conclude that their data are inconsistent with the random-network model of silica glass because, in theory, such a glass should produce no detail beyond 7 Å. Their data suggest that 90% or more of the atoms comprise parts of tridymite regions of dimensions up to 13 Å or greater. They suggest an analogy to a state obtained by a group of twinned crystals arranged in a large number of different orientations. The conclusions by Konnert and Karle, if correct, are extremely important because they give support to the crystallite-model of glass structure. In this model, a glass is composed of volumes of crystalline-like order bonded together in a random fashion by regions of lesser order. In a multicomponent glass, an essentially heterogeneous structure consisting of several types of crystallities, generally corresponding to the structures of the appropriate liquidus phases, is suggested (e.g., see Uhlmann and Kolbeck, 1976). This is in contrast to the random-network model, in which the glass structure is homogeneous with only a minimal amount of crystal-like ordering.

This model is challenged by Wright and Leadbetter (1976), who question whether the structure observed by Konnert and Karle (1972) at large interatomic distances is real or simply an artifact of the function chosen to represent the data. Nevertheless, it is clear that the short-range bond topology is crystal-like: the only remaining question concerns the extent of ordering. A more pressing problem is how to determine whether the structure of the glass is indeed the same as the melt. It is reasonable to infer that molten silica is less ordered than the glass. The small entropy of fusion of cristobalite (~ 1.0 cal/deg) is consistent with the interpretation that the melt structure is that of a random network but does not disprove the crystallite hypothesis (Bell and Dean, 1968). The question remains unresolved.

BINARY SILICATE MELTS

It has been known for some time that that the properties of silica-poor melts require the postulate that they contain silicate anionic polymers of finite size. Silica-rich melts, on the other hand, are believed to be essentially infinite networks of linked SiO_4 tetrahedra, with some of the bridging bonds replaced by non-bridging bonds. The fundamental challenge to silicate melt theory is to develop a model of silicate anion polymerization that describes the evolution of a SiO_2-poor melt into a SiO_2-rich melt. Unfortunately, polymer theory can describe the equilibria only among chain or branched silicate chains of finite length. Treatment of more complex equilibria involving infinite chains, ring, or network structures are limited by the mathematical complexities inherent in polymer theory itself. Since SiO_2-poor and SiO_2-rich melts are very different in their structural, physical, and thermodynamic properties, it is convenient to consider each individually.

SiO₂-Rich Melts

Si–O bonds have a strong directional character resulting in the "stiff", spacious, infinite network structure of pure molten silica. The structure gives rise to a liquid of extremely high viscosity, low density, low conductivity, and small thermal expansion. This same structure has an unusual sensitivity to defects. In silica glass, for example, traces of H_2O lower the viscosity by one order of magnitude: simply touching a pristine surface with a finger deposits sufficient alkali to cause the contaminated spot to devitrify (Weyl and Marboe, 1962).

This sensitivity to defects is also illustrated by the equilibrium relationships of cristobalite with a silicate melt. At the melting point, cristobalite, a network silicate, coexists with fused silica, a network structure characterized by long-range disorder. The addition of a small amount of alkali oxide to the equilibrium depresses the liquidus temperature, because the added components are concentrated into the liquid phase. These alkali ions form non-bridging bonds in the melt, thereby stabilizing the melt phase relative to cristobalite. In a previous section, it was argued that the most stable non-bridging bonds are formed by cations of low ionization potential. Such cations would most effectively stabilize the melt relative to cristobalite. For example, the addition of 5 mole % of Cs_2O (IP = 3.9 eV) depresses the liquidus by 210°C, whereas 5 mole % of Li_2O (IP = 5.5 eV) depresses the liquidus by only 80°C.

Moreover, the addition of a few mole % of oxides of the divalent cations with high ionization potential depresses the liquidus by only a few degrees and in most systems renders the single liquid unstable by the occurrence of silicate-liquid immiscibility. This indicates that non-bridging oxygens in these melts do not greatly stabilize the melt relative to either cristobalite crystallization or to silicate-liquid immiscibility.

In order to examine this phenomenon in greater detail and avoid the two-liquid regions, liquidus compositions of cristobalite at 1,500°C are plotted against the ionization potentials of the cations (Figure 4). Two

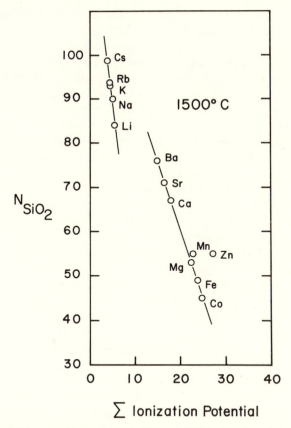

Figure 4. Liquidus compositions of cristobalite at 1,500°C plotted against the sum of the ionization potentials of the network modifying cations. Liquidi data from Levin et al. (1964, 1969).

trends are defined, one each for the univalent and divalent cations. In each trend, the liquidus composition shifts to more SiO_2-poor compositions with increasing ionization potential of the cation. The significance of this correlation is subtle but important. All the liquidus melts are isothermal, isobaric and coexist with cristobalite. To the extent that cristobalite can be treated as a pure SiO_2 phase, all melts are buffered at the same chemical potential of SiO_2.

Hess (1977) proved that

$$\mu_{SiO_2} = 2\mu_{O^0},$$

so that all melts are also buffered at the same chemical potential of the bridging oxygen. The magnitude of μ_{O^0} at constant T and P is a function of both the concentration and the activity coefficient of the bridging oxygen. Both of these quantities relate to the polymerized state of the melt.

The value of μ_{O^0} can therefore be used as a thermodynamic index of the state of polymerization of a melt. If the activity coefficient of the bridging oxygen equaled unity, the chemical potential of the bridging oxygen would be proportional to the concentration of the bridging oxygen. Therefore the value of μ_{O^0} would also be a structural index of the polymerized state of the melt, i.e. two melts characterized by the same μ_{O^0} would also be polymerized to the same degree. To the extent that this is true, it can be concluded for example, that the binary melt at Na_2O = .10 mole % is polymerized to the same degree as the binary melt at MgO = .47 mole %. Therefore, all the melts in Figure 4 may exist in approximately the same state of polymerization. Melts containing cations of high ionization potentials reach this state at much lower SiO_2 contents than melts containing cations of low ionization potentials.

However, the stability of a silicate melt is a function not only of the concentration of the oxygen species but also of their distribution within the melt. A random distribution maximizes the configurational entropy of the melt and at constant enthalpy would minimize the Gibbs free energy (see Equation (8)). However, a random distribution of non-bridging oxygen, O^-, provides each cation with only a small number of coordinating O^- in SiO_2-rich melts. The positive charges of these cations are poorly screened from other cations by the non-bridging oxygen: this would result in an increase in the enthalpy and in the free energy of the melt (see Equation (8)).

At high temperatures, a more random distribution of oxygen species is favored, because the entropy term in the free energy equation, the product $T \Delta S$, would outweigh the opposing and unfavorable enthalpy term. However, as temperatures are lowered, a state is reached at which the $T \Delta S$ term of a highly randomized melt no longer compensates for the enthalpy. The free energy of the melt is now minimized by concentrating non-bridging oxygens around the metal cations, thus providing coordination polyhedra of several oxygens around each cation. In the extreme case, this ordering results in the formation of two coexisting silicate liquids, one rich in non-bridging oxygens and metal cations, the other in silicon and bridging oxygens.

The temperature at which silicate-liquid immiscibility occurs in these simple systems should be a function of the screening demands of the cations. The order of critical temperatures is Ca > Sr > Ba > Li > Na > K > Rb, Cs (see Hess, 1977, and unpublished data); this parallels the order of their ionization potentials. Since the onset of silicate-liquid immiscibility requires that the single-liquid phase becomes not only metastable but also unstable relative to the two liquids, the temperature of the critical point and the extent of the two-liquid field is one of the best indices of the stability of a silicate melt. In lieu of experimental data, a knowledge of the ionization potentials of the modifying cations is a prerequisite to an understanding of the relative stabilities of silicate melts. This reinforces the conclusions obtained in the previous sections.

This discussion has stressed the importance of the electronegativity of the cation, as approximated by the ionization potential, in determining the thermochemical properties of a silicate melt. Such an approach, although remarkably successful, is an over-simplification of a complex problem. The incorporation of oxides of polyvalent cations of the form M_mO_n, where $n > m$, illustrates the nature of some of these complications. Reconsider the question of the distribution of non-bridging oxygen in a melt. In a melt characterized by the maximum degree of randomness consistent with the requirements of local electrical neutrality, a monovalent cation is co-ordinated by only one non-bridging oxygen, a divalent cation by two, and a polyvalent cation with a number equal to its formal charge. Whereas non-bridging oxygens may occur singly in melts containing only univalent cations, the non-bridging oxygen must be concentrated around polyvalent cations. Thus, melts containing polyvalent cations are characterized by an inherently higher degree of ordering of the non-bridging oxygen, a property which increases with the charge of the cation.

The ordered structure of these melts is different not only in degree but also in kind from melts containing either monovalent or divalent cations. The ionization potentials and therefore the electron-drawing power of the polyvalent cations of charge 3^+ or more are comparable to that of the silicon ion. The free oxygen bond, O–M–O, should then be as strong as the bridging bond. Thus, the free energy change and the equilibrium constant of the equilibrium

$$O^0 + O^{2-} = 2O^-$$

are both small. Such melts contain few non-bridging oxygens; they are solutions containing locally ordered regions of either free or bridging oxygen. These melts are less stable than those containing non-bridging oxygens, and are characterized by higher liquidus temperatures and by large fields of silicate-liquid immiscibility. Moreover, the low stability and the paucity of the non-bridging oxygen in the liquid implies that the non-bridging oxygen would not be very stable in the crystalline state. Therefore, these systems should contain few intermediate silicate compounds near the liquidus. These characteristics are indeed obtained in most systems. For example, the liquids in the TiO_2–SiO_2 and Cr_2O_3–SiO_2 systems have several features in common: (1) they are stable only above 1,550°C; (2) they are in equilibrium only with the crystalline phases of TiO_2, Cr_2O_3 and SiO_2 composition; (3) they are not in equilibrium with any intermediate silicate compounds; and (4) they become unstable relative to immiscible silicate liquids (Levin et al., 1964). The two-liquid fields extend over nearly the entire compositional range of the system and to temperatures above 1,780°C. Some or all of these properties are observed in binary silicate systems containing ZrO_2, ThO_2, Sc_2O_3, and the oxides of the rare-earth elements, among others.

A significant exception to these generalizations occurs in the P_2O_5–SiO_2 system. The addition of 5 mole % of P_2O_5 depresses the cristobalite liquidus by more than 400°C; a stable field of silicate liquid immiscibility does not occur in the range from $N_{SiO_2} = 100$–50 mole % and two silicate compounds, $2SiO_2 \cdot P_2O_5$ and $SiO_2 \cdot P_2O_5$ occur as liquidus phases.

These properties are clearly not consistent with the high ionization potential of P^{+5} and the rules described above. One possible explanation for this paradox is obtained from the observation that the fundamental structural unit in phosphate minerals is the PO_4^{-3} tetrahedron, which is only slightly smaller than the SiO_4^{-4} tetrahedron. Thus, the two species may substitute for and copolymerize with each other thereby enhancing the stability of the melt. These conclusions are in agreement with studies of silicophosphate glasses by paper chromatography, which indicate that SiO_4 groups do indeed enter phosphate chain structures (Finn et al., 1976).

SiO_2-Poor Melts

From the preceding discussion, it might be supposed that a silicate melt of orthosilicate ($N_{SiO_2} = .33$) or metasilicate ($N_{SiO_2} = .50$) composition would consist solely of the same isolated SiO_4 tetrahedra or infinite SiO_3 chains that occur in the corresponding crystalline silicates. A melt of intermediate composition would be a mixture of SiO_4 and SiO_3 anionic groups. There is now direct experimental evidence to prove that this is not correct. Some qualitative data come from the separation of silicate anions by chromatographic techniques. These have shown, for example, that a glass of Pb_2SiO_4 composition contains not only isolated SiO_4 tetrahedra, but also Si_2O_7, Si_3O_{10} chains and Si_4O_{12} rings (Baltă et al., 1976). Unfortunately, the abundance of these species cannot as yet be determined, because the experimental techniques required to separate the silicate anionic polymers from the glass cause both breakdown and polymerization of the silicate anions present (Richardson, 1974).

The most convincing data on the polydisperse nature of anionic silicate groups are derived from phosphate glasses. These glasses can be dissolved in appropriate solutions without significantly altering the nature of the phosphate anions, and, by chromatographic techniques, the different types of anionic species can be totally recovered (Richardson, 1974). These data are particularly intriguing because, as discussed below, crystalline phosphates bear a close resemblance to crystalline silicates. The crystalline structures of the inorganic phosphates are based on the PO_4^{-3} tetrahedral group, which by sharing corners polymerizes to form chain, ring, and sheet structures. The structures of metal-rich phosphates are very similar to those of orthosilicate and metasilicate phases. For example, $LiPO_3$ has an infinite pyroxene-type chain structure (Fraser, 1977) and $Ca_3(PO_4)_2$ has an almost complete solid solution series with Ca_2SiO_4 (see Levin et al., 1964). Polymorphs of $AlPO_4$ are isostructural with tridymite, cristobalite, and quartz, and Simpson (1977) synthesized aluminum phosphate variants of feldspar ($NaAl_2PSiO_8$, KAl_2PSiO_8). Phosphate and silicate ions therefore have very similar structures, the main difference being a double bond in the phosphate ion (Figure 5), which limits its cross-linking ability. Unlike silica, pure P_2O_5 does not form a strong cross-linked network structure, but rather occurs as a discrete molecule with a very low melting point. Thus, the comparison of phosphate and silicate glasses is justified only if the P_2O_5/MO ratio is not much greater than unity.

The anionic groups in phosphate glasses are typically straight chains without branches; the variations in chain lengths in different glasses is illustrated in Figure 6 (Meadowcroft and Richardson, 1965; Westman and Gartaganis, 1957). Rings of 3 or 4 phosphate tetrahedra are few, but become more abundant as the metaphosphate composition is approached. The range of chain lengths in the glasses is greater: chains can be either

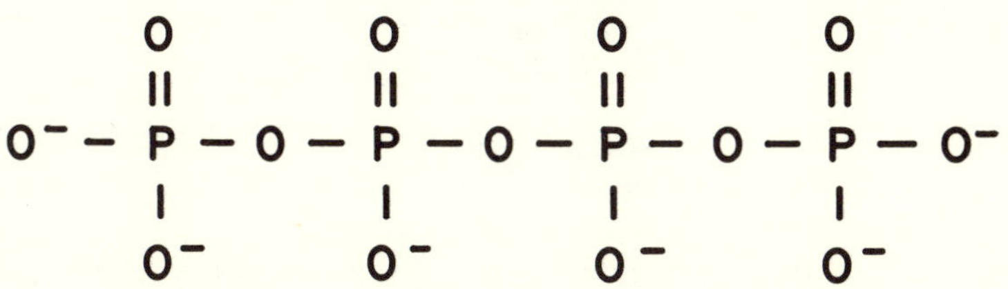

Figure 5. Diagram of chain of phosphate tetrahedra with indicated double bonded oxygen.

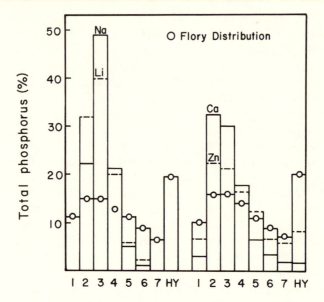

Number of phosphorus atoms per chain, n

Figure 6. The distribution of phosphorus in chains containing 1 to 7 phosphorus atoms in $5M_xO_y \cdot 3P_2O_5$ glasses where M_xO_y is Li_2O, Na_2O, CaO or ZnO. $-O-$ represents the Flory distribution (see text); Hy = chains containing more than 7 phosphorus atoms.

shorter or longer than those in the closest corresponding crystalline phosphate. However, the chain-length distribution has a more pronounced peak than that obtained for the condition where all polymerization reactions occur randomly. Under the latter condition, the chain-length distributions—known as the Flory (1953) distributions—are obtained if the polymerization reaction between chain species

$$2(P_nO_{3n+1})^{-(n+2)} = (P_{n+1}O_{3n+4})^{-(n+3)} + (P_{n-1}O_{3n-2})^{-(n+1)} \quad (11)$$

occurs with no free energy change

$$\Delta \bar{G}_R = 0, \quad (12)$$

and therefore

$$K_n = \frac{(P_{n+1})(P_{n-1})}{(P_n)^2} = 1. \quad (13)$$

According to the Flory model, the concentrations of chain species should vary only with the P_2O_5 content and are independent of the nature of the metal oxide. This is not obtained in phosphate glasses. The largest deviations from the Flory distribution occur in the concentrations of the smallest-sized ionic groups (Figure 6). Whereas in the Flory distribution the monomer should always be the most abundant species (on a mole basis), it is entirely absent in the alkali-phosphate glasses. However, the distribution of chain lengths becomes wider and more closely approaches the Flory distribution as one passes from the Na to Li, Ca and Zn-bearing phosphate glasses. The list of systems as given: Na, Li, Ca and Zn, is in order of increasing ionization potential of the cation and decreasing heats of mixing of the solutions (Figure 7). The deviations from the Flory

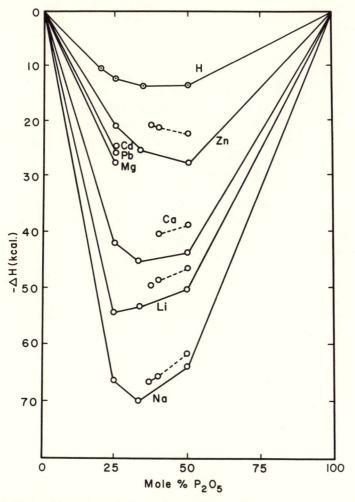

Figure 7. Heats of formation relative to the crystalline oxides for crystalline phosphates (solid lines) and relative to the metastable liquid oxides for the glassy phosphates (dashed lines) (Richardson, 1974; Jeffes, 1975).

distribution can thus be ascribed to energies involved in the reaction

$$O^0 + O^{2-} = 2O^-,$$

where the oxygen nomenclature is the same, except that the P_2O_5 rather than the SiO_2 is the polymeric component. The solution models discussed earlier apply here.

P_2O_5 melts containing cations of low ionization potential form strong non-bridging bonds because the electrons in the oxygen are drawn towards the P^{+5} cation. The enthalpy and the free energy change of reaction (11) should be strongly negative (Figure 7). This is in contrast to the Flory model, which assumes no free energy changes. Thus, the equilibrium constant

$$K_n = \frac{(P_{n+1})(P_{n-1})}{(P_n)^2}$$

and the abundances of the chain species will necessarily depart from the Flory distribution. Moreover, Richardson (1974) has shown that the distribution of phosphate chains will become narrower as the free energy change of reaction (11) grows more negative. Thus, only phosphate melts containing cations of high ionization potential should contain chain-length distributions approaching those predicted by the Flory model. These predictions are indeed confirmed by the chromatographic data.

The chain-length distribution of silicate polymeric species should follow similar rules. In lieu of experimental data, Masson and his coworkers (Masson et al., 1970) used the statistical principles of polymer theory to calculate the most probable size distribution of silicate ions as a function of the composition of the silicate melt. Masson (1972) has reviewed these results, so only the assumptions embodied in these calculations and some of the conclusions are described below.

In the Masson models, the building up of polymeric silicate ions occurs by a series of "condensation" reactions:

$$SiO_4^{-4} + Si_nO_{3n+1}^{-2(n+1)} = Si_{n+1}O_{3n+4}^{-2(n+2)} + O^{-2}, \qquad (14)$$

in which an O^{-2} ion is produced at each step.[1] The condensation reactions are each described by an equilibrium constant

$$K_n = \frac{(Si_{n+1}O_{3n+4})(O^{-2})}{(Si_nO_{3n+1})(SiO_4)}, \qquad (15)$$

in which the brackets refer to the appropriate activities of the silicate and oxygen ion species. The condensation reactions produce only straight or

---

[1] The O^{-2} refers to the oxygen ion and should not be confused with O^{2-}, which refers to the neutral M−O−M complex.

branched chain-silicate species. "Self-condensation" reactions yielding ring-silicate species, for example

$$Si_nO_{3n+1}^{-2(n+1)} = Si_nO_{3n}^{-2n} + O^{-2}, \qquad (16)$$

are not permitted, and therefore limit the applicability of these models to melts containing few ring silicate ions. Gaskell (1977) has considered this problem in detail and has concluded that the first significant appearance of ring ions occurs in silicate melts in the range $N_{SiO_2} = .3 - .4$.

Two fundamental assumptions are necessary to an application of these equations to actual silicate melts. First, it is assumed that the silicate chain and oxygen ions mix ideally on an anionic "lattice," and the cations mix ideally on a cationic lattice. In a binary silicate melt, the cation is derived from the complete dissociation of the metal oxide, MO, to give M^{+2} and O^{-2}. All silicon ions are considered to belong to silicate anions. Ion fractions are then substituted for activities in Equation (15). Second, it is postulated that the free energy of the condensation reactions in (14) is constant and independent of the chain length of the silicate ions. This means that all equilibrium constants in Equation (15) are equal. This concept, the "principle of equal reactivity" has found wide acceptance and success in the treatment of organic polymers and is a consequence of the strictly short-range bonding characteristics of organic molecules (Flory, 1953). This principle is probably not obeyed exactly in phases containing ionic bonds. However, Meadowcroft and Richardson (1965) have shown that this principle is indeed a good approximation in phosphate melts (and therefore probably also for silicate melts) provided that the polymer chains contain at least three or more phosphate ions.

Having made these assumptions, it is possible to derive an expression relating the activity of the metal oxide, a_{MO}, to the composition of the melt and to the value of K_n. For branched and straight silicate chains the expression is

$$\frac{1}{N_{SiO_2}} = 2 + \frac{1}{1 - a_{MO}} - \frac{3}{1 + a_{MO}\left(\dfrac{3 - K_n}{K_n}\right)}. \qquad (17)$$

The activity-composition curves calculated from Equation (17), where K_n is treated as an adjustable parameter, are shown in Figure 8 and are in reasonable agreement with the experimental data (see Masson et al., 1970). The predictions of the straight-chain model are adequate but less satisfactory (Masson, 1972). The activity curves determine the best value of K_n for a system at a fixed P and T. Once the K_n is given, it is possible to calculate the chain length distributions of the silicate chains as a function of melt composition. The reader is urged to consult the original reference (Masson et al., 1970) for the details of these calculations.

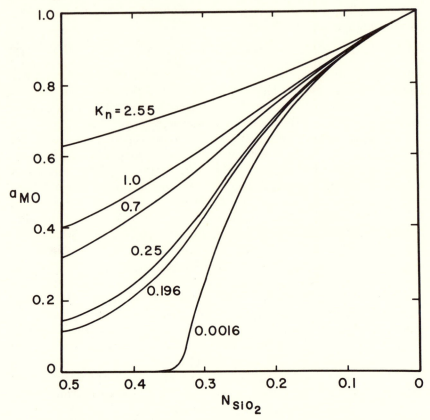

Figure 8. Activities of metal oxides relative to liquid standard states calculated for various values of K_n. The theoretical curves correspond closely to the activity curves of the following systems, given in descending order: SnO (1,100°C), FeO (1,600°C), FeO (1,300°C), MnO (1,600°C), PbO (1,000°C) and CaO (1,600°C) (Masson, 1972).

Calculated chain-length distributions for $CaO-SiO_2$ and $FeO-SiO_2$ melts are shown in Figure 9 (Masson, 1972). The smaller value of K_n for the $CaO-SiO_2$ melt indicates that the melt is less polymerized for a given composition than the $FeO-SiO_2$ melt.

The chain-length distribution for the $CaO-SiO_2$ system is strongly peaked near $N_{SiO_2} = .33$. The SiO_4 monomer is by far the most abundant silicate ion on a mole basis and is almost the sole silicate species present in SiO_2-poor melts. The latter appear to be composed predominantly of SiO_4 monomers, M^{+2} and O^{-2} ions. Large silicate chains become abundant near $N_{SiO_2} = .50$ and their sum is more than three times the SiO_4 abundance.

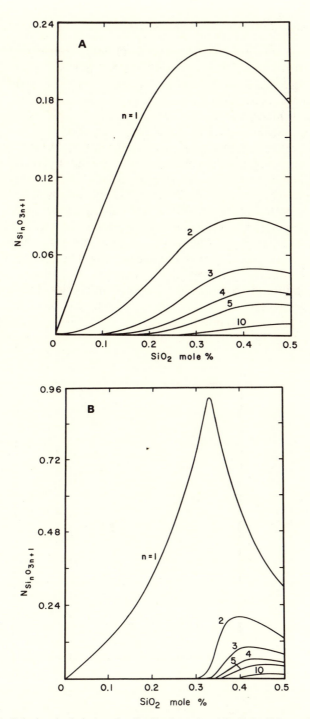

Figure 9. Calculated chain-length distributions for (A) FeO–SiO$_2$ melts at 1,600°C (K$_n$ = 1.0) and (B) for CaO–SiO$_2$ melts at 1,600°C (K$_n$ = .001) (Masson, 1972).

In contrast, the more polymerized $FeO-SiO_2$ melt contains a much broader distribution of silicate chains than the $CaO-SiO_2$ melts. Although the SiO_4 monomer is still the most abundant silicate ion, it never attains a concentration greater than 22 mole %. Large polysilicate ions occur even in SiO_2-poor melts, and are dominant in the more SiO_2-rich melts.

These conclusions parallel those obtained from the more general oxygen model of polymerization (see POLYMERIZATION MODEL). The polymerized state of the melt is indicated either by the value of K or K_n. A polymerized melt contains a large concentration of bridging oxygens which in the range $N_{SiO_2} = 0-50$ mole % is reflected in the existence of a broad distribution of silicate chains. It was argued previously, that the degree of polymerization of a melt is a function of the ionization potential of the modifying cation. Figure 10 is a plot of $\log K_n$ (Masson, 1972) against the ionization potentials of the cations. The correlation is satisfactory and indicates that melts containing Fe^{+2}, Co, Ni should contain a wide distribution of polysilicate anionic chain species. The highly polymerized state will no doubt also be expressed by an abundance of ring or more complex silicate anionic groups as $N_{SiO_2} = .50$ is neared. Melts containing Ca, Ba, Sr, K, Na, and Li, i.e. cations all of low ionization

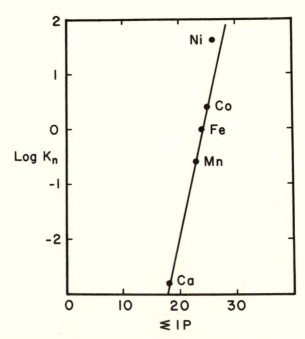

Figure 10. Correlation of $\log K_n$ with the ionization potentials of the divalent cations. A large value of K_n indicates that a melt is highly polymerized.

potential, are less polymerized and should contain distributions of silicate chains with a sharp peak near $N_{SiO_2} = .33$.

MULTICOMPONENT SYSTEMS

The concepts developed for binary silicate melts may be extended to multicomponent systems, although the situation is much more complex. The increase in complexity due to the presence of many cations is compounded by the occurrence of aluminate and phosphate ions which can copolymerize with the silicate ions, and by the varied coordination opportunities available to such cations as Fe^{+3} and Al^{+3}. These problems are considered separately in following sections.

General

Changes in the state of polymerization of a silicate melt may be monitored by determining whether the liquidus field of a unary crystalline phase is expanded or contracted owing to the addition of certain components to the melt (Kushiro, 1975). This is illustrated by examining the liquidus fields of forsterite and protoenstatite in the $MgO-SiO_2$ system. Hess (1977) has shown that in a binary silicate melt

$$2\mu_{O^0} = \mu_{SiO_2}, \tag{18}$$

$$\mu_{O^{2-}} = \mu_{MO}, \tag{19}$$

$$\mu_{O^-} = \tfrac{1}{4}\mu_{SiO_2} + \tfrac{1}{2}\mu_{MO}, \tag{20}$$

where $M = Mg$ in this example. For the assemblage forsterite-protoenstatite-melt, the conditions of heterogeneous and homogeneous equilibrium require that

$$
\begin{array}{cccc}
\text{FORSTERITE} & \text{MELT} & \text{MELT} & \\
\mu_{Mg_2SiO_4} & = \mu_{Mg_2SiO_4} & = 2\mu_{MgO} + \mu_{SiO_2}, & (21)
\end{array}
$$

$$
\begin{array}{cccc}
\text{PROTOENSTATITE} & \text{MELT} & \text{MELT} & \\
\mu_{MgSiO_3} & = \mu_{MgSiO_3} & = \mu_{MgO} + \mu_{SiO_2}. & (22)
\end{array}
$$

Equations (18), (19), (20) may be combined with (21) and (22), to show that

$$\mu_{Mg_2SiO_4} = 4\mu_{O^-}, \tag{23}$$

$$\mu_{MgSiO_3} = 2\mu_{O^-} + \mu_{O^0}. \tag{24}$$

These equations are instructive because they relate the thermodynamic properties of the equilibrium to the polymerized state of the melt as expressed by the chemical potentials of the bridging and non-bridging oxygen.

The addition of a cation (in oxide form) of low ionization potential to a melt initially in equilibrium with forsterite and protoenstatite has two effects. Free oxygens of high polarizability (strong Lewis bases) are added to the melt and displace the equilibrium

$$O^0 + O^{2-} = 2O^-$$

to the right thus forming more non-bridging oxygen. Moreover, the equilibrium constant, K, of the melt is also increased slightly. Thus, the net result of both these effects is to increase μ_{O^-} and to lower μ_{O^0}. From (23) and (24), the chemical potential of Mg_2SiO_4 in the melt will increase relative to the chemical potential of $MgSiO_3$. If the small temperature change is neglected as the cotectic is traced into the ternary system, forsterite should crystallize from the doped melt and the forsterite-protoenstatite cotectic will be shifted to higher SiO_2 contents. Therefore, the addition of a cation of low ionization potential to a melt containing cations of higher ionization potential shifts liquidus boundaries to higher SiO_2 contents. This effect (Figure 11) has been verified by Kushiro (1975).

The addition of a cation of high ionization potential to the melt has two opposing effects. Although free oxygens are added to the melt, the shift of the equilibrium

$$O^0 + O^2 = 2O^-$$

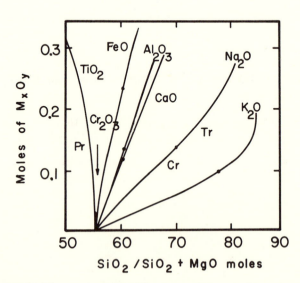

Figure 11. Shift of liquidus boundary between forsterite and protoenstatite at 1 atm with the addition of metal oxides (Kushiro, 1975).

to the right is mitigated or might even be reversed because the equilibrium constant of the ternary melt will be less than that of the binary melt. This effect is illustrated by the addition of TiO_2, which shifts the forsterite-protenstatite cotectic to lower SiO_2 compositions (Figure 11). These cations polymerize the liquid by forming coordination polyhedra with free oxygen. The oxygens needed to do this are stripped from the non-bridging complexes and result in the formation of additional bridging bonds.

In general, the results described in Figure 11 are consistent with the theory outlined above. The application of these ideas to a systematic examination of phase equilibria, however, must be done with caution. Any conclusions must consider the effect of solid solutions (for example, FeO in olivine and protoenstatite) and the effect of temperature or pressure on the thermodynamic properties of the coexisting phases.

Aluminum in Silicate Melts

On the basis of geometric packing arguments, aluminum has an ionic radius intermediate between that preferring four- and six-fold coordination with oxygen. In this sense, aluminum can assume the role either of a network-forming or network-modifying cation. To exist as a tetrahedrally coordinated cation within a silica-rich network, aluminum must be associated with a charge-balancing cation such as K^+, Na^+ or Ca^{+2}. However, not all cations assume the role of charge balancing cations in the crystalline state. It is therefore important to determine which cations couple effectively with aluminum and allow it to achieve tetrahedral coordination in the liquid state. In addition, a major and controversial question concerns the structural role of aluminum in silicate melts containing aluminum in excess of the charge balancing cations.

Certain physicochemical properties of soda alumino-silicate glasses (Day and Rindone, 1962) and melts (Riebling, 1966) develop either maxima or minima near the Na/Al mole ratio of 1. Viscosity data for these melts show this behavior particularly well (Figure 12). The substitution of Al_2O_3 for Na_2O in $Na_2O–SiO_2$ melts causes a steep viscosity increase that leads to a pronounced maximum close to the Al/Na = 1 line. These data suggest that, initially, Al enters into network formation in the form of AlO_4 tetrahedra. The binary $Na_2O–SiO_2$ melts are moderately depolymerized liquids for these silica contents and thus are of very low viscosity. The progressive replacement of Al for Na has two consequences. First, strong Al–O bonds form at the expense of weak Na–O bonds, thus tending to strengthen the anionic portion of these liquids. Moreover, for each AlO_4 tetrahedron entering into network formation, a Na ion must occupy nearby cavities within the aluminosilicate framework in order to

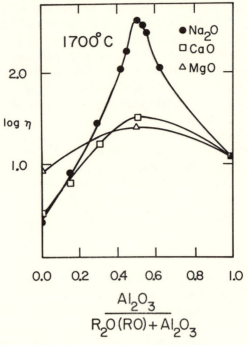

Figure 12. Viscosity data for Na_2O-, MgO- and CaO-aluminosilicate melts at 1,700°C (Riebling, 1966, 1964).

balance the formal charge of aluminum. Thus, in melts with Na/Al > 1, Na ions occur both as network modifiers and, in association with AlO_4 species, as network formers.

At the Na/Al = 1 composition, most if not all of the network-modifying Na ions have been eliminated. This viscous liquid is composed of a stiff anionic aluminosilicate network enclosing Na in irregular-shaped cavities. Further replacement of Na by Al not only removes a Na cation that is needed to charge balance the anionic framework but also adds an Al that somehow must be accommodated within the structure. Day and Rindone (1962) conclude that the Al ions in excess of the Na/Al = 1 ratio enter into six-fold coordination and thereby become network modifiers.[2] Since the Al–O bonds in AlO_6 octahedra are weaker than those in AlO_4 tetrahedra, the occurrence of Al in a network modifying role reduces the viscosity of the melt. If these coordination changes are continued to the

[2] Note, however, that laser Raman spectroscopic measurements of $NaAlSi_3O_8$ + 12 mole % Al_2O_3 glasses do not point to the presence of AlO_6 species, only AlO_4 (Mysen, pers. comm.).

Al_2O_3–SiO_2 binary system, it might be expected that all the Al ions would occur as AlO_6 octahedra in the silicate melt. This is probably not correct. At one atmosphere, the aluminosilicate phase on the liquidus is mullite, a mineral which contains Al in both four-fold and six-fold coordination. High pressures of tens of kilobars are required to bring kyanite, an Al_2SiO_5 polymorph containing Al solely in AlO_6 octahedra, to the liquidus. Thus it is highly probable that the Al_2O_3–SiO_2 melt at one atmosphere, and Al_2O_3-rich melts in general, would contain both AlO_4 and AlO_6 polyhedra.

In order for local charge balance to be maintained these Al polyhedra must not be distributed randomly but must segregate into Al_2O_3-rich microphase regions. The formation of microphase Al-rich melts suggests that macroscopic phase separation of two immiscible liquids should occur at some lower temperature. MacDowell and Beall (1969), and more recently Galakhov et al. (1976b), have determined that a metastable two-liquid field does indeed exist below 1,300°C in the Al_2O_3–SiO_2 system. Moreover, in the three-component glass Ga_2O_3–Al_2O_3–SiO_2, the metastable two-liquid field expands continuously from the Al_2O_3–SiO_2 side to the Ga_2O_3–SiO_2 side. The smooth increase in the critical temperature from 1,300°C in the Al_2O_3–SiO_2 glass to 1,640°C in the Ga_2O_3–SiO_2 glass indicates that the structural role of Ga is similar to that of Al. The relatively large ionic radii of Ga^{+3} ($r = .62$ Å in octahedral and $r = .47$ Å in tetrahedral coordination) indicates that its preference for octahedral coordination is higher than that of Al ($r = .53$ Å in octahedral and $r = .39$ Å in tetrahedral coordination). This observation would support the model in which a large percentage of the Al ions were also in octahedral coordination. The lower critical temperature in the Al_2O_3–SiO_2 system probably indicates that some of the Al ions are also in tetrahedral coordination, and therefore are more compatible with the silica network structure.

In contrast to the phase-equilibria studies of the Ga_2O_3–Al_2O_3–SiO_2 system, the metastable two-liquid field in the MgO–Al_2O_3–SiO_2 system (Figure 13) does not rise and expand smoothly from the Al_2O_3–SiO_2 side to the MgO–SiO_2 side (Galakhov et al., 1976a). Instead, it forms a saddle with a minimum in temperature and width occurring near the $MgO/Al_2O_3 = 1$ composition. The explanation of this extremum may be the same as that used for the maxima in the viscosity curves in the Na_2O–Al_2O_3–SiO_2 system. In melts containing $MgO/Al_2O_3 > 1$, all Al occurs as AlO_4 tetrahedra within the anionic framework. MgO in excess of that needed to charge-balance the AlO_4 negative charge occurs in the usual role as a network modifier. The inability to screen the charge of the Mg cation with non-bridging oxygen renders the liquid unstable, thereby causing silicate liquid immiscibility. At $MgO/Al_2O_3 = 1$, all Mg^{+2} ions are in

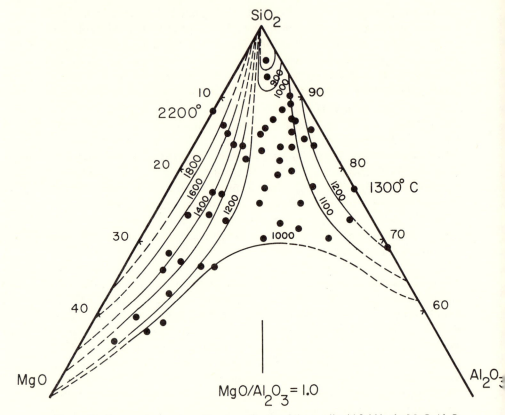

Figure 13. Temperature-composition coordinates of the two liquid field in the MgO-Al₂O₃-SiO₂ system. The isotherms below 1,700°C are contours of the metastable solvus.

structural cavities within the aluminosilicate network, and non-bridging oxygen are not required to form coordination polyhedra around the Mg^{+2} ion. Immiscibility occurs only at temperatures below 1,100°C. In this instance, immiscibility must arise from a general instability of an SiO_2-rich, aluminosilicate network. This problem will be considered shortly. For compositions $MgO/Al_2O_3 < 1$, some of the Al enters octahedral coordination; as a network modifier this Al requires a screen of non-bridging oxygen to shield its +3 charge from other cations. This makes the melt less stable. The result is an expansion of the two-liquid field to higher temperatures and more extreme compositions. The screening demand of Al^{+3} as measured by its ionization potential is more than twice that of Mg^{+2}. On this basis, it would be expected that the critical temperature in the Al_2O_3-rich melts would be comparable if not higher than that in MgO-rich melts. Obviously this is not true (Figure 13). The paradox is resolved if it is assumed that some of the Al occurs as AlO_4 tetrahedra, which in association with AlO_6 species lend greater stability to the liquid. Although the phase relations in the $MgO-Al_2O_3-SiO_2$ system (and probably in most other $N_2O-Al_2O_3-SiO_2$ or $MO-Al_2O_3-SiO_2$ systems) are generally consistent with the model for aluminum coordination, some important questions remain unresolved.

Viscosity maxima occur in the $MgO-$ and $CaO-Al_2O_3-SiO_2$ (Riebling, 1964) melts, but they are neither as pronounced nor as large as those in the Na_2O-bearing melts (Figure 12). One possible explanation for this phenomenon is that some Al^{+3} ions are in octahedral coordination even in melts where MO/Al_2O_3 is greater than 1. There is no evidence in the crystalline state that AlO_6 species are favored in SiO_2-rich compositions. The Al-bearing liquidus minerals, cordierite and anorthite, are both tectosilicates in which the Al^{+3} is in tetrahedral coordination. Since the AlO_6 complex conserves volume more efficiently than AlO_4, it is difficult to argue that it is stabilized at higher temperature in a melt phase of lower density. Thus, the existence of AlO_6 species in the MO-rich melts is a possible but unlikely explanation for the low viscosities.

A more attractive hypothesis is that the stability of the aluminosilicate network is less in MgO- and CaO-bearing melts than those containing alkalis. There is some evidence, admittedly speculative, to support this. In feldspar, the entrance of small cations in the irregularly shaped cavities of the aluminosilicate framework involves considerable deformation of the structure (Smith, 1974). An index of this deformation is obtained by comparing the variation of the length of the shortest cation-oxygen distance, denoted $M-O_A$, within this cavity for different cations. These distances are (in nanometers) Rb = .30; K = .27; Ba = .27; Na = .24; and Ca = .23 (Smith, 1974). The intertetrahedral angle $T-O_A-T$, where T = tetrahedral cation, decreases, and consequently the $T-O_A$ bond length increases with the distortion of the network around the small M cation. The small $T-O_A-T$ angle and longer $T-O_A$ bond lengths correlate with smaller bond-overlap populations and therefore are centers of generally weaker bonds (e.g., Brown et al., 1969, 1970; Gibbs et al., 1972). If this model is applicable to silicate melts, a distorted network should contain several weakened bonds which under stress give rise to the relatively low viscosity. The smaller the cation, the less rigid the network, and the more fluid the melt.

OXIDATION-REDUCTION EQUILIBRIA

It is well known that the oxidation states of such elements as Fe, Eu and Cr in a silicate melt are functions of the oxygen fugacity, temperature, and pressure of the ambient environment (e.g. Morris et al., 1974). Less well appreciated is the fact that the oxidation state is sensitive also to the composition and therefore the structure of the melt (Paul and Douglas, 1965a,b; Morris and Haskin, 1974). In this section, equilibria are derived that relate the redox ratio, M^{n+1}/M^n, of an element to the local structure of the melt as described by the concentration and distribution of the oxygen species.

These homogeneous equilibria may be written once the coordination states of these redox cations are determined. This analysis is fraught with complexity because of the varied coordination opportunities available to a cation within a silicate melt. For example, Fe^{+3} in a crystalline silicate typically occurs in a single coordination state with non-bridging oxygen. However, in a silicate melt, it is possible that Fe^{+3} occurs in several types and sizes of coordination polyhedra in which the oxygen species may be free and/or non-bridging. Notwithstanding these complications, structures of crystalline silicates will be used as a guide. The analysis will focus on the Fe^{+2}-Fe^{+3} equilibria, although this same approach can readily be applied to other redox equilibria.

In crystalline silicates, Fe^{+2} typically but not invariably occurs in six-fold coordination states with non-bridging oxygens. There are examples in which Fe^{+2} occurs in eight-fold (e.g. garnet) and four-fold (e.g. staurolite) coordinated states with non-bridging oxygens. These complexities will be ignored. Fe^{+3} is coordinated either by four or six non-bridging oxygen as in iron sanidine or andradite, respectively. When in the four-fold coordinated state, Fe^{+3} is associated with a charge balancing cation in much the same manner as Al^{+3}. It is assumed that these same coordination states are obtained in silicate melts. However, it was argued earlier that cations of high ionization potential such as Fe^{+3} may form bonding states with free as well as non-bridging oxygens. Since crystal field theory (Burns, 1970) predicts that Fe^{+3} has no preference for either six- or four-fold coordinated sites, it is assumed that the Fe^{+3} can be coordinated also to four and/or six free oxygens in a silicate melt.

The equations for the redox equilibria between Fe^{+2} coordinated with six non-bridging oxygen and the variously coordinated Fe^{+3} ions are given in Table 1. In order to write a balanced reaction, the following rules

Table 1. Homogeneous equilibria between Fe^{+2} – Fe^{+3} in Al_2O_3-free silicate melts. See text for discussion.

(1) $2SiO(\frac{1}{2}Fe^{+2}) + SiON + \frac{1}{2}SiOSi + \frac{1}{4}O_2 = 4[SiO\overset{*}{F}e^{+3}]^{-1/4} + N^+$

(2) $2SiO(\frac{1}{2}Fe^{+2}) + SiON + \frac{1}{4}O_2 = 2[\overset{*}{F}e^{+3}O\overset{*}{F}e^{+3}]^{-1/2} + N^+ + \frac{3}{2}SiOSi$

(3) $4SiO(\frac{1}{2}Fe^{+2}) + SiOSi + \frac{1}{2}O_2 = 6SiO(\frac{1}{3}Fe^{+3})$

(4) $4SiO(\frac{1}{2}Fe^{+2}) + \frac{1}{2}O_2 = 3(\frac{1}{3}Fe^{+3})O(\frac{1}{3}Fe^{+3}) + 2SiOSi$

are adopted. Fe^{+3} cations, denoted by an asterisk, and all Si, are counted as $\frac{1}{4}$ of an ion. The O^{-2}, Fe^{+2} and monovalent cations, $\overset{*}{N}{}^{+}$, are each counted as one ion.[3] For example, in the $SiO\overset{*}{Fe}{}^{+3}$ complex of Equation (1) both Si and Fe^{+3} are tetrahedrally coordinated and contribute $\frac{1}{4}$ of their charge to the non-bridging oxygen. The complex has a net charge of $-\frac{1}{4}$ $[O^{-2} - \frac{1}{4}(Si^{+4} + Fe^{+3})]$ which is charge balanced by $\frac{1}{4}\overset{*}{N}{}^{+}$. The $SiO(\frac{1}{2}Fe^{+2})$ complex contains a non-bridging oxygen which is bonded by $\frac{1}{4}Si^{+4}$ and $3Fe^{+2}$ cations. The Fe^{+2} cations are in six-fold coordination. Each of the Fe^{+2} cations contributes $\frac{1}{6}$ of its $+2$ charge, and therefore three such Fe^{+2} ions contribute a total charge of $+1$ and are equivalent to $\frac{1}{2}Fe^{+2}$. The $SiO(\frac{1}{2}Fe^{+2})$ complex has no net charge. Similar arguments can be applied to derive the other complexes in the equilibria in Table 1.

Equations (1) and (2) are equilibria in which the ferric iron is coordinated to four non-bridging and free oxygen respectively. The $\overset{*}{N}{}^{+}$ is a monovalent cation (or $\frac{1}{2}\overset{*}{M}{}^{+2}$ for a divalent cation) that is needed to balance the negative charge of the tetrahedrally coordinated Fe^{+3}. These monovalent cations are located within irregularly shaped cavities in the ferric-silicate network. Equations (3) and (4) are equilibria in which Fe^{+3} is coordinated to six non-bridging and free oxygen, respectively. All these equations predict that the redox ratio, Fe^{+3}/Fe^{+2}, is a complex function of the structure of the silicate melt. However, the validity of any or all of these equilibria can be determined only by testing them against experimental data. This is done below.

Douglas and his coworkers systematically measured the value of the redox ratio in alkali-oxide melts doped with about .6 wt % Fe (Paul and Douglas, 1965a,b). Since Fe occurs in very low concentrations, it may be considered as a probe ion which responds to but does not greatly influence the liquid structure. The melts were equilibrated in air, and were run at constant T, P, f_{O_2} and total Fe contents. Therefore, any observed changes in the redox state of iron are due solely to the variations of the alkali oxide content of the melt. In all the experiments, the redox ratio of iron increased with the alkali content (Figure 14). The oxidation shifts were greatest for the K_2O-SiO_2 melts and least for the Li_2O-SiO_2 melts.

The addition of alkali oxide to a silica-rich melt introduces a cation of low ionization potential and an oxygen ion that behaves as a strong Lewis base—that is, it is a good electron donor. These conditions lead to the formation of strong non-bridging oxygen, thereby increasing the activity of the non-bridging oxygen and decreasing the activity of the bridging bond. In reactions (2) and (4) in Table 1, the addition of alkali oxide to the silicate melt should drive the reactions to the right. In (2) this is caused by the increase in the activity of the non-bridging oxygen,

---

[3] Note that O^{-2} again refers to the oxygen ion and not to the free oxygen, O^{2-}.

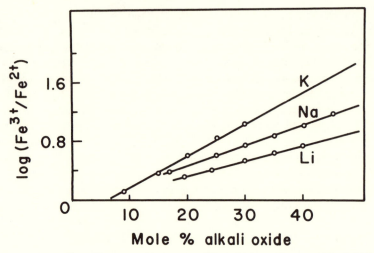

Figure 14. Log Fe^{+3}/Fe^{+2} versus mole fraction SiO_2 in binary silicate melts at 1,400°C.

that is, the activity of the Si–O–N complex and the decrease in the activity of the bridging oxygen, i.e. the activity of the Si–O–Si complex. In (4), the reaction is driven by a decrease in the activity of the bridging oxygen. The changes in the activities of the oxygen species result in an increase in the activity of the Fe^{+3} complex and a decrease in the activity of the Fe^{+2} complex. The redox rate, Fe^{+3}/Fe^{+2} should therefore be increased in agreement with the experimental data. The magnitude of this increase is controlled by the changes in the activities of the oxygen species. The activities of the non-bridging and bridging oxygen are most increased and decreased, respectively, by cations of low ionization potential. The order of ionization potentials for the alkali ions is K < Na < Li. From what was said above, the redox ratio in alkali silicate melts should be in the order K > Na > Li. This is what is in fact obtained (Figure 14).

Thus, the experimental data are consistent with the equilibria given in reactions (2) and (4). This conclusion is not true of reaction (3). The addition of alkali oxide to a silicate melt reduces the activity of the bridging oxygen and therefore increases and decreases the activities of the Fe^{+2} and Fe^{+3} complexes, respectively. The redox ratio should be lower in more alkali-rich melts, a result which is not obtained. This argues convincingly that Fe^{+3} cannot be coordinated *solely* by six non-bridging oxygen. An analysis of (1) is complicated by the fact that the redox ratio is increased by an increase in the activity of *either* the non-bridging or bridging oxygen.

An examination of the equilibria in Table 1 indicates that ferric iron may occur solely as a four-fold coordinated cation with free and non-bridging oxygen or as a six-fold coordinated cation with free oxygen. These equilibria, however, are not the only possible ones, because they do not admit the very likely possibility of finding Fe^{+3} in several co-ordination sites within the same melt. Support for this comes from the Mössbauer analysis of SiO_2–Na_2O–CaO–Fe–O glasses, also equilibrated in air, which indicate that Fe^{+3} exists in both four- and six-fold coordination in roughly equal concentrations (Levy et al., 1976). Although caution should be exercised in using such data (Waff, 1977), the Mössbauer spectra do suggest that the redox equilibria are more complex than those listed in Table I. These more complex equilibria can be evaluated by algebraically combining the simple reactions.

To illustrate this principle, Equations (2) and (3) are combined so that ferric iron occurs in four- and six-fold coordination in equal amounts. Thus

$$2SiO(\tfrac{1}{2}Fe^{+2}) + \tfrac{1}{2}SiON + \tfrac{1}{4}O_2 = [\overset{*}{Fe}^{+3}O\overset{*}{Fe}^{+3}]^{-1/2} + \tfrac{1}{2}N^+$$
$$+ \tfrac{3}{2}SiO(\tfrac{1}{3}Fe^{+3}) + \tfrac{1}{2}SiOSi. \quad (25)$$

This homogeneous equilibrium satisfies the following constraints. (1) Increasing the N_2O content of the silicate melt displaces the equilibrium to the right, and increases the redox ratio as required by the experimental data (Paul and Douglas, 1965a,b). (2) The log of the redox ratio Fe^{+3}/Fe^{+2} is, depending on the value of the activity coefficient ratio, $\gamma Fe^{+3}/\gamma Fe^{+2}$, proportional to $\tfrac{1}{4}\log f_{O_2}$, a dependence commonly observed in silicate melts. Actually, the pre-log f_{O_2} coefficient varies between .21–.25 (Waff and Weill, in preparation) and may be explained by a f_{O_2} dependence for the activity coefficient ratio term. (3) The ratio of Fe^{+3} in four-fold coordination to that in six-fold coordination is not compositionally dependent. This agrees with the work of Levy et al. (1976), who found that this ratio did not vary with SiO_2 or Na_2O/CaO content in the system CaO–Na_2O–Fe–Fe_2O_3–SiO_2. Thus, although the equilibrium in reaction (3) of Table 1 is not possible in simple silicate melts, it is viable when combined with some other equilibria.

The coupled reaction involving ferric iron in octahedral coordination with non-bridging oxygen and in tetrahedral coordination with free oxygen adequately satisfies the available Mössbauer experimental and theoretical constraints. In fact, all of the combinations of Equations (1) or (2) with (3) or (4) are equally satisfactory representations of the redox equilibria. None of these coupled equilibria can be eliminated until thermodynamic data become available to calculate the equilibrium constants of the equilibria. Until this is possible, one must accept the hypothesis that all the coordination states are likely in Al_2O_3-free melts. However, some additional information is obtained by coupling Equations

(1) and (2) and (3) and (4),

$$(1) + (2) \quad [\overset{*}{Fe}{}^{+3}O\overset{*}{Fe}{}^{+3}]^{-1/2} + SiOSi = 2[SiO\overset{*}{Fe}{}^{+3}]^{-1/4} \quad (26)$$

$$(3) + (4) \quad [(\tfrac{1}{3}Fe^{+3})O(\tfrac{1}{3}Fe^{+3})] + SiOSi = 2[SiO(\tfrac{1}{3}Fe^{+3})]. \quad (27)$$

The concentrations of Fe^{+3} in both four- and six-fold coordination with free oxygen increase at the expense of four- and six-fold coordinate complexes with non-bridging oxygens as the activity of bridging oxygens is decreased. The equilibria (1) and (4) in Table 1 thus become more important in SiO_2-poor melts.

DISCUSSION

In this concluding section, the polymerization model is used to examine two problems of igneous petrogenesis in which the silicate melt plays a central role: (1) the extent and origin of silicate liquid immiscibility in naturally occurring magmas; and (2) the effect of melt structure upon mineral-liquid trace element partitioning.

Silicate Liquid Immiscibility

Naturally occurring immiscible liquids of granitic and ferrobasaltic composition (Table 2), now preserved as partly devitrified glasses, are found

Table 2.

	1		2	
SiO_2	37.6	76.4	38.4	74.3
Al_2O_3	3.9	12.2	6.7	11.1
FeO	30.6	1.8	33.1	6.9
MgO	6.3	.5	.5	.1
CaO	11.0	1.1	11.3	2.1
Na_2O	.2	.5	.2	.4
K_2O	.3	6.1	.6	4.2
P_2O_5	3.1	.1	4.8	.2
TiO_2	6.0	.7	3.1	.7

Column Headings
(1) Naturally-occurring immiscible glasses in lunar basalt 14310 (Roedder and Weiblen, 1972).
(2) Immiscible melts produced experimentally as late stage liquids in lunar basalt 12038 at 987°C (Ryerson and Hess, 1978).

as both inclusions in minerals and within the mesostasis of lunar and terrestrial basalts (Roedder and Weiblen, 1970, 1971, 1972). Similar immiscible silicate melts (Table 2) have also been produced as residual liquids in fractional crystallization experiments of lunar and terrestrial basalts (Hess et al., 1975; Dixon and Rutherford, 1977). In both cases, the compositions of the coexisting silicate melts are analogues to immiscible silicate melts generated in the central two-liquid fields of the simpler K_2O–FeO–Fe_2O_3–Al_2O_3–SiO_2 and Na_2O–FeO–Fe_2O_3–Al_2O_3–SiO_2 systems (Roedder, 1951; Naslund, 1977). The normative composition of the granite immiscible melt is predominantly of quartz and feldspar and in this sense is a highly polymerized, tectosilicate liquid. The ferrobasalt melt contains abundant normative pyroxene and may be compared to a pyroxenitic melt containing a wide distribution of silicate chains. The miscibility gap between liquid pyroxenite and liquid tectosilicate thus mimics the miscibility gaps occurring between similar crystalline phases in the subsolidus regions. Naslund's (1977) results for the K_2O–FeO–Fe_2O_3–Al_2O_3–SiO_2 system are particularly instructive in this regard (Ryerson and Hess, 1978). At low oxygen fugacities ($f_{O_2} = 10^{-12}$), the stable two-liquid tielines lie on joins bounded by fayalite and a composition intermediate between silica and leucite. However, at high oxygen fugacities ($f_{O_2} = 10^{-.7}$) where fayalite is no longer stable with respect to magnetite-silica, the two liquid field widens such that the high FeO side of the solvus is now bounded by the magnetite composition. Tie lines joining the compositions of the coexisting immiscible melts are distinctly radial to the Fe_3O_4 composition. Hence the altered mineral stabilities are also reflected in changes in the melt structure. In this case an increase in the oxygen fugacity causes the homogeneous equilibrium

$$3FeSiO_3 \text{ (liquid)} + \tfrac{1}{2}O_2 \text{ (gas)} = Fe_3O_4 \text{ (liquid)} + 3SiO_2 \text{ (liquid)} \quad (28)$$

to be shifted to the right. In the oxygen nomenclature, non-bridging oxygen are converted to both free and bridging oxygen. The structure of the melt is transformed from one containing a wide distribution of silicate chains (that is, the structure of the ferrobasaltic immiscible liquid) at low oxygen fugacities, to a more polymerized and oxidized melt characterized by volumes rich in metal oxide and silica. The elimination of non-bridging oxygen and the oxidization of Fe^{+2} to the ferric state, a cation of high ionization potential, makes the melt unstable and causes the field of silicate liquid immiscibility to widen towards the Fe_3O_4 composition.

Such crystal-liquid analogies are now applied to determine whether other anhydrous magmas have the potential of developing silicate-liquid immiscibility. Crystalline silicates composed of polymerized structures intermediate to those of the pyroxenes, where one third of all Si–O bonds are bridging, and those of the tectosilicates, where all Si–O bonds are bridging (the tetrahedral Si–O–Al bond in the tectosilicate structures is roughly equivalent in strength to the Si–O–Si bond and is therefore treated here as a bridging bond), do not crystallize from anhydrous magmas. This suggests that anhydrous silicate melts of comparable intermediate states of polymerization, that is, magmas of dacite to andesite composition, should become immiscible at some temperature.

Whether silicate-liquid immiscibility does indeed develop stably in these magmas depends on the relative temperatures of the two-liquid field and their liquidi. Experiments indicate that the top of the two-liquid field of the lunar immiscible liquid lies near 1,100°C. Liquidus temperatures for andesites and dacites magmas at low pressures are above 1,200°C (see Ringwood, 1975). Moreover, a comparison of the composition of a typical andesite (or dacite) to the composition of a mixture of lunar immiscible liquids of comparable total SiO_2 content reveals that the andesites are significantly more Al_2O_3-rich (andesites $\sim 16\%$ versus 9% for mixture). Experiments in simple systems indicate that the addition of Al_2O_3 to magmas of low Al_2O_3 content sharply depresses the temperatures of the two-liquid field (e.g. Hess, 1977; Wood and Hess, 1977). Thus, it is anticipated that the two-liquid field of the andesite-dacite family of magmas is less than 1,100°C, and probably exists metastably below the solidus. These magmas should therefore not develop silicate-liquid immiscibility.

The onset of silicate-liquid immiscibility in natural anhydrous magmas is favored by a liquid line of descent characterized by an enrichment of FeO rather than SiO_2. These are residual liquids of ferrobasalt composition that have very low liquidus temperatures in the range of 1,100–1,000°C (Rutherford et al., 1974; Hess et al., 1975). A second requirement is that plagioclase crystallizes early to moderate the build-up of Al_2O_3 in the residual liquids. These magmas would be similar to those produced in the Na_2O–K_2O–FeO–Fe_2O_3–Al_2O_3–SiO_2 system and therefore have critical temperatures in excess of 1,000°C. The combination of low liquidus and high critical temperatures should produce silicate-liquid immiscibility in these late-stage magmas.

This trend of FeO enrichment has been experimentally reproduced in nearly all varieties of lunar basalt (Rutherford et al., 1974; Hess et al., 1975; Hess et al., 1977; McKay and Weill, 1977), and as anticipated above, nearly all lunar basalts have developed silicate liquid immiscibility in their residual magmas (Roedder and Weiblen, 1972; Dungan et al., 1975). The immiscible liquids are of granitic and ferrobasaltic composition (Table 2), and SiO_2-rich glasses of composition intermediate between these two extremes are not found in the mesostasis of any lunar basalts. Moreover, there are no examples of dacitic to andesitic volcanics in the lunar catalogue of rocks. These considerations suggest that nearly all samples of lunar granites were produced as immiscible liquids (Rutherford et al., 1976).

Differentiation trends characterized by strong FeO enrichments and relatively early crystallization of plagioclase, the "Fenner Trend," are obtained in many shallow terrestrial plutons of mafic compositions. Late stage granites or granophyres typically occur near the roof of many of these plutons. McBirney and Nakamura (1974) have shown by experiment that the granophyre of the Skaergard is likely to have originated as an immiscible liquid in late-stage ferrobasaltic magmas. These data and the preceding commentary suggest that many granophyres in such shallow mafic plutons have developed by silicate-liquid immiscibility. Moreover, the work of Dixon and Rutherford (1977) indicates that magmas of trondjemite composition found as segregations within ophiolite complexes may have a similar origin.

Mineral-Liquid Partitioning

The effect of silicate liquid structure upon mineral-liquid element partitioning can be determined from the element partitioning data for co-existing immiscible granite and ferrobasalt magmas (Ryerson and Hess, 1978; Watson, 1976). These immiscible melts may be in equilibrium with several minerals including olivine, pyroxene and plagioclase. According to the laws of phase equilibria, the chemical potential of any component, M, must be equal in all phases containing that component. Consider the example of a system containing pyroxene and two immiscible silicate melts. The chemical potential of FeO, μ_{FeO}, obeys the relation

$$\mu_{FeO}^{pyroxene} = \mu_{FeO}^{ferrobasalt} = \mu_{FeO}^{granite}. \tag{29}$$

The distribution coefficient, $D_{FeO}^{pyroxene-liquid}$, defined here as the weight % of the oxide in the pyroxene divided by the weight % of the oxide in the liquid is

$$D_{FeO}^{pyroxene-liquid} = \frac{FeO \text{ in pyroxene}}{FeO \text{ in liquid}}. \tag{30}$$

Now, the ratio of the two mineral-liquid distribution coefficients is equal to the value of the distribution coefficient between the two coexisting immiscible liquids,

$$\frac{D^{pyroxene-granite}}{D^{pyroxene-ferrobasalt}} = D^{ferrobasalt-granite}. \tag{31}$$

Thus, to determine the effects of melt structure on the values of $D^{mineral-liquid}$, it is necessary only to measure the value of the two-liquid distribution coefficient for the coexisting immiscible melts. Some implications of these data are reviewed below.

The patterns of element distribution between ferrobasalt and granitic immiscible melts were identical in every pair of melts studied (Ryerson and Hess, 1978; 1975; Watson, 1976). Elements can be grouped into three categories based on the value of the two-liquid distribution coefficients. A two-liquid distribution close to unity means that neither immiscible melt preferential accommodates the element. From Equation (31), the mineral-liquid distribution coefficients are equal

$$D^{\text{mineral-ferrobasalt}} = D^{\text{mineral-granite}}. \tag{32}$$

In terms of the degree of polymerization, the ferrobasalt and granite are approximate end members of the commonly occurring, volatile-free magmas. Since the mineral-liquid distribution coefficients are apparently unaffected by the state of polymerization of the melt, they can be assumed to remain roughly constant in all anhydrous magmas. This conclusion refers only to the influence of liquid structure and not to other important factors such as temperature, pressure or changing mineral composition. Elements that belong in this category are Ba, Sr and probably divalent Eu (Ryerson and Hess, 1978; Watson, 1976). These elements form cations of similar ionization potential, charge, and size.

The experimentally determined two-liquid distributions for Ti, Fe, Mn, Mg, Ca, P, Zr, Ta, Cr, Lu, Yb, Dy, Sm and La are all significantly greater than unity (Ryerson and Hess, 1978; Watson, 1976). The exact value of D is a function of the concentration of the given element, the concentration of other elements, and the proximity of the critical point for a given melt composition. The data for La illustrates the first two points (Figure 15). In low concentrations of La (less than $\sim .4$ wt % La_2O_3 for cases in which Henry's law applies), the two liquid distribution coefficient is 4 in P_2O_5-free immiscible melts. If the immiscible melts contain a few wt % P_2O_5, as is found in lunar immiscible melts (see Roedder and Weiblen, 1972), the distribution coefficient for La in the dilute concentration range is 15! Thus, it is not possible to list a single value for the two-liquid distribution coefficient for any element listed above.

However, the data do require that (Equation (31))

$$D^{\text{mineral-ferrobasalt}} < D^{\text{mineral-granite}}$$

for these elements. All these elements form cations of ionization potentials higher than Sr^{+2}. The two-liquid partitioning data indicate that, for cations of high ionization potential, the $D^{\text{mineral-liquid}}$ increases with increasing polymerization of the silicate melt. This is because cations of high ionization potential are not readily accommodated in a highly polymerized liquid structure. The argument was developed earlier and is not repeated here. This conclusion is verified by Watson (1977), who determined the partitioning of MnO between forsterite and liquids in the Mg_2SiO_4–$NaAlSi_3O_8$–$CaAl_2Si_2O_8$ system. He found that $D_{\text{Mn}}^{\text{forsterite-melt}}$

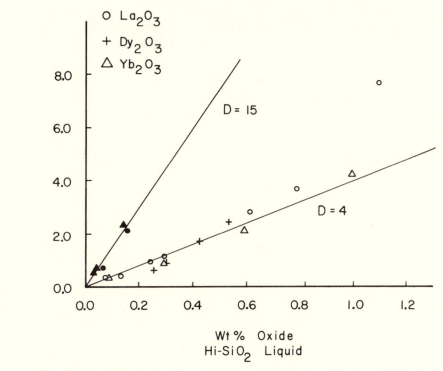

Figure 15. Distribution coefficients for La, Yb, Dy in coexisting immiscible melts at 1,000°C.
D = 15 for phosphorus-bearing melts.

increased with increasing polymerization of the melt at constant temperature.

Two-liquid distribution coefficients for Si, Al, K and Na are all less than unity. Al and Si, both cations of high ionization potential, are not partitioned into the ferrobasalt, but apparently copolymerize to form an aluminosilicate network in the granite melt. The alkalis are associated with the tetrahedrally coordinated Al in a charge-balancing capacity. In the granite immiscible melt, the mole ratio of Al to the total alkalis is greater than unity. This means that either some cations other than the alkalis charge-balance the network-forming Al, or some of the Al is coordinated to six-oxygens. Because the latter should be partitioned into the ferrobasalt immiscible melt, it is the author's opinion that most of the Al is tetrahedrally coordinated.

From Equation (31),

$$D^{\text{mineral-ferrobasalt}} > D^{\text{mineral-granite}}.$$

Thus, the magnitude of the mineral-liquid distribution coefficients for these elements should decrease with increasing polymerization of the anhydrous silicate melt. One possible exception should be noted. In per-alkaline magmas, where K + Na > Al (moles), the alkalis occur in concentrations in excess of that needed to charge-balance a network-forming Al ion. Since the alkalis form cations of low ionization potential, it is anticipated (but not yet proved) that the excess alkalis will be distributed equally between the immiscible silicate melts. The mineral-liquid distribution coefficients for the alkalis in peralkaline basalts should then be less than those in tholeiitic basalts.

CONCLUSION

An essential requirement for a successful theory is the reduction of the complexities of real systems to a few fundamental concepts. This is best done by a process of judicious simplification, in which certain aspects of a problem are ignored and others are emphasized. The wise execution of this method often makes the difference between a result which is purely academic and one which realistically grapples with the physical situation. Clearly, the polymerization model described in this paper is a relatively simple hypothesis that ignores the steric complications associated with the development of complex silicate anions. Moreover, the use of crystal-chemical concepts and crystalline silicate analogues to describe the structure of silicate melts is extremely simplistic. Nevertheless, the aim of this paper was to consider the present state of knowledge and extend these ideas to their theoretical limit, in hope of better constraining this wonderfully complex problem. The polymerization model is believed to obey the requirements of judicious simplification and may be the foundation from which fresh and novel approaches will rise.

REFERENCES

Baes, C. F., Jr., 1970. A polymer model for BeF_2 and SiO_2 melts: *J. Solid State Chem., 1*, 159–169.

Baltã, P., E. Baltã, and V. Ceausescu, 1976. Polymer mass distribution in silicate glasses: *J. Non-Cryst. Solids, 20*, 323–348.

Bell, R. J. and P. Dean, 1968. The structure of vitreous silica: Validity of the random network theory: *Phil. Mag., 25*, 1381–1398.

Bockris, J. O'M., J. A. Kitchener, S. Ignatowicz and J. W. Tomlinson, 1948. The electrical conductivity of silicate melts: systems containing Ca, Mn, Al: *Discuss. Faraday Soc., 4*, 281–286.

Bockris, J. O'M., and D. C. Lowe, 1954. Viscosity and structure of molten silicates: *Proc. Roy. Soc.* (London), A266, 423–435.

Bockris, J. O'M., J. D. MacKenzie and J. Kitchener, 1955. Viscous flow in silica and binary liquid silicates: *Trans. Faraday Soc., 51*, 1734–1748.

Bockris, J. O'M., J. W. Tomlinson and J. L. White, 1956. Structure of liquid silicates: Partial molar volumes and expansivities: *Trans. Faraday Soc.*, *52*, 299–310.

Brown, G. E., G. V. Gibbs and P. H. Ribbe, 1969. The nature and the variation in length of the Si–O and Al–O bonds in framework silicates: *Amer. Mineral.*, *54*, 1044–1061.

Brown, G. E. and G. V. Gibbs, 1970. Stereochemistry and ordering in the tetrahedral portion of silicates: *Amer. Mineral.*, *55*, 1587–1607.

Burns, R. G., 1970. *Mineralogical Applications of Crystal Field Theory*: Cambridge University Press, Cambridge.

Charles, R. J., 1969. The origin of immiscibility in silicate solutions: *Phys. Chem. Glasses*, *10*, 169–178.

Cotton, A. F. and G. Wilkinson, 1966. *Advanced Inorganic Chemistry*, John Wiley and Sons, New York.

Day, D. E. and G. E. Rindone, 1962. Properties of soda aluminosilicate glasses: I. Refractive index, density, molar refractivity and infrared absorption spectra: *J. Amer. Ceram. Soc.*, *45*, 489–496.

Dixon, S. J. and M. J. Rutherford, 1977. Trondhjemites as late stage immiscible liquids from fractionated oceanic basalts (Abs.): *EOS*. (*Trans. Amer. Geophys. Union*, *58*, 520).

Duffy, J. A. and M. D. Ingram, 1976. An interpretation of glass chemistry in terms of the optical basicity concept: *J. Non-Cryst. Solids*, *21*, 373–410.

Dungan, M. A., B. N. Powell and P. W. Weiblen, 1955. Petrology of some Apollo 16 feldspathic basalts: KREEP melt rocks? (Abs.): *Lunar Sci.* VI, Lunar Sci. Inst., Houston, 223–225.

Finn, C. W. F., D. J. Fray, T. B. King and J. G. Ward, 1976. Some aspects of structure in glassy silicophosphates: *Phys. Chem. Glasses*, *17*, 71–76.

Flory, P. J., 1953. *Principles of Polymer Chemistry*, Cornell University Press.

Fraser, D. G., 1977. Thermodynamic properties of silicate melts, in *Thermodynamics in Geology*, 301–325; D. Reidel Publishing Co., Dordrecht, Holland.

Galakhov, F. Ya., V. I. Aver'yanov, V. T. Vailonova and M. P. Aresher, 1976a. The metastable liquid phase separation region in the $MgO-Al_2O_3-SiO_2$ system: *Soviet Glass Phys. Chem.*, *2*, 405–408.

Galakhov, F. Ya., V. I. Aver'yanov, V. T. Vavilonova and M. P. Aresher, 1976b. The metastable two liquid region in the $Ga_2O_3-Al_2O_3-SiO_2$ and $Al_2O_3-SiO_2$ system: *Soviet J. Glass Phys. Chem.*, *2*, 127–132.

Gaskell, D. R., 1977. Activities and free energies of mixing in binary silicate melts: *Metall. Trans.*, *8B*, 131–145.

Gibbs, G. V., M. M. Hamil, L. S. Bartell and H. Yow, 1972. Correlations between Si–O bond length, Si–O–Si angle and bond angle overlap populations calculated using extended Huckel molecular orbital theory: *Amer. Mineral.*, *57*, 1578–1613.

Grieg, J. W., 1927. Immiscibility in silicate melts: *Amer. J. Sci.*, *13*, 1–44, 133–154.

Hess, P. C., 1971. Polymer model of silicate melts: *Geochim. Cosmochim. Acta*, *35*, 289–306.

Hess, P. C., 1977. Structure of silicate melts: *Can. Mineral.*, *15*, 162–178.

Hess, P. C., M. J. Rutherford, and H. W. Campbell, 1977. Origin and evolution of LKFM basalt: *Proc. 8th Lunar Sci. Conf.* 2357–2373.

Hess, P. C., M. J. Rutherford, R. N. Guillemette, F. J. Ryerson and H. A. Tuchfeld, 1975. Residual products of fractional crystallization of lunar magmas: an experimental study: *Proc. 6th Lunar Sci. Conf., 1*, 895–909.

Jeffes, J. H. E., 1975. The thermodynamics of polymeric melts and slags: *Silicates Ind., 40*, 325–340.

Konnert, J. H. and J. Karle, 1972. Tridymite-like structure in silica glass: *Nature Phys. Sci., 236*, 92–94.

Kushiro, I., 1975. On the nature of silicate melt and its significance in magma genesis: regularities in the shift of the liquids boundaries involving olivine, pyroxene and silica minerals: *Amer. J. Sci., 275*, 411–431.

Lentz, C. W., 1964. Silicate minerals as sources of trimethylsilyl silicates and silicate structure analysis of sodium silicate solutions: *Inorg. Chem., 3*, 574–578.

Levin, E. M., C. R. Robbins, and H. F. McMurdie, 1964. Phase diagrams for ceramists: *Amer. Ceram. Soc.*,

Levin, E. M., C. R. Robbins and H. F. McMurdie, 1969. Phase diagrams for ceramists 1969 Supplement: *Amer. Ceram. Soc.*

Levy, R. A., C. H. P. Lupis and P. A. Flinn, 1976. Mössbauer analysis of the valence and coordination of iron cations in SiO_2–Na_2O–CaO glasses: *Phys. Chem. Glasses, 17*, 94–103.

McBirney, A. R. and Y. Nakamura, 1974. Immiscibility in late-stage magmas of the Skaergaard Intrusion: *Carnegie Inst, Wash. Yearbook, 73*, 348–351

McKay, G. A. and D. F. Weill, 1977. KREEP petrogenesis revisited: *Proc. 8th Lunar Sci. Conf.*, 2339–2356.

MacDowell, J. F and G. H Beall, 1969. Immiscibility and crystallization in Al_2O_3–SiO_2 glasses: *J. Amer. Ceram. Soc. 52*, 17–25

Masson, C. R., 1965. An approach to the problem of ionic distribution in liquid silicates: *Proc. Roy. Soc. London, A287*, 201–221.

Masson, C. R., 1968. Ionic equilibria in liquid silicates: *J. Amer. Ceram. Soc., 51*, 134–143.

Masson, C. R., I. B. Smith and S. G. Whiteway, 1970. Activities and ionic distributions in liquid silicates: application of polymer theory: *Can. J. Chem., 48*, 1456–1464.

Masson, C. R., 1972. Thermodynamics and constitution of silicate slags: *J. Iron Steel Inst., 210*, 89–96.

Meadowcroft, T. R. and F. D. Richardson, 1965. Structural and thermodynamic aspects of phosphate glasses: *Trans. Faraday Soc., 61*, 54–70.

Morris, R. V. and L. A. Haskin, 1974. EPR measurement of the effect of glass composition on the oxidation states of europium: *Geochim. Cosmochin Acta, 38*, 1435–1445.

Morris, R. V., L. A. Haskin, G. M. Biggar and M. J. O'Hara, 1974. Measurement of the effects of temperature and partial pressure of oxygen on the oxidation states of europium in silicate glasses: *Geochim. Cosmochim. Acta, 38*, 1447–1459.

Mozzi, R. L. and B. E. Warren, 1969. The structure of vitreous silica: *J. Appl. Cryst., 2*, 164–172.

Naslund, H. R., 1976. Liquid immiscibility in the system $KAlSi_3O_8$–$NaAlSi_3O_8$–FeO–Fe_2O_3–SiO_2 and its application to natural magmas: *Carnegie Inst. Wash. Yearbook, 75*, 592–597.

Navrotsky, A., 1976. Silicates and related minerals: solid state chemistry and thermodynamics applied to geothermometry and geobarometry: *Progress in Solid State Chem.*, *11*, 203–264.

Navrotsky, A. and W. E. Coons, 1976. Thermochemistry of some pyroxenes and related compounds: *Geochim. Cosmochim. Acta*, *40*, 1281–1288.

Paul, A. and R. W. Douglas, 1965a. Ferrous-ferric equilibrium in binary alkali silicate glasses: *Phys. Chem. Glasses*, *6*, 207–211.

Paul, A. and R. W. Douglas, 1965b. Cerous-ceric equilibrium in binary alkali borate and alkali silicate glasses: *Phys. Chem. Glasses*, *6*, 212–215.

Pauling, L., 1960. *The Nature of the Chemical Bond*, 3rd Ed. Ithaca: Cornell Press.

Richardson, F. D., 1975. *Physical Chemistry of Melts in Metallurgy*: Academic Press, London.

Riebling, E. F., 1964. Structure of magnesium aluminosilicate liquids at 1,700°C: *Can. J. Chem.*, *42*, 2811–2821.

Riebling, E. F., 1966. Structure of sodium aluminosilicate melts containing at least 50 mole % SiO_2 at 1,500°C: *J. Chem. Phys.*, *44*, 2857–2865.

Ringwood, A. E., 1975. *Composition and Petrology of the Earth's Mantle*: McGraw-Hill.

Roedder, E., 1951. Low temperature liquid immiscibility in the system K_2O–FeO–Al_2O_3–SiO_2: *Amer. Mineral.*, *36*, 282–286.

Roedder, E. and P. W. Weiblen, 1970. Lunar petrology of silicate melt inclusions, Apollo 11 rocks: *Proc. Apollo 11 Lunar Sci. Conf.*, *1*, 801–837.

Roedder, E. and P. W. Weiblen, 1971. Petrology of silicate melt inclusions, Apollo 11 and Apollo 12, and terrestrial equivalents: *Proc. 2nd Lunar Sci. Conf.*, 507–528.

Roedder, E. and P. W. Weiblen, 1972. Petrographic features and petrologic significance of melt inclusions in Apollo 14 and 15 rocks: *Proc. 3rd Lunar Sci. Conf.*, 251–279.

Rutherford, M. J., P. C. Hess, and G. H. Daniel, 1974. Experimental liquid line of descent and liquid immiscibility for basalt 70017: *Proc. 5th Lunar Sci. Conf.*, *1*, 569–583.

Rutherford, M. J., P. C. Hess, F. J. Ryerson, H. W. Campbell and P. A. Dick, 1976. The chemistry, origin, and petrogenetic implications of lunar granite and monzonite: *Proc. 7th Lunar Sci. Conf.*, *2*, 1723–1740.

Ryerson, F. J. and P. C. Hess, 1975. The partitioning of trace elements between immiscible silicate melts (Abs.): *EOS Trans. Amer. Geophys. Union*, *56*, 470.

Ryerson, F. J. and P. C. Hess, 1978. Implications of liquid-liquid distribution coefficients to mineral-liquid partitioning: *Geochim. Cosmochim. Acta*, *42*, 921–932.

Simpson, D. R., 1977. Aluminum phosphate variants of feldspar: *Amer. Mineral.*, *62*, 351–355.

Smith, J. V., 1974. *Feldspar Minerals*, *2*: Springer-Verlag, New York.

Toop, G. W. and C. S. Samis, 1962a. Activities of ions in silicate melts: *Trans. Met. Soc. AIME*, *224*, 878–887.

Toop, G. W. and C. S. Samis, 1962b. Some new ionic concepts of silicate slags: *Can. Met. Quart.*, *1*, 129–152.

Uhlmann, D. R. and A. G. Kolbeck, 1976. Phase separation and the revolution in concepts of glass structure: *Phys. Chem. Glasses*, *17*, 146–158.

Veblen, D. R., P. R. Buseck and C. W. Burnham, 1977. Asbestiform chain silicates: New minerals and structural groups: *Science, 198*, 359–366.

Waff, H. S., 1977. The structural role of ferric iron in silicate melts: *Can. Mineral., 15*, 198–199.

Warren, B. E. and J. Biscoe, 1938. Fourier analysis of X-ray patterns of soda-silica glass: *J. Am. Ceram. Soc., 21*, 259–265.

Watson, E. B., 1976. Two liquid partition coefficients: experimental data and geochemical implications: *Contrib. Mineral. Petrol., 56*, 119–134.

Watson, E. B., 1977. Partitioning of manganese between forsterite and silicate liquid: *Geochim. Cosmochim. Acta, 41*, 1363–1374.

Westman, A. E. R. and P. A. Gartaganis, 1957. Constitution of sodium, potassium, and lithium phosphate glasses: *J. Amer. Ceram. Soc., 40*, 293–299.

Weyl, W. A. and E. C. Marboe, 1962. *The Constitution of Glass: A Dynamic Interpretation 1*, John Wiley, New York, 1–427.

Weyl, W. A. and E. C. Marboe, 1964. *Constitution of Glasses: A Dynamic Interpretation 2*, John Wiley, New York. 429–892.

Wood, M. I. and P. C. Hess, 1977. The role of Al_2O_3 in immiscible silicate melts (Abs): *EOS, Trans. Am. Geophy. Union 58*, 520.

Wright, A. C. and A. J. Leadbetter, 1976. Diffraction studies of glass structure: *Phys. Chem. Glasses, 17*, 122–145.

Zachariasen, W. H., 1932. Atomic arrangement in glass: *J. Amer. Ceram. Soc., 54*, 3841–3851.

Chapter 2

THE IGNEOUS SYSTEM
$CaMgSi_2O_6$—$CaAl_2Si_2O_8$—$NaAlSi_3O_8$:
VARIATIONS ON A CLASSIC THEME
BY BOWEN

D. F. WEILL and R. HON

Department of Geology, University of Oregon, Eugene

A. NAVROTSKY

Department of Chemistry, Arizona State University, Tempe

INTRODUCTION

It is a pleasure to contribute to a volume honoring N. L. Bowen's *The Evolution of the Igneous Rocks*. Since it was first published fifty years ago, Bowen's book has had a profound influence on several generations of geoscientists, who found in it a happy blend of introductory text and advanced treatise. The book continues to play an important part in the modern development of igneous petrology: even today it contains surprisingly little that is stale or badly dated. That this is so is not at all an indirect demonstration that petrologists have been idle since its publication. On the contrary, there has been a veritable explosion of data and understanding of igneous petrogenesis since Bowen's Princeton Lectures were first published, but it can truly be said that the seeds of much of the modern work are to be found in Bowen's own work, and in *The Evolution of the Igneous Rocks* in particular.

Bowen not only anticipated much of the modern work, but he did so in a lucid style that generously simplified difficult themes. Certainly one of Bowen's dominant motifs was that a fundamental understanding of crystal-liquid phase equilibria in silicate systems was essential to interpreting igneous rocks. It is difficult to improve on Bowen's own words (1928) in emphasizing this point.

> The accumulated data upon silicate systems offer us aid in two ways. They furnish us with definite facts concerning the behavior of certain silicates and their mixtures under varying conditions of temperature and pressure. They reveal the types of inter-relationships commonly displayed by silicates and the fundamental factors controlling these and thus permit us to deduce principles of more general applicability.

In wishing to "deduce principles of more general applicability" Bowen was voicing concern about the problem of extrapolation of phase equilibria from simple silicate systems to magmas. He had expressed this concern earlier in his well known paper of 1915 on the ternary system CaMgSi$_2$O$_6$(Di)–CaAl$_2$Si$_2$O$_8$(An)–NaAlSi$_3$O$_8$(Ab).

> The aim of experimental investigations of silicate melts is, of course, the explanation of some of the multitudinous, more or less disconnected facts concerning igneous rocks that have accumulated with the advance of descriptive petrography. It still is, and possibly always will be, a considerable extrapolation from the simple systems that can be investigated quantitatively to the more complex systems, or perhaps rather system, represented by magmas. Nevertheless, considerable progress is being made in the shortening of this extrapolation. It is not long since the discussion of the theory of the crystallization of igneous rocks was carried out by describing the crystallization of some simple binary eutectic mixture such as salt and water, and winding up with the statement that the crystallization of igneous rocks is in some measure analogous. A better grasp of the theoretical aspects of the problems and a considerable number of exact investigations of silicate melts themselves have now carried the matter well beyond this stage.

Since these words were written a large number of additional investigations of silicate melts have been carried out in "simple," "natural," and "dirty" systems. To be sure, these investigations have greatly expanded our catalog of empirical facts on silicate crystal-liquid phase equilibria. Each such investigation contributes to our understanding of a particular problem of petrogenesis, and the aggregate has carried our general sophistication about petrogenesis "well beyond this stage." Nevertheless, the "considerable extrapolation" from simple systems to magmas that Bowen alluded to remains elusive.

Although it is true that we are still far from being able to calculate phase equilibria in generalized magma systems even at one atmosphere, it is also true that the methodological and conceptual framework for beginning an attack on this important problem is available. Few would argue that classical thermodynamics is not the appropriate formalism or that the present technology of high temperature thermochemistry is inadequate for the task. Even fewer would argue against the proposition that the place to anchor a thermodynamic description of magmatic phase equilibria is with a simple, "magma-like" system; yet no such description exists.

Bowen selected the ternary system Di–An–Ab to illustrate the crystallization of basaltic and dioritic magmas reduced to their simplest form. The same system has also been used many times, following Bowen's lead, to illuminate for students of petrology the mysteries of experimental determination of phase equilibrium in glass-forming silicate systems, as well as the geometry of ternary phase equilibrium diagrams. For many of the same reasons we have also selected this system to illustrate some of the problems associated with achieving a thermodynamic description of a "simple" magma. At present, even such a simple system is far from being completely described, but substantial additions have now been made to the anatomy of Bowen's Di–An–Ab system, and we are happy to be able to present them in the appropriate setting of this commemorative volume. Much of our story involves a presentation of new data, but we also feel a strong obligation towards the pedagogical responsibilities of a book commemorating Bowen's. We can only hope that the mixture we have chosen to present will be palatable to students and practitioners alike of Bowen's modern petrology.

DIRECT MEASUREMENT OF PHASE EQUILIBRIUM

Experimental

Starting with Bowen's (1913) investigation of the melting of plagioclase feldspars, crystal-liquid equilibria in the Di–An–Ab system have been studied using the quench method by various investigators (Bowen, 1915; Osborn, 1942; Schairer and Yoder, 1960; Kushiro and Schairer, 1970; Kushiro, 1973; Murphy, 1977). A combined total of 146 data points defining the temperature-composition saturation surfaces of plagioclase and pyroxene have been assembled in Table 1, and the compositional coverage of the direct determination of phase equilibrium is illustrated

Table 1. A compilation of experimental data on pyroxene-plagioclase-liquid equilibrium in the bulk system $CaMgSi_2O_6(Di)$–$CaAl_2Si_2O_8(An)$–$NaAlSi_3O_8(Ab)$ and empirical equations for the pyroxene and plagioclase saturation surfaces. All compositions are given in mole fractions. T_m is the experimentally measured equilibrium temperature; T_c is the equilibrium temperature calculated according to the empirical equations given at the bottom of the table. All temperatures are given in °C.

| Data Point No. and [reference] | Pyroxene Saturation Surface | | | | | |
| | Liquid Composition | | | Temperature | | |
	Di	An	Ab	T_m	T_c	ΔT
1 [2]	.8371	.1629	.0000	1350	1344	6
2 [2]	.6583	.3417	.0000	1277	1281	−4
3 [2]	.7841	.0000	.2159	1340	1335	5
4 [2]	.5477	.0000	.5423	1285	1283	2
5 [2]	.2875	.0000	.7125	1210	1212	−2
6 [2]	.5575	.2951	.1474	1267	1259	8
7 [2]	.5552	.2247	.2201	1275	1269	6
8 [2]	.8151	.0813	.1036	1345	1343	2
9 [2]	.4517	.1846	.3637	1252	1249	3
10 [2]	.6406	.1352	.2242	1300	1300	0
11 [2]	.2899	.1354	.5747	1210	1209	1
12 [2]	.1883	.0825	.7292	1182	1177	5
13 [2]	.1305	.0369	.8326	1153	1153	0
14 [2]	.5203	.3079	.1718	1250	1246	4
15 [2]	.4700	.2821	.2479	1240	1237	3
16 [2]	.4156	.2533	.3311	1230	1227	3
17 [2]	.3602	.2150	.4248	1219	1218	1
18 [2]	.3414	.1993	.4593	1216	1215	1
19 [2]	.2655	.1296	.6049	1200	1201	−1
20 [2]	.1721	.0587	.7692	1176	1171	5
21 [2]	1.0000	.0000	.0000	1392	1393	−1
22 [3]	.9204	.0796	.0000	1368	1370	−2
23 [3]	.8371	.1629	.0000	1343	1344	−1
24 [3]	.7498	.2502	.0000	1313	1314	−1
25 [3]	.6583	.3417	.0000	1280	1281	−1
26 [3]	.6395	.3605	.0000	1274	1274	0
27 [4]	.9159	.0000	.0841	1368	1368	0
28 [4]	.8289	.0000	.1711	1347	1346	1
29 [4]	.7386	.0000	.2614	1323	1325	−2
30 [4]	.5967	.0000	.4033	1293	1294	−1
31 [4]	.4467	.0000	.5533	1263	1258	5
32 [4]	.1417	.0000	.8583	1158	1157	1
33 [4]	.1186	.0000	.8814	1140	1146	−6
34 [5]	.7442	.1241	.1317	1318	1324	−6
35 [5]	.5549	.2160	.2291	1268	1271	−3
36 [5]	.4744	.2550	.2706	1237	1243	−6
37 [5]	.4539	.2650	.2811	1233	1235	−2
38 [6]	.3947	.2226	.3827	1225	1227	−2
39 [6]	.4214	.2305	.3481	1230	1233	−3
40 [6]	.4534	.2471	.2995	1235	1239	−4
41 [6]	.4748	.2684	.2568	1240	1241	−1
42 [6]	.4960	.2852	.2188	1245	1243	2
43 [7]	.5540	.3540	.0920	1250	1249	1
44 [7]	.5570	.3570	.0860	1250	1249	1

Table 1 (*continued*)

Data Point No. and [reference]	Pyroxene Saturation Surface					
	Liquid Composition				Temperature	
	Di	An	Ab	T_m	T_c	ΔT
45 [7]	.5460	.3160	.1380	1250	1252	−2
46 [7]	.5120	.2540	.2340	1250	1253	−3
47 [7]	.4720	.1930	.3350	1250	1253	−3
48 [7]	.4340	.1500	.4160	1250	1250	0
49 [7]	.4180	.1110	.4710	1250	1250	0
50 [7]	.4000	.1060	.4940	1250	1245	5
51 [7]	.4270	.0000	.5730	1250	1253	−3
52 [7]	.4280	.0000	.5720	1250	1254	−4
53 [7]	.4200	.0000	.5800	1250	1251	−1
54 [7]	.7050	.2950	.0000	1300	1298	2
55 [7]	.6930	.2580	.0490	1300	1299	1
56 [7]	.6770	.2280	.0950	1300	1298	2
57 [7]	.6670	.1750	.1580	1300	1302	−2
58 [7]	.6570	.1390	.2040	1300	1304	−4
59 [7]	.6410	.1180	.2410	1300	1302	−2
60 [7]	.6230	.1070	.2700	1300	1299	1
61 [7]	.6160	.0560	.3280	1300	1299	1
62 [7]	.6190	.0000	.3810	1300	1298	2
63 [7]	.6180	.0000	.3820	1300	1298	2
	Plagioclase Saturation Surface					
64 [1]	.0000	.8333	.1667	1521	1525	−4
65 [1]	.0000	.6667	.3333	1490	1487	3
66 [1]	.0000	.5650	.4350	1465	1463	2
67 [1]	.0000	.5000	.5000	1450	1448	2
68 [1]	.0000	.3333	.6667	1394	1400	−6
69 [1]	.0000	.2750	.7250	1372	1375	−3
70 [1]	.0000	.2500	.7500	1362	1362	0
71 [1]	.0000	.2000	.8000	1334	1332	2
72 [1]	.0000	.0550	.9450	1205	1198	7
73 [1]	.0000	.1400	.8600	1287	1287	0
74 [1]	.0000	.1111	.8890	1265	1260	5
75 [1]	.0000	.0400	.9600	1175	1178	−3
76 [1]	.0000	.0300	.9700	1158	1165	−7
77 [2]	.5623	.4377	.0000	1328	1323	5
78 [2]	.4613	.5387	.0000	1390	1388	2
79 [2]	.2431	.7569	.0000	1485	1485	0
80 [2]	.1237	.7128	.1635	1482	1482	0
81 [2]	.2853	.5183	.1964	1404	1404	0
82 [2]	.5075	.3248	.1677	1257	1261	−4
83 [2]	.1805	.4121	.4074	1375	1373	2
84 [2]	.2943	.3849	.3208	1339	1342	−3
85 [2]	.3484	.3255	.3261	1298	1298	0
86 [2]	.4334	.2841	.2825	1251	1255	−4
87 [2]	.2917	.2361	.4722	1259	1259	0
88 [2]	.3461	.2155	.4384	1224	1231	−7
89 [2]	.1554	.2326	.6120	1305	1301	4
90 [2]	.2120	.1558	.6322	1242	1232	10
91 [2]	.1195	.1303	.7502	1254	1254	0

Table 1 (*continued*)

| Data Point No. and [reference] | Plagioclase Saturation Surface | | | | | |
| | Liquid Composition | | | | Temperature | |
	Di	An	Ab	T_m	T_c	ΔT
92 [2]	.1770	.0919	.7311	1200	1198	2
93 [2]	.1883	.0825	.7292	1182	1185	−3
94 [2]	.1189	.0463	.8348	1178	1184	−6
95 [2]	.1305	.0369	.8326	1153	1170	−17*
96 [2]	.5203	.3079	.1718	1250	1250	0
97 [2]	.1721	.0587	.7692	1176	1171	5
98 [2]	.2655	.1296	.6049	1200	1190	10
99 [2]	.3602	.2150	.4248	1219	1227	−8
100 [2]	.4700	.2821	.2479	1240	1246	−6
101 [2]	.4156	.2533	.3311	1230	1241	−11*
102 [2]	.3414	.1993	.4593	1216	1221	−5
103 [2]	.0000	1.0000	.0000	1553	1552	1
104 [3]	.6395	.3605	.0000	1274	1271	3
105 [3]	.1249	.8751	.0000	1533	1533	0
106 [3]	.2431	.7569	.0000	1488	1485	3
107 [3]	.3551	.6449	.0000	1438	1440	−2
108 [3]	.4613	.5387	.0000	1388	1388	0
109 [3]	.5623	.4377	.0000	1323	1323	0
110 [5]	.4539	.2650	.2811	1233	1240	−7
111 [6]	.3947	.2226	.3827	1225	1225	0
112 [6]	.4214	.2305	.3481	1230	1226	4
113 [6]	.4534	.2471	.2995	1235	1230	5
114 [6]	.4748	.2684	.2568	1240	1238	2
115 [6]	.4960	.2852	.2188	1245	1243	2
116 [7]	.0000	.1043	.8957	1250	1253	−3
117 [7]	.0000	.1015	.8985	1250	1250	0
118 [7]	.0663	.1060	.8277	1250	1253	−3
119 [7]	.1394	.1382	.7224	1250	1251	−1
120 [7]	.1625	.1339	.7036	1250	1238	12*
121 [7]	.2329	.1978	.5693	1250	1253	−3
122 [7]	.2365	.1997	.5638	1250	1253	−3
123 [7]	.2354	.2116	.5530	1250	1261	−11*
124 [7]	.3365	.2223	.4412	1250	1238	12*
125 [7]	.3410	.2285	.4305	1250	1241	9
126 [7]	.3974	.2479	.3547	1250	1241	9
127 [7]	.4019	.2558	.3423	1250	1245	5
128 [7]	.4915	.2860	.2225	1250	1244	6
129 [7]	.4838	.2930	.2232	1250	1249	1
130 [7]	.5107	.3124	.1769	1250	1254	−4
131 [7]	.5413	.3254	.1333	1250	1255	−5
132 [7]	.5540	.3540	.0920	1250	1268	−18*
133 [7]	.5570	.3570	.0860	1250	1269	−19*
134 [7]	.0000	.1407	.8593	1300	1287	13*
135 [7]	.0000	.1493	.8507	1300	1295	5
136 [7]	.0389	.1806	.7805	1300	1312	−12*
137 [7]	.1880	.2566	.5554	1300	1303	−3
138 [7]	.2895	.3039	.4066	1300	1301	−1
139 [7]	.4049	.3528	.2423	1300	1300	0
140 [7]	.4735	.3704	.1561	1300	1292	8
141 [7]	.5555	.4010	.0435	1300	1297	3
142 [7]	.5819	.4181	.0000	1300	1309	−9
143 [4]	.0953	.0000	.9047	1138	1141	−3
144 [4]	.0599	.0000	.9401	1148	1145	3
145 [4]	.0241	.0000	.9759	1143	1136	7
146 [4]	.0000	.0000	1.0000	1118	1121	−3

Table 1 (*continued*)

		Coexisting Pyroxene Compositions		
	$CaMgSi_2O_6$	$CaAl_2SiO_6$	$Mg_2Si_2O_6$	$NaAlSi_2O_6$
43 [7]	.8880	.0730	.0310	.0080
44 [7]	.8740	.0850	.0330	.0080
45 [7]	.8780	.0660	.0480	.0080
46 [7]	.9130	.0370	.0370	.0130
47 [7]	.9140	.0280	.0410	.0170
48 [7]	.9170	.0210	.0440	.0180
49 [7]	.9150	.0120	.0530	.0200
50 [7]	.9170	.0160	.0460	.0210
51 [7]	.9260	.0060	.0490	.0190
52 [7]	.9170	.0100	.0560	.0170
53 [7]	.9140	.0060	.0550	.0250
54 [7]	.9190	.0420	.0390	.0000
55 [7]	.9240	.0280	.0450	.0030
56 [7]	.9280	.0240	.0420	.0060
57 [7]	.9260	.0130	.0510	.0100
58 [7]	.9320	.0100	.0480	.0100
59 [7]	.9250	.0090	.0550	.0110
60 [7]	.9230	.0080	.0580	.0110
61 [7]	.9290	.0080	.0520	.0110
62 [7]	.9240	.0050	.0590	.0120
63 [7]	.9310	.0050	.0520	.0120

		Coexisting Plagioclase Compositions	
	$CaAl_2Si_2O_8$	$NaAlSi_3O_8$	$CaMgSi_3O_8$
64 [1]	.9400	.0600	nd
65 [1]	.8750	.1250	nd
66 [1]	.8333	.1667	nd
67 [1]	.8020	.1980	nd
68 [1]	.7100	.2900	nd
69 [1]	.6667	.3333	nd
70 [1]	.6450	.3550	nd
71 [1]	.5800	.4200	nd
72 [1]	.3333	.6667	nd
73 [1]	.5000	.5000	nd
74 [1]	.4500	.5500	nd
75 [1]	.2500	.7500	nd
76 [1]	.2250	.7750	nd
77 [2]	1.0000	.0000	nd
78 [2]	1.0000	.0000	nd
79 [2]	1.0000	.0000	nd
96 [2]	.9000	.1000	nd
97 [2]	.3333	.6667	nd
98 [2]	.5000	.5000	nd
99 [2]	.6667	.3333	nd
100 [2]	.8333	.1667	nd
101 [2]	.7130	.2870	nd
102 [2]	.6415	.3585	nd
103 [2]	1.0000	.0000	nd
104 [3]	1.0000	.0000	nd
105 [3]	1.0000	.0000	nd
106 [3]	1.0000	.0000	nd
107 [3]	1.0000	.0000	nd
108 [3]	1.0000	.0000	nd
109 [3]	1.0000	.0000	nd
112 [6]	.7182	.2818	nd

Table 1 (*continued*)

	Coexisting Plagioclase Compositions		
	$CaAl_2Si_2O_8$	$NaAlSi_3O_8$	$CaMgSi_3O_8$
116 [7]	.4195	.5805	.0000
117 [7]	.4205	.5795	.0000
118 [7]	.4699	.5243	.0058
119 [7]	.4915	.4987	.0098
120 [7]	.5165	.4727	.0108
121 [7]	.5771	.4072	.0157
122 [7]	.5798	.4055	.0147
123 [7]	.5852	.3991	.0157
124 [7]	.6167	.3646	.0187
125 [7]	.6412	.3391	.0197
126 [7]	.6681	.3102	.0217
127 [7]	.6779	.2944	.0277
128 [7]	.7329	.2354	.0317
129 [7]	.7389	.2303	.0308
130 [7]	.7681	.2011	.0308
131 [7]	.8022	.1629	.0349
132 [7]	.8409	.1232	.0359
133 [7]	.8472	.1128	.0400
134 [7]	.5130	.4870	.0000
135 [7]	.5170	.4830	.0000
136 [7]	.5132	.4868	.0000
137 [7]	.6438	.3483	.0079
138 [7]	.7166	.2666	.0168
139 [7]	.7852	.1889	.0259
140 [7]	.8310	.1431	.0259
141 [7]	.9119	.0550	.0331
142 [7]	.9658	.0000	.0342
146 [4]	.0000	1.0000	nd

Equation of Pyroxene Saturation Surface**

$$T(°C) = 1,085.6 + 575.2(x) + 72.5(y) - 544.7(x^2) - 848.4(y^2) + 276.8(x^3) + 1,846.1(xy^3)$$

63 points used in regression; $\Sigma(\Delta T)^2 = 616$; standard error of estimate $= 3.3°$

Equation of Plagioclase Saturation Surface**

$$T(°C) = 1,121.2 + 788.4(x) + 1,551.6(y) - 7,481.8(x^2) - 6,640.0(xy) - 3,044.6(y^2)$$
$$+ 15,769.2(x^3) + 37,941.1(x^2y) + 9,212.1(xy^2) + 3,080.4(y^3) - 7,823.1(x^4)$$
$$- 74,614.5(x^3y) - 32,077.7(x^2y^2) - 3,025.0(xy^3) - 1,156.5(y^4) + 32,215.8(x^4y)$$
$$+ 48,865.1(x^3y^2)$$

74 points used in regression; $\Sigma(\Delta T)^2 = 1,332$; standard error of estimate $= 5.0°$

Equations of Cosaturation Curve (Cotectic)**,

$$T(°C) = 1,139.5 + 688.0(y) - 3,059.0(y^2) + 8,683.5(y^3) - 799,360(y^7) + 1,711,691(y^8)$$
$$x = 0.104 + 1.236(y) - 1.510(y^2) + 591.108(y^5) - 2,201.148(y^6) + 5,272.268(y^8)$$

* These points were not included in the regression analysis for the plagioclase saturation surface.

** x = mole fraction of Di and y = mole fraction of An

Valid for $0 \leq y \leq 0.3639$

References: [1] Bowen, 1913; [2] Bowen, 1915; [3] Osborn, 1942; [4] Schairer and Yoder, 1960; [5] Kushiro and Schairer, 1970; [6] Kushiro, 1973; [7] Murphy, 1977

in Figure 1. For those points where the information is available we have also included in Table 1 the compositions of the solid solution phases.

Before the development of *in situ* microanalysis of quench specimens by electron-beam induced X-rays, phase-equilibrium data points—that is, compositions of coexisting phases—were mostly determined by "bracketing." The appearance or disappearance of a phase was noted as phase boundaries were crossed by varying the temperature at a constant bulk composition or vice versa. (See Bowen, 1915, for a good illustration of this approach applied to the Di–An–Ab system.) The electron microprobe now allows for a more direct approach, because a single run from which equilibrated phases are directly analyzed yields the same information that previously had to be determined from a series of bracketing runs. In addition to having this obvious advantage, direct analysis of coexisting phases also eliminates one of the necessary (but not always

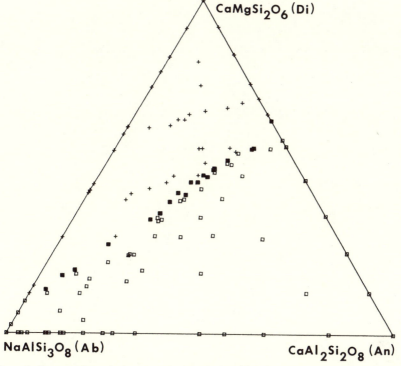

Figure 1. Compositions (moles) on the liquidus surfaces in the system CaMgSi$_2$O$_6$–CaAl$_2$Si$_2$O$_8$–NaAlSi$_3$O$_8$ as reported from quenching experiments. See text and Table 1 for references and details of composition. Crosses on pyroxene liquidus, open squares on plagioclase liquidus, and solid squares on the pyroxene + plagioclase cotectic.

justified) assumptions of the bracketing technique, that is, that phase compositions are limited to certain predetermined stoichiometric combinations. In the context of the Di–An–Ab ternary diagram this means that pre-microprobe determinations by the bracketing technique had to assume that plagioclase was a solid solution within the limits CaAl$_2$Si$_2$O$_8$– NaAlSi$_3$O$_8$, that diopside had the exact composition CaMgSi$_2$O$_6$, and, consequently, that residual liquids had compositions within the triangle CaMgSi$_2$O$_6$–CaAl$_2$Si$_2$O$_8$–NaAlSi$_3$O$_8$. Fortunately, these assumptions, while not strictly correct, were valid to a good first approximation, and the earlier work is reliable, subject to the above limitations. Nevertheless, it is interesting to consider some of the compositional details which have been brought out recently.

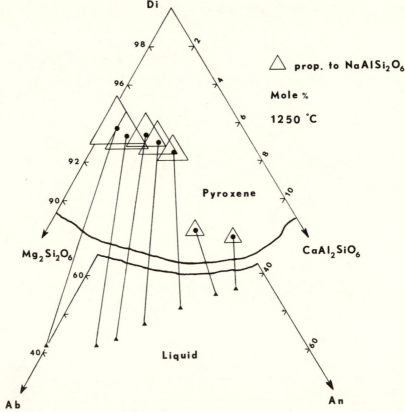

Figure 2. Compositions (moles) of pyroxene (solid solution) and liquids in equilibrium at 1,250°C. Dots represent pyroxene compositions projected on the composition triangle CaMgSi$_2$O$_6$–Mg$_2$Si$_2$O$_6$–CaAl$_2$SiO$_6$. The size (length of side) of each associated triangle is equal to the concentration of NaAlSi$_2$O$_6$. Coexisting liquid compositions are indicated by the tie-line and solid triangle plotted in the composition triangle Di–An–Ab. Data from Murphy (1977).

COMPOSITIONS OF THE SOLID SOLUTIONS

Osborn (1942) was the first to point out that pyroxenes crystallized from binary Di–An liquids did not correspond exactly to the composition $CaMgSi_2O_6$ and that, therefore, liquids from which such pyroxenes had precipitated could not be considered strictly binary. Substitution of Mg^{2+} for Ca^{2+} and the coupled substitutions of $Na^+ + Al^{3+}$ for $2M^{2+}$ and $2Al^{3+}$ for $M^{2+} + Si^{4+}$ on the pyroxene lattice account for most of the departures from the idealized diopside formula. These substitutions can be conveniently represented in terms of the pyroxene components $Mg_2Si_2O_6$, $NaAlSi_2O_6$, and $CaAl_2SiO_6$ (Kushiro and Schairer, 1970; Kushiro, 1973; Murphy, 1977). Even at one atmosphere the concentration of "jadeite" component ($NaAlSi_2O_6$) is not negligible. Murphy's 1977 results on pyroxene compositions are incorporated in Table 1, and the systematics of the pyroxene vs. coexisting liquid compositions at 1,250°C are illustrated in Figure 2.

Plagioclase solid solutions crystallized from ternary liquids contain "excess" Mg that may be considered to enter the structure via a coupled substitution of $Mg^{2+} + Si^{4+}$ for $2Al^{3+}$ and can be represented as the

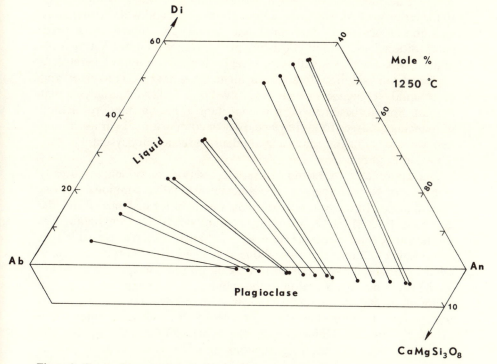

Figure 3. Compositions (moles) of plagioclase (solid solution) and liquids in equilibrium at 1,250°C. After Murphy (1977).

component CaMgSi$_3$O$_8$ (Kushiro and Schairer, 1970; Bryan, 1974; Bruno and Facchinelli, 1975; Longhi et al., 1976; Murphy, 1977). The concentration of this component as determined by Murphy (1977) is given in Table 1. Systematic trends in plagioclase vs. coexisting liquid compositions at 1,250°C are illustrated in Figure 3.

COMPOSITIONS OF THE LIQUIDS

Although any ternary liquid which has crystallized a finite amount of pyroxene or plagioclase thereby becomes non-ternary, there are good reasons for wanting to plot liquid compositions in the Di–An–Ab triangle. In the experimental determination of saturation surfaces, the liquids are usually not subject to much crystallization: that is, the bulk composition is chosen to be only slightly oversaturated with either pyroxene or plagioclase at the temperature of a run, so that the residual liquid compositions are usually not far off the ternary composition plane. Furthermore, the earlier bracketing work did not consider departures from ternary composition, and, if we are to combine old and new data, it is useful to consider the most efficient way to reduce "inexact ternary" compositions to the ternary plane. The problem is a general one, and will become more common as phase equilibria in simple systems are studied using direct compositional analyses of coexisting phases rather than bracketing techniques. Direct analyses will often reveal a limited amount of heretofore neglected solid solution. In addition, all analyses are subject to error, and even liquids (or glasses) which are truly ternary will yield analyses that are not. For all these reasons it is important to have a standard method of reducing analyses to a preselected set of compositional variables, and here we briefly illustrate such a method, using the ternary system Di–An–Ab as an example.

One approach to reducing analyses to a set of components is already well known to petrologists who calculate normative components from rock analyses. In the example before us we might for instance "form" the components CaMgSi$_2$O$_6$, CaAl$_2$Si$_2$O$_8$, and NaAlSi$_3$O$_8$ from an analysis of the liquid and normalize the sum to unity. The trouble with that procedure, of course, is the usual one associated with norm calculations: what order should be followed in making up the normative molecules? While the order of formation in the standard norm recipe or its various adaptations may have some significance in specific applications, norms are calculated, more often than not, according to what amounts to an arbitrary sequence. Unfortunately, the sequence of calculation can significantly affect the resulting fractions of normative components. The most straightforward method of getting around this problem is to use a procedure that incorporates a neutral criterion instead of a sequential recipe, and calculate what we might call a "least square norm."

In the general case we will have analytically measured concentration factors for a number of constituents that we wish to distribute among preselected normative components in a way that minimizes $\Sigma(C_m - C_n)^2$, where the sum is taken over all constituents, C_m is the measured concentration of a constituent, and C_n is its concentration calculated from the normative composition. For the particular case of an analysis of a quenched glass nominally in the system Di–An–Ab, we will have measured concentration factors (C_m) for each of the five constituents SiO$_2$, Al$_2$O$_3$, MgO, CaO, and Na$_2$O. Taking the constituent SiO$_2$ as an example, we may calculate C_n from the mass balance relation; $C_n =$ .50Di + .50An + .75Ab, where Di, An, and Ab represent molar norm fractions. Similar expressions can easily be worked out for the four other constituents. It remains only to set the partial derivatives of $\Sigma(C_m - C_n)^2$ with respect to Di, An, and Ab equal to zero, and solve for the three normative components. (The solution is not constrained by Di + An + Ab $\equiv$ 1 at this stage; normalization is the last step.)

$$Di = .32SiO_2 - 1.64Al_2O_3 + 2.36MgO + 1.00CaO - .28Na_2O$$

$$An = -.32SiO_2 + 2.64Al_2O_3 - 1.36MgO + 2.00CaO - .72Na_2O$$

$$Ab = 1.28SiO_2 - .56Al_2O_3 - .56MgO - 2.00CaO + .88Na_2O$$

The equations above yield the optimum mole fractions of Di, An, and Ab from the measured concentrations of SiO$_2$, Al$_2$O$_3$, and so on. If desired, the problem can, of course, also be worked out in terms of weight fractions by substituting the appropriate numerical coefficients. The compositions listed in Table 1 from Murphy (1977) have been calculated in this fashion.

REPRESENTATION OF THE SATURATION SURFACES

Ternary T–X saturation surfaces are conventionally represented by temperature contours drawn on the ternary composition triangle. The details of contour interval and shape are perforce determined by the distribution of data points. For some applications it is sufficient to "eyeball" the contours. In cases where there are relatively few data points this may well be the only way to derive the surface. Indeed, when it is done by the experimentalist who has developed a feel for the phase diagram, the method is preferable to any mathematical approach, because it is usually very difficult to estimate departures from ideal thermodynamic behavior and, hence, to predict the curvature of liquidus surfaces. Nevertheless, without some quantitative approach it is difficult to obtain any idea about the accuracy involved in interpolating between data points. Fortunately, the large number of data points on the plagioclase and pyroxene liquidus surfaces in the Di–An–Ab system (Figure 1) provides much better coverage than usual, and we can take advantage of this by fitting the points to some analytical function suitable for interpolation.

An empirical fit of the plagioclase and pyroxene liquidus surfaces is also very useful in that it makes it possible to compare the thermodynamically derived liquidus with the experimental liquidus over a continuum, rather than only at discrete points. As long as the fitting function is to be used for interpolative purposes only and has no physical significance, it is best to use a simple mathematical formalism that lends itself easily to regression analysis. We have fitted the liquidus surfaces to polynomials in two independent composition variables by stepwise regression. The coefficients of the two equations are given in Table 1. The differences between experimentally measured and calculated temperatures are also given for each data point in Table 1. The fit is excellent (standard errors of estimate are 3.3 and 5.0°C for the pyroxene and

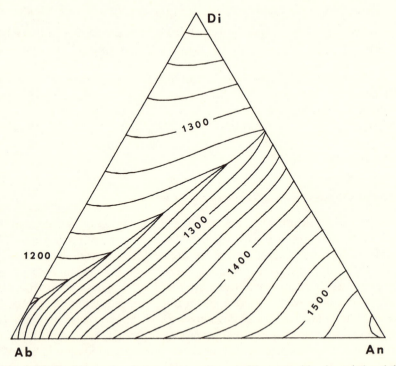

Figure 4. The liquidus surface (temperature contours in °C; composition in moles) and the cotectic line in the system Di–An–Ab according to the equations listed in Table 1.

plagioclase surfaces, respectively). Since the actual liquidus surfaces can be assumed to be smooth, the fit makes it possible to recognize anomalous experimental points by their large ΔT. Such points are also identified in Table 1. Considering the number of separate laboratories and methods involved in developing the data set in over sixty years, it is remarkably self-consistent. The equations in Table 1 and their illustration in Figure 4 can be considered with a high degree of confidence to represent the true saturation surfaces. In order to complete the description of the experimental phase-equilibrium results, we also include in Table 1 the equations for temperature and composition along the cosaturation line.

Such equations have a variety of applications. For example, they can be combined with models for the T–X dependence of silicate liquid densities (Bottinga and Weill, 1970) and viscosities (Bottinga and Weill, 1972; Shaw, 1972) in order to derive expressions for density and viscosity along the liquidus surfaces. The results are shown in Figures 5 and 6. Using these equations, it is possible to calculate the solid-liquid density contrast

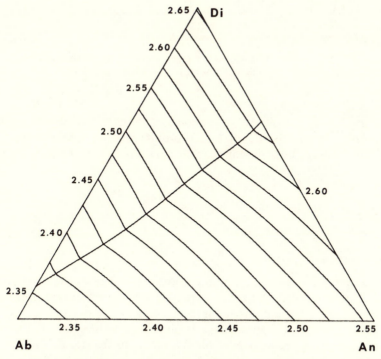

Figure 5. Density of liquids (contours in grams/cm^3) along the liquidus surface (moles). The pyroxene + plagioclase cotectic is also shown for reference.

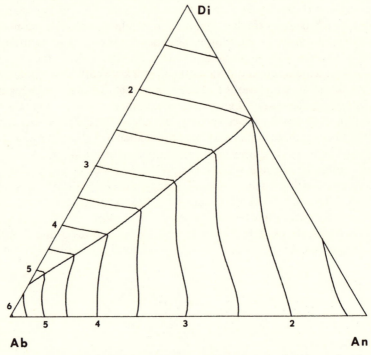

Figure 6. Viscosity of liquids (contours in log$_{10}$poises) along the liquidus surface (moles). The pyroxene + plagioclase cotectic is also shown for reference.

and the liquid viscosity as continuous functions of any crystallization path in the "haplobasaltic" system. It is quite realistic to look forward to the day when such calculations can be carried out for real basalts.

THERMOCHEMISTRY

ENTHALPY OF FUSION

The enthalpies of fusion of diopside, anorthite, and albite ($\Delta_f H_{Di}$, $\Delta_f H_{An}$, and $\Delta_f H_{Ab}$, respectively) are important quantities which are needed in thermal energy calculations and in developing thermodynamic equations to represent crystal-liquid phase equilibrium. Values for $\Delta_f H_{Di}$ and $\Delta_f H_{Ab}$ have been recommended recently by Carmichael et al. (1977). Both values were chosen by these authors to be compatible with the slopes of experimentally determined P–T fusion curves (Boyd and England, 1963) and the equilibrium relation $dP/dT = \Delta_f H/T\Delta_f V$ for these two minerals. Enthalpies of fusion have also been derived from calorimetry (enthalpies of solution combined with high temperature enthalpy), and from melting-point lowering curves in simple binary systems. The latter method usually depends on certain assumptions about the ideal behavior of liquid and solid solutions. Since one of the objects of this paper is to demonstrate that it is best not to make such assumptions, we will not further consider estimates of $\Delta_f H_{Di}$ and $\Delta_f H_{Ab}$ derived from melting-point lowering curves.

We have recently measured the enthalpies of solution of crystalline diopside, high albite, and anorthite in molten $2PbO \cdot B_2O_3$ at $712 \pm 1°C$. When combined with the enthalpies of solution of glasses of identical compositions (see section below on solution calorimetry) these yield enthalpies of "vitrification," $\Delta_v H$, which can be converted to $\Delta_f H$ according to

$$\Delta_f H(T) = \Delta_v H(T_{sc}) + \int_{T_{sc}}^{T_g} (C_{p,g} - C_{p,c}) \, dT + \int_{T_g}^{T} (C_{p,l} - C_{p,c}) \, dT, \quad (1)$$

where T_{sc} is the temperature of solution calorimetry, T_g is the glass transition temperature, and $C_{p,g}$, $C_{p,c}$, and $C_{p,l}$ are the isobaric heat capacities of glass, crystalline solid, and supercooled liquid, respectively. From Equation (1) it is obvious that the magnitude of the difference between $\Delta_f H$ and $\Delta_v H$ is dependent on T, T_g, and the differences in heat capacities between the crystalline and glass or liquid phases. These relations are illustrated in Figure 7 for the case of CaMgSi$_2$O$_6$, and it is clear that $\Delta_v H$ and $\Delta_f H$ (at liquidus temperatures) cannot be considered equivalent without introducing serious error.

A listing of $\Delta_f H$ for diopside and albite derived from various sources is given in Table 2 along with other necessary thermochemical data. For our purposes the agreement among the various estimates of $\Delta_f H$ is quite good. In all calculations carried out in this paper we have used the values of $\Delta_f H_{Di}$ and $\Delta_f H_{Ab}$ recommended in Table 2. In the case of anorthite there is no satisfactory agreement on $\Delta_f H$, and the estimates based on various combinations of calorimetric measurements range from 40 kilocalories/mole (Ferrier, 1969a,b) to 19–32 kilocalories/mole (Kracek and Neuvonen, 1952; this work) at the melting point. Much of the uncertainty is due to the large interval between the melting point of anorthite (1,830°K) and the temperatures at which heats of solution have been measured. In addition there is a large uncertainty in the heat capacity of liquid CaAl$_2$Si$_2$O$_8$. At present it must be concluded that the enthalpy of fusion of anorthite is a poorly known quantity (see Ludington, 1979).

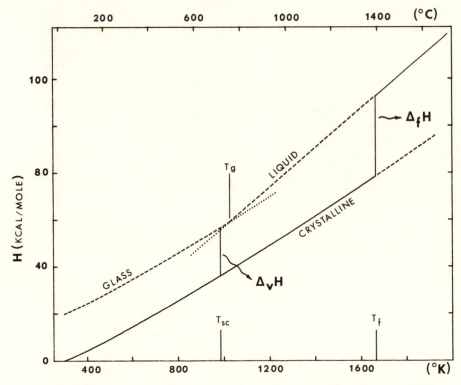

Figure 7. Enthalpies of CaMgSi$_2$O$_6$ phases relative to crystalline diopside at 298°K.
T_{sc} = 985°K; T_g = 1,026°K; T_f = 1,665°K; and $\Delta_v H$(@ 985°K) = 20,459;
$\Delta_f H$(@ 1,665°K) = 34,085 calories/mole. See text and Table 2 for details.

PREPARATION OF SAMPLES FOR SOLUTION CALORIMETRY

The enthalpy of solution, $\Delta_s H$, of glasses with compositions in the three
binary systems, Di–An, An–Ab, and Di–Ab have been measured at
712°C, using 2PbO·B$_2$O$_3$ as solvent. Enthalpies of solution have also
been measured for crystalline diopside, anorthite, and albite. Details of
the Calvet type twin calorimeter, sample assembly, and run procedures
have been described by Navrotsky and Kleppa (1968), Navrotsky (1973),
and Navrotsky and Coons (1976). Crystalline albite used for these measurements was prepared by heat treating Amelia albite at 1,050°C for
55 days. Diopside and anorthite were crystallized from synthetic glasses
at 1,350°C for 150 hours. The three end-member glass components were
prepared from reagent grade SiO$_2$, Al$_2$O$_3$, and MgO (fired progressively
to 950°C until constant weight was achieved), Na$_2$CO$_3$ and CaCO$_3$
(350–400°C until constant weight). Stoichiometric proportions of these
components were mixed under alcohol and then slowly decarbonated by

Table 2. Thermochemical data for the compounds $CaMgSi_2O_6$, $CaAl_2Si_2O_8$, and $NaAlSi_3O_8$. All quantities are expressed in units of calories, moles, and °K.

$$CaMgSi_2O_6$$

$C_{p,c} = 52.87 + 7.84 \times 10^{-3}T - 15.74 \times 10^5 T^{-2}$	Kelley (1960)
$C_{p,g} = 51.32 + 10.30 \times 10^{-3}T - 13.24 \times 10^5 T^{-2}$	Kelley (1960)
$C_{p,l} = 82.52$ to $84.98*$	Ferrier (1968b), Carmichael et al. (1977)
$T_f = 1,665$	
$T_g = 1,026**$	Briggs (1975)

$\Delta_f H(@\ 1,665°K) = 34,385$ (Tamman, 1903; HF solution calorimetry @ 293°K)[#]; 30,700 (Ferrier, 1968a,b; HF/HCl solution calorimetry @ 353°K and drop calorimetry); 34,626 (Navrotsky and Coons, 1976; $Pb_2B_2O_5$ solution calorimetry @ 985°K)[#]; 31,043 (Carmichael et al., 1977; from dP/dT of melting curve and $\Delta_f V$); 34,085 (this work; $Pb_2B_2O_5$ solution calorimetry @ 985°K)[#]

Adopted for calculations in this work:
$\Delta_f H(T) = 30.88T - 3.92 \times 10^{-3}T^2 - 15.74 \times 10^5 T^{-1} - 5,518$

$$CaAl_2Si_2O_8$$

$C_{p,c} = 64.42 + 13.70 \times 10^{-3}T - 16.89 \times 10^5 T^{-2}$	Kelley (1960)
$C_{p,g} = 66.46 + 11.84 \times 10^{-3}T - 18.22 \times 10^5 T^{-2}$	Kelley (1960)
$C_{p,l} = 85.54$ to $101.8*$	Ferrier (1969b), Carmichael et al. (1977)
$T_f = 1,830$	
$T_g = 1,085**$	Arndt and Häberle (1973)

$\Delta_f H(@\ 1,830°K) = 18,924$ to $30,924$ (Kracek and Neuvonen, 1952; HF solution calorimetry @ 348°K)[#]; 39,900 (Ferrier, 1969a,b; HF/HCl solution calorimetry @ 353°K and drop calorimetry); 20,098 to 32,098 (this work; $Pb_2B_2O_5$ solution calorimetry @ 985°K)[#]

$$NaAlSi_3O_8$$

$C_{p,c} = 61.70 + 13.90 \times 10^{-3}T - 15.01 \times 10^5 T^{-2}$	Kelley (1960)
$C_{p,g} = 61.31 + 18.00 \times 10^{-3}T - 16.16 \times 10^5 T^{-2}$	Kelley (1960)
$C_{p,l} = 85.69*$	Carmichael et al. (1977)
$T_f = 1,391$	
$T_g = 1,018**$	Arndt and Häberle (1973)

$\Delta_f H(@\ 1,391°K) = 14,263$ (Kracek and Neuvonen, 1952; HF solution calorimetry @ 348°K)[#] 14,971 (Holm and Kleppa, 1968 and Hlabse and Kleppa, 1968; Pb–Cd–B–O system solution calorimetry @ 964–971°K)[#]; 16,153 (Waldbaum and Robie, 1971; HF solution calorimetry @ 323°K)[#]; 15,620 (this work; $Pb_2B_2O_5$ solution calorimetry @ 985°K)[#]

Adopted for calculations in this work:
$\Delta_f H(T) = 23.99T - 6.95 \times 10^{-3}T^2 - 15.01 \times 10^5 T^{-1} - 3,224$

---

* Due to lack of precise data we are forced to consider $C_{p,l}$ in the supercooled liquid range to be constant, with negligible temperature dependence—clearly a rough approximation. The lower limits listed for $CaMgSi_2O_6$ and $CaAl_2Si_2O_8$ are probably better estimates in the temperature range immediately above T_g, whereas the upper limits listed are probably more representative of the region near the melting point. In our calculations we have used the average of the range indicated. In the case of $CaAl_2Si_2O_8$ the uncertainty is large, making it impossible to recommend with any confidence an expression for $\Delta_f H$.

** Also see Table 5 and the discussion of T_g in the text.

[#] All values of $\Delta_f H(@\ T_f)$ that are based on solution calorimetry of glasses, that is, on $\Delta_v H(@\ T_{sc})$, have been corrected according to Equation (1) using the thermochemical data given in this table.

gradually increasing temperature to 1,000°C. Each composition was fused at 1,650°C, quenched, and ground for a minimum of three cycles. All intermediate compositions were synthesized directly from appropriate mixtures of the three end-member glasses. Again, these mixtures were fused at 1,650°C, quenched and ground for three cycles.

All liquids in the ternary system are good glass formers and no problems were encountered with devitrification. Final glass compositions were all checked microscopically for homogeneity in oil immersions (index of refraction constant to within .0002). All glass compositions were

Table 3. Weighed-in (ideal) compositions of glasses for calorimetry compared with electron microprobe analyses*, and typical results of microprobe check for glass homogeneity**.

					Glass Compositions				
Ideal Composition (mole fraction)						Weight Fractions			
Di	An	Ab		SiO$_2$	Al$_2$O$_3$	MgO	CaO	Na$_2$O	Σ
1.0000	.0000	.0000	weighed-in	.5549	.0000	.1861	.2590	.0000	1.0000
			microprobe	.5469	.0000	.1858	.2632	.0000	.9959
.7500	.0000	.2500	weighed-in	.5930	.0559	.1326	.1845	.0340	1.0000
			microprobe	.5893	.0564	.1296	.1884	.0329	.9966
.7500	.2500	.0000	weighed-in	.5180	.1099	.1303	.2418	.0000	1.0000
			microprobe	.5189	.1077	.1259	.2444	.0000	.9969
.6250	.0000	.3750	weighed-in	.6107	.0818	.1078	.1500	.0497	1.0000
			microprobe	.6049	.0812	.1072	.1518	.0518	.9969
.6250	.3750	.0000	weighed-in	.5014	.1595	.1051	.2340	.0000	1.0000
			microprobe	.4963	.1570	.1038	.2353	.0011	.9935
.5000	.0000	.5000	weighed-in	.6275	.1065	.0842	.1171	.0647	1.0000
			microprobe	.6281	.1058	.0834	.1212	.0637	1.0021
.5000	.5000	.0000	weighed-in	.4857	.2061	.0815	.2267	.0000	1.0000
			microprobe	.4854	.2025	.0779	.2287	.0000	.9945
.2500	.0000	.7500	weighed-in	.6588	.1525	.0402	.0559	.0927	1.0000
			microprobe	.6594	.1530	.0418	.0587	.0882	1.0011
.2500	.7500	.0000	weighed-in	.4573	.2910	.0383	.2134	.0000	1.0000
			microprobe	.4577	.2893	.0388	.2178	.0012	1.0048
.1000	.9000	.0000	weighed-in	.4417	.3373	.0148	.2061	.0000	1.0000
			microprobe	.4422	.3319	.0147	.2096	.0018	1.0002
.0000	.0000	1.0000	weighed-in	.6874	.1944	.0000	.0000	.1182	1.0000
			microprobe	.6847	.1964	.0000	.0000	.1164	.9975
.0000	.2500	.7500	weighed-in	.6206	.2394	.0000	.0527	.0873	1.0000
			microprobe	.6202	.2372	.0000	.0516	.0850	.9940
.0000	.5000	.5000	weighed-in	.5559	.2830	.0000	.1038	.0573	1.0000
			microprobe	.5614	.2794	.0000	.1052	.0559	1.0019
.0000	.7500	.2500	weighed-in	.4930	.3254	.0000	.1534	.0283	1.0000
			microprobe	.5014	.3182	.0000	.1530	.0272	.9998
.0000	.9000	.1000	weighed-in	.4561	.3502	.0000	.1825	.0112	1.0000
			microprobe	.4570	.3440	.0000	.1848	.0120	.9978
.0000	1.0000	.0000	weighed-in	.4319	.3665	.0000	.2016	.0000	1.0000
			microprobe	.4289	.3602	.0000	.2051	.0000	.9942

Table 3 (*continued*)

Ideal Composition (mole fraction)				Glass Homogeneity			
				Average Counts	Variance	Sigma	99% Inhomogeneity
Di	An	Ab	Element	($\bar{N}$)	(s^2)	Ratio	Limits (% Relative)
1.0000	.0000	.0000	Si	43788	54438	1.12	.00− .53
			Mg	47889	57735	1.10	.00− .49
			Ca	78795	178710	1.51	.23− .63
.5000	.5000	.0000	Si	36943	33552	0.95	.00− .41
			Al	7305	7885	1.04	.00−1.13
			Mg	21081	30004	1.19	.00− .86
			Ca	69366	134022	1.39	.17− .60
.5000	.0000	.5000	Si	56215	91385	1.28	.06− .58
			Al	3995	4816	1.10	.00−1.71
			Mg	24516	38061	1.25	.00− .85
			Ca	41063	65710	1.27	.02− .68
			Na	3807	4045	1.03	.00−1.55
.0000	1.0000	.0000	Si	31559	41809	1.15	.00− .66
			Al	13269	15775	1.09	.00− .93
			Ca	61390	88697	1.20	.00− .51
.0000	.5000	.5000	Si	52004	56835	1.05	.00− .43
			Al	160678	369421	1.52	.17− .44
			Ca	80592	213444	1.63	.29− .68
			Na	3602	5224	1.20	.00−2.11
.0000	.0000	1.0000	Si	92191	203943	1.49	.20− .57
			Al	103193	287403	1.67	.27− .62
			Na	7396	7656	1.02	.00−1.08

* Analytical parameters for quantitative analysis: ARL-EMX-SM; 15 kev; 50 na specimen current on brass; 30-second integration on 4 spots; for Na only (6 na specimen current on brass and 30-second integration on 10 spots).

** Homogeneity test: 20-second integration on 50 spots at 50 na specimen current on brass for Si, Al, Mg and Ca; for Na only (30-second integration on 50 spots at 6 na specimen current on brass). Some "anomalies" in counts vs. concentration of elements are due to use of different diffracting crystals. See text for explanation of inhomogeneity limits.

analyzed with the electron microprobe. The results are given in Table 3. In all cases the probe analysis and the weighed-in composition agree within analytical error. Our general experience in glass synthesis indicates that it is best to consider the microprobe analysis as a check against serious discrepancies, and to adopt the weighed-in composition as the "true" value if there is agreement within analytical error.

The enthalpy of solution of glasses in this system is generally a non-linear function of composition. In addition to verifying the average composition of the glasses with the microprobe it is obviously important to monitor their homogeneity. X-ray counts for all elements except oxygen were accumulated at 50 randomly selected spots. Typical results are incorporated in Table 3. The data may be interpreted as follows. If a glass specimen is perfectly homogeneous the dispersion from the mean counts ($\bar{N}$) will be determined by the statistics of X-ray production (assuming, of course, that there are no other significant sources of variance), so that we might expect the observed standard deviation to be approximately equal to the standard "counting error" or $s^2 \approx \sigma_c^2 = \bar{N}$. Accordingly, the "sigma ratio", s/σ_c, is often used as an indicator of homogeneity, with the expectation that it should not differ greatly from unity for a homogeneous specimen. Sigma ratios are given in Table 3, and it can be seen that the glasses are relatively homogeneous according to this loosely defined criterion.

A more objective way of defining homogeneity is achieved by the following procedure. The parent population, from which we have drawn 50 sample counts, is assumed to have a Gaussian distribution whose total variance is due to a contribution from X-ray production statistics (σ_c^2) and a contribution from inhomogeneity (σ_i^2). The ratio, $s^2/(\sigma_c^2 + \sigma_i^2)$, may be expected to follow the chi-square probability function (49 degrees of freedom). Using the values of chi-square corresponding to the integral probabilities of 0.995 and 0.005, we can solve for the lower and upper limits of inhomogeneity at the 99% probability level. It can be seen from Table 3 that the glasses are homogeneous within the typical accuracy limits of the electron microprobe for each element analyzed.

Synthesis of a glass may precede calorimetry by a considerable period of time. For protection against contamination or reaction with the atmosphere, the glasses are sealed under vacuum as coarse pieces in plastic vials immediately after synthesis. The vials are opened and the samples crushed and ground to $-200/+350$ mesh size only immediately prior to loading in the calorimeter. Selected glass compositions have been thoroughly tested to determine that no devitrification occurs during the 3–4 hours necessary to equilibrate the samples in the calorimeter at 712°C before dissolution.

ENTHALPY OF SOLUTION, $\Delta_s H$

The enthalpy of solution of sixteen glasses in the three binary systems as well as the three crystalline solids in the system are reported in Table 4. The heat effects are much larger for the crystalline phases than for the glasses. Values for the latter range from endothermic ($\Delta_s H_{Ab} = 4{,}778$ calories/mole) through near-zero ($\Delta_s H_{Di} = -45$ calories/mole) to exothermic ($\Delta_s H_{An} = -3{,}211$ calories/mole). The enthalpy of solution reported for each composition is an average of a substantial number of replicate runs (6 to 32). Our best estimate of the instrumental precision (s.d.) of individual runs during the series of runs in the calorimeter is approximately ± 1.4 calories/gram, and this has been incorporated into the estimated error of the mean listed for each composition in Table 4.

The enthalpy of solution data can be put into more convenient analytical form for subsequent calculations by means of stepwise linear regression, using a polynomial equation of the general form indicated below (see p. 196 in Guggenheim, 1967):

$$\Delta_s H = A_1(x) + A_2(y) + A_3(z) + xy[A_4 + A_5(y) + A_6(y^2) + A_7(y^3)]$$
$$+ yz[A_8 + A_9(y) + A_{10}(y^2) + A_{11}(y^3)]$$
$$+ xz[A_{12} + A_{13}(x) + A_{14}(x^2) + A_{15}(x^3)], \tag{2}$$

where x, y, and z are mole fractions of Di, An, and Ab, respectively. After eliminating terms of low significance the regression analysis (weighted least squares fit; weighting factor $= 1/s_m^2$) yields the optimal equation for the enthalpy of solution,

$$\Delta_s H = -46(x) - 3,201(y) + 4,780(z) + xy[4,966 + 3,054(y)]$$
$$+ yz[7,256 + 3,893(y^2)] - xz[5,664]. \tag{3}$$

Table 4. Enthalpy of solution, $\Delta_s H$, of glasses and crystalline solids in Pb$_2$B$_2$O$_5$ at 712 ± 1°C. Compositions are given in mole fractions and enthalpies in calories/mole.

Composition			$\Delta_s H$		
			Measured ± (s.d.)$_m$	Calculated	
Di	An	Ab	(no. of runs)	(Eqn. 3)	Difference
	Glasses				
1.0000	.0000	.0000	−45 ± 61 (25)	−46	−1
.7500	.0000	.2500	179 ± 130 (6)	98	−81
.7500	.2500	.0000	158 ± 133 (6)	239	81
.6250	.0000	.3750	235 ± 134 (6)	436	201
.6250	.3750	.0000	131 ± 127 (7)	203	72
.5000	.0000	.5000	975 ± 101 (11)	951	−24
.5000	.5000	.0000	135 ± 110 (10)	−1	−136
.2500	.0000	.7500	2,537 ± 106 (11)	2,512	−25
.2500	.7500	.0000	−1,277 ± 123 (9)	−1,052	225
.1000	.9000	.0000	−2,113 ± 102 (14)	−2,191	−78
.0000	.0000	1.0000	4,778 ± 65 (32)	4,780	2
.0000	.2500	.7500	4,198 ± 108 (12)	4,191	−7
.0000	.5000	.5000	2,848 ± 109 (12)	2,847	−1
.0000	.7500	.2500	525 ± 121 (10)	565	40
.0000	.9000	.1000	−1,396 ± 129 (9)	−1,466	−70
.0000	1.0000	.0000	−3,211 ± 75 (27)	−3,201	10
	Crystals				
1.0000	.0000	.0000	20,414 ± 352 (4)		
.0000	1.0000	.0000	15,373 ± 242 (10)		
.0000	.0000	1.0000	17,180 ± 218 (9)		

Equation (3) is valid along the three binary systems Di–An ($x + y = 1$, $z = 0$), An–Ab ($y + z = 1$, $x = 0$), and Di–Ab ($x + z = 1$, $y = 0$). Enthalpies of solution calculated according to Equation (3) are compared to the measured values in Table 4 where it can be seen that the quality of the fit is good (standard error of estimate = 133 calories/mole).

ENTHALPY OF MIXING, $\Delta_m H$

The excess enthalpies of mixing in each of the binary systems can be derived from Equation (3) as illustrated in the sequence of reactions given below for the An–Ab system.

$$y\text{An (glass)} \rightarrow y\text{An (solution)}; \qquad \Delta H(4) = y\Delta_s H_{\text{An}} \qquad (4)$$

$$z\text{Ab (glass)} \rightarrow z\text{Ab (solution)}; \qquad \Delta H(5) = z\Delta_s H_{\text{Ab}} \qquad (5)$$

$$\text{An}_y\text{Ab}_z \text{ (glass)} \rightarrow \text{An}_y\text{Ab}_z \text{ (solution)}; \quad \Delta H(6) = \Delta_s H_{\text{An}_y\text{Ab}_z} \quad (6)$$

Combining these three reactions according to (4) + (5) − (6) we get

$$y\text{An (glass)} + z\text{Ab (glass)} \rightarrow \text{An}_y\text{Ab}_z \text{ (glass)};$$
$$\Delta H(7) = \Delta H(4) + \Delta H(5) - \Delta H(6) \qquad (7)^1$$

The enthalpy change for the process expressed by reaction (7), i.e., the mixing of y moles of CaAl$_2$Si$_2$O$_8$ and z moles of NaAlSi$_3$O$_8$ in the glassy state can easily be derived by calculating $\Delta H(4)$, $\Delta H(5)$, and $\Delta H(6)$ from Equation (3) and substituting in Equation (7). When this is carried out, the enthalpy of mixing in the An–Ab binary is thus seen to be given by the last three terms of Equation (3) (with their signs reversed) evaluated at compositions such that $y + z = 1$ and $x = 0$:

$$\Delta_m H_{\text{An–Ab}} = -yz[7{,}256 + 3{,}893(y^2)]. \qquad (8)$$

Similary, for the Di–Ab and Di–An system we obtain

$$\Delta_m H_{\text{Di–Ab}} = xz[5{,}664] \qquad (9)$$

$$\Delta_m H_{\text{Di–An}} = -xy[4{,}966 + 3{,}054(y)]. \qquad (10)$$

Equations (8–10) give the enthalpy of mixing in the binary systems for the glassy state at the temperature of solution calorimetry ($T_{sc} = 712°C$). For purposes of thermodynamic calculations involving liquid-crystal equilibrium we need the mixing properties in the liquid or supercooled liquid state at temperatures corresponding to the saturation surfaces in the system. Taking into account the change of heat capacity

[1] The enthalpy change due to reaction (7) is equal to the sum shown only if the solutions resulting from reactions (4) and (5) are sufficiently dilute in An and Ab that no measurable enthalpy change occurs when the two solutions are mixed. This is the case for all the calorimetry runs discussed in this paper.

at the glass transition, the general expression for the temperature dependence of $\Delta_m H$ is given by

$$\Delta_m H(@\ T) = \Delta_m H(@\ T_{sc}) + \int_{T_{sc}}^{T_g} (C_{p,g})\,dT + \int_{T_g}^{T} (C_{p,l})\,dT$$

$$- \sum_i X_i \int_{T_{sc}}^{T_g} (C_{p,i,g})\,dT - \sum_i X_i \int_{T_g}^{T} (C_{p,i,l})\,dT, \quad (11)$$

where X is a mole fraction, the summations are taken over the i end-member components of the mixed phase, and all other symbols are as previously defined. The nature and general magnitude of the correction to $\Delta_m H(@\ T_{sc})$ necessitated by increasing the temperature through the glass transition and into the supercooled liquid range can conveniently be discussed by reference to Equation (11). To start with we can observe that if variations of T_g with glass composition were negligible, and the heat capacities of glasses and liquids were linear with respect to composition, that is, if

$$T_g = \text{constant}, \quad (12)$$

$$C_{p,g} = \sum_i X_i C_{p,i,g}, \quad (13)$$

$$C_{p,l} = \sum_i X_i C_{p,i,l}, \quad (14)$$

the correction is zero, and $\Delta_m H(@\ T) = \Delta_m H(@\ T_{sc})$. Existing data on the heat capacities of silicate glasses (Bacon, 1977) and liquids (Carmichael et al., 1977) indicate that Equations (13–14) are valid first approximations, but there is no justification for adopting Equation (12) even as a gross approximation in the present context.

The exact thermodynamic nature of the glass transformation is complex. For present purposes it is sufficient to note that glass-forming silicate compositions of the type we are dealing with here exhibit discontinuities in thermal expansion and heat capacities at reasonably well-defined and reproducible temperatures. Ideally, we need only concern ourselves with the high-temperature enthalpy, $H_T - H_{T_{sc}}$, of the mixed composition and of the end-member component compositions, since these would give us, respectively, the sum of the second and third terms and the sum of the fourth and fifth terms in Equation (11). If such high-temperature enthalpy data were directly available there would be no need to consider either the sequential integration in Equation (11) or the glass transition at all. However, since such data are not available, we use the glass transition temperature simply as a convenient reference point to mark the end of the "glass" integration regime and the beginning of the "liquid" integration regime for purposes of calculating enthalpy.

The change in heat capacity at the glass transition in silicates averages about 8 calories/°K mole (Bacon, 1977). Provided the glass transition temperatures and the heat capacities used in Equation (11) are not systematically and grossly in error, it is unlikely that the total uncertainty in calculating the temperature correction to $\Delta_m H(@\ T_{sc})$, that is, the last four terms of Equation (11), will exceed about ± 100 calories/mole. Arndt and Häberle (1973) made a systematic study of the variations of T_g with composition in the An–Ab system by measuring the change of linear thermal expansion below and above the glass transition. Although their results are subject to different interpretations in detail, the general trend strongly suggests an approximately linear relation between composition and T_g. Using differential thermal analysis (DTA), Briggs (1975) measured T_g for the compositions Di, Di$_{50}$An$_{50}$, and An. The only compositional overlap between the determinations of Briggs and that of Arndt and Häberle is at the An composition, where Briggs recommends a value appreciably higher than that of Arndt and Häberle. We have adopted the latter here because it is part of a consistent set in which composition was systematically varied by small increments. In Table 5 we have listed the published values of T_g at various compositions in the systems Di–An

Table 5. Glass transition temperatures, T_g, for compositions along the Ab–An and Di-An binary systems. Compositions* in mole fractions and temperature in °K.

Ab(z)	An(y)	T_g	Reference
1.0000	.0000	1036**	Arndt and Häberle (1973)
.8960	.1040	1025	"
.8050	.1950	1029	"
.7370	.2630	1036	"
.6050	.3950	1041	"
.5130	.4870	1051	"
.4150	.5850	1053	"
.3140	.6860	1065	"
.2150	.7850	1069	"
.1160	.8840	1078	"
.0180	.9820	1086	"
Di(x)	An(y)		
1.0000	.0000	1028	Briggs (1975)
.5000	.5000	1051	"
.0000	1.0000	1153[#]	"

Recommended equation: $T_g = 1,026(x) + 1,085(y) + 1,018(z)$

* Compositions listed for the Ab–An system are averages of the two compositions listed for each glass by Arndt and Häberle

** Not used in linear regression because sample contained bubbles.

[#] Not used in linear regression because value is anomalously high relative to measurements of Arndt and Häberle and An–Ab. Only the glass transition temperature for the Ab composition departs significantly from an otherwise linear trend. Arndt and Häberle (1973) reported the presence of bubbles in the glass of Ab composition, and this may have significantly affected their measurement of thermal expansion. With the present set of data, and for the purposes of this paper, there is no justification for adopting any model more elaborate than a linear one for $T_g = f$(composition):

$$T_g(\text{in } °\text{K}) = 1{,}026(x) + 1{,}085(y) + 1{,}018(z). \tag{15}$$

Equations (13–15) and the appropriate data from Table 2 may now be substituted in Equation (11) to arrive at solutions for $\Delta_m H(@ \ T)$. The resulting analytical solutions are long and awkward. For example, the solution in the An–Ab system is

$$
\begin{aligned}
\Delta_m H_{\text{An–Ab}}(@ \ T \geq 1{,}085°\text{K}) = {} & 20{,}874(y) + 13{,}904(z) - 29{,}523(y^2) \\
& - 24{,}819(z^2) \\
& - 61{,}408(yz) + 6{,}969(y^3) + 9{,}327(z^3) \\
& + 23{,}673((y^2z) \\
& + 26{,}017(yz^2) - 3{,}893(y^3z) \\
& + \frac{[1{,}822(y) + 1{,}616(z)] \times 10^3}{[1{,}085(y) + 1{,}018(z)]}. \tag{16}
\end{aligned}
$$

Instead of retaining such cumbersome expressions for the enthalpy of mixing we have performed linear regression analyses of discrete points from them (25 points at .04 mole fraction intervals in each binary system) to obtain an optimal fit using the formalism for composition dependence expressed in Equation (2). The resulting expressions for $\Delta_m H(@ \ T \geq 1{,}085°\text{K})$ are of comparable precision and much simpler to handle. For example, the agreement between Equations (16) and (17) is better than a few calories/mole over the entire range of compositions in the An–Ab system:

$$\Delta_m H_{\text{An–Ab}}(@ \ T \geq 1{,}085°\text{K}) = -yz[6{,}667 - 15(y) + 3{,}893(y^2)] \tag{17}$$

$$\Delta_m H_{\text{Di–Ab}}(@ \ T \geq 1{,}026°\text{K}) = xz[5{,}788] \tag{18}$$

$$\Delta_m H_{\text{Di–An}}(@ \ T \geq 1{,}085°\text{K}) = -xy[5{,}367 + 3{,}054(y)]. \tag{19}$$

Within the limitations of the thermochemical data employed and the models implied in Equations (13–15), we have now derived analytical expressions for the enthalpy of mixing in the systems An–Ab, Di–Ab, and Di–An. In the next section of this paper we will use these expressions (Equations (17–19)) to calculate crystal-liquid phase equilibrium. At this point, we note with interest that Bowen (1928), preparatory to an analysis of the thermal effects of assimilation of rocks by magma, made what amounts to the reverse calculation for the same binary systems.

Using a model for the configurational entropy of mixing in the liquid
and solid solutions which assumed completely random mixing of units
of $CaMgSi_2O_6$, $CaAl_2Si_2O_8$, and $NaAlSi_3O_8$, and combining this
entropy with his experimentally determined crystal-liquid phase equilibrium diagrams, Bowen was able to calculate differential mixing heats

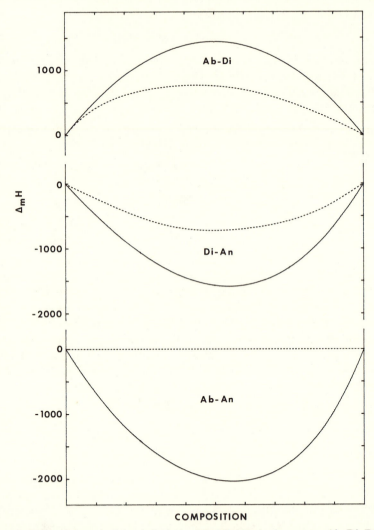

Figure 8. The enthalpy of mixing (calories/mole) of liquids in the systems Ab–Di, Di–An,
and Ab–An. The solid curves are drawn according to Equations (17–19), and the
dashed curves are the estimates of Bowen (1928). The left-right order of components in the system label on the figure corresponds to the composition scale
(moles).

(i.e., partial molar enthalpies of mixing, or $\Delta_m\bar{H}$, in the notation employed in this paper), and hence reconstruct the integral heats of mixing (our $\Delta_m H$) along the three binary systems. In Figure 8, Bowen's estimates of $\Delta_m H$ are shown for comparison with the graphical representations of Equations (17–19).

Bowen's phase diagram for the system An–Ab has long been used in introductory igneous petrology classes as a profitable illustration of a thermodynamic calculation of phase equilibria between liquid and solid solutions for which the mixing of the two end-member components could be considered ideal in both phases. Bowen himself (1913) was the first to point out the satisfactory agreement between the thermodynamically calculated solidus and liquidus curves and his experimental results. Hence, Bowen's estimate of $\Delta_m H_{An-Ab}$ was implied to be zero, or, at least negligible. Our calorimetric measurements indicate that mixing of liquids in this system is considerably exothermic, and that the agreement between the experimentally determined and the calculated phase diagram in this case is a fortuitous result of the assumptions built into the calculations rather than a fundamental confirmation of ideal mixing of CaAl$_2$Si$_2$O$_8$ and NaAlSi$_3$O$_8$ units in both liquid and solid solution. A corollary of our results for the liquids is that the mixing of the above units in crystalline plagioclase at near-liquidus temperatures is non-ideal and gives rise to a significant heat effect.

In the other two systems, Bowen's reconstructions of the enthalpy of mixing agree with our measurements in sign, but they are significantly different in magnitude. Since Bowen's calculations were based on essentially correct phase diagrams for the binaries, these differences in $\Delta_m H$ are already a clue that our $\Delta_m H$ data cannot be used, in combination with an entropy of mixing model that is ideal in terms of a random distribution of "Di," "An," and "Ab" units, to calculate the correct phase diagrams.

CALCULATIONS OF CRYSTAL-LIQUID PHASE EQUILIBRIUM

EQUATION OF THE LIQUIDUS SURFACE

The Gibbs function, G, of a solution phase containing n_i moles of i components may be expressed as

$$G = \sum_i n_i \left(\frac{\partial G}{\partial n_i}\right)_{T,\, n \neq i} = \sum_i n_i(\mu_i^\circ + \Delta_m\bar{H}_i - T\Delta_m\bar{S}_i), \qquad (20)$$

where μ_i° is the molar Gibbs free energy or the chemical potential of pure component i, and $\Delta_m\bar{H}_i$ and $\Delta_m\bar{S}_i$ are the partial molar enthalpy and entropy of mixing of component i, defined respectively by

$$\Delta_m\bar{H}_i = \left(\frac{\partial\Delta_mH}{\partial n_i}\right)_{T,\,n\neq i,} \tag{21}$$

$$\Delta_m\bar{S}_i = \left(\frac{\partial\Delta_mS}{\partial n_i}\right)_{T,\,n\neq i.} \tag{22}$$

Equation (20) implies that the chemical potential of component i is

$$\mu_i \equiv \left(\frac{\partial G}{\partial n_i}\right)_{T,\,n\neq i} = \mu_i^\circ + \Delta_m\bar{H}_i - T\Delta_m\bar{S}_i. \tag{23}$$

Since the equilibrium state in a heterogeneous system is defined by an equality of chemical potentials for each component throughout all phases, it is clear that no progress in calculating the conditions of crystal-liquid equilibria can be made without data or reasonable model assumptions about $\Delta_m\bar{H}_i = f$(composition, T) and $\Delta_m\bar{S}_i = f$(composition, T) in all phases. The discussion of thermochemical data in the previous section was concerned primarily with determining $\Delta_mH = f$(composition, T) in the binary liquids. Unfortunately, the enthalpy of mixing for plagioclase crystalline solutions is not known, and, in view of the fact that the liquid components of this system do not mix ideally, it should certainly not be assumed that there is no excess enthalpy of mixing for the solid solutions. For the time being the lack of appropriate data rules out the possibility of calculating the plagioclase saturation surface. Accordingly, we concentrate here on calculating "diopside" equilibrium, neglecting the relatively small amount of solid solution in the pyroxene phase.

The basic equation for diopside-liquid equilibrium is

$$\mu_{\mathrm{Di}}^{\circ,s} = \mu_{\mathrm{Di}}^l = \mu_{\mathrm{Di}}^{\circ,l} + \Delta_m\bar{H}_{\mathrm{Di}} - T\Delta_m\bar{S}_{\mathrm{Di}}, \tag{24}$$

which may be combined with

$$\frac{(\mu_{\mathrm{Di}}^{\circ,l} - \mu_{\mathrm{Di}}^{\circ,s})}{T} = \int_T^{T_f}\frac{\Delta_fH_{\mathrm{Di}}}{T^2}\,dT \tag{25}$$

to yield a form of the equilibrium equation which is useful for calculating the diopside saturation surface:

$$\int_T^{T_f}\frac{\Delta_fH_{\mathrm{Di}}}{T^2}\,dT + \frac{\Delta_m\bar{H}_{\mathrm{Di}}}{T} - \Delta_m\bar{S}_{\mathrm{Di}} = 0. \tag{26}$$

The integral term is easily evaluated after substituting the expression for Δ_fH_{Di} that is given in Table 2. The partial enthalpy of mixing in the second term is obtained by partial differentiation of Equations (18, 19):

$$\Delta_m \bar{H}_{Di}(\text{in the Di–Ab system}) = (1 - x)^2[5,788] \tag{27}$$

$$\Delta_m \bar{H}_{Di}(\text{in the Di–An system}) = (1 - x)^2[-8,421 + 6,107(x)]. \tag{28}$$

ENTROPY OF MIXING, $\Delta_m S$

Finally, we must adopt a model for the entropy of mixing so that the last term of Equation (26) may be evaluated. Obviously, any simple model that considers only the change in configurational entropy upon mixing of selected species can be nothing more than a rough approximation, but the exercise can be rewarding nevertheless. Of the large number of possible models involving various types of units for mixing, we have selected three for purposes of illustration. The first simply involves mixing of the three end-member components, $CaMgSi_2O_6$, $CaAl_2Si_2O_8$, and $NaAlSi_3O_8$; for convenience, it can be labeled the molecular mixing model. This is a simple extension of the ideal mixing model adopted by Bowen (1913) to calculate the An–Ab phase diagram. As previously discussed, it is not a realistic model, because of the significant enthalpy of mixing of liquids measured along the three binaries. Furthermore, it is difficult to reconcile this model with what is known about the structural roles of elements in silicate melts. We include it here because past practice has given it some claim to legitimacy, and because it can be used to illustrate the utility of Equation (26) in evaluating entropy models.

Flood and Knapp (1968) carried out a thermodynamic analysis of several feldspar-silica liquidus phase diagrams and suggested some structural models for the liquids. They concluded that the liquidus curves along the $NaAlSi_3O_8$–SiO_2 binary could be reproduced by calculations if the configurational entropy of mixing was assumed to be due solely to a randomizing of SiO_4 and AlO_4 tetrahedra. Such a scheme is compatible with a liquid structure in which most of the oxygens form bridges between tetrahedra, but in which Al^{3+} (coupled with Na^+) can exchange freely with Si^{4+} within a polymerized framework. The model is numerically equivalent to randomly mixing units of $NaAlO_2$ and SiO_2. For compositions along the $CaAl_2Si_2O_8$–SiO_2 binary, Flood and Knapp found excellent agreement with the melting curves by postulating random mixing of $Ca_2Al_4Si_3O_{14}$ and SiO_2 units. The large aluminosilicate unit may be considered as a polymer of SiO_4 and AlO_4 tetrahedra, held together by bridging oxygens. The Ca atoms are bound within the aluminosilicate unit. For compositions rich in the An component ($2CaAl_2Si_2O_8 = Ca_2Al_4Si_3O_{14} + SiO_2$), the model states that about three-fourths of the Si atoms are tied up in aluminosilicate polymer groups, while the rest form independent SiO_4 tetrahedra. Furthermore, the model implies that, within the aluminosilicate polymer unit, the Al and Si positions are specific and these atoms are ordered. Again, for compositions rich in the An component, for which the Al/Si ratio approaches unity, this ordering of Al and Si within a $Ca_2Al_4Si_3O_{14}$ unit inherent in the Flood and Knapp model may be considered a manifestation in the liquid state of the tendency for tetrahedrally coordinated Al and Si to be ordered in plagioclase feldspars of comparable composition (see, for example, Loewenstein (1954) and Smith (p. 79, 1974) for a discussion of the "aluminum avoidance rule"). It seems to us that the choice of $Ca_2Al_4Si_3O_{14}$ and SiO_2 as randomly mixed units runs into serious trouble (in addition to that caused by their large volume difference) as compositions become increasingly rich in SiO_2. As the concentration of this component increases, we would expect an increase in the fraction of bridging oxygens in the melt, that is, an increase in the general polymerization of the melt, until pure SiO_2 is reached, with all tetrahedra bridged by all oxygens. Unfortunately, the random mixing of SiO_2 and $Ca_2Al_4Si_3O_{14}$ can only be conceived in the context of a melt structure that, contrary to these expectations, requires that as the SiO_2 concentration increases, there is a concomitant increase in the fraction of Si in independent SiO_4 tetrahedra free to mix with $Ca_2Al_4Si_3O_{14}$ units. This particular shortcoming of the model does not occur if a component such as $CaMgSi_2O_6$ is added to either $CaAl_2Si_2O_8$ or $NaAlSi_3O_8$, because the addition of this component increases the value of $O/(Si + Al)$ in the melt, and would be expected in a general way to depolymerize the melt. We have considered, therefore, as a marginally permissible extension of the Flood and Knapp proposal, an entropy-of-mixing model for the Di–An–Ab system in which there is random mixing of $Ca_2Al_4Si_3O_{14}$, $NaAlO_2$, SiO_2, CaO, and MgO units.

The two mixing models discussed above consider the liquid to be a quasi lattice, on which the number of sites are simply equal to the number of chosen mixing units; the models are essentially defined by the choice of mixing units. The test of an entropy of mixing model is its compatibility with measured liquidus phase relations and enthalpy of mixing (as expressed in Equation (26)), and with what is known about silicate melt structure in general. Such models obviously can be deficient due to an unrealistic choice of mixing units. Indeed, it is questionable whether any single set of mixing units can remain totally realistic over the range of compositions represented by the Di–An–Ab system, because the melt structure and its dominant species are generally expected to change with composition. About the best that we can hope for is that, if such configurational entropy-of-mixing models are based on a reasonable average view of the melt structure within defined composition limits, the models will enable us to calculate the main contributions to the entropy increase when Di, An, and Ab components are mixed.

Perhaps the most critical aspect of melt structure to keep in mind when considering models for the calculation of the entropy of mixing is the fundamental difference in structural roles played by Si and Al, which form oxygen tetrahedra (more or less polymerized depending on O/(Si + Al) and the types of other cations in the melt), and the larger cations of much weaker ionic potential such as Na, Ca, and Mg, which are interstitial to the tetrahedral units. Since all liquids within the system Di–An–Ab have a relatively large fraction of bridging oxygens (2 $\leq$ O/(Si + Al) $\leq$ 3) it is unreasonable to consider that Si and Al atoms, tied up in polymeric species, are capable of exchanging with interstitial cations such as Na, Ca, and Mg. Perhaps a more realistic mixing model for aluminosilicate melts with a relatively high fraction of bridging oxygens would be to consider two independent sublattices, with Al and Si randomly distributed on one, and Na, Ca, and Mg on the other. Although we might expect such a two-lattice model to be more successful within the system Di–An–Ab than were the other models, we emphasize that there are also obvious deficiencies associated with it. For instance, it neglects possible ordering in the Si–Al distribution, as well as any tendency for clustering by the formation of Na–AlO$_4$, possibly AlO$_4$–Ca–AlO$_4$, and even AlO$_4$–Mg–AlO$_4$ groupings, in which the nonframework cation is located near the Al to provide local charge balance. Also, because the two-lattice model is based on the simple approximation that a large fraction of oxygens are bridging, it fails to take into account any change in the ratio of bridging to non-bridging oxygens. Therefore, the model necessarily ignores any possible entropy effects due to the mixing of units from one sublattice with units from the other sublattice. In spite of these difficulties, this model is more generally representative than the previous two of the realities of liquid structure in the Di–An–Ab system, and we have adopted this two-lattice model as our third illustrative example.

The expressions for configurational entropy of mixing corresponding to the three models are given in Table 6. The derivations are straightforward and follow directly from consideration of the configurational complexions (for example, see Denbigh, 1966) of the system in the mixed and unmixed states. We confine ourselves here to one illustrative example, that of the two-lattice model. Consider the mixing of X moles of CaMgSi$_2$O$_6$, Y moles of CaAl$_2$Si$_2$O$_8$, and Z moles of NaAlSi$_3$O$_8$. After mixing, the total number of cation sites on the tetrahedral sublattice of this model is simply equal to the total number of Al and Si atoms, that is, $(2XN + 4YN + 4ZN)$, where N is Avogadro's number. The number of Si and Al atoms to be randomly distributed on this sublattice are $(2XN + 2YN + 3ZN)$ and $(2YN + ZN)$ respectively. Similarly, the total number of cation sites on the interstitial sublattice is equal to $(2XN + YN + ZN)$, into which $(XN + YN)$ Ca atoms, (XN) Mg atoms, and (ZN)

Table 6. Configurational entropy of mixing (in calories/°K mole) for the three mixing models discussed in the text. Mole fractions of CaMgSi$_2$O$_6$, CaAl$_2$Si$_2$O$_8$, and NaAlSi$_3$O$_8$ are represented by x, y, and z respectively. The partial molar mixing entropies are obtainable directly from $\Delta_m S = x \Delta_m \overline{S}_{Di} + y \Delta_m \overline{S}_{An} + z \Delta_m \overline{S}_{Ab}$.

Model and Comments	Mixing Units	Entropy of Mixing ($\Delta_m S$)
"molecular" model; used by Bowen (1913) to calculate the plagioclase melting diagram	CaMgSi$_2$O$_6$, CaAl$_2$Si$_2$O$_8$, NaAlSi$_3$O$_8$	$-R[x \ln x + y \ln y + z \ln z]$
extension of model proposed by Flood and Knapp (1968) to explain the feldspar-silica binary systems	CaAl$_4$Si$_3$O$_{14}$, SiO$_2$, NaAlO$_2$, CaO, MgO	$-R\left[x \ln \dfrac{64x^2(3-x-2.5y)^2}{(4-3y)^4} + y \ln \dfrac{y\sqrt{2(3-x-2.5y)}}{(4-3y)\sqrt{y}} \right.$ $\left. + z \ln \dfrac{256(1-x-y)(3-x-2.5y)^3}{27(4-3y)^4} \right]$
"two-lattice" model; Al and Si on one sublattice; Na, Ca, and Mg on the other; only model to correctly predict an activity greater than zero for CaAl$_2$Si$_2$O$_8$ in liquids of the Di-Ab system.	SiO$_2$, AlO$_{1.5}$, and (separately) NaO$_{.5}$, CaO, MgO	$-R\left[x \ln \dfrac{x(3-x-y)^2(x+y)}{(2-x)^2(1+x)^2} + y \ln \dfrac{(3-x-y)^2(1-x+y)^2(x+y)}{(2-x)^4(1+x)} \right.$ $\left. + z \ln \dfrac{16z(3-x-y)^3(1-x+y)}{27(2-x)^4(1+x)} \right]$

Na atoms are to be randomly distributed. The total configurational complexity, Ω_m, of the mixed state is equal to

$$\Omega_m = \frac{(2XN + 4YN + 4ZN)!}{(2XN + 2YN + 3ZN)!(2YN + ZN)!} \cdot \frac{(2XN + YN + ZN)!}{(XN + YN)!(XN)!(ZN)!}. \tag{29}$$

Similarly, considering the configurations possible for the three pure components before mixing, we obtain

$$\Omega_0 = \frac{(2XN)!}{(XN)!(XN)!} \cdot \frac{(4YN)!}{(2YN)!(2YN)!} \cdot \frac{(4ZN)!}{(ZN)!(3ZN)!}. \tag{30}$$

The change of configurational entropy due to mixing is given by $k \ln (\Omega_m/\Omega_0)$ where k is the Boltzmann constant. Applying Stirling's approximation for large numbers ($\ln N! \simeq N \ln N - N$) the molar entropy of mixing reduces to

$$\Delta_m S = -Rx \ln \frac{x(3 - x - y)^2(x + y)}{(2 - x)^2(1 + x)^2}$$

$$- Ry \ln \frac{(3 - x - y)^2(1 - x + y)^2(x + y)}{(2 - x)^4(1 + x)}$$

$$- Rz \ln \frac{16z(3 - x - y)^3(1 - x + y)}{27(2 - x)^4(1 + x)}, \tag{31}$$

where R is the gas constant and the lower case x, y, and z denote, as before, the mole fractions of Di, An, and Ab, respectively. The partial molar entropies of mixing are readily obtained by taking the partial derivatives of $\Delta_m S$ according to Equation (22), or by applying $\Delta_m S = x \Delta_m \bar{S}_{Di} + y \Delta_m \bar{S}_{An} + z \Delta_m \bar{S}_{Ab}$ directly to Equation (31). For example,

$$\Delta_m \bar{S}_{Di} = -R \ln \frac{x(3 - x - y)^2(x + y)}{(2 - x)^2(1 + x)^2}. \tag{32}$$

The reader is referred to Table 6 for the complete set of equations from all three mixing models.

The Diopside Liquidus Curves in the Binary Systems Di-Ab and Di-An

Once the choice of entropy model is made, Equation (26) can be solved for temperature at any composition along the diopside liquidus in the Di–Ab and Di–An systems. Results using the two-lattice entropy calculation are illustrated in Figure 9. The agreement with experiment is excellent over the total range of composition (unusually large in the case of Di–Ab). The difference between calculated and experimentally measured liquidus temperature is shown for each of the three entropy-of-mixing models in Figure 10. The success of the two-lattice model relative to the "Flood and Knapp" and "molecular" models is apparent. Agreement between experiment and calculation confirms the idea that, for compositions in the systems under consideration, the two-lattice model is more representative of liquid structure; however, this agreement does not shed any light on the obvious shortcomings of the model, discussed in the previous section. At this stage we can only note that the model is compatible with the experimentally observed phase equilibrium and with the measured enthalpies of mixing in the two binaries, and so is the logical choice if we wish to extend our calculations into the ternary system.

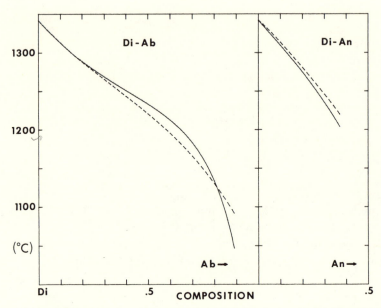

Figure 9. Diopside liquidus curves (composition in mole fraction) in the Di–Ab and Di–An systems. The solid curves are calculated according to Equation (26) and the two-lattice entropy model. The dashed curves are based on direct experimental results (see data and equation in Table 1).

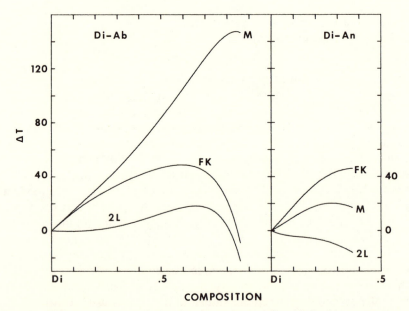

Figure 10. Difference between calculated and experimental diopside liquidus temperatures ($\Delta T = T_{calculated} - T_{measured}$ in °C) for three entropy-of-mixing models; M = molecular, FK = Flood and Knapp, and 2L = two-lattice.

EXTRAPOLATION INTO THE TERNARY SYSTEM

Equations (17–19) give the enthalpy of mixing along the three binary systems. An approximate representation of excess-mixing properties in a ternary system in terms of the excess-mixing properties along its three limiting binary systems was proposed by Kohler (1960). The Kohler equation calculates the excess property within the ternary by assuming that binary interactions causing the excess property at any particular composition along a limiting binary are "ideally diluted" by the third component as composition is extended into the ternary while preserving the concentration ratio of the binary components. The resulting expression for the molar enthalpy of mixing of a ternary composition x, y, z (mole fractions of Di, An, and Ab, respectively) is

$$\Delta_m H_{Di-An-Ab} = (1 - x)^2 \Delta_m H_{An-Ab} + (1 - y)^2 \Delta_m H_{Di-Ab}$$
$$+ (1 - z)^2 \Delta_m H_{Di-An}, \tag{33}$$

where $\Delta_m H_{An-Ab}$ is the molar enthalpy of mixing in the An–Ab binary system evaluated at the y/z (or An/Ab) ratio of the ternary composition, $\Delta_m H_{Di-Ab}$ is the molar enthalpy of mixing in the Di–Ab binary system evaluated at the x/z (or Di/Ab) ratio of the ternary composition, and $\Delta_m H_{Di-An}$ is the molar enthalpy of mixing in the Di–An binary system evaluated at the x/y (or Di/An) ratio of the ternary composition. The expression for ternary enthalpy of mixing given in Equation (33) is thus seen to be the sum of three binary enthalpy-of-mixing terms, each multiplied by the appropriate "$(1 - x)^2$" factor to account for the effect of dilution of the binary interactions by the addition of the third component. Combining Equations (33) and (17–19), we obtain the following expression for the molar enthalpy of mixing in the ternary system:

$$\Delta_m H_{Di-An-Ab} = xy[-5,367 - 3,054(y)(x + y)^{-1}] + xz[5,788]$$
$$+ yz[-6,667 + 15(y)(y + z)^{-1} - 3,893(y)^2(y + z)^{-2}]. \tag{34}$$

The corresponding enthalpy of mixing surface in the ternary system is illustrated in Figure 11.

The partial molar enthalpy of mixing of $CaMgSi_2O_6$ in the ternary system is obtained from

$$\Delta_m \bar{H}_{Di} = \left(\frac{\partial [(X + Y + Z) \Delta_m H_{Di-An-Ab}]}{\partial X} \right)_{Y,Z}, \tag{35}$$

where X, Y, and Z are the number of moles of Di, An, and Ab, respectively, and $x = X/(X + Y + Z)$, $y = Y/(X + Y + Z)$, and $z = Z/(X + Y + Z)$.

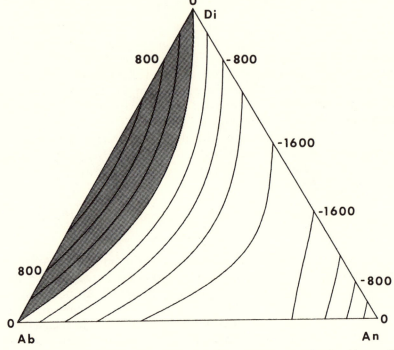

Figure 11. Enthalpy of mixing (calories/mole) for liquids in the system Di–An–Ab according to Equation (34). The mixing of CaMgSi$_2$O$_6$, CaAl$_2$Si$_2$O$_8$, and NaAlSi$_3$O$_8$ to form liquids within the shaded area is endothermic.

The solution to Equation (35) is given below, and the $\Delta_m \bar{H}_{Di}$ ternary surface is illustrated in Figure 12.

$$\Delta_m \bar{H}_{Di} = -5{,}367(y)(1-x) - 3{,}054(y)^2(y - x + xz)(x + y)^{-2}$$
$$+ 5{,}788(z)(1-x) + 6{,}667(yz) - 15(y)^2(z)(y + z)^{-1}$$
$$+ 3{,}893(y)^3(z)(y + z)^{-2}. \tag{36}$$

We are now in a position to extend the use of Equation (26) into the ternary system, and it is instructive to consider two alternative approaches. (1) Using the temperature and composition of the experimentally determined diopside liquidus surface (as defined by the empirical equation in Table 1), we can solve Equation (26) for $\Delta_m \bar{S}_{Di}$. This allows us to gain some perspective on the entropy function that is compatible with the experimental phase equilibrium results and with the enthalpy of mixing, which we have just extrapolated into the ternary system according to Kohler's approximation. The results of this exercise are illustrated in Figure 13,

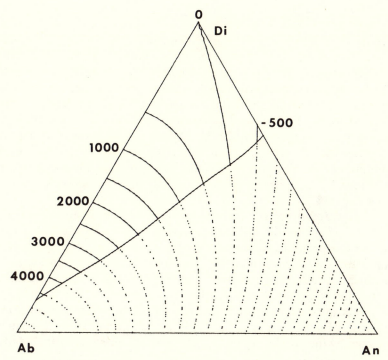

Figure 12. Partial molar enthalpy of mixing of Di (calories/mole) in liquids of the system Di–An–Ab according to Equation (36). The limit of diopside saturated liquids is shown for reference.

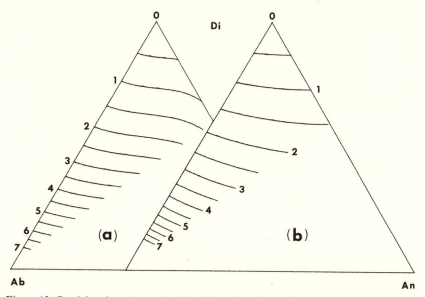

Figure 13. Partial molar entropy of mixing of Di (calories/°K mole) in liquids of the system Di–An–Ab. (a) Calculated along the diopside saturation surface from Equation (26) and the experimentally determined saturation surface. (b) Calculated directly from the two-lattice model according to Equation (32).

where we have also plotted, for comparison, the $\Delta_m\bar{S}_{Di}$ surface as calculated directly from the two-lattice model of Equation (32). The agreement is excellent along the entire liquidus surface. (2) Encouraged by the agreement between the "back-calculated" entropy and the two-lattice model entropy, we can adopt the latter model to calculate the entire ternary liquidus surface of diopside. The results are shown in Figure 14, where they may be compared with the experimentally measured liquidus surface. When all the uncertainties in the measurements and model assumptions leading up to the comparison of Figure 14 are considered, the agreement between experiment and thermodynamic calculation is very satisfying.

We are currently extending our calorimetric analysis to include glass compositions within the Di–An–Ab ternary. Preliminary results indicate that, for compositions in the primary field of crystallization of diopside, enthalpies of mixing predicted by the Kohler approximation are in fair agreement with the experimentally determined enthalpies, lending further support to the two-lattice model. For compositions in the primary field of crystallization of plagioclase, and rich in the An component, measured enthalpies of solution imply more exothermic mixing enthalpies than are predicted by the Kohler equation; this region is receiving further study.

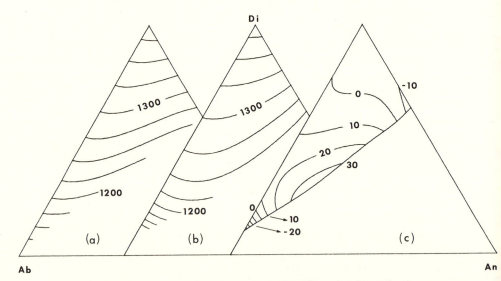

Figure 14. Comparison of calculated and measured diopside saturation surface (temperature contours in °C; composition in moles). (a) Experimentally determined, drawn according to equation recommended in Table 1. (b) Calculated from Equation (26) and the two-lattice entropy model. (c) $\Delta T = T_{\text{calculated}} - T_{\text{measured}}$.

CONCLUDING REMARKS

A key measure of progress in any branch of science is its ability to quantify and generalize its subject matter. The recent literature is an accurate indicator of the emphasis that geochemists and petrologists are placing on quantitative generalizations about silicate melts (for example, Wasserburg, 1957; Shaw, 1964; Thompson, 1967; Bottinga and Weill, 1970, 1972; Hess, 1971; Burnham, 1975; Fraser, 1976; Carmichael et al., 1977; Bottinga and Richet, 1978).

In this paper we have dissected a portion of the thermodynamic anatomy of the system Di–An–Ab, a system which was recognized long ago by Bowen as a valuable guide to interpreting phase equilibria in silicates, and, hence, central to understanding igneous processes and the origin of igneous rocks. The importance of achieving even a limited thermodynamic description of phase equilibria in a magma-like system will be obvious to the reader. Direct and specific experimental investigation of equilibrium and rate-controlled igneous phenomena will continue to be necessary for a long time. However, given the extremely broad ranges of both chemical composition and physical conditions involved in igneous processes, a comprehensive description will never be achieved without a theoretical framework suitable for extrapolation and interpolation. Thermodynamic data are necessary input for modeling many kinetic processes, and thermo-dynamic models *per se* are certainly the best hope for systematizing magmatic phase equilibria.

A general thermodynamic description of magmatic phase equilibria is a distant goal, and it is still too early to sort out from the various approaches those which hold the most promise for future generalization. Indeed, the problem is sufficiently difficult to render it unlikely that a single thermo-dynamic model will ever be successful for calculating phase equilibria in magmas. In our opinion it is more likely that a combination of direct experimental determination of phase equilibria, structural information on silicate melts, and a systematic program of calorimetry will lead to a manageable number of predictive models, each valid within a restricted range of composition, temperature and pressure.

Faced with such a long-term challenge, it is critical that we stimulate more than just a passing interest in these problems among today's students of igneous petrology. Several generations of students have benefited from the clear lessons that Bowen put in his *"Evolution of the Igneous Rocks."* We regard the thermodynamic description of a simple ternary silicate system that is attempted here as a modest extension of Bowen's com-bination of science and pedagogy, and we hope that we have provided a sufficient number of "exercises left to the student" to kindle interest in the development of thermodynamic models for magmatic phase equilibria.

ACKNOWLEDGMENTS

The debt we owe to N. L. Bowen is large, obvious, and a distinct pleasure to acknowledge here. We have profited from discussions with S. Banno, Y. Bottinga, I. Carmichael, and R. Robie. It is also a pleasure to acknowledge the financial assistance of Grants NGL-38-003-020 from the National Aeronautics and Space Administration and DMR 77-18235 from the National Science Foundation.

REFERENCES

Arndt, J. and Häberle, F., 1973. Thermal expansion and glass transition temperatures of synthetic glasses of plagioclase-like compositions, *Contrib. Mineral. Petrol. 39*, 175–183.

Bacon, C. R., 1977. High temperature heat content and heat capacity of silicate glasses: experimental determination and a model for calculation, *Am. J. Sci. 277*, 109–135.

Bottinga, Y. and Richet, P., 1978. Thermodynamics of liquid silicates, a preliminary report, *Earth Planet. Sci. Lett. 40*, 382–400.

Bottinga, Y. and Weill, D. F., 1970. Densities of liquid silicate systems calculated from partial molar volumes of oxide components, *Am. J. Sci. 269*, 169–182.

Bottinga, Y. and Weill, D. F., 1972. The viscosity of magmatic silicate liquids: a model for calculation, *Am. J. Sci. 272*, 438–475.

Bowen, N. L., 1913. The melting phenomena of the plagicoclase feldspars, *Am. J. Sci., 4th Series, 35*, 577–599.

Bowen, N. L., 1915. The crystallization of haplobasaltic, haplodioritic and related magmas, *Am. J. Sci., 4th Series, 40*, 161–185.

Bowen, N. L., 1928. *The Evolution of the Igneous Rocks*, Princeton University Press, Princeton, 333 pp.

Boyd, F. R. and England, J. L., 1963. Effect of pressure on the melting of diopside, CaMgSi$_2$O$_6$, and albite, NaAlSi$_3$O$_8$, in the range up to 50 kilobars, *J. Geophys. Res. 68*, 311–323.

Briggs, J., 1975. Thermodynamics of the glass transition temperature in the system CaO–MgO–Al$_2$O$_3$–SiO$_2$, *Glass Ceramics Bull. 22*, 73–82.

Bruno, E. and Facchinelli, A., 1975. Crystal-chemical interpretation of crystallographic anomalies in lunar plagioclases, *Bull. Soc. Franc. Mineral. Crist. 98*, 113–117.

Bryan, W. B., 1974. Fe–Mg relationships in sector-zoned submarine basalt plagioclase, *Earth Planet. Sci. Lett. 24*, 157–165.

Burnham, C. W., 1975. Water and magmas: a mixing model, *Geochim. Cosmochim. Acta 39*, 1077–1084.

Carmichael, I. S., Nicholls, J., Spera, F. J., Wood, B. J., and Nelson, S. A., (1977). High temperature properties of silicate liquids: applications to the equilibration and ascent of basic magma, *Phil. Trans. Roy. Soc. London, A, 286*, 373–431.

Denbigh, K., 1966. *The Principles of Chemical Equilibrium*, 2nd Ed., Cambridge University Press, 494 pp.

Ferrier, A., 1968a. Mesure de la chaleur de dévitrification du diopside synthétique, *Compt. Rend., Paris, Serie C, 266,* 161–163.

Ferrier, A., 1968b. Mesure de l'enthalpie du diopside synthétique entre 298 et 1,885°K, *Compt. Rend., Paris, Serie C, 267,* 101–103.

Ferrier, A., 1969a. Mesure de l'enthalpie de dévitrification de l'anorthite synthétique, *Compt. Rend., Paris, Serie C, 269,* 185–187.

Ferrier, A., 1969b. Étude expérimentale de l'enthalpie de l'anorthite synthétique entre 298 et 1,950°K, *Compt. Rend., Paris, Serie C, 269,* 951–954.

Flood, H. and Knapp, W. J., 1968. Structural characteristics of liquid mixtures of feldspar and silica, *J. Am. Ceram. Soc. 51,* 259–263.

Fraser, D. G., 1976. Thermodynamic properties of silicate melts, in *Thermodynamics in Geology*, edited by D. G. Fraser, Reidel, Dordrecht, 301–325.

Guggenheim, E. A., 1967. *Thermodynamics, 5th Ed.,* North-Holland, Amsterdam, 390 pp.

Hess, P. C., 1971. Polymer model of silicate melts, *Geochim. Cosmochim. Acta 35,* 289–306.

Hlabse, T. and Kleppa, O. J., 1968. The thermochemistry of jadeite, *Am. Mineralogist 53,* 1281–1292.

Holm, J. L. and Kleppa, O. J., 1968. Thermodynamics of the disordering process in albite, *Am. Mineralogist 53,* 123–133.

Kelley, K. K., 1960. *High-temperature heat-content, heat-capacity, and entropy data for the elements and inorganic compounds*, U.S. Bureau of Mines Bull. 584, 232 pp.

Kohler, F., 1960. Zur Berechnung der thermodynamischen Daten eines ternären Systems aus den zugehörigen binären Systemen, *Monatschefte für Chemie 91,* 738–740.

Kracek, F. C. and Neuvonen, K. J., 1952. Thermochemistry of plagioclase and alkali feldspars, *Am. J. Sci.,* Bowen volume, 293–318.

Kushiro, I., 1973. The system diopside-anorthite-albite: determination of compositions of coexisting phases, *Carnegie Inst. Wash. Yrbk. 72,* 502–507.

Kushiro, I. and Schairer, J. F., 1970. Diopside solid solutions in the system diopside-anorthite-albite at 1 atm and at high pressures, *Carnegie Inst. Wash. Yrbk. 68,* 222–226.

Loewenstein, W., 1954. The distribution of aluminum in the tetrahedra of silicates and aluminates, *Am. Mineralogist 39,* 92–96.

Longhi, J., Walker, D. and Hays, J. F., 1976. Fe, Mg, and silica in lunar plagioclases, *Lunar Science VII*, The Lunar Science Institute, Houston, 501–503.

Ludington, S., 1979. Thermodynamics of melting of anorthite deduced from phase equilibrium studies, *Am. Mineralogist, 64,* 77–85.

Murphy, W. M., 1977. "An experimental study of solid-liquid equilibria in the albite-anorthite-diopside system," M. S. Thesis, University of Oregon.

Navrotsky, A., 1973. Ni SiO$_4$-enthalpy of the olivine-spinel transition by solution calorimetry at 713°C, *Earth Planet. Sci. Lett. 19,* 471–475.

Navrotsky, A. and Coons, W. E., 1976. Thermochemistry of some pyroxenes and related compounds, *Geochim. Cosmochim. Acta 40,* 1281–1288.

Navrotsky, A. and Kleppa, O. J., 1968. Thermodynamics of formation of simple spinels, *J. Inorg. Nucl. Chem. 30*, 479–498.

Osborn, E. F., 1942. The system CaSiO$_3$-diopside-anorthite, *Am. J. Sci. 240*, 751–788.

Schairer, J. F. and Yoder, H. S., 1960. The nature of residual liquids from crystallization, with data on the system nepheline-diopside-silica, *Am. J. Sci., 258A*, 273–283.

Shaw, H. R., 1964. Theoretical solubility of H$_2$O in silicate melts, quasi-crystalline models, *J. Geology 72*, 601–617.

Shaw, H. R., 1972. Viscosities of magmatic silicate liquids: an empirical method of prediction, *Am. J. Sci. 272*, 870–893.

Smith, J. V., 1974. *Feldspar Minerals, vol. 1*, Springer-Verlag, New York, 627 pp.

Tamman, G., 1903. *Kristallisieren und Schmelzen*, Barth, Leipzig, 577 pp.

Thompson, J. B. Jr., 1967. Thermodynamic properties of solid solutions, in *Researches in Geochemistry, vol. 2*, edited by P. Abelson, Wiley, New York, 340–361.

Waldbaum, D. R. and Robie, R. A., 1971. Calorimetric investigations of Na–K mixing and polymorphism in the alkali feldspars, *Z. Kristallogr. 134*, 381–420.

Wasserburg, G. J., 1957. The effects of H$_2$O in silicate systems, *J. Geology 65*, 15–23.

Chapter 3

VISCOSITY, DENSITY, AND STRUCTURE OF SILICATE MELTS AT HIGH PRESSURES, AND THEIR PETROLOGICAL APPLICATIONS

IKUO KUSHIRO

Geological Institute, University of Tokyo and Geophysical Laboratory, Carnegie Institution of Washington

INTRODUCTION

Viscosity and density are two properties of magma that have an important effect on igneous processes. Bowen (1934) emphasized the importance of viscosity as a dominant controlling factor during the passage of magma through the Earth, the pouring-out of lava upon the surface, and the sinking of crystals and the rise of gases within a magma body. A number of viscosity measurements have been made by both field observations of lava flows and experiments in laboratories (e.g., Kani, 1934; Kozu and Kani, 1935; Nichols, 1939; Krauskopf, 1948; Minakami, 1951; Shaw et al., 1968; Murase and McBirney, 1973; Bottinga and Weill, 1972). Bowen himself (1934) measured the viscosities of albite and orthoclase melts, as well as that of the join diopside-enstatite. Most of these measurements were made at atmospheric pressure, and the results are applicable to magmatic processes on or near the surface. For understanding deep magmatic processes, such as separation of partial melt from the source, ascent of magma, and crystal fractionation at depth, we need to know the viscosities of magmas at high pressures. Several such high-pressure measurements have been made on basaltic, andesitic and granitic melts, as well as on synthetic silicate melts of geologic interest (e.g., Shaw, 1963;

Burnham, 1963; Scarfe, 1973; Kushiro et al., 1976; Kushiro, 1976; Khitarov et al., 1976).

The density of magma has also been measured at atmospheric pressure, but no direct measurements of density have been made at high pressures. Recently, the density of basaltic and hydrous andesitic melts has been measured at pressures to 20 kb (Fujii and Kushiro, 1977; Kushiro, 1978b).

These viscosity and density measurements at high pressures are useful for understanding deep magmatic processes. In this paper, the results of these measurements are briefly reviewed and are applied to the crystal fractionation of magma in deep-seated magma chambers and to the rate of ascent of magma.

A knowledge of the structure of silicate melts is fundamental to our understanding of the viscosity, density, and liquidus relations of rock-forming silicate minerals. This paper, therefore, concerns to some extent the structural changes of silicate melts or magmas at high pressures in relation to the viscosity and density changes.

CHANGES IN THE VISCOSITY AND STRUCTURE OF MELT OF $NaAlSi_2O_6$ COMPOSITION WITH PRESSURE

Changes in the viscosity and structure of silicate melts with pressure are discussed first for melts of relatively simple compositions. Among the several different compositions of melt examined, the melt of jadeite composition ($NaAlSi_2O_6$) was selected, because of the possible structural change of the melt during the solid state phase change which results from the reaction nepheline + albite $\rightleftharpoons$ jadeite.

Viscosity

The viscosity of a melt of jadeite composition has been determined at pressures up to 24 kb (Kushiro, 1976) at 1,350°C, using the falling-sphere method in the solid-media, piston-cylinder apparatus. The results are shown in Figure 1. As shown in the figure, the viscosity of the melt at 1,350°C decreases with increasing pressure from 6.8×10^4 poises at 1 atm to 5.3×10^3 poises at 24 kb.[1] The rate of decrease with pressure is greatest

---

[1] Possibility of diffusion of H_2O into the container during the run has been checked using unsealed graphite and sealed Pt containers. The results obtained with both the containers were the same within the uncertainty of measurement for both albite and jadeite melts (Kushiro, 1978a and unpublished data), indicating that the effect of H_2O is not significant for the viscosity values obtained in this and other experiments.

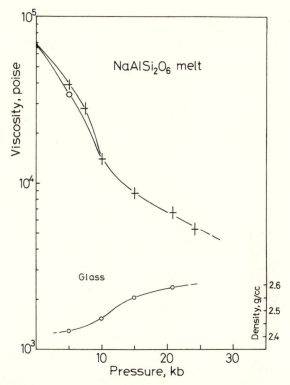

Figure 1. Viscosity change of $NaAlSi_2O_6$ melt with pressure at 1,350°C and density change of quenched melt (glass) of the same composition determined by Kushiro (1976).

between 7.5 and 10 kb. The decrease of viscosity of this melt must be associated with some structural change. Kushiro (1976) discussed such changes on the basis of Waff's (1974) suggestion that the coordination change of Al from four to six would result in the depolymerization of (Si, Al)–O–(Si, Al) bonding and a decrease of viscosity of the melt.

The structural change of the melt of jadeite composition is also indicated by the change in activation energy for viscous flow. Figure 2 shows log η (viscosity) plotted against $1/T°K$. The slope of the curve, which gives the activation energy for viscous flow, is steeper at 1 atm and 7.5 kb than at 20.7 kb. Activation energies obtained at 1 atm and 7.5 kb are nearly identical (80 kcal/mole), whereas at 20.7 kb activation energy decreases to about 45 kcal/mole. According to Riebling (1966) and Hess (1971), melts with lower activation energies are less polymerized in the $Na_2O–Al_2O_3–SiO_2$ system. If their suggestion is accepted, the melt of jadeite composition is less polymerized at 20.7 kb than at 1 atm and 7.5 kb. However, the

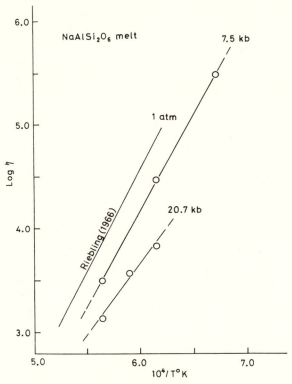

Figure 2. Viscosity change of $NaAlSi_2O_6$ melt with temperature at 1 atm, 7.5 kb and 20.7 kb. Data at 1 atm are from Riebling (1966).

results of recent Raman spectroscopic studies on jadeite glass (Sharma et al., 1978) do not agree with this discussion, as will be shown later. Sharma et al. have suggested that the structure of the jadeite glass quenched at high pressures is similar to that of coesite, whereas the same glass quenched at 1 atm has a structure similar to that of cristobalite. Recently Mysen et al. (1979) suggested that the jadeite melt as well as other highly polymerized melts [(number of non-bridging oxygens)/(number of tetrahedrally coordinated cations) < 1] undergo network collapse and weakening with pressure, thereby decreasing the viscosity.

DENSITY CHANGE OF QUENCHED MELT

To increase our understanding of the structure of a melt of jadeite composition at high pressures, the density of glass quenched from such a melt at different pressures has been determined (Kushiro, 1976). The density increases with increasing pressure of quenching (Figure 1); the rate of increase is largest at pressures between 10 and 15 kb. It is probable that the structure of the glass changes to a more close-packed configuration at higher pressures. Although the density of the glass is different from that of a melt of the same composition, the glass's density and structural changes would be similar to those of the melt of jadeite composition.

X-RAY FLUORESCENCE SPECTROSCOPY

To examine the structural change suggested above, X-ray fluorescence spectroscopy has been applied to the quenched melt of jadeite composition. De Kempe et al. (1961), Day and Rindone (1962), and Leonard et al. (1964) have shown that the wave length of Al $K\alpha$ and $K\beta$ radiations shift with coordination change of Al in amorphous silico-aluminas. Velde and Kushiro (1978) measured the wave lengths of Al $K\alpha$ and $K\beta$ radiations on quenched melt of jadeite composition, and found that both shift with increasing pressure of quenching. Figure 3A shows the shift of Al $K\alpha$ wave length with pressure. At 1 atm the wave length of Al $K\alpha$ of the glass is the same as that of albite; that is, Al in the glass is in 4-fold co-ordination. With increased pressure of quenching, the wave length for the glass becomes shorter, approaching the wave length of Al $K\alpha$ for jadeite, in which Al is all in the 6-fold coordination. A similar wave-length shift of Al $K\alpha$ has been observed for the glass of $CaAl_2Si_2O_8$ (anorthite) and $CaAl_2SiO_6$ (Ca-Tschermak's pyroxene) composition; the wave length of Al $K\alpha$ shifts from that of plagioclase at 1 atm toward that of corundum with increasing pressure, at least to 18 kb (Kushiro, unpublished data). Figure 3B shows the shift with pressure of Al $K\beta$ for the glass of jadeite composition. Again, the wave length shifts from that for albite at 1 atm to that for jadeite at high pressures. The shift is not constant, but is largest between 1 atm and 10 kb. These results accord with the suggestion that Al in the glass of jadeite composition changes from 4-fold coordination at low pressures to 6-fold coordination at high pressures. As mentioned before, the study with Raman spectroscopy (Sharma et al., 1978) suggests no significant coordination change of Al in the jadeite glass. If their results and discussion are accepted, the wave length shift must be explained by other mechanisms. One possibility may be that the (Si, Al)–O bond length in the melt changes with pressure.

INFRARED SPECTROSCOPY

Infrared transmission spectra, however, support depolymerization of Si–O bonds in the melt of jadeite composition quenched at high pressures

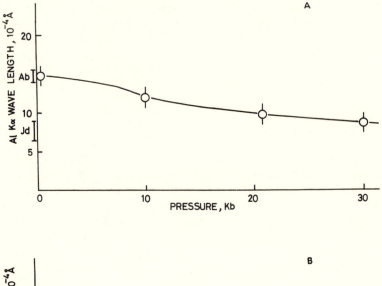

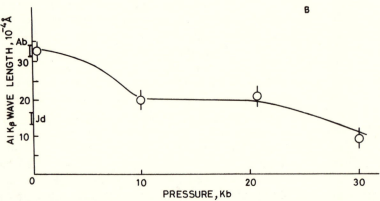

Figure 3 (A and B). Shift of wave length of Al $K\alpha$ (3A) and Al $K\beta$ (3B) for quenched melt (glass) of NaAlSi$_2$O$_6$ composition as a function of pressure (Velde and Kushiro, 1978).

(Velde and Kushiro, 1976). Figure 4 shows infrared transmission spectra of the quenched melt produced at 1 atm, and at 10, 15, 21, 24, and 30 kb. The intensity of the absorption band (given by depth of the valley) near 700 cm^{-1} decreases with increasing pressure of quenching. This absorption is attributed to the Si–O–Si stretch mode, and the decrease of its intensity indicates depolymerization of the Si–O–Si bonds. By addition of H$_2$O to the melt, the same absorption for the quenched melt is much reduced, as is shown by the lowest spectrum in Figure 4. It is believed that H$_2$O in the silicate melt breaks the Si–O–Si bonds very efficiently

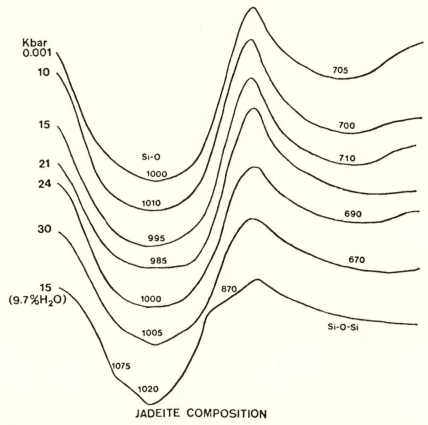

Figure 4. Infrared transmission spectra of quenched melt of $NaAlSi_2O_6$ composition produced at various pressures (Velde and Kushiro, 1976).

(Wasserburg, 1957; Burnham, 1975), and this result supports the above arguments.

Melting Curve Inflection

If the structure of melt changes with pressure, as suggested above, the melting relations may also change. Bell and Roseboom (1969) determined the melting relations for jadeite composition up to 43 kb (Figure 5). The liquidus curve determined by them, when drawn according to their data points, shows an inflection at pressures below 10 kb. The pressure of inflection is in accord with the pressure range for rapid changes in viscosity of the melt and in density of the quenched melt (Figure 1). Rapid density

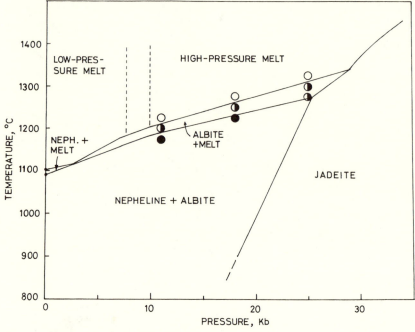

Figure 5. Melting relations for NaAlSi$_2$O$_6$ (jadeite) composition determined by Bell and Roseboom (1969). A possible structural change of melt is suggested on the basis of the various measurements (see text). Dashed lines indicate a region where melt may change effectively its structure.

increase of the melt results in a rapid decrease in ΔV (volume change) for melting, thereby decreasing the slope of the liquidus curve.

RAMAN SPECTROSCOPY

Recently, Sharma et al. (1978) made Raman spectroscopic studies on a glass of jadeite composition, synthesized at pressures between 1 atm and 40 kb. They observed a small shift of a weak band in 900–1,200 cm^{-1} toward lower frequencies, and of a strong peak at 472 cm^{-1} toward higher frequencies, with increasing pressure of quenching. They also found that all the bands are strongly polarized in the high pressure glass, compared with those in the 1 atm glass. These changes in Raman spectrum must be associated with some structural changes of the jadeite glass with pressure. However, major features of the spectra for high pressure glass are similar to those for 1-atm glass. Sharma et al. (1978) concluded from the above observations that all the glasses synthesized at pressures between 1 atm and 40 kb has a 3-dimensional network structure, in which Al^{3+} occurs predominantly in 4-fold coordination; and that the local symmetry of Si, Al in the network is lowered, compared with that in vitreous SiO_2, so that the glass structure at high pressure may in part resemble that of coesite. They also suggested that the network structure becomes more ordered, and the SiO_4 and AlO_4 polyhedra less distorted, with increasing pressure.

These conclusions are in direct contrast to those suggested from the Al K radiation and infrared spectra data. The Mössbauer spectroscopic study on the quenched melts (glass) of $NaAl_{1-x}Fe_xSi_2O_6$ ($X = .1–.5$) compositions by Virgo and Mysen (1977) also suggests the possibility of a coordination change of Al in these glasses at pressures higher than 15 to 20 kb. The density change of the melt and the inflection on the liquidus curve can, however, be explained by the structural change suggested by Raman spectroscopy.

All of the above-mentioned observations of the melt or the glass of jadeite composition strongly indicate that some pressure-induced structural change would take place in the melt of jadeite composition at high pressures, although no conclusive statement can be made at present on the nature of this structural change. Coordination change of Al would explain most of the observations. However, the results obtained by the Raman spectroscopy suggest that the coordination change of Al is not significant for the structural change of jadeite melt. If this is so, the structural change must be associated with change in the configuration of the SiO_4 and AlO_4 tetrahedra, the Si–O and Al–O bond length, or both.

STRUCTURAL CHANGES OF MELT OF $NaAlSi_3O_8$ COMPOSITION

Viscosity

For comparison with the data on jadeite composition, the melt of albite composition ($NaAlSi_3O_8$) has been studied at high pressures (Kushiro, 1978a). Figure 6 shows the viscosity of the anhydrous albite melt at 1,400°C in the pressure range between 1 atm and 20 kb. The viscosity drops from about 1.13×10^5 poises at 1 atm to about 1.8×10^4 poises at 20 kb at 1,400°C. The decrease is by a factor of 6, compared with a factor of 10 for the jadeite composition. The rate of decrease in the viscosity of the melt of albite composition is not constant, but is more rapid in the pressure range between 12 and 15 kb than in other pressure ranges. This pressure range for rapid viscosity decrease is higher than that for the melt of jadeite composition.

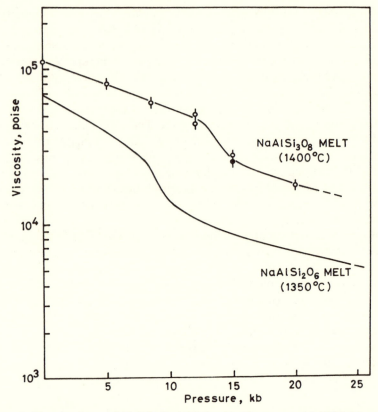

Figure 6. Viscosity change of NaAlSi$_3$O$_8$ (albite) melt with pressure at 1,400°C determined by Kushiro (1978a). That of NaAlSi$_2$O$_6$ melt at 1,350°C is shown for comparison.

DENSITY CHANGE OF QUENCHED MELT

The density of the glass of albite composition quenched at different pressures has been also determined (Kushiro, 1978a) and is shown in Figure 7. The density of the glass increases with increasing pressure of quenching from 2.38 g/cc at 1 atm to 2.525 g/cc at 25 kb. The rate of increase in density is, again, not constant, but is largest at pressures between 15 and 20 kb. The density of the glass of jadeite composition, shown in Figure 7 for comparison, increases at a greater rate between 10 and 15 kb than it does in the other pressure ranges examined. The pressure at which the greatest increase in density of the glass occurs is higher for the albitic glass than for the jadeitic glass by 5–6 kb. It is interesting that the univariant curve for the breakdown of albite to jadeite + quartz (Birch and LeComte, 1960) lies about 6 kb higher than the univariant curve for the

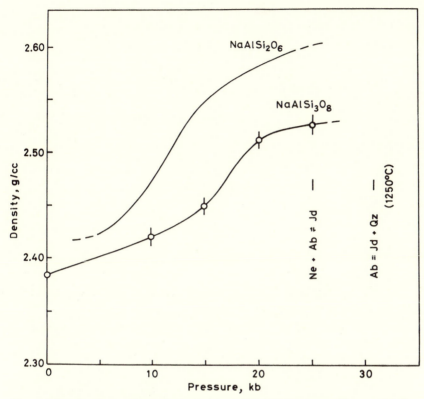

Figure 7. Density change of quenched melt (glass) of $NaAlSi_3O_8$ composition as a function of pressure of quenching (Kushiro, 1978). That of $NaAlSi_2O_6$ composition is shown for comparison. Two short vertical bars indicate the pressures for the reactions nepheline (Ne) + albite (Ab) $\rightleftharpoons$ 2 jadeite (Jd) and albite $\rightleftharpoons$ jadeite + quartz (Qz) at 1,250°C.

reaction nepheline + albite $\rightleftharpoons$ jadeite (Robertson et al., 1957) near the solidus temperatures, as indicated in Figure 7.

Melting Curve Inflection

The melting curve of albite also suggests that structural changes occur in the melt. Boyd and England (1963) carefully determined the melting curve of albite up to 32 kb. Their data are reproduced in Figure 8. If the melting curve is drawn according to their data points, the curve clearly shows an inflection at pressures near 15 kb. Rapid viscosity changes take place between 13 and 16 kb, as is shown in Figure 7; this pressure range co-incides with that of the inflection of the melting curve similarly to the

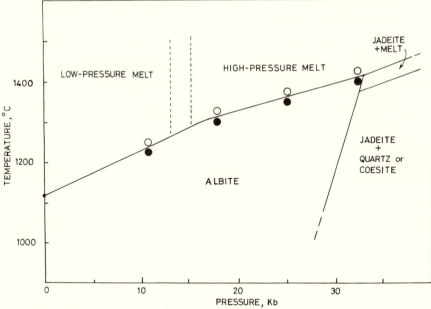

Figure 8. Melting relations for NaAlSi₃O₈ (albite) composition determined by Boyd and
England (1963) and Bell and Roseboom (1969). A possible structural change of
melt is suggested on the basis of viscosity and density changes. Dashed lines
indicate a region where melt may change effectively its structure.

melt of jadeite composition. Such inflection is due to a decrease in ΔV
(volume change) for melting, and indicates a density increase of albite
melt at about 15 kb. It is concluded that the albite melt effectively changes
to a denser structure near 15 kb, at least between the liquidus temperature
and 1,400°C.

Solubility Change of CO_2

The structural change of the melt of albite composition has been also
suggested by Mysen (1976) on the basis of the solubility of CO_2. He
observed an increase of $CO_3^{2-}/(CO_2 + CO_3^{2-})$ in the melt by about
100% at pressures between 20 and 30 kb. He suggested that, if coordination of Al changes from 4-fold to 6-fold, then the (Al, Si)–O–(Al, Si) bonds
are broken, resulting in an increase of $a_{O^{2-}}$ in the melt, and at fixed f_{CO_2},
increased $a_{O^{2-}}$ yields high $X_{CO_3^{2-}}$ and hence greater solubility. Other
explanations, based on structural changes that do not include coordination change of Al (e.g., cristobalite-type → coesite-type structural change),
have not been given so far.

DIFFERENCE IN VISCOSITY CHANGE BETWEEN JADEITE AND ALBITE MELTS

To examine the structural relations between albite and jadeite melts at high pressures, the viscosities of these melts at different pressures were compared. Figure 9 shows the viscosities of melts of nepheline, jadeite and albite compositions at 1,500°C and 1 atm, determined by Riebling (1966). In these melts, Al is virtually all in 4-fold coordination. It is remarkable that log η of these three melts lie on a straight line, indicating that log η is proportional to the mole percent of SiO_2. Figure 9 also shows the viscosities of jadeite and albite melts at 20 kb at 1,400°C. It is interesting that both the lines converge with increasing SiO_2 content. This suggests that the viscosity of pure silica melt changes only slightly with increasing pressure, at least to 20 kb. The viscosity of pure silica melt may decrease at high pressures due to the change of Si from 4-fold to 6-fold coordination, but this will take place at much higher pressures. If such a linear relation holds among melts of jadeite, albite and SiO_2-rich compositions, and if the viscosity of silica-rich melt does not change significantly with pressure, at least to 20 kb, the observation that the viscosity drop of albite melt from 1 atm to 20 kb is smaller than that of jadeite melt is understandable.

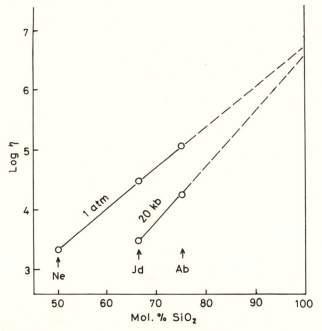

Figure 9. Relations in viscosity of melts of $NaAlSiO_4$ (Ne), $NaAlSi_2O_6$ (Jd), and $NaAlSi_3O_8$ (Ab) compositions at 1 atm at 1,500°C taken from Riebling (1966) and 20 kb at 1,400°C.

CHANGES IN REACTIONS INVOLVING MELTS WITH PRESSURE

On the basis of the changes in the viscosity of melts and the density of quenched melts, as well as the data of X-ray fluorescence, Raman spectroscopy and infrared spectroscopy, it is suggested that the structural changes or phase transformations of melts of albite and jadeite compositions take place at high pressures. If the structure of melt changes with pressure, reactions involving melt may also change with pressure. Such changes are well demonstrated in the system $NaAlSiO_4$–Mg_2SiO_4–SiO_2 (Figure 10).

The liquidus relations for the composition B25, which is at the intersection of the joins Mg_2SiO_4–$NaAlSi_3O_8$ and $MgSiO_3$–$NaAlSi_2O_6$

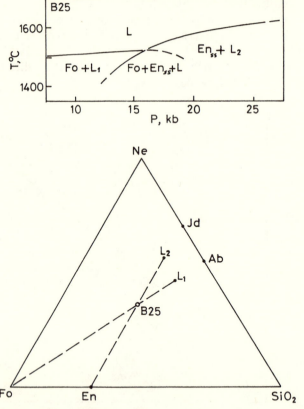

Figure 10. Liquidus relations for B25 (forsterite 35 albite 65 by weight) composition determined by Kushiro (1968) and stable tie lines involving liquid at low pressures (<16 kb) and high pressures (>16 kb).

have been determined (Kushiro, 1968). Below 16 kb, forsterite is on the liquidus; above 16 kb, enstatite is on the liquidus. In the region Fo + L, the liquid must lie on the extension of the Fo-B25 join (e.g., L_1), whereas in the region En_{ss} + L, the liquid must lie on or near the extension of the En − B25 join (e.g., L_2).

Subsolidus phase relations pertinent to this system have been determined by Schairer and Yoder (1961) and Yoder and Tilley (1962). At low pressures the join forsterite-albite is stable, but at high pressures, at least at 30 kb, the join forsterite-albite is broken and the join enstatite-jadeite becomes stable.

Shift of the stable tie line from the join Fo − L_1 to the join En − L_2 at high pressure is similar to the shift from the join Fo − Ab to the join En − Jd. The molar volume decreases from the Fo + L_1 assemblage to the En + L_2 assemblage with a constant crystal/liquid ratio by about 4.5%: this is calculated on the assumption that the density difference between L_1 and L_2 is the same as that between albite glass and jadeite glass at 20 kb. The amount of decrease is smaller than that for the subsolidus reaction Fo + Ab = 2En + Jd (14%); however, the calculation suggests that En + L_2 is stable relative to Fo + L_1 at high pressures. Such similarity in the shift of stable tie lines is also observed in the system forsterite–$CaAl_2SiO_6$–SiO_2, where the forsterite-liquid tie lines at low pressures also change to the enstatite-liquid tie lines at high pressures. The melt containing anorthite component would also change to a denser structure at high pressures.

CHANGE IN TRACE-ELEMENT PARTITIONING

If the structure of melt changes with pressure, the partitioning of trace elements may also change. To test this possibility, partitioning of Ni between forsterite and melt has been determined in the system Mg_2SiO_4–$NaAlSi_2O_6$ at pressures between 1 atm and 20 kb (Mysen and Kushiro, 1978). This system is suitable for testing the effect of pressure, because the liquidus temperature of forsterite changes little with pressure. In particular, at 1,300°C the composition of liquid coexisting with forsterite is nearly identical between 1 atm and 20 kb, so that the compositional effect on partitioning can also be eliminated (Figure 11A). In these experiments, ^{63}Ni was added to compositions $Fo_{20}Jd_{80}$ and $Fo_{40}Jd_{60}$, and the experiments were made at 1,300°C at 1, 5, 10, 15, and 20 kb. The run products, consisting of forsterite and glass, were analyzed with the β-track method (Mysen and Seitz, 1975). The partition coefficient of Ni between forsterite and melt (glass) as a function of pressure is shown in Figure 11B. The partition coefficient D_{Ni}^{ol-liq}(Ni_{ol}/Ni_{liq}) increases from 22.5 $\pm$ 1.4 at 1 kb to

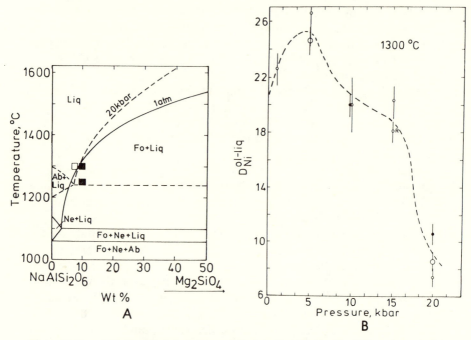

Figure 11 (A and B). Equilibrium diagram of the join $NaAlSi_2O_6$–Mg_2SiO_4 at 1 atm (solid lines) and 20 kbar (dashed lines) (11A), and change of D_{Ni}^{ol-liq} (Ni in olivine/Ni in liquid) with pressure at 1,300°C (11B) (Mysen and Kushiro, 1978).

26.5 $\pm$ 2.7 at 5 kb and decreases to 7.4 at 20 kb. The K_D value $[(Ni/Mg)_{ol}/(Ni/Mg)_{liq}]$ increases from 1.9 at 1 kb to 2.3 at 5 kb and decreases to .6 at 20 kb. Such changes in the partition coefficient would probably be related to structural changes of the melt; however, no clear explanation of the relation between the structural change of jadeite-rich melt and the change in partition coefficient of Ni can be given at present.

CHANGES IN VISCOSITY AND DENSITY OF BASALTIC MELT

VISCOSITY CHANGES

The viscosity changes of melts at high pressures have been discussed above in relation to melts of relatively simple compositions. Similar changes would be expected in natural rock melts or magmas at high pressures. The

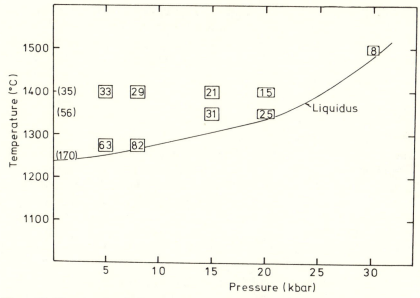

Figure 12. Viscosity of melt of Kilauea 1921 olivine tholeiite determined by Kushiro et al. (1976) and Fujii and Kushiro (1977).

viscosity of a melt of Kilauea 1921 olivine tholeiite has been measured at pressures to 30 kb (Kushiro, Yoder and Mysen, 1976; Fujii and Kushiro, 1977). The results are shown in Figure 12 (numbers in boxes signify viscosity in poise). The viscosity decreases slightly with increasing pressure at constant temperature; for example, the viscosity drops from 35 poises at 1 atm to 15 poises at 20 kb at 1,400°C, and from 55 to 25 poises at 1,350°C. The magnitude of the viscosity drop at constant temperature is much smaller than that of jadeite melt and is closer to that of albite melt. Similar results have been obtained for melt of an aphyric abyssal tholeiite collected near the Mid-Atlantic Ridge during the DSDP Leg 45 cruise. The viscosity decreases from about 160 poises at 1 atm to about 80 poises at 12 kb at 1,300°C (Fujii and Kushiro, 1977).[2]

Along the liquidus of basalt, however, viscosity decreases more significantly; the viscosity of a Kilauea 1921 olivine tholeiite melt decreases from

---

[2] Recently Scarfe et al. (1979) found that the viscosity of melts of diopside and sodium metasilicate compositions increases with increasing pressure at constant temperature. They discussed that the viscosity of highly polymerized melts [defined as (number of non-bridging oxygens)/(number of tetrahedrally coordinated cations) < 1] such as albite and jadeite melts (at least at 1 atm) decreases with increasing pressure, whereas that of the depolymerized melts increases. In the light of their experiments, small rates of decrease of viscosity of relatively depolymerized basaltic melts are understandable.

about 170 poises at 1 atm to about 8 poises at 30 kb. Such a large decrease is due to the effects of both pressure and temperature.

The viscosity change along the liquidus is directly applicable to the problem of the generation and ascent of magmas, and a large viscosity decrease of basaltic melts along the liquidus is significant, as discussed below.

DENSITY CHANGES

The density of a Kilauea 1921 olivine tholeiite melt has been directly determined along the liquidus temperature, using the falling-sphere method (Fujii and Kushiro, 1977). Two different spheres of different densities were used to determine simultaneously the viscosity and density of the melt. The density of the melt obtained by this method is shown in Figure 13. The density of the glass quenched from the same melt at high pressures was also determined (Kushiro et al., 1976; Fujii and Kushiro, 1978). The glass has a higher density than the melt has at pressures lower than 10 kb, but the difference becomes smaller at pressures higher than 15 kb, as is shown in Figure 13.

The density of the olivine tholeiite melt increases with increasing pressure, but the rate of increase is larger in the pressure range from 10 to

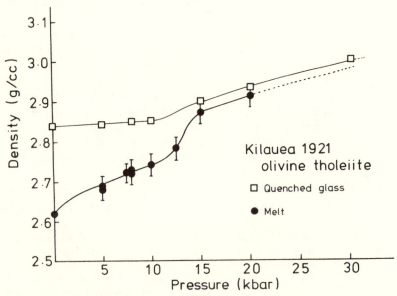

Figure 13. Density of melt of Kilauea 1921 olivine tholeiite (solid circle) and its quenched melt (glass) (open square) determined by Fujii and Kushiro (1977).

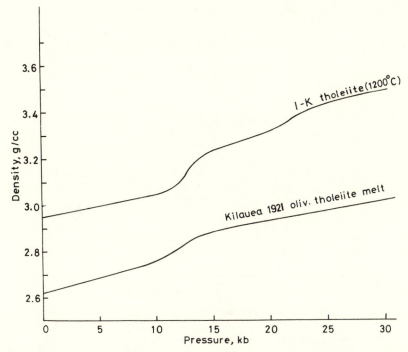

Figure 14. Comparison of density change of the Kilauea 1921 olivine tholeiite melt (Fujii and Kushiro, 1977) and that of crystalline olivine tholeiite (Ito and Kennedy, (1971).

15 kb than in other pressure ranges.[3] The density increase is about 15% from 1 atm to 30 kb, and about 7% between 10 and 15 kb.

For comparison, the density change of rock of an olivine tholeiite composition at 1,200°C, estimated by Ito and Kennedy (1971), is shown in Figure 14. The density increase from 1 atm to 30 kb is about 20%, compared to 15% for the melt. The pattern of density change for an olivine tholeiite melt is similar to that for the subsolidus change of olivine tholeiite to eclogite, although there is another slope change at higher pressures (19 to 22 kb). It is most probable that the melt of basaltic composition transforms effectively to a melt with a higher density structure in the pressure range where density increase is largest. A possible phase equilibrium diagram for a basaltic composition is constructed in Figure 15.

---

[3] The variation of density with temperature at constant pressure is within the uncertainty of measurements (± 0.03 g/cc), at least in the temperature range between 1,250 and 1,400°C (Fujii and Kushiro, 1977). The same observation has been made by Murase and McBirney (1973) for a Columbia River basalt melt at 1 atm (i.e., 2.60 g/cc at 1,250°C and 2.58 g/cc at 1,450°C).

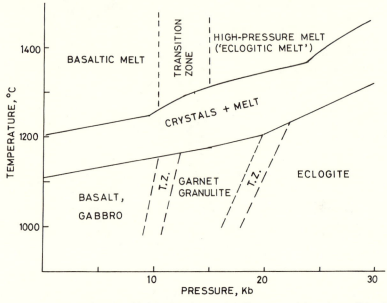

Figure 15. A suggested phase diagram for olivine tholeiite composition. A possible phase change of the melt is suggested on the basis of the density change. The subsolidus phase transformation is based on the data by Ito and Kennedy (1971).

Yoder and Tilley (1962) used the term "eclogitic melt" for melts of basaltic composition equilibrated at high pressures. The high-density basaltic melt may be called "eclogitic melt", although no direct evidence has been available so far for the presence of a significant amount of octahedrally coordinated Al in the high density basaltic melt.

CRYSTAL FRACTIONATION AT DIFFERENT PRESSURES

PLAGIOCLASE FLOATING AND FORMATION OF ANORTHOSITE

Because of the density increase of basaltic melt with pressure, the trend of crystal fractionation may change with pressure. The most interesting consequence of the density increase of basaltic melt is the density reversal between basaltic melt and plagioclase. Figure 16 shows the densities of plagioclase of compositions An_{90} and An_{65}, and of a melt of Kilauea 1921 olivine tholeiite. Thermal expansivity and compressibility of plagioclase have been taken into consideration in drawing the curves for the density changes of these plagioclases.

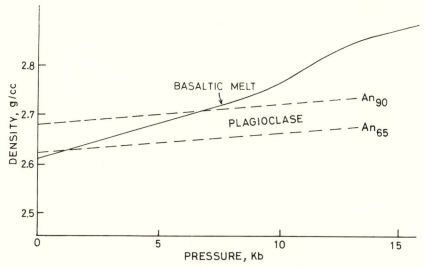

Figure 16.   Density change of Kilauea olivine tholeiite melt and plagioclase of An_{90} and An_{65}
with pressure.

The density of An_{90} is larger than that of the olivine tholeiite melt at
pressures below about 6 kb, but it becomes smaller above 6 kb. Consequently, this plagioclase sinks in the olivine tholeiitic melt at shallow
depths (<20 km), but it floats at greater depths. Floating and sinking
experiments have actually been done for An_{90} plagioclase at pressures
between 3 and 10 kb, using the same technique employed in the viscosity
and density measurements. At 6, 7.5, and 10 kb, plagioclase of An_{90}
initially placed at the bottom of the charge floated, and at 3 kb, plagioclase
sank during the $\frac{1}{2}$ hr run, confirming the above suggestion.

The pressure for such density reversal depends on the compositions
of plagioclase and melt. For plagioclase of An_{65} and the olivine tholeiite
melt, the density reversal occurs at much lower pressures ($1 \sim 2$ kb).
However, if coexisting melt is more enriched in salic components and has a
lower density, the density reversal should occur at higher pressures. On
the other hand, if the melt is more enriched in ferromagnesian components
such as lunar ferrobasalts, plagioclase would float at lower pressures,
even at 1 atm.

The results of these experiments may be relevant to the formation of
some terrestrial anorthosite (Kushiro and Fujii, 1977). If a basaltic magma
chamber exists at shallow levels in the crust, plagioclase sinks with
ferromagnesian minerals such as olivine and pyroxene, forming gabbro,
as shown in Figure 17. On the other hand, if the magma chamber exists at
deeper levels (>20 km), plagioclase floats, whereas olivine and pyroxene

CONTINENTAL CRUST

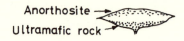

UPPER MANTLE

Figure 17. Contrasting cumulates in basaltic magma chambers at shallow and deep levels
in the continental crust.

sink. In consequence, anorthosite will be formed near the top of the magma
chamber and ultramafic cumulates will be formed at the bottom. If this is
the case, anorthosite should preferentially form in the lower parts of the
continental crust. Floating of plagioclase at high pressures as a mechanism
of genesis of anorthosite has been suggested by Morse (1968), and recently
by Goode (1977) in the layered gabbroic Kalka Intrusion of central
Australia. Myers and Platt (1977) described a layered anorthosite complex
at Majorqap qava, Southwest Greenland, which consists of a thick
anorthosite unit in the upper part and an ultramafic unit near the bottom,
although the complex might have been formed by two successive intrusions. According to these authors, the presence of a pyroxene-spinel
symplectite indicates that the complex was formed at pressures between
6 and 9 kb. These two examples would strongly support the present
hypothesis. Massif-type anorthosite is often associated with metamorphic
rocks of granulite facies. This evidence is also consistent with the present
hypothesis. Perhaps the reason why massif-type anorthosite bodies are
found almost exclusively in the Precambrian shield areas may be that only
here has erosion extended to the lower crust or the level of anorthosite
formation.

RATE OF SINKING OF CRYSTALS AT DIFFERENT PRESSURES

The rate of sinking of crystals in basic magma at different levels can be
calculated using the viscosity and density of melt of basaltic composition
and assuming that basaltic or "eclogitic" melts behave as a Newtonian
fluid near the liquidus temperatures. Figure 18 shows the velocity of

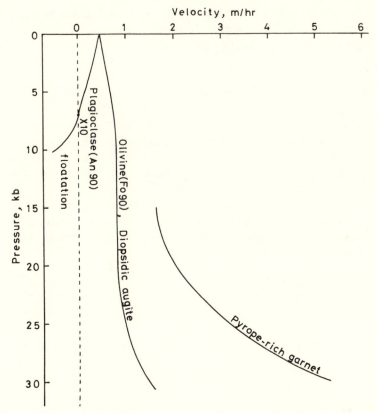

Figure 18. Rates of sinking (or floating) of crystals (2 mm in diameter) in olivine tholeiite melt as a function of pressure. Viscosity and density values used for the calculation are given in Figure 19.

sinking of garnet, diopsidic augite, and olivine crystals with a diameter of 2 mm in the melt of olivine tholeiitic composition near the liquidus temperature. It is striking that the velocities of sinking of garnet and diopsidic clinopyroxene are very large at pressures higher than 25 kb; the velocity is greater than 3 m/hr and 1 m/hr for garnet and clinopyroxene, respectively. This is due to the low viscosity of the melt along the liquidus at high pressures. Eclogite fractionation, stressed by O'Hara and Yoder (1967), would be extremely effective at depths greater than 80 km. It is also interesting that the velocity of sinking of olivine and clinopyroxene is nearly constant between 22 kb and 8 kb (a depth range from 65 km to 25 km), due to the compensation of density decrease and viscosity increase of the melt. Further decrease of velocity at lower pressure is due to the increase of viscosity of the melt. Plagioclase floats at pressures greater than about 6 or 7 kb. The velocity ($\frac{1}{10}$ of the values shown in Figure 24) progressively increases away from the pressure point at which the densities of both the melt and plagioclase coincide.

RATE OF ASCENDING MAGMA

From the viscosity and the density of basaltic melt at high pressures, the velocity of ascending magma can also be estimated. Figure 19 shows the changes with pressure in the viscosity and density of an olivine tholeiite melt along the liquidus temperature. These were used to estimate the velocity of sinking of crystals shown above.

The velocity of ascending magma was calculated, assuming that magma ascends along a vertical crack of constant width as a consequence of a pressure gradient (the result of the density difference between melt and

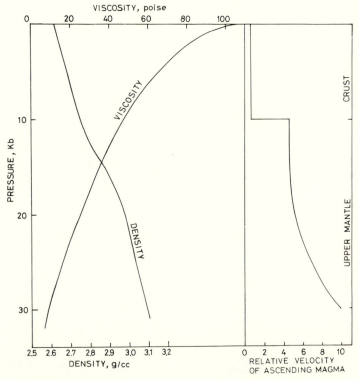

Figure 19. Viscosity and density variations of an olivine tholeiite (Kilauea 1921) melt with pressure (left) and relative ascending velocity of the magma (right). See text for the method of calculation and the assumptions made.

surrounding rocks) in the crack. In this case, the velocity of ascending magma is approximated by the following equation:

$$v = \frac{(\rho - \rho')g\,\Delta x^2}{12\eta},$$

where ρ and ρ' are the densities of the surrounding rocks and melt, g is the gravitational constant, Δx is the width of the crack, and η is the viscosity of the melt (Takeuchi et al., 1972). The calculated relative velocity of ascending magma is shown in Figure 19. The velocity at 30 kbar is set at a value of 10, and those at different pressures are given as relative values. As shown in the figure, the velocity is greater at deeper levels. It is interesting that the velocity suddenly drops at the crust-mantle boundary, due to the large rock-density change which takes place there. If this mechanism really operates, magma chambers should tend to form near the Moho discontinuity. In the upper mantle, particularly in the relatively deep levels, however, magma may not ascend by this mechanism, because the hypothesized cracks are unlikely to exist in relatively low Q upper-mantle materials. In such levels magma may ascend by zone melting (Harris, 1957) or partial zone melting, which is a combination of stoping and partial melting processes (Kushiro, 1968). The velocity change shown here, therefore, may be applicable only in the relatively shallow levels (< 50 km).

CONCLUDING REMARKS

Pressure-induced structural changes or phase changes in simple silicate melts and in a melt of basaltic composition are suggested on the basis of the viscosity, density and melting curve inflections, and on measurements by X-ray fluorescence spectroscopy, infrared absorption, Mössbauer spectroscopy and Raman spectroscopy on the quenched melt of jadeite composition. The nature of the structural change of the melt is, however, not fully understood at present. Most observations of the melt and the glass of jadeite and albite compositions suggest that the structural changes are mainly due to the shift of Al from 4-fold to 6-fold coordination; however, Raman spectroscopy on the jadeite glass does not support such coordination change, but suggests a cristobalite → coesite type structural change. Further studies are needed to settle this problem.

Because of the structural changes of silicate melts with pressure, partitioning of trace elements between crystals and magma and diffusion rate of elements in magma may also change with pressure.

The pattern of fractional crystallization would also change with pressure. For example, plagioclase sinks in basaltic magma at shallow levels in the crust, but it floats in the same magma at deep levels in the continental crust, forming anorthosite near the top of the magma chamber. Rates of sinking of ferromagnesian minerals in basic melts (eclogitic magma) are greater at depths greater than 80 km. The rate of ascent of magma also changes with depth.

These consideration and suggestion are believed to be important for magma genesis and should be extended in more quantitative ways.

ACKNOWLEDGMENTS

The author wishes to thank Drs. T. Fujii, R. B. Hargraves, B. O. Mysen, S. K. Sharma, B. Velde, D. Virgo, and H. S. Yoder, Jr. for their discussion and criticism on various parts of the manuscript.

REFERENCES

Bell, P. M. and E. H. Roseboom, 1969. Melting of jadeite and albite to 45 kilobars with comments on melting diagrams of binary systems at high pressures, *Mineral, Soc. Amer. Special Publ.* No. 2, 151–161.

Birch, F. and P. LeComte, 1960. Temperature-pressure plane for albite composition, *Amer. J. Sci.*, 256, 209–217.

Bottinga, Y. and D. F. Weill, 1972. The viscosity of magmatic silicate liquids: model for calculation, *Amer. J. Sci.* 272, 438–475.

Bowen, N. L., 1934. Viscosity data for silicate melts, 15th Annual Meeting, 1, *Eos Trans. AGU* 249–255.

Boyd, F. R. and J. L. England, 1963. Effect of pressure on the melting of diopside, $CaMgSi_2O_6$, and albite, $NaAlSi_3O_8$, in the range up to 1,750°C, *J. Geophys. Res. 68*, 311–323.

Burnham, C. W., 1963. Viscosity of water-rich pegmatite melt at high pressure (abstract), *Geol. Soc. Amer. Spec. Pap. 76*, 26.

Burnham, C. W., 1975. Water and magmas; a mixing model, *Geochim. Cosmochim. Acta 39*, 1077–1084.

Day, D. and G. Rindone, 1962. Properties of soda aluminosilicate glasses, III. Coordination of aluminum irons, *J. Amer. Ceram. Soc. 45*, 579–581.

DeKempe, C. M., C. Gastuche and G. W. Brindley, 1961. Ionic coordination in alumino-silicic gels in relation to clay mineral formation, *Amer. Mineral. 46*, 1370–1381.

Fujii, T. and I. Kushiro, 1977. Density, viscosity and compressibility of basaltic liquid at high pressures, *Carnegie Inst. Washington, Year Book 76*, 419–424.

Goode, A. D. T., 1977. Floatation and remelting of plagioclase in the Kalka intrusion, central Australia: petrological implications for anorthosite genesis, *Earth Planet. Sci. Letters 34*, 375–380.

Harris, P. G., 1957. Zone refining and the origin of potassic basalts, *Geochim. Cosmochim. Acta 12*, 195–208.

Hess, P. C., 1971. Polymer model of silicate melts, *Geochim. Cosmochim. Acta 35*, 289–306.

Ito, K. and G. C. Kennedy, 1971. An experimental study of the basalt-garnet granulite-eclogite transition. in *The Structure and Physical Properties of the Earth's Crust* J. G. Heacock, editor, Amer. Geophys. Union, Washington, D. C., 303–314.

Kani, K. 1934. The measurement of the viscosity of basalt glass at high temperatures, 1 and 2, *Proc. Imp. Acad. Tokyo 10*, 29–32, 79–82.

Khitarov, N. I., E. B. Lebedev, A. M. Dorfman, and A. B. Slutsky, 1976. The pressure dependence of the viscosity of basalt melt. *Geokhimiya, No. 10*, 1489–1497.

Kozu, S., and K. Kani, 1935. Viscosity measurements of the ternary system diopside-albite-anorthite at high temperatures, *Proc. Imp. Acad. Tokyo 11*, 383–385.

Krauskopf, K. B., 1948. Lava movement at Paricutin volcano, Mexico, *Bull. Geol. Soc. Amer. 59*, 1267–1283.

Kushiro, I., 1968. Compositions of magmas formed by partial zone melting of the earth's upper mantle, *J. Geophys. Res. 73*, 619–634.

Kushiro, I., 1976. Changes in viscosity and structure of melt of $NaAlSi_2O_6$ composition at high pressures, *J. Geophys. Res. 81*, 6347–6350.

Kushiro, H. S. Yoder, and B. O. Mysen, 1976. Viscosities of basalt and andesite melts at high pressures, *J. Geophys. Res. 81*, 6351–6356.

Kushiro, I., 1978a. Viscosity and structural changes of albite ($NaAlSi_3O_8$) melt at high pressures. *Earth Planet. Sci. Letters 41*, 87–90.

Kushiro, I., 1978b. Density and viscosity of hydrous calc-alkalic andesite melt at high pressures. *Carnegie Inst. Washington, Year Book, 77*, 675–677.

Kushiro, I., and T. Fujii, 1977. Floatation of plagioclase in magma at high pressures and its bearing on the origin of anorthosite, *Proc. Japan Acad. 53, Ser. B.*, 262–266.

Leonard, A., S. Suzuki, J. J. Fripiat, and C. De Kempe, 1964. Structure and properties of amorphous silicoaluminas. I. Structure from X-ray fluorescene spectroscopy and infrared spectroscopy, *J. Phys. Chem. 68*, 2608–2617.

Minakami, T., 1951. On the temperature and viscosity of the fresh lava extruded in the 1951 Oo-sima eruption, *Bull. Earthq. Res. Inst. 29*, 487–498.

Morse, S. A., 1968. Layered intrusions and anorthosite genesis, in *Origin of Anorthosite and Related Rocks*, N.Y. Isachsen, ed., N.Y.S. Museum Mem. 18, 175–187.

Murase, T., and A. R. McBirney, 1973. Properties of some common igneous rocks and their melts at high temperatures, *Geol. Soc. Amer. Bull. 84*, 3563–3592.

Myers, J. S. and R. G. Platt, 1977. Mineral chemistry of layered Archean anorthosite at Majorqap qava, near Fiskenaesset, southwest Greenland, *Lithos 10*, 59–72.

Mysen B. O., 1976. The role of volatiles in silicate melts: solubility of carbon dioxide and water in feldspar, pyroxene, and feldspathoid melts to 30 kb and 1,625°C, *Amer. J. Sci. 276*, 969–996.

Mysen, B. O. and M. G. Seitz, 1975. Trace element partitioning determined by beta-track mapping: an experimental study using carbon and samarium as examples, *J. Geophys. Res. 80*, 2627–2635.

Mysen, B. O. and I. Kushiro, 1978. Pressure dependence of nickel partitioning between forsterite and aluminous silicate melts, *Earth Planet. Sci. Letters 42*, 383–388.

Mysen, B. O., D. Virgo and C. M. Scarfe, 1979. Viscosity of silicate melts as a function of pressure: Structural interpretation. *Carnegie Inst. Washington, Year Book 78* (in press).

Nichols, R. L., 1939. Viscosity of lava, *J. Geol. 47*, 290–302.

O'Hara, M. J., and H. S. Yoder, Jr., 1967. Formation and fractionation of basic magmas at high pressures, *Scot. J. Geol. 3*, 67–117.

Riebling, E. F., 1966. Structure of sodium aluminosilicate melts containing at least 50 mole % SiO_2 at 1,500°C, *J. Chem. Phys. 44*, 2857–2865.

Robertson, E. C., F. Birch, and G. J. F. MacDonald, 1957. Experimental determination of jadeite stability relations to 25,000 bars, *Amer. J. Sci. 255*, 115–137.

Scarfe, C. M., 1973. Viscosity of basic magmas at varying pressure, *Nature 241*, 101–102.

Scarfe, C. M., B. O. Mysen and D. Virgo, 1979. Changes in viscosity and density of melts of sodium disilicate, sodium metasilicate and diopside composition with pressure. *Carnegie Inst. Washington, Year Book 78* (in press).

Schairer, J. F., and H. S. Yoder, 1961. Crystallization in the system nepheline-forsterite-silica at 1 atmospheric pressure, *Carnegie Inst. Washington, Year Book 60*, 141–144.

Sharma, S. K., D. Virgo and B. O. Mysen, 1978. Raman study of structure and coordination of Al in $NaAlSi_2O_6$ glasses synthesized at high pressures. *Carnegie Inst. Washington, Year Book, 77*, 658–62.

Shaw, H. R., 1963. Obsidian-H_2O viscosities at 1000 and 2000 bars in the temperature range 700° to 900°C, *J. Geophys. Res. 68*, 6337-6343.

Shaw, H. R., 1969. Rheology of basalt in the melting range, *J. Petrology 10*, 510–535.

Shaw, H. R., T. L. Wright, D. L. Peck, and R. Okamura, 1968. The viscosity of basaltic magma: An analysis of field measurements in Makaopuhi lava lake, Hawaii, *Amer. J. Sci. 266*, 225–264.

Takeuchi, H., N. Fujii and M. Kikuchi, 1972. Mechanism of magma ascent (in Japanese), *Zishin, ser. 2, 25*, 266–268.

Velde, B., and I. Kushiro, 1976. Infrared spectra of high-pressure quenched silicate liquid, *Carnegie Inst. Washington, Year Book 75*, 618–620.

Velde, B., and I. Kushiro, 1978. Structure of sodium alumino-silicate melts quenched at high pressures; infrared and aluminum K-radiation data, *Earth Planet. Sci. Letters 40*, 137–140.

Virgo, D. and B. O. Mysen, 1977. Influence of structural changes in $NaAl_{1-x}Fe_xSi_2O_6$ melts on $Fe^{3+} - \Sigma Fe$ ratio, *Carnegie Inst. Washington, Year Book 76*, 400–407.

Waff, H. S., 1975. Pressure-induced coordination changes in magmatic liquids, *Geophys. Res. Lett. 2*, 193–196.

Wasserburg, G. J., 1957. The effect of H_2O in silicate systems, *J. Geol. 65*, 15–23.

Yoder, H. S., and C. E. Tilley, 1962. Origin of basalt magmas: An experimental study of natural and synthetic rock systems, *J. Petrol. 3*, 342–532.

Chapter 4

TRACE-ELEMENT CONSTRAINTS ON MAGMA GENESIS

S. R. HART

Department of Earth and Planetary Sciences, Massachusetts Institute of Technology

C. J. ALLÈGRE

Institut de Physique du Globe de Paris, Université Paris VI

INTRODUCTION

Perhaps the greatest change since Bowen's day in our approach to magma genesis is in the fields of trace-element and isotopic geochemistry. Mass spectrometers, although first operated in 1918, did not make an appearance in the earth sciences until the classic studies of Nier (1938, 1939). The field of trace-element geochemistry was initiated in 1937 by the pioneering works of Goldschmidt (1937) and Fersmann (1934), but the major impact of trace-element studies on igneous petrogenesis was not felt until the early 1950's, when the development of the optical spectrograph allowed useful trace element analysis of geologic specimens. The most notable of the studies initiated at that time where those of Wager and Mitchell (1951) and Nockolds and Allen (1953, 1954). And yet, despite the enormous activity in this field, especially in the past decade, little new ground has actually been broken compared to that covered by Bowen in his classic treatise (Bowen, 1928). We have of course added enormous detail to our understanding of source compositions, P-T facies of partial melting, and quantitative liquid lines of descent, but the basic concepts of igneous petrogenesis have changed very little. Partial melting

© 1980 by Princeton University Press
Physics of Magmatic Processes
0-691-08259-6/80/0121-39$01.95/0 (cloth)
0-691-08261-8/80/0121-39$01.95/0 (paper)
For copying information, see copyright page

is now thought to have a greater role in producing the observed chemistries of eruptive rocks, relative to fractional crystallization. And the processes of interaction between magmatic rocks (especially sialic rocks) and their host are now known to be very important in altering primary chemistries.

In this paper, we will review the nature of the information that trace elements and isotopes provide us in understanding the petrogenesis of the igneous rocks. We will start by considering some basic kinetic constraints, and will follow that with a discussion of mafic rocks, tracing their history from melting through fractional crystallization to wall-rock interactions and the like. Finally, the sialic igneous rocks will be discussed in light of trace-element and isotopic constraints. These rocks present a more complex problem: while our understanding of mafic rock petrogenesis is fairly sound, our grasp of the processes that produce sialic rocks is much more tenuous. The decade to come will probably see many advances in this area.

KINETIC CONSIDERATIONS

Before discussing the role of trace elements in igneous processes from an equilibrium point of view, it is helpful to understand some of the kinetic considerations that may limit the applicability of such discussions. Many of these kinetic limitations have recently been discussed by Hofmann and Hart (1975, 1978) and Nelson and Dasch (1976). Based on rather abundant data for element diffusion in basalt and obsidian liquids (Hofmann and Magaritz, 1977; Hofmann, this volume), and rather sketchy data for element diffusion in silicate minerals, the following generalizations can be made: at basaltic magmatic temperatures of 1150–1250°C: (1) diffusion coefficients for elements such as Ca, Sr, Ba, Co, Eu, and Gd in a basalt melt will be in the range $10^{-6} - 10^{-8}$ cm^2/sec. The maximum variation between any of these elements at a given temperature is about a factor of four (Eu and Gd are slower than Co and Ca); (2) diffusion coefficients for Sr and Ba in obsidian melts are lower than in basalt melts by a factor of five to ten (at a given temperature); (3) the effect of pressure on diffusion coefficients is relatively small. For example, the diffusion coefficient for Ca in silicate melt at 1,300°C decreases by a factor of three for a pressure increase of 30 kb (Watson, 1979); (4) elements such as Sr, U, Al, and Ni in mantle-type minerals probably have diffusion coefficients in the range $10^{-11} - 10^{-13}$ cm^2/sec.

From these values, we can estimate the kinetic constraints involved in various geologic processes at temperatures of 1,150–1,250°C (using the lower estimates of diffusion in melt and solid of 10^{-8} and 10^{-13} respectively).

(1) Sub-solidus equilibration: exchange of a trace element between two 1-cm grains in contact will take about .5 m.y. Separation of the grains by 1 cm of similar material will only increase equilibration time to 1 m.y. Transport distances will reach 60 cm in 1 b.y. Equilibration times of 100 m.y. or more can be expected only if the grains are separated by a "barrier" phase which has a low partition coefficient (solubility) of the element in question.

(2) Equilibration between solids in the presence of melt: if the melt wets all grain boundaries, equilibration between 1 cm grains will occur in less than .04 m.y. If melt occurs only as interconnected tubes (Bulau and Waff, 1977), equilibration times may approach the .5 m.y. given above for the subsolidus case.

(3) Equilibration within melts: characteristic transport distances will be .5 cm in one year, 5 m in 1 m.y., and .5 km in 1 b.y.

From these considerations, we may expect all phases in a partially molten mantle to be in elemental and isotopic equilibrium (a mantle diapir rising at a rate of 10 cm/yr will only travel 4 km between the time it starts to melt and the time complete intermineral equilibration is achieved; it is unlikely that significant melt separation can take place in this short time interval). Thus the production of a partial melt should very closely approximate an equilibrium process. However, the migration of this melt through overlying mantle may result in partial re-equilibration with that mantle, and this process may be one of nonequilibrium. Modeling of such a process will depend on a much more detailed understanding of conduit dimensions and migration velocities than is currently at hand.

In the case of mineral crystallization, it is clear that kinetic considerations are frequently important (witness the common occurrence of zoned crystals). The problem arises in several ways. First, once a volume of solid is crystallized, will it have any tendency to re-equilibrate with a changing melt composition (i.e., is crystallization adequately described in terms of a surface equilibrium or Rayleigh-distillation model)? Second, during crystallization, is diffusion in the melt rapid enough to keep the melt composition adjacent to the growing crystal identical to that in the bulk melt reservoir? If not, equilibrium partition coefficients are not applicable; rather, one must use an "effective" partition coefficient that will depend on the relative magnitudes of the crystal growth rate versus element diffusion coefficients in the melt (Smith, Tiller and Rutter, 1955; Albarede and Bottinga, 1972; Flemings, 1974). Taking the first question, it may be noted that a 5 mm phenocryst will reach 90% of equilibrium with a melt in about 3,500 years (for $D = 10^{-13}$ cm^2/sec). A large (2 km) convecting magma chamber will reach subsolidus temperatures throughout in about 2,000 years, and "magma-exposure" times of phenocrysts in such a body may permit considerable re-equilibration between crystals and melt. Thus, Rayleigh crystallization models may be strictly applied only for considerably smaller cooling units, where magma-exposure times of less than a few hundred years can be expected. Considering the second question, it has been shown by Smith, et al. (1955) that a chemical boundary layer is formed adjacent to a solid as it crystallizes, and that the concentration of a trace element in the surface of the solid (C_s) is related to the initial concentration in the melt (C_0), for a one-dimensional case, by:

$$\frac{C_s}{C_0} = \frac{1}{2}\left\{1 + \text{erf}\left[\frac{\sqrt{(R/D)x}}{2}\right]\right.$$
$$\left. + (2k - 1)\exp\left[-k(1 - k)\left(\frac{R}{D}\right)x\right]\text{erfc}\left[\frac{(2k - 1)}{2}\sqrt{(R/D)x}\right]\right\},$$

where k is the solid/liquid partition coefficient,
   R is the crystal growth rate,
   D is the diffusion coefficient of the element, and
   x is linear dimension of the solid.

C_s/C_0 is thus the "effective" partition coefficient, which, for small k ($\lesssim .3$), is approximately equal to: $k(1 + Rx/D)$. From this it can be seen that the effective partition coefficient will be within 20% of the real partition coefficient as long as $Rx/D \lesssim .2$. While growth rates of phenocrysts are not well known, a value of 4×10^{-10} cm/sec was reported for plagioclase in the Kilauea lava lakes by Kirkpatrick (1977). With $D = 1 \times 10^{-8}$ cm^2/sec (basalt at 1,150°C) partition coefficients will closely approach "true" values for crystal sizes of 5–10 cm and smaller. Thus, except for cases of unusually rapid phenocryst growth, elements of low partition coefficient (< 1) in basalt may be modelled very adequately with equilibrium partition coefficients. (On the other hand, crystallization of granitic melts at lower temperature, where D's may be 10^{-10} cm^2/sec or less, may lead to significant departures between "effective" and "true" partition coefficients.)

Finally, for trace elements of large parition coefficient, the problem is considerably more serious. Solution of the complete equation above (again using a growth rate of 4×10^{-10} cm/sec and diffusion coefficient of 1×10^{-8} cm^2/sec) leads to effective partition coefficients which may be in error by more than 20% for crystal sizes greater than .01 cm ($k = 10$) or .2 cm ($k = 3$). Note the strong dependence on k; note also that the effective partition coefficient will always be smaller than the true partition coefficient. If we consider the partitioning of Ni between olivine and melt, where k ranges from 5 to 20 (Hart and Davis, 1978) and where growth rates may exceed those for plagioclase, it is clear that modeling of olivine crystallization as a simple Rayleigh fractionation process may lead to serious errors.

EQUILIBRIUM TRACE-ELEMENT MODELS

The general conclusion based on kinetic considerations is that, in magmatic circumstances, an equilibrium approximation is not too bad. In fact, much important progress has been made using this first-order approximation. Without giving a complete discussion of this approach (for a good review, see Allègre and Minster, 1978), we want to emphasize that in most cases authors work with two models:

(1) *Rayleigh fractionation model*:

$$C_l^i = C_{0,l}^i f^{(D^i - 1)},$$

where

C_l^i = concentration of element i in the melt,

$C_{0,l}^i$ = initial concentration of element i in the melt,

f = fraction of melt

D^i = bulk partition coefficient of element i between solid and melt (which is the sum of the products of the individual mineral partition coefficients and their weight fractions,

(2) *Berthelot-Nernst equilibrium model*:

$$C_l^i = \frac{C_0^i}{D + F(1 - D)},$$

where

C_l^i = concentration of element i in melt,

C_0^i = initial concentration of element i in solid,

F = fraction of melt

D = bulk partition coefficient of element i between residual solid and melt.

The Rayleigh equation is commonly used to model fractional crystallization in a closed system, such as a magma chamber; the Berthelot-Nernst equation is used to model partial melting processes.

A number of more complicated models have been proposed (Allègre and Minster, 1978; O'Hara, 1977; Langmuir et al., 1977) but have not yet been extensively used. In many cases, our knowledge of partition coefficients is too limited to justify the use of more complex models. A review of partition coefficient data may be found in Irving (1978).

INVERSION TECHNIQUES

A major step forward in the use of trace elements to constrain igneous processes has been taken by the introduction of inversion techniques (Allègre et al., 1977; Minster et al., 1977; Allègre and Minster, 1978). Rather than taking the earlier approach of calculating a series of models and testing them against the data to find the most appropriate one, the inverse technique starts with the data itself (and some limiting constraints regarding the types of processes involved) and from it generates a single model which is, in essence, the "best-fit" model. This approach has the advantage of using all of the data, of treating the data quantitatively and objectively, and of allowing statistical estimates of modeling uncertainties and information quantity. The technique spotlights those samples and data which are most critical to the model (and thus identifies those elements which need the most accurate analyses), the partition coefficients which are most in need of experimental verification, and the extent to which various processes are statistically distinguishable from each other.

ISOTOPIC TRACERS

One category of trace element investigation which is particularly useful in petrogenetic studies involves both radiogenic and light stable isotopes.

Some light elements like hydrogen or oxygen have minor isotopes (respectively D or ^{18}O) and the ratios between these minor isotopes and the major isotopes (D/H, $^{18}O/^{16}O$) vary in nature. These variations are small and are usually expressed in δ notation:

$$\delta^{18}O = \left[\frac{(^{18}O/^{16}O)_{sample} - (^{18}O/^{16}O)_{reference}}{(^{18}O/^{16}O)_{sample}} \right] \times 1,000$$

As pointed out by Taylor and Epstein (1962) these isotopic variations can be used as petrogenetic tracers. Other elements such as Sr, Pb and Nd have isotopes which are decay products of long-lived radioactivities. Therefore, isotopic ratios such as $^{87}Sr/^{86}Sr$, $^{206}Pb/^{204}Pb$ and $^{143}Nd/^{144}Nd$ vary in nature as a consequence of the radioactive decay of the parent isotopes ^{87}Rb, ^{238}U and ^{147}Sm. The magnitude of the variation depends on the time-integrated history of the ratios $^{87}Rb/^{86}Sr$, $^{238}U/^{204}Pb$ and $^{147}Sm/^{144}Nd$. Unlike the light stable isotopes, these radiogenic isotope ratios are not fractionated in any significant way by physico-chemical processes. An excellent discussion of the principles of isotope geochemistry may be found in Faure (1978). We want to emphasize only a few points here:

(1) Secondary processes such as weathering and alteration can markedly alter the isotopic compositions of rocks. Consequently it may be more difficult to interpret isotopic variations in old rocks than in young ones. Furthermore, the response of an isotopic system to secondary processes can vary markedly from one system to another.

(2) Pb isotopes are particularly useful as petrogenetic tracers because there are two radioactive uranium parent isotopes, of different half-life, which both decay to the same element, Pb.

(3) Isotopic ratios are most efficient as "tracers" when there is a large contrast between different environments. For example, Rb/Sr ratios are markedly higher in the continental crust than in the mantle, and thus interactions between these two reservoirs are ideally traced using Sr isotopes. On the other hand, the crust and mantle have fairly similar U/Pb ratios, so Pb isotopic ratios may be of less use in tracing crust-mantle interactions.

(4) It is important to use a combination of as many isotopic systems as possible in trying to understand complex petrogenetic processes, as a single isotopic system may frequently give ambiguous results. For example, volcanic rocks with $^{87}Sr/^{86}Sr = .705$ are frequently claimed to be "mantle-derived." However, there are many reservoirs which may contain Sr of this isotopic composition—young continental crust, old deep-crustal granulites, young volcanogenic sediments, altered oceanic crust, and so forth.

PETROGENESIS OF MAFIC ROCKS

SOURCE CHARACTERISTICS

Trace elements are helpful in specifying a number of things about a magma source. For incompatible trace elements (that is, those with crystal/liquid partition coefficients considerably less than one, such as Rb, Zr, Ba, U, light REE), even relatively small degrees of melting (10%) of the source produce a liquid which incorporates most of these elements from the source, so that ratios of these elements will be very similar to those in the source (Gast, 1968; Schilling, 1971). In other words, if the Rb/Sr ratio of a mantle source is .03, the ratio in a derived melt will also be close to .03. Only in the case of very small degrees of melting ($<5\%$), where the residue contains a phase which is effective at fractionating Rb and Sr (such as amphibole or phlogopite) will the Rb/Sr of the melt be significantly different than that in the original source (and even then, expected differences are not large—about 30 to 50%).

Isotope ratios will be similarly carried over into the melt, so that the $^{87}Sr/^{86}Sr$ or the $^{206}Pb/^{204}Pb$ ratio of a primary magma may be equated with that of its source. The only exception to this is a case in which equilibrium between the melt and the residual phases is not maintained during the melting episode. Based on the kinetic arguments discussed above, it seems likely that equilibrium will be the normal situation, and that disequilibrium melting can occur only in very special circumstances (where a melt is formed in a material which has been stored sub-solidus for an appreciable time [$>10^8$ years], and very quickly [$<10^6$ years] extracted after being formed). It should also be emphasized that "isotope fractionation" cannot be appealed to in order to explain isotopic differences. In the case of both Sr and Nd, fractionation corrections based on the presence of several non-radiogenic isotopes are routinely applied to all data. Pb has only one nonradiogenic isotope, and thus natural fractionation effects cannot be corrected for; however, Pb isotopic differences in nature are large (5–10%), and natural fractionation effects are likely to be quite small ($<.1\%$) due to the large mass of the isotopes.

From these considerations, it has been possible to prove that basalt magmas are derived from mantle sources which are heterogeneous in composition, at least with respect to the isotopic composition of Sr, Nd, and Pb (see references in Hofmann and Hart, 1978) and to the elemental ratios of numerous incompatible elements (Rb/Cs, K/Ba, Zr/Nb, Sm/Nd, U/Pb, Rb/Sr, to mention just a few). It is also possible to show that these differences in mantle composition are of long duration—probably of the order of 1–2 b.y. (Brooks, et al., 1976; Richard et al., 1976; De Paolo and Wasserburg, 1976; Tatsumoto, 1978). It is *not* possible to specify as yet exactly how these differences arose—whether some of them are relicts of original heterogeneity built in during formation of the earth, or the result of differences caused by single or multiple episodes of melt extraction, or a result of metasomatic effects within the mantle. It is possible to "map" the mantle based on these differences, at least areally, and to show that the mantle supplying ocean ridge basalt is similar the world around, but different from that supplying oceanic island or continental basalts (Hofmann and Hart, 1978; Tatsumoto, 1978; O'Nions, Hamilton and Evensen, 1977). This mapping has been used to support various kinematic models of the mantle, such as the mantle-plume concept (Schilling, 1975; White, Schilling and Hart, 1976), but as yet there are insufficient constraints on the depth distribution of these differences to unambiguously define the scale or motions of these mantle heterogeneities.

In addition to specifying some of the trace element and isotopic characteristics of various mantle magma sources directly, these elements can also be used to specify something about the mineralogy and major element character of the mantle. This is done by using trace elements that are "compatible"—that is, that have crystal/melt partition coefficients greater than one (for example, the heavy REE and Sc in garnet, Ni in olivine, Cr in clinopyroxene, Sr in plagioclase, Rb and Ba in phlogopite). In these cases, the presence of one of these phases in a melting residue will tend to "buffer" the concentration of that element in the melt, somewhat independently of the degree of melting. Thus, alkali basalts tend to have REE patterns which show wide variation in light-REE abundances but relatively constant heavy-REE abundances—that is, the patterns tend to converge toward the heavy-REE end. This is strong support for the presence of garnet as a residual phase in the mantle from which alkali basalts are derived (Kay and Gast, 1973; Sun and Hanson, 1975). Conversely, mid-ocean ridge basalts (MORB) have REE patterns that are are parallel to each other, but have variable abundances of total REE. This is strong evidence that garnet was not a component of the residue from which ridge basalts were derived (Kay, Hubbard and Gast, 1970; Schilling, 1975). Note that this does not prove that melting did not take place in the garnet-peridotite facies, it only implies that the melt last equilibrated with a garnet-free assemblage. Thus these basalts were either derived at shallow depths (< 30 kb), or at greater depths but by a large degree of melting which consumed all of the garnet. On the other hand, the relative constancy of Sr in MORB, relative to other incompatible elements, suggests plagioclase as a residual phase—thus it is quite likely that MORB are derived from shallow depths, within the plagioclase peridotite facies (Kay, Hubbard, and Gast, 1970). Similarly, the wide variations of Rb and Ba in MORB suggest the absence of phlogopite as a residual phase following melting.

In principle, other compatible elements could be used in the same way. Unfortunately, the concentrations of compatible elements are also very sensitive to fractional crystallization, so that relatively minor amounts of shallow crystallization of a mineral such as olivine may tend to obscure any information that Ni might carry regarding presence of olivine in a mantle source. Thus, although the nickel content of basalts varies widely, this is due to olivine crystallization and does not indicate the absence of olivine in the mantle source (Hart and Davis, 1978). Since each potential mantle phase seems to have at least one trace element which is "compatible" for that phase alone, it should ultimately be possible to specify the mineralogy of various mantle magma sources with some confidence.

PARTIAL MELTING CONSTRAINTS

The concentration of an incompatible element in a primary magma is a function of the original concentration in the source and the degree or extent of melting. Therefore, if two magmas can be assumed to come from the same source, a "relative" degree of melting for the two magmas can be specified quite closely. For this reason, tholeiites and alkali basalts that are erupted in close geographic proximity and have similar incompatible element ratios and isotopic ratios may be presumed to come from the same or a similar mantle source. For example, tholeiites and alkali basalts from a given volcano on Hawaii seem to show similar Pb isotope ratios (Tatsumoto, 1978). Any difference in overall absolute abundances of incompatible elements in these basalts, then, is related to the different degrees of melting involved. Similarly, the alkali basalts of the Azores were probably derived from mantle similar to that supplying tholeiites to the ridge transecting the island platform (White, Schilling and Hart, 1976): the concentrations of incompatible elements such as Rb, Ba and Cs are about twice as high in the alkali basalts as in the tholeiites—thus the tholeiites probably represent *twice* the degree of melting that the alkali basalts do (White, 1977). If the tholeiites are 25% melts, the alkali

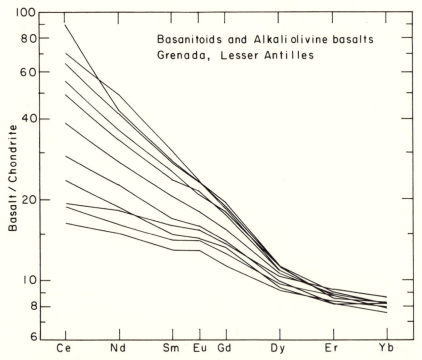

Figure 1. Rare-earth patterns in basanitoids and alkali olivine basalts from Grenada (Shimizu and Arculus, 1975). Normalizing factors from Sun and Hanson (1975). Total variation in Yb value is only 14%, compared to the variation of more than a factor of 5 in Ce.

basalts are 12%; if the tholeiites are 10%, the alkali basalts are 5%, etc. Unfortunately, the *absolute* degree of melting is difficult to determine by trace element arguments alone, since the absolute concentrations of trace elements in the mantle is poorly constrained at present.

Another example of the use of incompatible elements to determine relative degrees of partial melting was reported for a series of basanitoids and alkali olivine basalts from Grenada, Lesser Antilles (Shimizu and Arculus, 1975). The fan-shaped REE pattern, shown in Figure 1, strongly suggests garnet as a residual phase in the production of all of the basalts; the variation in light-REE (Ce) abundances is about a factor of 5.5. This suggests a relatively large variation in degree of melting for the various basalts of this series; numerical modeling suggests values of 17% melting for one end member and 2.2% for the other (Shimizu and Arculus, 1975). It must be remembered, however, that it is the relative difference in degree of melting that is well constrained—the absolute values are subject to considerably more uncertainty due to model-sensitive parameters.

Crystal Fractionation Constraints

Crystallization of various basaltic minerals is not effective in fractionating the incompatible elements from each other; thus the ratios of various incompatible elements (K/Rb, U/Pb, Ba/Ce, etc.) will not be altered, even by large amounts of crystallization of minerals such as plagioclase, clinopyroxene, or olivine. The concentration levels will of course be affected, with concentrations in the melt phase increasing as crystallization of these minerals proceeds (90% crystallization will increase melt concentrations of incompatible elements by ∼ 10 times, and so on).

In contrast to this, the melt concentrations of compatible elements may be drastically changed during crystal fractionation, so that melt concentrations of elements such as Ni, Cr, Sr, and Sc serve as excellent indicators of the extent of crystallization of the minerals olivine, clinopyroxene, plagioclase, and garnet, respectively. Because each common basaltic mineral has at least one trace element which is characteristically compatible for that mineral alone, the specification of relative mineral proportions during crystallization can theoretically be done with considerable precision. The behavior of both compatible and incompatible elements has been modelled in several natural differentiation series with considerable success. For example, Zielinski and Frey (1970) used major element data to determine relative proportions of crystallizing minerals in an alkali-basalt/trachyte series on Gough Island. Using these mineral proportions and the relevant partition coefficients for the elements Sr, Ba, Cr, Ni and the REE, they found excellent agreement between predicted and observed trace element trends in this volcanic series. A similar study, carried out on an olivine-basalt/trachyte sequence from Reunion Island, was successful in modeling the observed variations of Sr, Ba, Cr, Ni, REE, U, and Th (Zielinski, 1975).

In addition to the detailed modeling of accepted differentiation sequences, compatible trace elements have had a major role in constraining more general petrogenetic models. For example, the model of basalt genesis proposed by O'Hara (1968) assumes a large amount ($\geq 50\%$) of

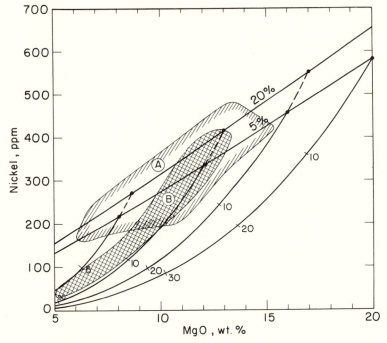

Figure 2. Ni–MgO relationships for partial melting and fractional crystallization models. Partial melting curves labeled 5% and 20% are batch partial melts from a peridotite mantle containing 2,300 ppm Ni (assuming melts of a wide range of MgO can be produced). Four curves for fractional crystallization of olivine are shown, starting from the 5% melting line with liquids of 8, 12, 16 and 20% MgO. Amounts of olivine crystallized are indicated by numbers along curves. Ni partition coefficient used is given by:

$$\left[D = \frac{124}{MgO} - 0.9 \right], \text{ see Hart and Davis, 1978.}$$

No allowance is made here for possible pressure effects on the partition coefficient or for non-Henry's Law behavior (Mysen, 1978; Mysen and Kushiro, 1979). Enclosed areas are (A) fields of natural basalts containing high-pressure xenoliths, and (B) fields enclosing basalts and differentiates from five volcanic series, accumulative rocks not included (Hart and Davis, 1978).

eclogite and olivine fractionation. Such a process would be expected to severely deplete the residual liquids in the heavy-REE (due to garnet crystallization) and in Ni (due to olivine crystallization). The lack of observed depletion of these elements has been cited as evidence against O'Hara's model (Gast, 1968; Hart and Davis, 1978; Shimizu and Arculus, 1975). For example, in Figure 2, the field of Ni–MgO contents of natural basalts (which are non-accumulative with respect to olivine) is compared to trends for partial melting and fractional crystallization (Hart and Davis, 1978). It is clear that, although this field does appear to outline a fractional crystallization trend (it is parallel to a fractional crystallization trajectory), the primary melts have MgO contents of 10–12% rather than the 20% or so postulated by the O'Hara model. Fractional crystallization of a primary melt of 20% MgO to produce an 8% MgO melt ($\sim 35\%$ olivine crystallization) would result in Ni contents only one quarter of those observed in natural series.

Modeling of trace element behavior in the Skaergaard, the classic example of a differentiated layered intrusion, has also been done successfully (Paster, Schauwecker, and Haskin, 1974), though it was necessary to include the effects of trapped liquid in order to bring about a reasonable fit of the model to the observations.

Miscellaneous Processes

Under this heading we include such diverse processes as wall-rock reaction, contamination, volatile transfer, and alteration. The operation of each of these processes in natural systems will have important consequences in understanding the overall petrogenetic process—however, their operation does not always have an unambiguous fingerprint on the products generated by the process. A case in point is wall-rock reaction, which might, for example, refer to the continued interaction of a magma with the walls of the conduit during magma ascent. For incompatible elements, this process will lead to a steady increase in their concentration, with the limit being given by the bulk partition coefficient between the wall-rock assemblage and the magma. In effect, the liquid will tend, in terms of the incompatible elements, to approach in appearance a liquid formed by very small degrees of melting. (As the degree of melting approaches zero, the concentration of an incompatible element in a melt approaches C_0/D—the original solid composition divided by the partition coefficient.) This is one possible way to explain magmas—such as, for instance, nephelinites—showing very high concentrations of incompatible elements. Interpretation of these magmas as very small degrees of melting ($<1\%$) creates a problem as to how such small amounts of melt are extracted from their source. The physics of this process are very poorly understood, but conventional wisdom suggests extraction of such small amounts of melt may be difficult (Yoder, 1977). By allowing for wall-rock reaction (or zone melting), an initially large degree of melt (say 10%) can be equilibrated with a mantle volume at least 10 times the original volume (perhaps by diffusion through interconnected grain boundary melt channels), resulting in a melt which resembles a 1% melt. In simplest terms, such a process will be difficult to identify, since the resulting melt will look identical to a small degree batch melt. It is possible, however, that such a process will not take place at equilibrium (since melt migration may be rapid relative to diffusion times), and that therefore the final melt may bear some characteristic trace element signatures related to the different diffusion rates of elements in a silicate melt. For example, in a basalt melt at 1,300°C, sodium diffusion is about 20 times as fast as REE (Gd or Eu) diffusion (Hofmann, this volume; Magaritz and Hofmann, 1977). The Na/Eu ratio of a melt involving strong wall rock reaction may therefore be significantly higher than the Na/Eu ratio expected for a melt derived by equilibrium partial melting. Clearly, real understanding of how much wall-rock reaction is involved in magma genesis must await further progress in studies of melt migration paths and velocities.

Magma contamination might be considered as a process similar to wall-rock reaction, but more commonly it is considered in terms of actual addition to and digestion of foreign material by a magma. It can be most easily recognized, then, when the foreign materials have a different and characteristic trace element or isotopic imprint. Thus the presence of oceanic sediment contamination, with high Ba and Cs contents, has been postulated in some of the tholeiitic basalts of Japan (Hart, et al., 1970), while the absence of such sediment contamination has been argued for basalts from the Tonga and Mariana arcs, based on Pb-isotope evidence, which did not show any of the high 207/204 signatures characteristic of oceanic sediment (Oversby and Ewart, 1972; Sinha and Hart, 1972; Meijer, 1976).

Another interesting example of crustal contamination of a basaltic magma has been documented by Fratta and Shaw (1974) and Dostal and Fratta (1977), who compared the trace element character of a single diabase dike with respect to its emplacement in two markedly different country rocks. The dike, 10 meters in width, was found to have significantly higher levels of K, Rb, Ba, Li and Tl where it intruded a granite-gneiss country rock, as compared to the same dike where it intruded basic-volcanic country rock. Thus, while contamination of magmas may occur during their upward passage through the crust, it is clear that such contamination may also occur at the final site of emplacement. In this circumstance, the contamination probably did not occur by bulk digestion of incorporated wall rock, since other trace elements such as REE, U, Zr, Nb, Sc, Co, and Cr showed no such effects.

Frequently, where a differentiated basalt sequence is modeled using both major and trace elements, an excellent fit can be achieved for the major elements and the compatible and refractory incompatible elements (such as Zr, the heavy REE, Sr, Th, Nb, and so on), whereas the volatile incompatible elements (such as K, Rb, and Cs) may show abnormally high concentrations in the late-stage differentiates as compared to the predictions of the model. This has sometimes been taken as evidence that the magmas were not all derived from the same source or that a complex melting process in the source produced a variety of magma compositions (Langmuir, et al., 1977). The discrepancy might equally well be ascribed to a process—for instance, volatile transfer—occurring in the magma chamber or in the cooling unit itself. An example of this may be the study of chemical variations across a single 10 meter basalt flow (Watkins, et al., 1970; Hart, et al., 1971). Variations in the concentrations of K, Rb, Cs, U, and Th of up to a factor of five were noted, without any significant variation in the major elements or trace elements such as Sr, Ba, Ni, or Cr. The variations were interpreted in terms of the upward movement of a very small amount (1%) of residual fluid (magma or volatile phase?) after crystallization of the flow was essentially complete. There is abundant evidence for migration of volatiles in salic rocks, such as that based on the oxygen isotope data, to be discussed below. There are fewer examples in mafic rocks, but the widespread oxygen exchange noted in the Muskox and Skaergaard layered intrusions (Taylor, 1968) and the anomalously low $\delta^{18}O$ values for many Icelandic basalts (Muehlenbachs, et al., 1974) is clear proof of the effects of volatile or aqueous-phase movements during mafic-rock petrogenesis. Also, carbon-isotope measurements on mafic rocks show clearly that a CO_2-rich phase has separated from most magmas during their ascent (Pineau, et al., 1976), leaving a magma depleted in $\delta^{13}C$ relative to its mantle source.

Because there are relatively few well-documented examples of trace-element movement by volatile transfer, the significance of this mechanism is difficult to evaluate. Needless to say, with the increasing refinement and complexity of partial-melting and crystallization models (Langmuir, et al., 1977; O'Hara, 1977), it is becoming more and more imperative that late stage effects such as contamination and volatile-phase transfer be understood, so that their effects can be accounted for before modeling of more "primary" processes is undertaken.

PETROGENESIS OF GRANITIC ROCKS

In comparison with the tremendous amount of data accumulated on basaltic rocks, the data on granitic rocks are rather meager. Recently, however, granitic rocks have come under focus and sufficient data are now available to provide a first approach to this problem. If we are now still in the basalt decade, we predict a granite decade next for geochemistry. Interest in granitic rocks derives from two independent sources. First, the problem of granite genesis is related to the problem of the development of continental crust. Second, many ore deposits are associated with granitic rocks and the urgent problem of finding mineral resources demands a careful understanding of the granite problem.

From the point of view of modern geochemistry (i.e., a quantitative modeling approach), the problem of granite petrogenesis is far more difficult than that of basalt petrogenesis, for several reasons. (1) The petrology of granites is in a rather primitive stage and we lack convenient classification systematics. For example, we do not have the equivalent of the Yoder-Tilley basalt tetrahedron and corresponding phase diagrams for granitic systems. (2) While it is now well established that basalt comes from the mantle by partial melting of ultramafic materials, the origin of granites is still under great debate. From differentiation of mantle derived melts to processes of crustal anatexis, including also metasomatic processes, there is considerable room for argument. Bowen's hypotheses on these problems are still quite alive. (3) As we shall see later, partial melting and fractional crystallization are not the only processes important to granite genesis: other phenomena involving fluid transfer, including fluid extraction during melting and fluid expulsion after crystallization, also play a role. Geochemical data on partitioning between fluid and solid for most elements are still scarce. (4) Complex and multiple mixing processes between heterogeneous materials are probably involved in granite genesis, and these very often lead to considerable scatter in the data, in contrast to the generally regular behavior of elements involved in basalt petrogenesis.

SOURCE REGION CHARACTERISTICS

The problem of the source material for granites is certainly more difficult than for basalts. The basic question is the same as it was in Bowen's time. Are granites the product of prolonged fractionation of mantle-derived melts, or do they represent remelting of sediments or gneisses? Both processes are feasible experimentally and both have been argued by field geologists. This problem, which is one of the central issues of the granite controversy, can be attacked quite efficiently with isotopes and trace elements.

Because they are one of the products of Rb–Sr dating, initial $^{87}Sr/^{86}Sr$ ratios are probably the most common type of geochemical data on granitic rocks. The principle of their use is quite simple. The Rb/Sr ratio of crustal rocks (granites, sediments, etc.) is far higher than the Rb/Sr of mantle material (for example, 3 compared to .03). Thus the $^{87}Sr/^{86}Sr$

isotopic ratio grows at a rate of 10^{-4}/m.y. in a granitic environment, as opposed to a maximum growth of only 10^{-6}/m.y. in a mantle reservoir. Assuming that the continental crust is mostly granitic in composition, we can see that after several hundred million years a continental crust reservoir will be very different from a mantle reservoir in $^{87}Sr/^{86}Sr$. A granite formed by remelting of existing continental crust will therefore have a significantly higher $^{87}Sr/^{86}Sr$ ratio than a granite which is derived from mantle material by magmatic fractionation (Hurley et al., 1962; Hedge and Walthall, 1963). By plotting the initial $^{87}Sr/^{86}Sr$ ratio versus time for a large number of granites of different ages, we see a large spread of values tending to diminish with time in the past. Most of the values are above mantle values, but the lowest values are very close to the mantle trend.

Young Granites (Cenozoic)

The initial Sr isotope ratios range from .760 (Palung Himalayas, Hamet and Allègre, 1976b) to .703 (California, Kistler and Peterman, 1973). Values above .707 seem to have a definite crustal reworking or contamination signature because oceanic volcanic rocks invariably have $^{87}Sr/^{86}Sr$ lower than .707 (see review in Hofmann and Hart, 1978). Values above .715 may infer that complete crustal anatexis is involved.

The work done in the western United States (Kistler and Peterman, 1973; Earley and Silver, 1973; Armstrong, et al., 1977) shows remarkable results. The $^{87}Sr/^{86}Sr$ initial ratios of the granitic rocks of ages between 190 m.y. and 60 m.y. are independent of the actual ages but are dependent on the nature of the environment and basement in which the plutons are emplaced. We can define roughly two types of provinces: one in which the ($^{87}Sr/^{86}Sr$) initial ratios are low ($\leq$ .7043) and one in which the ($^{87}Sr/^{86}Sr$) initial ratios are high ($\geq$ .7055). Rocks with low ratios occur when the plutons intrude Paleozoic or Mesozoic eugeosynclinal rocks. All of the rocks with high ratios occur within known or inferred Precambrian rocks. A clear boundary can be drawn between the provinces.

Similar results are found using Pb isotopic ratios of igneous rocks in the western United States (Doe, 1967; Zartman, 1974). One province includes the west coast of the United States (California, Oregon, and Washington), where the basement in largely Phanerozoic. The other province is the continental interior, where the basement is largely Precambrian. The provinces are distinctly different in Pb isotopic signature. The Phanerozoic basement province rocks show radiogenic (J-type) $^{206}Pb/^{204}Pb$ ratios which correspond well with mantle values. For the continental interior province, the spread in $^{206}Pb/^{204}Pb$ ratios is large, and the majority of the samples plot in the nonradiogenic (B-type) lead field. We should note that at present no known mantle-derived magmas fall in this field. On the other hand, Gray and Oversby (1972) have suggested that the lower crust has a low U/Pb ratio and would therefore be a possible source material for igneous rocks containing B-type lead. (A general discussion of J-type and B-type leads can be found in Doe, 1970.) We can conclude from this that the plutons have assimilated or interacted with their country rocks, and that the observed ratios are derived, at least in part, from the crustal environment. This correlation of isotopic pattern with type of basement seems to be characteristic of the linear orogenic zones that represent volcanic arcs built on continental borders.

A second study involves the Himalayan granites in the Indian plate, along the main central thrust (Le Fort, 1975; Mattauer, 1975; studied by Hamet and Allègre, 1976b). Here, leucogranites occur in a Precambrian environment. The $^{87}Sr/^{86}Sr$ ratios are extremely high (.740 to .760) and seem to represent a derivation almost wholly from the Precambrian environment. However, more typical orogenic granodiorites occur along the Tsanpo suture, with lower initial ratios (.710). Leucogranites seem to be a common facet of collision-type orogenic belts; their origin by palingenesis of old crust seems to be clearly demonstrated.

European Hercynian Granites

Recently, systematic Rb–Sr studies have been done on Hercynian granites from Europe by Brewer and Lippolt (1974), Vitrac and Allègre (1975), Hamet and Allègre (1976a) and Duthou (1976). All of these studies showed initial $^{87}Sr/^{86}Sr$ ratios that are higher than the values for usual mantle sources, and thus support the idea of involvement of older crustal material (Figure 3). Following the early suggestion of Allègre and Dars (1965), each granite may be considered as a mixture between an "intrusive component" and an "upper-crustal component". The intrusive component may be of mantle or lower-crust derivation. *Maximum* values calculated for the mantle component in this model vary between zero and thirty percent (Vitrac and Allègre, 1975). Hamet and Allègre (1976a) and Vitrac and Allègre (1975) attempted to test this conclusion using Rb and Sr as trace elements. They observed a close correspondence in Rb and Sr content between the granites and their host rocks. This "correlation" is systematic and is in agreement with the observations made with the radiogenic isotopes. Using the available partition coefficients, they reported quantitative models that fit the observed Rb–Sr ratios of the granites. In principle, the derivation of these granites from basic precursors is possible, but implies considerable crystallization of plagioclase. In the absence of evidence for such a large mass of plagioclase cumulate, the only way to derive these granites is by remelting of the crust, followed by local fractional crystallization or mixing between several components.

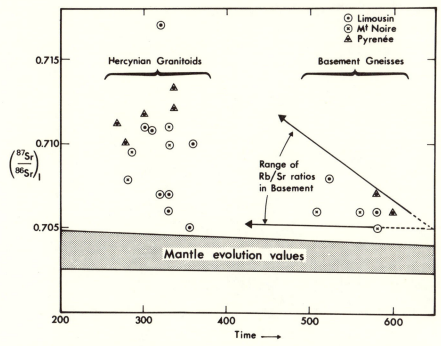

Figure 3. Initial ^{87}Sr/^{86}Sr ratios for Hercynian granites from France. Data are from Hamet and Allègre (1976a), Vitrac and Allègre (1975) and Duthou (1977). Initial ratios and evolution vectors (Rb/Sr ratios) for basement gneisses are also plotted. It is apparent that all of the Hercynian granite points fall within the range of evolved basement values at 300 m.y.

In addition to the Sr isotope studies described above, Pb isotope studies have also been made on the granitic rocks from the Hercynian of France (Allègre et al., 1979). While these showed a relatively large variation in ^{206}Pb/^{204}Pb, all values correspond to U/Pb ratios higher than mantle values and place these granites well inside the J-type field (relative to their 300 m.y. age). These observations will be discussed elsewhere, but they argue in favor of significant contamination by old continental crust or formation of these granites by complete remelting of such a crust.

The question of source materials for granitic rocks has also been studied using trace elements as petrogenetic tracers. Fourcade and Allègre (1980) studied the 280 m.y. old Querigut massif of France in great detail using the REE. Data for whole rock samples from the Querigut massif are plotted on a Ce/Yb–Eu/Yb correlation plot (Figure 4). The Ce/Yb ratio indicates the variation in the light versus the heavy rare earth elements and the Eu/Yb ratio indicates the importance of feldspar behaviour. On

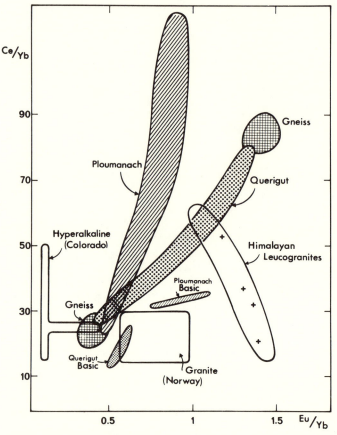

Figure 4. Ce/Yb versus Eu/Yb diagram for various granitic rocks. Mixing processes on this
diagram will generate linear trends. The gneisses bordering the Querigut rocks
fall at both ends of the Querigut trend, suggesting a mixing process. The Norwegian
mangerites (not shown) fall well outside the field of associated granites.

this diagram a mixing process is a straight line, while fractional crystallization or partial melting processes are nonlinear. Using the available
partition coefficients, we can easily see that fractionation or melting
processes move the points more strongly in a horizontal direction than
in a vertical direction. On this plot, we see that the various granitic facies
of the Querigut massif form a linear band, with wide variations in the
Ce/Yb ratios. The two ends of the band are represented by the host
gneisses presumed to represent the lower crust. The basic to dioritic rocks
of the massif are clustered, but are not on the main trend; their position
is such that it is impossible to derive the granitic facies from the basic

facies by fractional crystallization, or even by direct mixing. This observation is also supported by the study of the Ploumanach granite in Brittany (Fourcade, 1980).

Interesting results relating to the source materials of volcanic rocks have been reported for the ignimbrite province of Italy. Turi and Taylor (1976a,b) and Javoy (1970) found ignimbrites with very high $\delta^{18}O$ (+12 to +13). These values are comparable with those of sedimentary and metasedimentary rocks, and can be understood only if these ignimbrites are mainly the remelting of crustal materials. Sr and Pb isotope results support this conclusion (Vollmer 1976, 1977). In addition, K/Rb ratios studied by Dupuy and Allègre (1971) varied between 50 and 190 and showed a negative correlation with Rb concentration. Quantitative modeling, using measured partition coefficients, showed that these rocks could not be produced from a basalt parent because such a process would require the precipitation of at least 50% hornblende. Two processes were shown to be consistent with the data: one is the partial melting of continental crustal material, and the other, proposed by Turi and Taylor (1976) using our data, is a mixing between mafic magmas and continental crust.

Neodymium-Isotope Constraints on Source Materials

The use of this isotopic tracer in terrestrial studies is quite new, with the first results for granitic rocks appearing in the paper by Richard et al. (1976). However, a systematic approach to this problem (Allègre et al., 1979; Ben Othman et al., 1979) has only recently been developed. The approach is similar to that used for Sr isotopes, but in this case continental crustal materials have lower Sm/Nd ratios than the mantle has, so that derivation from a mantle reservoir will give more radiogenic $^{143}Nd/^{144}Nd$ isotope ratios than a recycling process from pre-existing crust. One particular advantage of Nd isotopes compared with Sr isotopes is the lower geochemical mobility of the rare earths compared to Rb and Sr. For example, the rare earths are almost inert during alteration, metamorphism and fluid transport. Thus their isotopic signature will not be strongly affected by hydrothermal circulation or remobilization.

The use of Nd as an isotopic tracer is best illustrated by use of a Nd–Sr correlation diagram (Figure 5). For mantle materials there is a negative correlation between Sr and Nd isotope ratios. Young granites all show negative ε_{Nd} values (defining $\varepsilon_{Nd} = 0$ as that in a system which has evolved with chondritic Sm/Nd ratios), and when initial ratios of these young granites are plotted (Figure 5), they fall distinctly to the right of the mantle trend, in the direction of old crustal material. These results argue for a significant crustal contribution in the genesis of these granites, with the

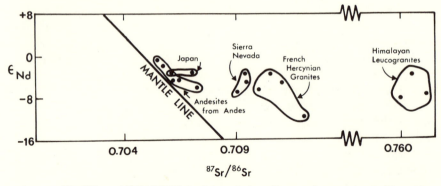

Figure 5. $^{87}Sr/^{86}Sr - {}^{143}Nd/^{144}Nd$ (ε_{Nd}) correlation diagram for young granitic rocks (data are from Beth Othman, et al., 1978). Note that most of the rocks plot to the right of the mantle evolution line defined by oceanic volcanic rocks (DePaolo and Wasserburg, 1976).

extent depending on the type of granite; it is large for the leucogranites of the Himalayas and relatively small for granodiorites from Japan.

GRANITIC SOURCE REGIONS IN THE PAST

The general conclusion which can be claimed without doubt is that most post-Precambrian granitic rocks are not direct products of mantle magmatism. Reworking of, or contamination with, older continental crust must play an important role in their genesis. This conclusion is not so obvious for Precambrian granites, however. Are they the same? Do some of them come directly from the mantle? Or are they also products of melting or interaction with crustal materials?

A recent well-documented study is that by Moorbath and Pankhurst (1976), who distinguished two types of granitic rocks (Figure 6). One is a calc-alkaline type which has very low strontium-isotope ratios, compatible with derivation from a mantle reservoir. The other is "true" granite, which appears to represent pure reworking of continental crust. However, we do not know if the first type is purely mantle-derived, or if it is derived from mantle materials with some crustal reworking. For example, graywackes that are subjected to crustal anatexis may give a mantle isotopic signature even though the petrogenetic process is purely crustal. Peterman et al. (1967) showed the eugeosynclinal sediments (graywackes) have $^{87}Sr/^{86}Sr$ ratios consistent with those required for granite sources.

The general result from Sr isotope data, that most Precambrian granites fall near a mantle source line, is also supported by Nd isotope data. DePaolo and Wasserburg (1976a,b), Allègre et al. (1978a), and McCulloch

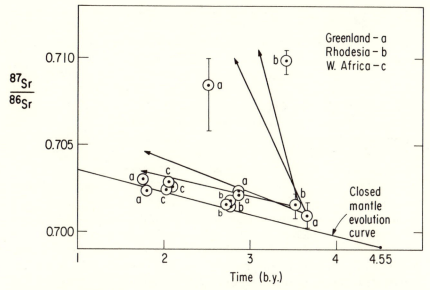

Figure 6. Initial $^{87}Sr/^{86}Sr$ ratios for some Precambrian granites from Greenland (a), Rhodesia (b) and West Africa (c). Data are from Moorbath (1977) and Allegre (unpublished). Vectors show evolutionary trends of these granites as a function of the average Rb/Sr ratio of each rock unit. The mantle evolution line is deduced from the Nd–Sr correlation diagram (DePaolo and Wasserburg, 1976).

and Wasserburg (1978) have reported initial Nd isotope ratios on a number of Precambrian granitic rocks; most of these are consistent with derivation from a mantle that has evolved with chondritic Sm/Nd ratios (as opposed to the lower Sm/Nd ratios expected for evolution of crustal materials).

PARTIAL-MELTING AND FRACTIONAL-CRYSTALLIZATION CONSTRAINTS

For a long time, partial melting processes were considered to be in contrast with fractional crystallization processes in the question of granite genesis; the former was associated with crustal anatectic hypotheses, the latter with derivation from basaltic melts. However, we now see that the evidence for crustal involvement in granite genesis is quite impressive: the question therefore becomes, how do these two processes interact?

There is abundant evidence regarding the operation of fractional crystallization processes during granite genesis. As an example, consider again the Querigut massif (Fourcade, 1980), discussed above. The minerals from each facies of this granite show a Rayleigh distillation process for the partitioning of rare-earth elements between phases. The most remarkable examples are the biotites, which have, in some facies, large negative europium anomalies, and which in other facies have almost no europium anomalies (Figures 7A, 7B). These observations have several important consequences: (1) we cannot simply use the ratio of bulk concentrations to calculate the partition coefficients between phases; (2) if the minerals do represent Rayleigh fractionation, then the formation of the granite represents an in-situ fractional crystallization process.

Fourcade also studied the behaviour of hornblende throughout the whole massif. He found that for the basic (dioritic) members, hornblende evolution follows a regular curve analogous to a fractional crystallization process. This regularity is not present, however, in the granitic members. On the scale of a single crystal, Shimizu et al. (1978) have shown that the hornblendes are zoned in rare earths and that this zoning corresponds to the range of hornblende patterns observed for a large variety of rock types. This observation clearly shows the existence of local fractional crystallization in these granite massifs, Total rocks from the basic through granodioritic facies also follow a fractional crystallization trend, though this pattern breaks down on going to the granitic facies (Fourcade and Allègre 1980).

A slightly different behaviour is observed for granite of alkaline associations. These rocks (Barker et al., 1975) have very large negative europium anomalies, and thus plot almost on the Ce/Yb axis of the Ce/Yb–Eu/Yb diagram (Figure 4). A similar position is also found for some Corsican granites (unpublished). The general trend shows two branches: one is vertical and seems to be the result of mixing (as suggested by Barker et al., 1975); the other is almost horizontal and suggests a significant feldspar fractionation process (and thus indicates fractionation in a magma chamber).

While these studies demonstrate the effectiveness of fractional crystallization processes, such processes may not operate on a single homogeneous starting material, as there is also isotopic evidence that suggests considerable heterogeneity in the initial melts. The most spectacular variations in this sense are certainly those reported by Taylor and Silver (1978) for the southern California batholith, where they found a systematic variation from west to east for both oxygen and strontium isotopes. Similar variations have been reported for Sr isotopes in the Sierra Nevada batholith (Kistler and Peterman, 1973) and the Idaho batholith (Armstrong, et al., 1977). Even though the data are not yet very abundant, close examination of the ε_{Nd} values from a single massif also show significant variations (see, for example, the Sierra Nevada batholith, or Querigut granite, Figure 5). These results seem to confirm that most granitic plutons are not initially isotopically homogeneous. The variations found

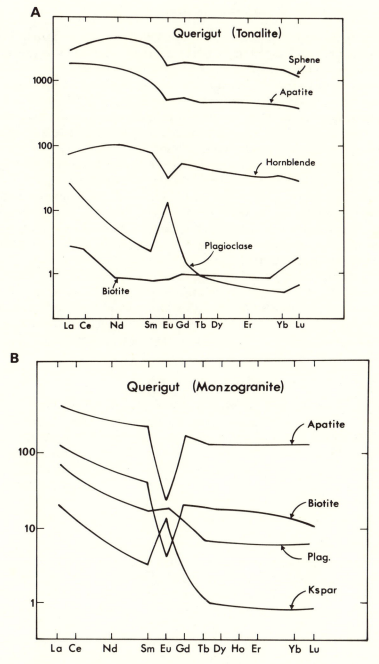

Figure 7. REE patterns (chondrite normalized) in minerals from a tonalite and a monzo-granite, Querigut massif, French Pyrénées (Fourcade, 1979). Note the marked differences between the patterns for the same mineral (e.g., biotite, plagioclase, apatite) from different rocks.

in the southern California batholith or Querigut granite correlate with the petrologic rock type and support the idea of a mixing process between a precursor coming from depth and a crustal component.

The evidence for partial melting processes in granite genesis was partly given earlier in the discussion of source characteristics. However, we can examine the process from a more physico-chemical point of view. In their detailed studies of French Hercynian granites, Fourcade and Allègre (1980) have shown that the mafic (basic) "precursor" does not mix with the surrounding environment (Figure 4). On the other hand, several types of country rocks seem to participate in the granite forming process itself. Similar conclusions were reached by Barker et al. (1975) in their study of alkaline granites. Trace-element modeling of this process implies a very large degree of melting of the country rocks (almost 100%). It is possible that the mafic precursor acts only as a heat source, generating melts at various crustal levels that are then injected into shallower levels.

A very different type of melting seems to occur with the Himalayan anatectic granites discussed earlier. Fourcade (1980) has proposed that these granites are essentially in-situ melts, formed by a combination of fluids circulating in shear zones, and shear heating related to the shear zone. The melts do not appear to migrate very far and thus they retain their anatectic characteristics.

THE ROLE OF HYDROTHERMAL CIRCULATION

The extensive oxygen and hydrogen isotope studies carried out on granitic rocks by H. P. Taylor and his associates in recent years is without a doubt the most advanced and well documented aspect of the geochemistry of granite (Taylor, 1968, 1974, 1978). The results are also extremely important because they point out the significance of water-rock exchange processes and the existence of low ^{18}O magma types. We will present first the theoretical basis for the work, and then provide a few typical examples and discuss the general consequences.

Theoretical Basis

Most igneous rocks have typical $\delta^{18}O$ values between 5.5 and 7 per mil (relative to standard mean ocean water). As shown by Taylor, magmatic fractionation at high temperature produces very small oxygen isotope fractionations between phases. For example, if granitic melts are the fractional crystallization products of "normal" basaltic magmas, they would have maximum $\delta^{18}O$ values of $+10$. For δD, normal values are between -50 and -85 per mil. On the other hand, surface-derived meteoric waters, such as ocean water or fresh water, have completely distinct $\delta^{18}O$ and δD isotopic ratios (Epstein and Mayeda, 1953; Craig, 1961). Figure 8 illustrates the various fields for these isotopes in meteoric

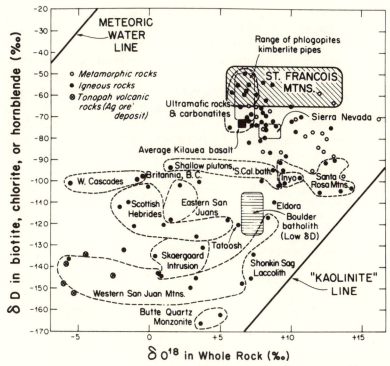

Figure 8. Plot of δD in biotite, chlorite or hornblende versus $\delta^{18}O$ in whole rock for a variety of igneous and metamorphic rocks (reproduced from Taylor, 1974b). The meteoric water line (Craig, 1961) and sedimentary kaolinite line are shown for reference. Standard mean ocean water is taken as δD and $\delta^{18}O = 0$. Mantle materials and mantle-derived magmas have a very restricted range of $\delta^{18}O$ ($+5.5$ to $+6$) and δD (-50 to -75). Note the rough correspondence between low δD values and low $\delta^{18}O$ values for hydrothermally altered igneous rocks.

waters, normal igneous rocks, and a variety of hydrothermally altered igneous rocks. Because the isotopic ratios of meteoric water are created by the meteorological cycle, such ratios have a very definite geographic pattern (this pattern seems to have been preserved over most of North America during recent geological time; see Lawrence and Taylor, 1974). Thus, the interaction between magmatic and meteoric reservoirs can produce numerous abnormal effects. This interaction between water and rock will produce different results depending on geography, temperature and the conditions under which it occurs.

Because the fractionation factor between phases decreases with increasing temperature, the isotopic ratios of rock and water tend to converge at high temperatures to give similar values. However, the amount of variation in the values will depend on whether the system is open or closed, and also, of course, on kinetic factors. At high temperatures, because the kinetics are faster, we can expect to achieve equilibrium, especially in aqueous conditions. In general, such water-rock exchanges will lower the $\delta^{18}O$ of granitic rocks. At low temperature, the effects are quite different. First, the interaction of a granite with water drives the major minerals out of their stability fields, leading to alteration reactions in which new minerals are formed. This will cause a change in both mineral and whole-rock isotopic ratios. In addition to this, the fractionation factor between minerals and water can be quite large at low temperature. Thus the low-temperature alteration minerals will generally have high $\delta^{18}O$ values.

Following Taylor (1978), we can classify plutonic granitic rocks into three categories:

Low $\delta^{18}O$ granites $\delta^{18}O < +6$
Normal $\delta^{18}O$ granites $+6 < \delta^{18}O < +10$
High $\delta^{18}O$ granites $\delta^{18}O > +10$

In general, the low $\delta^{18}O$ granites also exhibit low δD (as low as -130 or lower); see Figure 8.

Young Low-$\delta^{18}O$ Granitic Plutons

Taylor divides the problem of low $\delta^{18}O$ granitic plutons into two parts: (a) the interaction between crystallizing rocks and a convecting hydrothermal system, and (b) so-called low-$\delta^{18}O$ magmas.

The first phenomenon appears to be extremely common for epithermal plutons, and reinforces the point of view of Autran et al. (1970) that the petrology of granites is largely controlled by their depth of emplacement. The low $\delta^{18}O$ intrusions are always subvolcanic types and show effects largely of hydrothermal reactions, with the formation of alteration minerals. Isotopically, the feldspars are the most depleted in $\delta^{18}O$, in accord with laboratory experiments, which show that feldspars are the most easily exchangeable minerals for oxygen. Detailed studies of the Skye plutons, Skaergaard intrusion, the Idaho batholith, and many porphyry-copper quartz-monzonites have demonstrated the hydrothermal interactions quite convincingly and provide a good picture of the convection patterns outside the plutons (Taylor, 1978). A comparative study in the North American Coast Range, as well as measurements in the Nevada and Southern California batholith, have shown that the hydrothermal exchange is confined to the epithermal part of the pluton and decreases with depth (Taylor, 1978). This evidence from young granitic rocks is in complete agreement with present-day geothermal systems like, for example, those in Yellowstone and New Zealand. Assuming a simple equilibrium exchange process, Taylor has calculated the minimum water/rock ratios from the observed data: the (W/R) ratio varies from .1 to 2.

In addition to this problem, there are several cases of low $\delta^{18}O$ magmas. Evidence for these is given by the isotopic ratio of quartz, a mineral resistant to isotopic exchange: the ratio is low and decreases with time in a single plutonic (Taylor, 1974) or volcanic (Friedman, et al., 1974) sequence. Taylor argues that this feature may be the result of meteoric water that has penetrated the magma chamber and diffused into the magma. Because of physical difficulties created by both the limited solubility of water in magma and the relatively slow diffusion rate of water in magma, Taylor proposes an alternate mechanism: assimilation of the magma chamber roof, which has already been altered by water circulation and has low $\delta^{18}O$ values. This mechanism implies contamination of the magma with all species which are contained in the roof material.

Secondary Alteration

The effects discussed above are synchronous with pluton formation. Additional complications, involving secondary alterations, can also occur much later. These effects, which are associated with relatively low temperature exchanges, will show relatively large isotopic fractionations. The results of these secondary phenomena are complex and can affect element redistribution. However, Taylor has shown that these late exchanges produce brick-red feldspars (hematite exsolution) and thus can be detected by petrographic inspection. Chronologically, these effects can be very late compared to the crystallization age of the granite. In the St. Francois mountains, Missouri, at least 300 m.y. separate the secondary events and the primary formation age (Wenner and Taylor, 1976). In the Bushveld layered intrusion, the time difference is probably more than 500 m.y. However, a clear separation between the different types of alteration is not always so obvious. For example, when several episodes of plutonism occur in one area, the hydrothermal convection regime of the youngest one can cause low temperature alteration of the older one.

These hydrothermal effects have far reaching implications for two aspects of granite formation—the temperature conditions, and the scale of element migrations. Using two minerals for which we know the variation of partition coefficient with temperature (Clayton and Epstein, 1958; O'Neil and Clayton, 1974) we can calculate the temperature of equilibration, assuming that isotopic equilibrium was reached between the minerals (or the two minerals and fluid). This assumption can be tested in a rigorous way by using several minerals (Javoy, et al., 1970). Use of this technique, after recalibration of the thermometer by a critical evaluation of the partition data (Bottinga and Javoy, 1973), has revealed that in most cases the minerals in granites are not in equilibrium. The temperatures obtained are often between 600° and 400°C, and thus probably record late stage hydrothermal processes or mineral "closure" temperatures. The temperature of alteration processes has been determined on veins by the analysis of alteration products (Taylor, 1974), and gives results in the range 350° to 450°C, in agreement with other estimates.

When we do have an estimate of temperature, the application of the mineral-water calibration curve gives the isotopic composition of the hydrothermal water. Such results can be obtained for both $\delta^{18}O$ and δD, and can be checked by direct analysis of fluid inclusions. The results can then be compared with the data bank of isotopic compositions of known waters. The typical result is that the hydrothermal water circulating in crystallized plutons has more or less exchanged with the surrounding rocks (remember that this exchange will affect $\delta^{18}O$ but will have little effect on δD) but is clearly not juvenile or magmatic water. These results, which appear to apply for most hydrothermal systems, including ore deposits, are in agreement with the discovery made by Craig (1971) twenty years ago that most present-day hydrothermal systems are composed of meteoric water.

This hydrothermal exchange may also produce redistribution of certain elements, as for example, the alkalies. K, Rb and Cs partitioning has been studied extensively by Iyama and his group under hydrothermal conditions (Iyama, 1972). Despite several experimental difficulties in obtaining equilibrium, calibration curves were determined which permit the construction of feldspar-mica and feldspar-feldspar geothermometers. Such curves have been used by Caron and Lagache (1971) on natural granites in order to determine their temperatures of formation. In every case, although they observed reasonable regularity, the resulting temperatures are very low ($\approx 400°$). Since we know the temperature of granite formation is around 600–750°C, these low temperatures are attributed to re-equilibration during a late fluid stage or to continuous re-equilibration during the slow cooling of the granite. As discussed above for oxygen isotopes, the most probable interpretation is re-equilibration in the presence of fluids (even with structural modification of major minerals): these fluids strongly accelerate the diffusion and exchange of trace elements (especially in the presence of appropriate anions like Cl^-, Wyart and Sabatier, 1962). If this interpretation is true, even the total rock chemistry can be strongly modified by such hydrothermal interaction, and some of the observed regularities for O, Sr and trace elements as discussed above may be secondary effects. Albarede (1975) has discussed a quantitative model for such an open system, and showed that the observed facts are compatible with his model. These observations suggest caution in the use of mobile elements at this stage, and suggest that a larger role be given to studies involving such non-mobile elements as the rare earths.

Hydrothermal processes such as those discussed above may also be intimately related to metallogenesis associated with granite formation.

As we have emphasized, late stage fluids are extremely important in granitic-rock petrogenesis. The search for distinctive metal anomalies around "productive" granites is not always a successful prospecting guide. In part, this paradox was illuminated by Holland (1972), when he showed that element partitioning between aqueous fluid and granitic melt depends strongly on the anion content of the fluid, and that this dependency increases with the valency of the cations. We can then understand how concentrations of metals occur, if we remember that local fractional crystallization will ultimately lead to a fluid extremely rich in metals. This phenomenon can also occur during initial partial melting, where the metal is preconcentrated in a fluid phase and moves in it to the apex of intrusion.

The role of the fluid phase and its metallogenic effects in granite genesis should be investigated more deeply in the near future. Considerable promise in this direction is shown by the studies of sulfur and carbon isotopes. Though the data is sparse, large variations of these isotopes exist in granites, as shown by Shima et al. (1965) for sulfur and Pineau (1978) for carbon. Ohmoto (1972) showed that the coupling of the two isotopic pairs $^{32}S/^{34}S$ and $^{13}C/^{12}C$ may provide more favorable opportunities to reconstruct the redox conditions and the composition of the gas phase, and to provide constraints about the initial composition of sulfur and carbon. This method can probably be applied to the problem of granite genesis. It is our opinion now that (1) natural variations do exist, as shown by the early observations; (2) we do have a methodological approach, as developed by Ohmoto for ore deposits; and (3) this approach should be applied to granites. These studies will be particularly important if we remember that the volatile-phase and the redox conditions play an important role in determining the process of ore-fluid extraction from granitic reservoirs.

CONCLUSIONS REGARDING BASALT AND GRANITE GENESIS

Though our knowledge of granite genesis is at a rather primitive stage, we can draw some tentative conclusions comparing the petrogenesis of basalts and granites.

Basalts are derived by partial melting of peridotite in the upper mantle. The upper mantle is certainly heterogeneous, at least with respect to trace elements and isotopes such as Sr, Nd and Pb. Tholeiitic basalts have probably undergone their last "equilibration" with mantle material at relatively shallow depths, possibly in the plagioclase-peridotite facies. Alkali basalts originate at greater depth and probably last "equilibrated" with garnet-facies peridotite.

The dominant source of young granitic rocks is probably within the crust, though some mixing with differentiated mafic mantle-derived precursors is possible. The source region of older (Archean) granites is debatable—in many cases, the evidence is compatible with a mantle origin.

Partial melting plays a major role in both basalt and granite genesis. For basalts, variable degrees of partial melting are responsible for the difference between alkali basalt and tholeiite. For granites, the degree of melting is probably large in all cases, and the differences between various granites is therefore related to compositional heterogeneities in the source region. The presence or absence of water during this melting probably plays a significant role in the composition of the final melts, but this effect is poorly understood at present. The nature of the host environment is obviously important in granite genesis; basalts, particularly continental basalts, may also bear the imprint of their emplacement environment, though this question is currently under considerable debate.

Compared to partial melting effects, fractional crystallization processes appear to be less important in basalt and granite petrogenesis. As emphasized by O'Hara (1968), most basalts have undergone some fractional crystallization, and "primary" basalt magmas are rare. However, this process is not responsible for the major chemical distinctions between basalt types. Similarly, granitic magmas also certainly undergo differentiation (which may be responsible for the production of late-stage metallogenic fluids), but their major chemical characteristics are probably inherited from their source regions.

Finally, hydrothermal fluids (largely derived from meteoric waters) play a very important role in granite petrogenesis at shallow levels; similarly, juvenile CO_2–H_2O rich fluids probably play an important role in basalt petrogenesis, insofar as they control the conditions of partial melting at mantle depths.

Geochemical studies since Bowen's time have supplied important constraints on many petrogenetic models, but, as is common in science, they have also raised many new questions. We trust that these questions will generate as many exciting advances in the field of igneous petrogenesis as did those raised by Bowen fifty years ago.

REFERENCES

Albarede, F., and Bottinga, Y., 1972. Kinetic disequilibrium   trace-element partitioning between phenocrysts and host lava, *Geochim. Cosmochim. Acta* *36*, 141–156.

Albarede, F., 1975. The heat flow/heat generation relationship: an interaction model of fluids with cooling intrusions, *Earth and Planet. Sci. Lett. 27*, 73–78.

Allègre, C. J., and Dars, R., 1965. Chronologie Rb–Sr et granitologie, *Geol. Rundschau 55*, 226–237.

Allègre, C. J., Treuil, M., Minster, J. F., Minster, B., and Albarede, F., 1977. Systematic use of trace elements in igneous processes in volcanic suites, *Contrib. Mineral. Petrol. 60*, 57–75.

Allègre, C. J. and Minster, J. F., 1978. Quantitative models of trace element behavior in magnetic processes, *Earth and Planet. Sci. Lett. 38*, 1–25.

Allègre, C. J., Ben Othman, D., and Juery, A., 1978. Limitations on granite formation inferred by Nd, Sr and Pb isotopic systematics, *Trans. Am. Geophys. U. 59*, 392.

Allègre, C. J., Ben Othman, D., Polve, M., and Richard, P., 1979. The Nd-Sr isotopic correlation in mantle materials and geodynamic consequences, *Physics of the Earth and Planetary Interiors 19*, 293–306.

Armstrong, R. L., Taubeneck, W. H., and Hales, P. O., 1977. Rb–Sr and K–Ar geochronometry of Mesozoic granitic rocks and their Sr isotopic composition, Oregon, Washington, and Idaho, *Geol. Soc. Am. Bull. 88*, 397–411.

Autran A., Fonteilles, M., and Guitard, G., 1970. Relations entre les intrusions de granitoides, l'anatexie et le metamorphisme regional consideres principalement du point de vue de role de l'eau, *Bull. Soc. Geol. France XII 4*, 673–731.

Barker, F., Wones, D. R., Sharp, W. N., and Desborough, G. A., 1975. The Pikes Peak batholith, Colorado Front Range, and a model for the origin of the gabbro-anorthosite-syenite-potassic-granite suite, *Precambrian Res. 2*, 97–160.

Ben Othman, D., Richard, P., and Allegre, C. J., 1979. Neodymium and strontium isotopes in granites, in press.

Bottinga, Y., and Javoy, M., 1973. Comments on oxygen-isotope geothermometry, *Earth and Planet. Sci. Lett. 20*, 250–265.

Bowen, N. L., 1928. The Evolution of the Igneous Rocks, Princeton: Princeton University Press.

Brewer, M. S. and Lippolt, H. J., 1974. Petrogenesis of basement rocks of the Upper Rhine region elucidated by rubidium-strontium systematics, *Contrib. Mineral. Petrol. 45*, 123–141.

Brooks, C., Hart, S. R., Hofmann, A. W., and James, D. E., 1976. Ancient lithosphere: its role in young continental volcanism, *Science 193*, 1086–1094.

Bulau, J. R. and Waff, H. S., 1977. Thermodynamic considerations of liquid distribution in partial melts, *Trans. Am. Geophys. U. 58*, 535.

Carron, J. P. and Lagache, M., 1971. La distribution des elements alcalins Li, Na, K, Rb dans les mineraux essentiels des granites et granodiorites du sud de la Corse, *Bull. Soc. Mineral. Cristall. 94*, 70–80.

Clayton, R. N. and Epstein, S., 1958. The relationship between $^{18}O/^{16}O$ ratios in coexisting quartz, carbonate, and iron oxides from various geological deposits, *J. Geol. 66*, 352–371.

Craig, H., Isotopic variations in meteoric waters, *Science 133*, 1702–1703.

DePaolo, D. J. and Wasserburg, G. J., 1976a. Nd isotopic variations and petrogenetic models, *Geophys. Res. Lett. 3*, 249–252.

DePaolo, D. J. and Wasserburg, G. J., 1976b. Inference about magma sources and mantle structure from variations of $^{143}Nd/^{144}Nd$, *Geophys. Res. Lett. 3*, 743–746.

Didier, J., 1973. *Granites and their enclaves*, Amsterdam: Elsevier Publishing Co.

Doe, B. R., 1970. *Lead Isotopes*, New York: Springer-Verlag.

Doe, B. R. and Hart, S. R., 1963. The effect of contact metamorphism on lead in potassium feldspars near the Eldora stock, Colorado, *J. Geophys. Res. 68*, 3521–3530.

Doe, B. R., Tilton, G. R., and Hopson, C. A., 1965. Lead isotopes in feldspars from selected granitic rocks associated with regional metamorphism, *J. Geophys. Res. 70*, 1947–1968.

Doe, B. R., 1967. The bearing of lead isotopes on the source of granitic magma, *J. Petrol 8*, 51–83.

Dostal, J. and Fratta, M., 1977. Trace element geochemistry of a Precambrian diabase dike from western Ontario, *Can. J. Earth Sci. 14*, 2941–2944.

Dupuy, C. and Allegre, C. J., 1971. Fractionnement K/Rb dans les suites ignimbritiques de Toscane: un exemple de rejuvenation crustale, *Geochim. Cosmochim. Acta 106*, 427–458.

Duthou, P., 1976. "Chronologie Rb–Sr et geochemie des granitoids du Nord Limousin," Ph.D. thesis, University of Clermont Ferrand, France.

Earley, T. O. and Silver, L. T., 1973. Rb–Sr isotopic systematics in the Peninsular Range batholith in Southern and Baja California, *Trans. Am. Geophys. U. 54*, 494.

Epstein, S. and Mayeda, T., 1953. Variation of ^{18}O content of waters from natural sources, *Geochim. Cosmochim. Acta 4*, 213–224.

Faure, G., 1977. *Principles of Isotope Geology*, New York: Wiley and Sons.

Fersmann, A. E., 1934. *Geochemistry, vol. 2*, Leningrad.

Flemings, M. C., 1974. *Solidification processing*, New York: McGraw-Hill, Inc.

Fratta, M. and Shaw, D. M., 1974. "Residence" contamination of K, Rb, Li, and Ti in diabase dikes, *Can. J. Earth Sci. 11*, 422–429.

Friedman, I., Redfield, A. C., Schoen, B., and Harris, J., 1964. The variation in the deuterium content of natural waters in the hydrologic cycle, *Geophys. Rev. 2*, 177–124.

Fourcade, S., 1980. The phenocryst problem in granites, *Contrib. Mineral. Petrol.*, in press.

Fourcade, S. and Allegre, C. J., 1980. Trace elements in granite genesis, *Geochim. Cosmochim. Acta*, in press.

Goldschmidt, V. M., 1937. The principles of distribution of chemical elements in minerals and rocks. *J. Chem. Soc. 140*, 655–673.

Gray, C. M. and Oversby, V. M., 1972. The behavior of lead isotopes during granulite facies metamorphism, *Geochim. Cosmochim. Acta 36*, 939–952.

Hamet, J. and Allègre, C. J., 1976a. Hercynian orogeny in the Montagne Noire, France: application of ^{87}Rb–^{87}Sr systematics, *Geol. Soc. Am. Bull. 87*, 1429–1442.

Hamet, J. and Allègre, C. J., Rb–Sr systematics in granite from central Nepal (Manaslu): significance of the Oligocene age and high $^{87}Sr/^{86}Sr$ ratio in Himalayan orogeny, *Geol. 4*, 470–472.

Hart, S. R., Brooks, C., Krogh, T. E., Davis, G. L., and Nava, D., 1970. Ancient and modern volcanic rocks: a trace element model, *Earth and Planet. Sci. Lett. 10*, 17–28.

Hart, S. R., Gunn, B. M., and Watkins, N. D., 1971. Intralava variation of alkali elements in Icelandic basalt, *Am. J. Sci. 270*, 315–318.

Hart, S. R. and Davis, K. E., 1978. Nickel partitioning between olivine and silicate melt, *Earth and Planet. Sci. Lett. 40*, 203–219.

Hedge, C. E. and Walthall, F. G., 1963. Radiogenic strontium 87 as an index of geological processes, *Science 140*, 1214–1217.

Hofmann, A. W. and Magaritz, M., 1977. Diffusion of Ca, Sr, Ba, and Co in a basalt melt: implications for the geochemistry of the mantle, *J. Geophys. Res. 80*, 5432–5440.

Hofmann, A. W. and Hart, S. R., 1978. An assessment of local and regional isotopic equilibrium in the mantle, *Earth and Planet. Sci. Lett. 38*, 44–62.

Hofmann, A. W., 1978. Diffusion in natural silicate melts: a critical review, in *Physics of Magmatic Processes*, R. B. Hargraves, ed., Princeton: Princeton University Press.

Holland, H. D., 1972. Granites, solutions, and base-metal deposits, *Econ. Geol. 67*, 281–301.

Hurley, P. M., Hughes, H., Faure, G., Fairbairn, H. W. and Pinson, W. H., 1962. Radiogenic strontium-87 model of continental formation, *J. Geophys. Res. 67*, 5315–5334.

Hurley, P. M., Fairbairn, H. W., and Pinson, W. H., 1966. Rb–Sr isotopic evidence in the origin of potash-rich lavas of western Italy, *Earth and Planet. Sci. Lett. 1*, 301–306.

Irving, A. J., 1978. A review of experimental studies of crystal/liquid trace-element partitioning, *Geochim. Cosmochim. Acta 42*, 743–770.

Iyama, J. T., 1972. Fixation des elements alcalino-terreux Ba, Sr, Ca dans les feldspaths: etude experimentale, *24th Int. Geol. Cong.* Montreal, Sect. 10, 122–130.

Javoy, M., 1970. "Utilisation des isotopes de l'oxygene en magmatologie," unpublished thesis, Universite Paris VII, Paris, France.

Javoy, M., Fourcade, S., and Allegre, C. J., 1970, Graphical method of examination of $^{18}O/^{16}O$ fractionations in silicate rocks, *Earth and Planet. Sci. Lett. 10*, 12–16.

Kay, R. W., Hubbard, N. J., and Gast, P. W., 1970. Chemical characteristics and origin of oceanic ridge volcanic rocks, *J. Geophys. Res. 75*, 1585–1613.

Kay, R. W. and Gast, P. W., 1973. The rare earth content and origin of alkali-rich basalts, *J. Geol. 81*, 653–682.

Kirkpatrick, R. J., 1977. Nucleation and growth of plagioclase, Makaopuhi and Alae lava lakes, Kilauea Volcano, Hawaii, *Geol. Soc. Am. Bull. 88*, 78–84.

Kistler, R. W. and Peterman, Z. E., 1973. Variation in Sr, Rb, K, Na and initial $^{87}Sr/^{86}Sr$ in Mesozoic granitic rocks and intruded wall rocks in central California, *Geol. Soc. Am. Bull. 84* 3489–3512.

Kushiro, I., Yoder, H. S., and Mysen, B. O., 1976. Viscosity of basaltic and andesitic liquids at high pressure, *Carnegie Inst. of Washington Yearbook 75*, 614–618.

Langmuir, C. H., Bender, J. F., Bence, A. E., Hanson, G. N., and Taylor, S. R., 1977. Petrogenesis of basalts from the Famous area, Mid-Atlantic Ridge, *Earth and Planet. Sci. Lett. 36*, 133–156.

Lawrence, J. R. and Taylor, H. P., 1971. Deuterium and oxygen correlation: clay minerals and hydroxides in Quaternary soils compared to meteoric waters, *Geochim. Cosmochim. Acta 35*, 993–1003.

LeFort, P., 1975. Himalayas: the collided range and present knowledge of the continental arc, *Am. J. Sci. 275A*, 1–44.

Ludwig, K and Sliver, L. T., 1977. Lead isotope inhomogeneity in Precambrian igneous K-feldspars, *Geochim. Cosmochim. Acta 41*, 1457–1471.

McCulloch, M. T. and Wasserburg, G. J., 1978. Sm–Nd and Rb–Sr chronology and continental crust formation, *Science 200*, 1003–1011.

Magaritz, M. and Taylor, 1976a. $^{18}O/^{16}O$ and D/H studies of igneous and sedimentary rocks along a 500-km traverse across the Coast Range batholith into Central British Columbia at latitude 54°–55°N, *Can. J. Earth Sci. 13*, 1514–1536.

Magaritz, M. and Taylor, H. P., 1976b. Oxygen, hydrogen and carbon isotope studies of the Franciscan formation, Coast Ranges, California, *Geochim. Cosmochim. Acta 40*, 215–234.

Mattauer, M., 1975. Sur le mecanisme de formation de la schistosite dans l'Himalaya, *Earth and Planet. Sci. Lett. 28*, 144–154.

Meijer, A., 1976. Pb and Sr isotopic data bearing on the origin of volcanic rocks from the Mariana island-arc system, *Geol. Soc. Am. Bull. 87*, 1358–1369.

Minster, J. F., Minster, J. B., Allègre, C. J., and Treuil, M., 1977. Systematic use of trace elements in igneous processes: II. Inverse problem of the fractional crystallization process in volcanic suites, *Contrib. Mineral. Petrol. 61*, 49–77.

Moorbath, S. and Pankhurst, R. J., 1976. Further Rb–Sr age and isotope evidence for the nature of the late Archean plutonic event in West Greenland, *Nature 262*, 124–126.

Moorbath, S., 1977. Ages, isotopes, and evolution of Precambrian continental crust, *Chem. Geol. 20*, 151–187.

Muehlenbachs, K., Anderson, A. T., Jr., and Sigvaldason, G. E., 1974. Low O^{18} basalts from Iceland, *Geochim. Cosmochim. Acta 38*, 577–588.

Nelson, D. O., and Dasch, E. J., 1976. Disequilibrium of strontium isotopes between mineral phases of parental rocks during magma genesis—a discussion, *J. Volc. Geoth. Res. 1*, 183–191.

Nier, A. O., 1938. Variations in the relative abundances of the isotopes of common lead from various sources, *J. Am. Chem. Soc. 60*, 1571–1576.

Nier, A. O., 1939. The isotopic constitution of radiogenic leads and the measurement of geologic time, *Phys. Rev. 55*, 153–163.

Nockolds, S. R. and Allen, R., 1953, The geochemistry of some igneous rock series, I, *Geochim. Cosmochim. Acta 4*, 105–142.

Nockolds, S. R. and Allen, R., 1954. The geochemistry of some igneous rock series, II, *Geochim. Cosmochim. Acta 5*, 245–285.

O'Hara, M. J., 1968. The bearing of phase equilibrium studies on the origin and evolution of basic and ultrabasic rocks, *Earth Sci. Rev. 4*, 69–133.

O'Hara, M. J., 1977. Geochemical evolution during fractional crystallization of a periodically refilled magma chamber, *Nature 266*, 503–507.

Ohmoto, H., 1973. Systematics of sulfur and carbon isotopes in hydrothermal ore deposits, *Econ. Geol. 67*, 551–578.

O'Neil, J. R. and Clayton, R. N., 1964. Oxygen-isotope geothermometry, in *Isotopic and Cosmic Chemistry*, H. Craig, S. and G. Wasserburg, eds., The Netherlands: North-Holland, Amsterdam.

O'Nions, R. K., Hamilton, P. J., and Evensen, N. M., 1977. Variations in ^{143}Nd ^{144}Nd and ^{87}Sr/^{86}Sr ratios in oceanic basalts, *Earth and Planet. Sci. Lett., 34,* 13–22.

Oversby, V. M. and Ewart, A., 1972. Lead isotopic compositions of Tonga-Kermadec volcanics and their petrogenetic significance, *Contr. Mineral. Petrol, 37,* 181–210.

Oversby, V. M., 1978. Lead isotopes in Archean plutonic rocks, *Earth and Planet. Sci. Lett. 38,* 237–248.

Paster, T. P., Schauwecker, D. S., and Haskin, L. A., 1974. The behavior of some trace elements during solidification of the Skaergaard layered series, *Geochim. Cosmochim. Acta, 38,* 1549–1577.

Patterson, C. C. and Tatsumoto, M., 1964. The significance of lead isotopes in detrital feldspar with respect to chemical differentiation within the earth's mantle, *Geochim. Cosmochim. Acta 28,* 1–22.

Peterman, Z. E., Hedge, C. E., Coleman, R. G., and Snavely, P. D., Jr., 1967. ^{87}Sr/^{86}Sr ratios in some geosynclinal sedimentary rocks and their bearing on the origin of granitic magma in orogenic belt, *Earth and Planet. Sci. Lett. 2,* 433–439.

Pineau, F., 1978. "Geochemical studies using carbon isotopes," unpublished Ph.D. thesis, University of Paris.

Pineau, F., Javoy, M., and Bottinga, Y., 1976. ^{13}C/^{12}C ratios of rocks and inclusions in popping rocks of the mid-Atlantic ridge and their bearing on the problem of isotopic composition of deep-seated carbon, *Earth Planet. Sci. Letts. 29,* 413–421.

Richard, P., Shimizu, N., and Allègre, C. J., 1976. ^{143}Nd/^{144}Nd, a natural tracer: an application to oceanic basalt, *Earth Planet. Sci. Letts. 31,* 269–278.

Roddick, J. C. and Compston, W., 1977. Strontium isotopic equilibrium: a solution to a paradox, *Earth and Planet. Sci. Lett. 34,* 238–246.

Sakai, H., 1968. Isotopic properties of sulfur compounds in hydrothermal processes, *Geochem. J. 2,* 29–49.

Schilling, J. G., 1971. Sea-floor evolution: rare earth evidence, *Philos. Trans. R. Soc. Lond., A. 268,* 663–706.

Schilling, J. G., 1975. Rare-earth variations across "normal segments" of the Reykjanes Ridge, 60–63°N, Mid-Atlantic Ridge, 29°S, and East Pacific Rise, 2–19°S, and East Pacific Rise, 2°–19°S, and evidence on the composition of the underlying low-velocity layer, *J. Geophys. Res. 80,* 1459–1473.

Shima, M., Gross, W. H., and Thode, H. G., 1965. Sulfur-isotope abundances in basic sills, differentiated granites and meteorites, *J. Geophys. Res. 68,* 2835–2848.

Shimizu, N. and Arculus, R. J., 1975. Rare earth concentrations in a suite of basanitoids and alkali olivine basalts from Grenada, Lesser Antilles, *Contrib. Mineral. Petrol. 50,* 231–240.

Shimizu, N., Semet, M. P. and Allègre, C. J., 1978. Geochemical applications of quantitative ion microprobe analysis, *Geochim. Cosmochim. Acta 42,* 1321–1334.

Sinha, A. K., and Hart, S. R., 1972. A geochemical test of the subduction hypothesis for generation of island-arc magmas, *Carnegie Institution of Washington Yearbook 71*, 309–312.

Sinha, A. K., and Tilton, G. R., 1973. Isotopic evolution of common lead, *Geochim. Cosmochim. Acta 37*, 1823–1849.

Smith, V. G., Tiller, W. A., and Rutter, J. W., 1955. A mathematical analysis of solute redistribution during solidification, *Can. J. Physics 33*, 723–745.

Sun, S. S. and Hanson, G. N., 1975. Origin of Ross Island basanitoids and limitations upon the heterogeneity of mantle sources for alkali basalts and nephelinites, *Contrib. Mineral. Petrol. 52*, 77–106.

Tatsumoto, M., 1978. Isotopic composition of lead in oceanic basalt and its implication to mantle evolution, *Earth Planet. Sci. Lett. 38*, 63–87.

Taylor, H. P. and Epstein, S., 1962. Relationship between $^{18}O/^{16}O$ ratios in coexisting minerals of igneous and metamorphic rocks, Part I., *Geol Soc. Am. Bull. 75*, 461–480.

Taylor, H. P. and Epstein, S., 1966. Deuterium-hydrogen ratios in coexisting minerals of metamorphic and igneous rocks, *Trans. Am. Geophys. U. 47*, 213.

Taylor, H. P., 1968. The oxygen isotope geochemistry of igneous rocks, *Contrib. Mineral. Petrol. 19*, 1–71.

Taylor, H. P. and Silver, L. T., 1978. Oxygen isotope relationships in plutonic igneous rocks of the Peninsular Ranges batholith, southern and Baja California, in *Short Papers of 4th Internat. Conf. Geochron. Cosmochron. and Isotope Geology*, U.S. Geol. Surv. Open File Rept. 78–701, 423–426.

Taylor, H. P., 1974a. The application of oxygen and hydrogen isotope studies to the problem of hydrothermal alteration and ore deposition, *Econ. Geol., 69*, 843–853.

Taylor, H. P., 1974b. Oxygen and hydrogen isotope evidence for large-scale circulation and interaction between ground waters and igneous intrusions, with particular reference to the San Juan volcanic field, Colorado, in *Geochemical Transport and Kinetics*, Hoffmann et al., eds., Carnegie Institution of Washington Publication *634*, Washington, 299–324.

Taylor, H. P. and Magaritz, M., 1975. Oxygen and hydrogen isotopic studies of 2.6–3.4 b.y. old granites from the Barberton Mountain Land, Swaziland and the Rhodesian craton, *Geol. Soc. Am. Abs. with Program 7*, 1293.

Taylor, H. P., 1978. Oxygen and hydrogen isotope studies of plutonic granitic rocks, *Earth and Planet. Sci. Lett. 38*, 177–210.

Tilton, G. R. and Steiger, R. H., 1969. Mineral ages and isotopic composition of primary lead at Manitouwadge, Ontario, *J. Geophys. Res. 74*, 2118–2132.

Turi, B. and Taylor, H. P., 1976a. High ^{18}O igneous rocks from the Tuscan magmatic province, Italy, *Contrib. Mineral. Petrol. 55*, 33–54.

Turi, B. and Taylor, H. P., 1976b. Oxygen isotope studies of potassic volcanic rocks of the Roman Province, Central Italy, *Contrib. Mineral. Petrol. 55*, 1–31.

Vitrac-Michard, A. and Allègre, C. J., 1975. A study of the formation of a piece of continental crust by $^{87}Rb-^{87}Sr$ method: the case of the French Oriental Pyrenees, *Contrib. Mineral. Petrol. 50*, 257–285.

Vollmer, R., 1976. Rb–Sr and U–Th–Pb systematics of alkaline rocks: the alkaline rocks from Italy, *Geochim. Cosmochim. Acta 40*, 283–295.

Vollmer, R., 1977. Isotopic evidence for genetic relations between acid and alkaline rocks in Italy, *Contrib. Mineral. Petrol. 60*, 109–118.

Wager, L. R. and Mitchell, R. L., 1951. Distribution of trace elements during strong fractionation of basic magma—a further study of the Skaergaard intrusion, *Geochim. Cosmochim. Acta 1*, 129–208.

Watkins, N. D., Gunn, B. M., and Coy-yll, R., 1970. Major and trace element variations during the initial cooling of an Icelandic lava, *Am. J. Sci. 268*, 24–49.

Watson, E. B., 1979. Calcium diffusion in a simple silicate melt to 30 kbar, *Geochim. Cosmochim. Acta. 43*, 313–322.

Wenner, D. B. and Taylor, H. P., 1976. Oxygen and hydrogen isotope study of a Precambrian granite-rhyolite terrane, St. Francois Mountains, southeastern Missouri, *Bull. Geol. Soc. Am. 87*, 1587–1598.

White, W. M., Schilling, J. G. and Hart, S. R., 1976. Evidence for the Azores mantle plume from strontium isotope geochemistry of the Central North Atlantic, *Nature 264*, 659–663.

White, W. M., 1977. "Geochemistry of igneous rocks from the central North Atlantic: the Azores and the Mid-Atlantic Ridge," Ph.D. thesis, University of Rhode Island, Rhode Island.

Wyart, J. and Sabatier, G., 1962. L'equilibre des feldspaths et des feldspathoides en presence de solutions sodi-potassiques, *Norsk Geol. Tidssk.* bd. *42*, halvbd 2, 317–329.

Yoder, H. W., Jr., 1976. *Generation of Basaltic Magma*, Nat. Acad. Sci. Washington, D.C..

Zartman, R. E. and Wasserburg, G. J., 1969. The isotopic composition of lead in potassium feldspars from some 1 b.y. old North American igneous rocks, *Geochim. Cosmochim. Acta 33*, 901–942.

Zartman, R. E., 1974. Lead isotopic provinces in the cordillera of the western United States and their geologic significance, *Econ. Geol. 69*, 792–805.

Zielinski, R. A. and Frey, F. A., 1970. Gough Island evolution of a fractional crystallization model, *Contrib. Mineral. Petrol. 29*, 242–254.

Zielinski, R. A., 1975. Trace element evaluation of a suite of rocks from Reunion Island, Indian Ocean, *Geochim. Cosmochim. Acta 39*, 713–734.

Chapter 5

HEAT FLOW AND MAGMA GENESIS

E. R. OXBURGH

Department of Mineralogy and Petrology, Cambridge University

INTRODUCTION

One of the most pleasureable aspects of contributing to this volume
was the stimulus it provided for the re-reading of *The Evolution of the
Igneous Rocks*, something which this author is ashamed to confess to
not having done for fifteen years. On re-reading Bowen's words today,
one is struck first by Bowen's almost uncanny insight in discriminating
among the plethora of hypotheses for magma generation prevalent at
that time, and second, by how little our understanding of this field has
changed in fifty years.

Bowen recognized the fundamental importance of basalt. He saw
that it was probably generated by the partial melting of peridotite
within the mantle, and that once peridotite had yielded basalt it was in
some respects "barren." He recognized that basalt had been available
throughout much of the history of the Earth, and concluded that partial
melting probably occurred by some kind of pressure release mechanism.
Although these views are commonplace today, they were very far from
commonplace at the time he advanced them.

Bowen also recognized two major gaps in our understanding of magma
genesis: one in our poor control on the abundance and distribution of
radioactivity within the Earth, and the other in our ignorance of the
relative importance of conductive and convective processes of heat
transfer in cooling the Earth. Although we have perhaps made some
progress in answering these questions, they remain central to our study
of processes of magma genesis, and to the subject matter of this article.

The writer was invited to review the present status of heat-flow studies and to discuss their bearing on problems of magma genesis. Accordingly, in the first part of this article some of the general considerations governing temperature distribution in the outer parts of the Earth are outlined, and attention is drawn to some of the interesting but poorly understood patterns in the distribution of world heat flow that have been recognized over the last ten years. It is then shown that, insofar as magmas are most commonly generated in active tectonic regions, heat flow studies provide relatively little information on magma genesis, because they depend on some kind of approach to conductive equilibrium. Finally, there is a brief discussion of the various dynamic processes that may lead to melting in the Earth. The most important of these—that which is believed to be responsible for melting at mid-ocean ridges—is examined in detail, and some of its petrogenetic consequences are considered.

The subject of this article is itself so vast that, in many cases, it has not been possible to follow up points of interest or to explain all aspects of the background. For this reason an attempt has been made to provide a list of those references that, in the author's opinion, provide the best relevant reviews, without necessarily being either the most recent or, for that matter, the earliest pronouncement on any particular topic.

As in many aspects of Earth Science the evidence is in places still fragmentary; the pieces of the jigsaw puzzle are so few that they may be fitted together in different ways to hint at different pictures. The choice between the arrangements is at present often a matter of preference or judgement; indeed, what for one worker seems to be a crucial part of the pattern may be regarded by another as comprising a piece from a different puzzle.

CONDUCTIVE AND CONVECTIVE REGIMES

The thermal diffusivity, κ, of a material is given by $\kappa = k/\rho C_p$, where k is thermal conductivity, ρ is density and C_p is specific heat. Most rocks are characterized by low values of thermal diffusivity, so that they respond slowly to any change in ambient temperature, This low κ has two main consequences:

(1) If the crust undergoes any kind of thermal perturbation, the time for return to thermal equilibrium is long. A rough indication of how long is given by the conductive time constant τ:

$$\tau = \frac{l^2}{\kappa}, \qquad (1)$$

where l is a characteristic length (typically the distance across which heat must be transferred) and κ is the thermal diffusivity. (For example, this relationship shows that a time of the same order as the age of the Earth is required for significant conductive cooling to a depth of 400 km.)

(2) If the crust undergoes tectonic movements at rates of more than a few millimeters per year, there is a tendence for crustal blocks to carry their thermal structure with them as they move. Thus, in such circumstances, there is *convective* transfer of heat, and conductive solutions for the thermal structure are not applicable until the movements have ceased or become so slow as to be unimportant.

Whether conductive or convective processes dominate in any particular situation may be judged from the value of the Peclet number, P:

$$P = \frac{uL}{\kappa}, \tag{2}$$

where u is the velocity of motion and L a characteristic distance for the motion. In general convection dominates for $P > 1$ and conduction for $P < 1$. It is therefore useful to divide the surface of the Earth into $P > 1$ regions and $P < 1$ regions. The former essentially comprise active plate margins and zones of mid-plate igneous activity and rifting; the latter make up the greater part of plate interiors.

MEASUREMENTS OF HEAT FLOW

Heat is continuously conducted from the Earth's interior and lost at its outer surface. Measurement of the conductive heat flux through the crust provides an important constraint on temperatures at depth. We first consider how the flux is measured.

In a steady state, one-dimensional situation, the conductive flow of heat through a body is given by

$$q = k\beta, \tag{3}$$

where q is the heat flow, and k and β are respectively the thermal conductivity of the medium and the temperature gradient across it. This relationship is used both to estimate the steady flow of heat from the Earth's interior, and to extrapolate downwards temperature profiles measured at shallow depth.

For the calculation of heat flow, the thermal gradient must be measured at least several hundreds of meters beneath the surface in order to avoid the temperature perturbations associated with the movement of ground water, and with seasonal and recent climatic variation in surface temperature. With slight variations in technique, the measurements may be made in boreholes, mines or tunnels. In each case, however, there may be problems of temperature perturbation by the drilling operation itself. With the use of modern thermistor probes, temperatures may be measured to almost any degree of accuracy desired: temperature differences of .001°C are readily observed. If, as is commonly the case, radioactive heat generation is insignificant over the depth interval across which the gradient is measured, the steady state heat flow will not vary as a function of depth down the hole. The thermal gradient however, will vary inversely as the conductivity. It is nevertheless quite common to find gradient fluctuations which are not related to measured variation in conductivity and which are orders of magnitude too large to be explained by variations in the concentrations of heat-producing elements. Such fluctuations can result from lateral variation in thermal conductivity, and/or from conductivity anisotropy, both of which may lead to a flow of heat that, on a local scale, is not one dimensional. In this way the vertical thermal gradient may vary laterally over short distances by amounts which are large by comparison with the errors associated with its measurement at any one place. In addition, if the surface topography in the region of the measurement is steep and irregular, it is necessary to modify the measured gradient with a topographic correction (Birch, 1950).

It is standard practice when making heat flow measurements to attempt to avoid the effects of circulating ground water through choice of the area where the measurements are made (e.g. Roy et al., 1972). Even slight flows may carry amounts of heat which are large by comparison with the conductive flux (Bullard and Niblett, 1951). More serious, however, because they cannot be detected, are the effects of water movements in aquifers or fracture zones deeper than the bottom of the borehole or tunnel in which the temperature measurements are made. Even slow movements with characteristic velocities of only a few meters per year can have significant effects; if the water movement has an upward component the heat flow will tend to be enhanced, and it will be reduced if the motion is downward. The importance of such effects is clearly seen in the occasional hole that is deep enough to pass through a waterflow horizon (Figure 1A). It follows that it is not possible to interpret heat flow measurements unambiguously in regions of thick sedimentary cover without reference to the local hydrology.

For the purposes of calculating heat flow, thermal conductivity is normally measured on samples from the bore hole or tunnel in which the temperature gradient was measured. Again, the problems are geological rather than experimental. In consolidated rocks the measurements are generally carried out on short, water-saturated cylinders or discs between 1 inch (2.5 cm) and $1\frac{1}{2}$ inches (3.75 cm) in diameter and between $\frac{1}{4}$ inch (0.6 cm) and $1\frac{1}{4}$ inches (3.2 cm) thick (Birch, 1950; Roy et al., 1968; Sass et al., 1971). If the rocks are homogeneous on a fine scale there is no difficulty. But if, as is commonly the case, the rocks are inhomogeneous on a scale of more than a few inches, it may be necessary

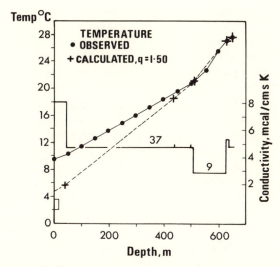

Figure 1A. A heat flow discontinuity in a 600 m borehole in Eastern England; below about 500 m the heat flow is about 1.50 μcal/cm^2 sec. and above that depth it is about 1.0 μcal/cm^2 sec. At about 500 m the hole (which is sealed from the water flow) intersects an inclined aquifer with a downward component of water flow. The water convects approximately 0.5 μcal/cm^2 sec. of heat flow down with it. The solid line with points gives the observed temperature profile, and the dashed line the profile corresponding to a heat flow of 1.5 μcal/cm^2 sec up to the surface. Measured conductivities and numbers of values are also given (from Richardson and Oxburgh, 1978).

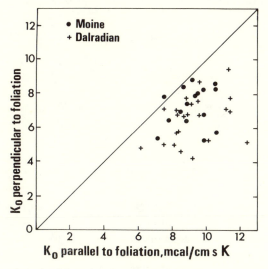

Figure 1B. Thermal conductivity measured both normal to and parallel to the foliation in a series of schists and gneisses from northwest Scotland (Richardson and Powell, 1977). Most of the rocks are anisotropic, some by more than a factor of two.

to make a large number of measurements in order that the thermal properties of the unit be adequately characterized. Unconsolidated samples may be handled with the needle-probe technique (Von Herzen and Maxwell, 1959).

Relatively few rocks are isotropic with respect to thermal conductivity, and in rocks with any kind of layered alternation of quartz and oriented micas, the conductivity measured normal to the foliation may be significantly less than that measured parallel to it (Figure 1B). The effects of such anisotropy upon the thermal structure of an area become important when the rocks are inclined or folded.

Because laboratory measurements of thermal conductivity are made on small samples, it is not possible to make allowance for macroscopic structures in the rocks, such as joint systems coated by moisture films. The few experiments which have been done, however, to measure the large scale thermal conductivity of rocks *in situ*, suggest that such effects are not generally important.

Several thousands of heat-flow measurements have now been carried out both on land and at sea (e.g., Lee, 1965; Sass and Munroe, 1974). Away from active igneous or hydrothermal areas, heat flow ranges from about 20 to 120 mW/m^{-2}, with a world average in the range of 55 to 60 mW/m^{-2}. (e.g., Lee, 1965). Such measurements may be used to estimate deeper crustal temperatures.

In order to predict the steady state temperature at any greater depth than that for which measurements have been made, use is made of the relationship

$$T_z = T_0 + q \int_0^z \frac{dz}{k},$$ (4)

where T_z is the temperature at z, the depth of interest, and T_0 is the surface temperature. The integral represents the thermal resistance down to depth z. In practice, the integral is solved by summation:

$$\int_0^z \frac{dz}{k} = \sum \frac{\Delta z_i}{k_i}$$ (5)

where Δz_i are depth intervals that may be characterized by a single value of conductivity k_i.

If heat production due to radioactivity is significant over the depth interval for which a temperature distribution is to be calculated, expression (5) may still be used; the temperature contribution from radioactivity at any depth is calculated separately and simply added to the conduction profile. The effects of all one-dimensional distributions of internal heat production may easily—if tediously in complex cases—be calculated from the expressions given in Carslaw and Jaeger (1959, p. 78–80).

Homogeneous internal production over a depth interval will give rise to a steady-state geothermal gradient that is steepest near the surface and decreases with depth; the gradient at any depth is proportional to the total amount of heat generated in the region below it, and at the base of the zone of heat production, the thermal gradient falls to zero. It follows that the effect of a given amount of internal heat production on temperatures at depth is least when heat production is concentrated near the surface, greater when it is homogeneously distributed through a larger depth, and, of course, greatest when it is concentrated at depth. As discussed below, this last distribution can probably only come about through tectonic burial.

We now consider the reliability of temperatures estimated in this way. Equations (4) and (5) show that, however well the local heat flow has been determined, T_z cannot be estimated unless the variation in rock thermal conductivity down to the depth of interest can be constrained. Rock conductivities may commonly vary from one lithology to another by factors of two or even three (Table 1) and this will be the order of uncertainty on the temperature estimate in the absence of such information. Similarly, the distribution of radioactivity must be specified. If the region for which a temperature estimate is required has a complex structure in which there is significant lateral variation in thermal conductivity, or conductivity anisotropy (Figure 1B), Equation (4) is no longer applicable, because the one-dimensional heat-transfer assumption upon which it is based is not satisfied. In such cases more complex models must be constructed by numerical methods (see Jones and Oxburgh, 1979).

The application of Equation (4) is limited in two other ways. First, is necessary that T_z lies within the zone of conductive heat transfer. There are various cases where, on the evidence of a high surface heat flow and a well-known conductivity structure, holes were drilled for the extraction of geothermal waters; drilling then revealed that at a relatively shallow depth there was a transition from conductive heat transfer to convective transfer by circulating water. The effect of such convective zones on the geothermal profile is similar to that of a rock layer of equivalent thickness but with a very high thermal conductivity and, in consequence, a very low thermal gradient (Equation (3)). The expected high values of T_z are therefore not realized. Similarly, on a much larger scale, Equation (4) may not be used to estimate temperatures in the Earth at depths greater than the thickness of the lithosphere. Within the lithosphere we may expect some approach to conductive equilibrium (see below), but in the asthenosphere the mantle must circulate to permit subduction and the generation of new plates, and convective heat transfer is important.

Table 1. Some useful quantities and relationships.

Thermal Conductivity: rock conductivity depends on mineralogy, grainsize, texture, and porosity; inasmuch as all of these may vary considerably within a single "rock type", unique values of conductivity cannot be given for a lithology unless it is very well characterized. (Data from Richardson and Oxburgh (1978) and unpublished results of Oxford Heat Flow Group.)

Mean Conductivities of Some English Stratigraphic Units

Lithostratigraphic type	*Wm/K*
Chalk	1.92
Kimmeridge Clay	1.28
Lower Lias Clay	1.38
Rhaetic	2.22
New Red Sandstone	3.31
New Red Marl	1.97
Rock Salt	5.75
Coal Measures:	
Sandstones	3.31
Siltstones	2.22
Mudstones	1.49
Coal	.31
Carboniferous Limestone + Magnesian Limestone	3.47
Igneous Rocks:	
Granite	2.6–3.6
Garnet Lherzolite	3.3–4.2
Basalt*	1.3–2.9
Diabase	1.7–2.5

Adiabatic gradient of temperature at 100 km $\sim$ .7°K/km

$$200 \text{ km} \sim .6°\text{K/km}$$

$$400 \text{ km} \sim .4°\text{K/km}$$

Solidus gradient for the mantle 40–100 km $\sim$ .4°K/km

Thermal diffusivity $= \kappa = \dfrac{k}{\rho C_p}$, commonly $\sim 10^{-6}$ m^2 s^{-1} K^{-1}

(k = thermal conductivity, C_p = specific heat, ρ = density)

C_p for rocks is $\sim$1.1–1.2 joule/gm up to $\sim$1,300°K

1 heat flow unit $= 10^{-6}$ cal cm^{-2} s^{-1} = 41.84 mW m^{-2}

1 heat generation unit $= 10^{-13}$ cal cm^{-3} s^{-1} = .4184 μW m^{-3}

1 conductivity unit $= 1$ m cal cm^{-1} s^{-1} k^{-1} = .4184 W m^{-1} K^{-1}

* very sensitive to abundance and nature of vesicles.

The second main constraint on the application of Equation (4) is the assumption of thermal equilibrium. Existing steady temperature distribution within the crust may be abruptly disturbed in many ways, as discussed above: igneous bodies may be emplaced, rapid tectonic movements may occur, or an active plate boundary (a $P > 1$ region) may become inactive. In all of these cases there has been a change in the thermal regime such that after the change the area may return to conductive equilibrium. As shown by Equation (1), the return may be protracted, and during that interval the conductive temperature gradient and the surface heat flow it sustains will change with time. For that reason these equations may not be used to calculate T_z unless z is small by comparison with l, where l is equal to $(\kappa t)^{1/2}$ (compare Equation (1)), and t is the time of the last change in thermal regime. In practice, there is no part of the Earth in true thermal equilibrium because of long term changes in internal heat production, continuous plate motion, and changes in erosion, climate and tectonics at its surface. Nevertheless, where t for a particular area is large, the assumption of equilibrium heat conduction is reasonable over a considerable depth interval.

THE THREE EMPIRICAL "LAWS" OF HEAT FLOW

During the last ten years or so, three important empirical relationships have been recognized for surface heat flow. If these relationships can be properly substantiated, they will provide important additional constraints on the processes controlling temperature distribution in the outer parts of the Earth. They are:

(1) The 'Birch-Lachenbruch law' for the dependence of surface heat-flow on near-surface radiogenic heat production in certain kinds of continental region.
(2) The '$t^{-1/2}$ law' for oceanic areas according to which the oceanic heat flow is thought to be inversely proportional to the square root of the age of the crust.
(3) The 'Polyak-Smirnov law' for an age dependence of heat flow in continental regions.

THE BIRCH-LACHENBRUCH LAW

Against all reasonable expectation, it was shown independently for two different areas—the Sierra Nevada (Lachenbruch; 1968) and New England (Birch et al., 1968)—that there is an empirical correlation between surface heat flow measured in eroded plutonic bodies, and the near-surface concentration of heat-producing elements in those bodies (Figure 2). Subsequently, the same relationship has been recognized in other plutonic regions, and even in some nonplutonic regions, although in the latter it tends to be less well defined (e.g., Rao and Jessop, 1975). Such regions, characterized by this coherent relationship between heat-flow and heat-production, are known as heat-flow provinces.

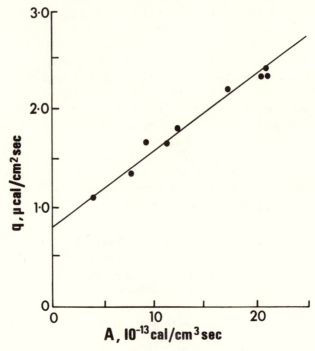

Figure 2. Heat production, A, plotted against surface heat flow, a, for a series of plutonic bodies in New England (Roy, Blackwell & Birch, 1968). See text for discussion.

The relationship shown in Figure 2 may be expressed as

$$q = q_m + Ah \qquad (6)$$

where q_m is the value of q on the zero intercept, A is the value of heat surface heat production, and h is a constant defined by the slope of the line and having units of length. As discussed in the original papers, there are various possible physical interpretations of this relationship.

The simplest interpretation is that, in any particular heat flow province, there is a surface layer of thickness h (commonly with a value of about 8 km) within which radioactive heat production varies from place to place. This near-surface contribution to the heat flow is superimposed on a uniform and deeper contribution, q_m, which is characteristic of the province as a whole, and is presumed to come from the mantle and lower crust. The quantity q_m is known as the reduced heat flow. In spite of its simplicity, such a model has no general geological plausibility. Many objections could be raised; the most serious of these being that there seems to be no way such a radioactive layer of uniform thickness could survive the effects of differential erosion.

Lachenbruch (1968) pointed out that the observations could also be satisfied by an exponential decrease in heat production with depth according to the relation

$$A(z) = A_0 \exp(-z/h) \tag{7}$$

where A_0 is the surface value of heat production and z is a depth beneath the surface (or a depth to which erosion may have occurred), and the exponential constant h is defined by the slope of the line in Equation (6). In this model the heat-flow/surface-heat production could be maintained throughout a region despite differential erosion. Furthermore, although there is no satisfactory geochemical explanation of such an exponential decline of heat production with depth in plutonic bodies, such a distribution is not unreasonable.

Lachenbruch went further in his interpretation of the Sierra Nevada observations. He suggested that within a particular geological province the variation in surface heat production could provide a measure of the degree of differential erosion (Figure 3); the plutons with the lowest heat production being the most deeply eroded. This interpretation is possible

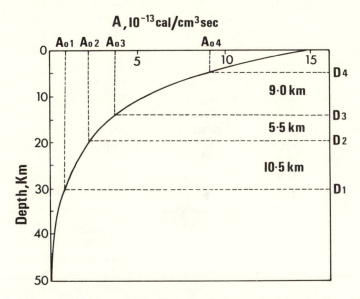

Figure 3. The exponential dependence of heat production on depth for the Sierra Nevada on the evidence of a q/A plot (e.g. Figure 2). Four heat-flow sites, 1, 2, 3, 4 are characterized by the values of near surface heat production shown. Lachenbruch (1968) pointed out that they could represent an erosional sequence, with site 1 being about 25 km more deeply eroded than site 4. The sites form a west-east progression 1 to 4, see the text for discussion.

only if an additional assumption is made—namely, that initially there was the same value of heat production everywhere at each level within the region. Given this assumption, and the depth-dependence function for distribution of heat production, it is easy to calculate differential erosion.

In the Sierra Nevada, however, there is a strong indication that the decrease in the heat production of the plutons in a east-west direction is simply a reflection of a change in their bulk chemistry (more silicic to less silicic) in that direction. Independent geological evidence for depth of burial also suggests that the regional gradient of erosion is in the opposite direction to that suggested by Lachenbruch.

Thus the Sierra Nevada observations can be explained by a series of plutons, each having internal heat production fractionated vertically according to the same exponential law (same h-value), but each giving a different mean value for heat production.

As mentioned above, a similar exponential relationship has been proposed for metamorphic areas. Such a relationship would be consistent with the known depletion in K, U and Th in high grade metamorphic basement regions (e.g. Hyndman et al., 1968) and with the higher concentrations of these elements found in regions of lower grade. Direct measurements of heat production in deep boreholes and the like (e.g., Lachenbruch and Bunker, 1971; Hawkesworth, 1974), although consistent with the exponential model, do not require it—the data do, however, establish some kind of decline in radioactivity with depth.

The h-value is a measure of the degree of upward concentration of radioactivity. Roughly two-thirds of the crustal radioactivity occurs within a depth interval of thickness h. Values of h vary from region to region but commonly range from 7 to 10 km. The very lowest values come from very high-grade metamorphic terrains and the highest from low-grade metamorphic areas (Table 2).

This problem of crustal heat production and its distribution and re-distribution by physical and chemical processes during crustal evolution is of fundamental importance, and is at present little understood.

Taken at their face values the linear A versus q plots seem to indicate the operation of a crustal process that is not only capable of establishing a vertical exponential distribution of K, U, and Th, but of distributing each of them according to the same exponential law (same h-value). This seems so improbable that one must ask whether the observations are good enough to distinguish between a truly exponential heat-production distribution, and a quasi-exponential distribution that is the sum of three different exponentials (K/h, Th/h and U/h) with different h values. The observations could tolerate significantly different values of h for different elements.

Table 2. Published values of reduced heat flow
and h (characteristic depth) for various
provinces (compilation of Sass and
Lachenbruch 1978 with additions from
Richardson and Oxburgh, 1978 and
Oxford Heat Flow Group unpublished
data for Rhodesia).

Continent Province	qm (mw/m^2)	h (km)
North America		
Basin and Range	59.02	9.4
Sierra Nevada	16.74	10.1
Eastern U.S.A.	33.07	7.5
Canadian Shield	22.19	12.4
Europe		
Baltic Shield	22.19	8.5
Ukranian Shield	25.12	9.1
England and Wales	23.44	16.0
Africa		
Niger	10.88	8
Zambia	40.19	(11)*
Rhodesia	<20	?
Asia		
Indian Shield	38.51	14.8
Australia		
Western Shield	26.37	4.5
Central Shield	26.79	11.1
Eastern Australia	57.35	(11.1)*

* value of D assumed for calculation q_m

In general, the concept of upward concentration of heat production in the crust is most understandable in partial-melting situations. K, Th, and U tend to be concentrated in partial melts, and these may migrate upwards under the influence of gravity. In individual plutons, concentration could occur during either crystallization or post-crystallization hydrothermal activity. However, in a terrain characterized by the injection of a number of individual plutons rising to different levels in the crust and having different mean values of heat production, it is very difficult to see how a well ordered regional distribution of heat sources could come about.

If the exponential distribution genuinely holds for metamorphic areas as well, some other process must be sought. And it may then be asked whether this is not the process which operates in igneous areas too. No clear model for such a process exists, but transport by upward-migrating or even circulating aqueous fluids during regional tectono-thermal activity is conceivable.

The opportunity remains for much work on almost every side of this problem.

THE "$t^{-1/2}$ LAW" FOR OCEANIC HEAT FLOW

It was not until the early nineteen-sixties that heat flow observations in the oceans became sufficiently abundant and sufficiently reliable for it to be clear that the mid-ocean ridges were characterized by significantly higher heat flow than the deeper parts of the ocean basin on their flanks. Subsequently, two broad classes of interpretation have been advanced to explain this pattern, which is now well established.

In the first interpretation, oceanic heat flow is viewed simply as the cooling of a thermal boundary layer formed on top of the up-welling and diverging mantle flow required beneath mid-ocean ridges by sea-floor spreading. At a relatively short distance from the ridge (e.g. 50 km or so) where the oceanic crust and underlying mantle are beyond the influence of the complex ridge-crest processes, they move laterally away from the ridge and undergo steady conductive cooling to the sea floor. It can be shown (e.g. Turcotte and Oxburgh, 1967) that the boundary layer (i.e. the cooled zone) thickens with time (t) in proportion to $t^{1/2}$, while the surface heat flows falls proportionally to $t^{-1/2}$. The cooling of the boundary layer involves thermal contraction, and reasonable values for the thermal properties of the oceanic crust and upper mantle show that cooling and contraction can account for most of the topographic difference between ridges and adjacent ocean basins. The subsidence of the ocean floor is also proportional to $t^{1/2}$, and the solutions for heat flow, topography and boundary-layer thickness become asymptotic at large values of t. The boundary layer is interpreted physically as the tectonic plate.

The alternative approach is to interpret the ocean floor as resulting from the cooling of a horizontally-moving slab between 50–100 km thick (e.g., McKenzie, 1967) and with a constant temperature maintained beneath it. For ocean floor younger than about 80 m.y., the two models give identical solutions for the two observable quantities, heat flow and topography. For ocean floor of greater age, however, there are differences, because in the slab model cooling is not permitted to extend deeper than the prescribed base of the slab, and hence both the depth of the sea floor and the heat flow reach constant values. In this sense the slab model may be regarded as the boundary layer model with an arbitrarily applied lower depth limit beyond which there is no further cooling or thickening.

A full discussion of the merits of these two models has been given by Parsons and Sclater (1977). Here we consider only some of the main points of issue. From a physical point of view the first or "boundary-layer model" has the attraction of simplicity; it involves the fewest assumptions about mantle behavior, beyond the observed fact of sea-floor spreading. The difficulty, however, is that it seems to predict ocean depths which are too great, and heat flow values which are too low, for the oldest parts of the oceans. Both of these problems are met by the second or "slab model," in which there is no thickening of the boundary layer beyond a certain depth. As pointed out by Sclater and Parsons (1976), however, the topography and heat-flow should depart from the $t^{1/2}$ law at different times if they are both responding to some fixed boundary condition at the base of the plate (in the Sclater and Parsons model the times were approximately 65 m.y. and 130 m.y., respectively). Part of the difficulty in demonstrating such a difference is that allowance must be made both for the elevation of the sea floor resulting from sedimentation and for the additional subsidence which has occurred in response to the sedimentary load. Furthermore it is clear (e.g. Sclater et al., 1975) that there are other long-wavelength characteristics of sea floor elevation that are not related to lithospheric cooling, and that sea floor of the same age may differ in elevation by as much as 1,500 m in different parts of the world.

Anderson et al. (1977) have discussed the problems of securing reliable heat-flow measurements at sea, and it is fair to conclude that at present there is so much "noise" in the measurements that the heat-flow field is less well known than the topography. The observational data are, therefore, not good enough at present to distinguish clearly between the two models, although the unmodified boundary-layer model does seem to be inadequate to explain the characteristics of old ocean floor. One possible explanation is that dynamic effects associated with the return flow under the plates counteract the subsidence in older parts of the oceans (e.g. Schubert et al. 1978). In any case, departures from the strict $t^{1/2}$ dependence are to be expected when account is taken in the boundary layer model of such features as phase changes within the cooling plate and physical properties that are temperature and pressure dependant. The slab model requires some kind of vigorous mantle circulation beneath the slab, such that new hot material is continually brought against the base of the plate to prevent it from cooling and thickening (e.g. Richter and Parsons, 1975).

Although the choice between these two models may appear to be somewhat esoteric and of only academic importance, it does have a significant bearing on estimates of the temperature distribution in the upper mantle and, indirectly, on our understanding of volcanic processes away from plate margins. Perhaps the feature to emphasize, however, is the agreement of the two models in explaining both the topography and heat flow in the younger parts of plates; it is hard to believe that this part of the explanation is not correct.

In the youngest parts of plates near ridge crests, simple conductive cooling models require considerable modification. The heat flow observations (e.g. Anderson et al., 1977) show a standard deviation which is larger than the mean; with increasing distance from the crest the mean value decreases, as does, to a greater extent, the standard deviation. Thus although high values of heat flow are restricted to crestal regions, low values occur there as well. Various reasons for this variation have been suggested (e.g. Oxburgh and Turcotte, 1969; Lister, 1972; Anderson et al, 1977); it is likely, however, that the main cause is the penetration of sea water into the newly formed oceanic crust to form convective circulations that exploit fracture permeability generated by thermal contraction. In such regions, conductive heat flow measurements can record only a small fraction of the total heat transfer. At greater distances from the ridge crest—corresponding to an age of some tens of millions of years—the circulation largely ceases, partly because the thermal gradients available to drive it are much less strong, and partly because of the sealing of fractures by the precipitation of phases from the hydrothermal fluids. The depth of penetration of the circulating fluids is not well established, but Spooner et al. (1974) estimate 4 km on the evidence of their oxygen-isotope studies.

Thus near-ridge crest investigations have established that the total heat loss from the oceans is very much greater than was estimated on the basis of conventional heat-flow measurements, so much greater that it is necessary to increase the estimate for the total world heat flux by about 20% (Oxburgh

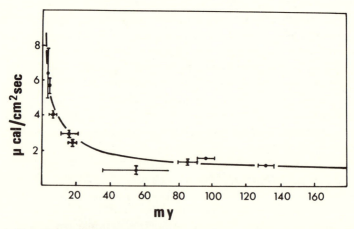

Figure 4. Heat flow dependence on age for the Pacific. Sclater's 'Reliable Means' plotted. Solid curve is the relationship predicted by the cooling of plate 125 km thick (after Parsons and Sclater, 1977).

and Turcotte, 1969; Williams and Von Herzen, 1974). This discovery in turn has implications for the world abundances of the heat producing elements. The same hydrothermal convection studies have also shown that deuteric alteration of oceanic basalts and modification of their elemental abundances may be more widespread than was previously anticipated.

THE POLYAK-SMIRNOV LAW

Polyak and Smirnov (1968) published compilations of continental heat flow observations classed according to geological province and crustal age. They showed that, on the average, "younger" regions have higher values of heat flow than older ones (Figure 5). This relationship has been broadly substantiated by subsequent work (e.g. Sclater and Francheteau, 1970; Pollack and Chapman, 1977; Sass and Lachenbruch, 1978).

It is clear that the dependence of heat flow upon age is different from that recognized in the oceans. In the first place the concept of age in continental crust is much less well defined than in the oceans, where it is possible in most cases to assign each area of crust a specific age of formation. On the continents, Polyak and Smirnov distinguish young volcanic provinces, active rift zones, Precambrian shields and platforms, and Phanerozoic folding provinces of different ages. If the folding provinces are considered in the context of accompanying regional metamorphism and possible igneous intrusion, it is clear that the age of the crust is taken to be the age of the last major thermal event to affect it. Such thermal events tend to be diffuse in both space and time, however, and there is considerable subjectivity in classifying the heat flow values in this way.

Since the major compilation of Polyak and Smirnov a great many new heat-flow observations have become available, but at the time of writing, no new compilation incorporating these, and applying better-defined age criteria, has been published.* In particular, it should now be possible to make a much better subdivision of the Precambrian, which is important because, as discussed below, the form of the dependence of heat flow on age for very old regions is critical to the interpretation of the relationship. It would also be important to establish whether there is a single universal curve, or whether different continental masses behave differently. The differences shown in Figure 5B between the Polyak and Smirnov compilation and the North American data were attributed by Sclater and Francheteau to a possible systematic error in thermal conductivity determination.

It may well be premature to look for a coherent physical interpretation of this pattern of continental heat flow, but it is possible to make several

---

* Note added in proof: this statement is no longer true. See Sclater et al., 1980.

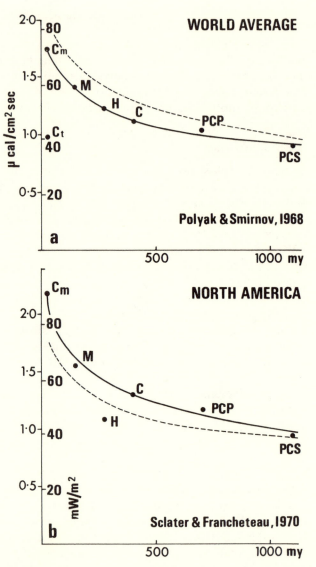

Figure 5. The variation of mean heat flow value with age (tectonic province) based on
Polyak & Smirnov (1968) and Sclater & Francheteau (1970); Cm—Cenozoic
miogeosymclines (a) and Basin and range (b)

M—Mesozoic fold belts
H—Hercynian orogenic belts
C—Caledonian orogenic belts
PEP—Precambrian platforms
PES—Precambrian shields.

NB. Error bars are omitted (see Sclater & Francheteau, 1970) for clarity but the
errors are very great and in the case of the Precambrian points almost impossible
to evaluate properly; see Figure 6. The solid curve in (a) is shown as the dashed
curve in (b) and vice versa.

preliminary observations. The time dependence of continental heat flow cannot reflect a *simple* conductive cooling of the crust since the most recent thermal event to affect it. The evidence for this assertion is to be found in Figure 6 where the slope of the q/t curve for the oceans is compared with that for the continents. The slope of the curve for the continents is clearly different from the $t^{1/2}$ dependence, characteristic of the oceans and required by the analytical solution for conductive cooling. If the continental relationship were also dependent on simple conductive cooling, the slopes should be the same (Oxburgh and Parmentier, 1978).

It may be helpful to consider the time dependence of continental heat flow in the context of the Birch-Lachenbruch relationship described earlier. It is evident that the substantial variation in near-surface radioactivity within any one particular age-province will give rise to a wide dispersion of heat flow values about the mean for the province. It is also clear that, because crustal thermal events are nearly always associated with uplift, they will be generally followed by rapid erosion. In many parts of the world, Caledonian mountain belts are, even today, relatively elevated above their surroundings, and are still being worn down. A consequence of

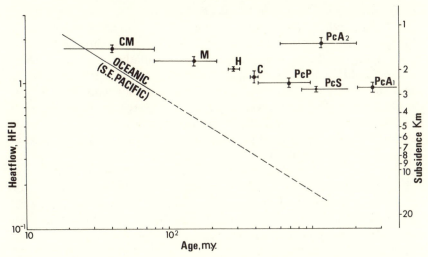

Figure 6. The observed dependence of oceanic heat flow and subsidence on ocean floor age in the south-east Pacific (solid curve); the dashed extrapolation shows the heat flow and subsidence that would be expected if a conductive boundary layer were able to continue cooling for a long period (after Oxburgh & Parmentier, 1978). The points give mean continental heat flow as a function of age after Polyak & Smirnov (1968) and Sass & Lachenbruch (1978). CM—Cenozoic miogeosynclines; M—Mesozoic fold belts; H—Hercynian orogenic belts; C—Caledonian orogenic belts; PcP—Precambrian platforms; PcS—Precambrian Shields; PcA$_2$—Precambrian Central Australia; PcA$_1$—Precambrian Western Australia. The decay of continental heat flow with time is very different from that observed in the oceans.

prolonged erosion of a crust, within which heat-producing elements are fractionated upwards, is that the mean value of surface heat production, and thus heat flow, should decrease with increasing age. This effect has been analysed by Richardson (1975), who showed that with a reasonable erosion rate, decreasing exponentially with time, it was possible to reproduce the earlier part (the first 500 m.y. or so) of the Polyak-Smirnov curve.

Erosion also has a second and opposite effect. As demonstrated by Clark and Jaeger (1969), if mean erosion rates excede more than a millimeter or so per year, temperatures in the lower part of the eroding mass are unable to readjust fully to the rapid removal of the overburden. In consequence, there is a transient steepening of the near-surface thermal gradient in regions of rapid erosion and an accompanying transient enhancement of the surface heat flow. Thus a comparison of heat-flow values from young orogenic belts with those from older regions, within which erosion has been negligible for some time, is not a comparison of like with like.

It follows that erosional history and the associated reduction in mean surface-heat production are both likely to contribute to the relationship shown in Figure 5. In principle, the effects of the differential removal of near-surface radioactivity might be isolated by examining the variation of reduced heat flow q_m with time, but attempts in this direction are not encouraging (Figure 7). It appears that there are other major perturbing influences affecting q_m, particularly in young areas. There must also be some question of the validity of some of the values for reduced heat flow; the observational points on which they are based are in some cases widely dispersed about the straight line drawn through them to derive the intercept value for reduced heat flow. Also, in a number of cases there must be serious question about the statistical significance of the rather small number of analyses used to characterize near-surface radioactivity.

Erosional phenomena, however, are not expected to have an important influence on continental heat flow at times greater than 500 m.y. after the last major thermal event. Unfortunately, the shape of the Polyak-Smirnov curve for times greater than 500 m.y. is very poorly constrained, and although heat flow may continue to fall beyond that point, it is not certain that it does so. The reduced heat-flow curve is even less well constrained. The shapes of the curves for very old stable crustal regions must reflect the long term evolutionary processes affecting the continental lithosphere (e.g. Oxburgh and Parmentier, 1978), with the q_m curve containing information about the time-dependence of the conductive heat flux to the base of the lithosphere, but further discussion of this interesting question is beyond the scope of this article.

A less conservative approach to these data has been taken by Pollack and Chapman (1977), who assume that conductive equilibrium obtains

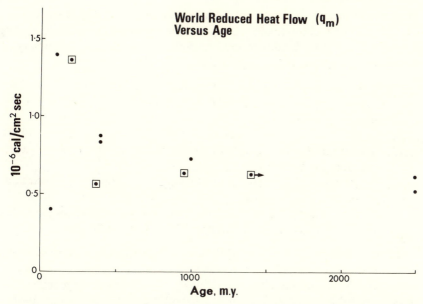

Figure 7. World continental reduced heat flow versus age after Hamza (1976) with additional points (squares) from Sass and Lachenbruch (1978) and Richardson and Oxburgh (1978). See text for discussion.

within the continents, and that the reduced heat flow for a heat-flow province is approximately 60% of the mean observed surface heat flow for the province. This assumption allows reduced heat flow to be calculated on a world-wide basis regardless of whether heat-producing element analyses are available. The reduced heat flow is assumed to be largely a conductive heat flow from the mantle and is used to estimate a thermal gradient in the subcontinental mantle. By assuming a particular dependence of mantle rheology on temperature and pressure, Pollack and Chapman calculate the thickness for the lithosphere at any point; thus, in essence, lithospheric thickness in the continents is regarded as inversely proportional to reduced heat flow. Although the general conclusion of this argument may well be correct, (i.e., that old continental lithosphere is thicker than average), the steps by which that conclusion is reached involve so many at present unwarranted assumptions that for the time being the conclusion must be sustained on other grounds.

DEEP GEOTHERMAL GRADIENTS

In estimating temperatures within the Earth today, we may therefore make use of these three empirical relationships, only the second of which can be regarded as having a sound physical interpretation. There are, in addition, observations of other kinds which may be required in order to make temperature extrapolations to greater depth, or can be used to constrain such extrapolations. These include the following:

(1) measurements of the thermal properties of rocks and minerals at high temperatures and pressures (e.g., Schatz and Simmons, 1972; Soga et al., 1972),

(2) variation of electrical conductivity with depth, (e.g., Tozer, 1959)

(3) velocity gradients and attenuation behavior of elastic waves (e.g., Spetzler and Anderson, 1968; Anderson, 1967),

(4) the chemical characterization of erupted magmas (e.g., Green and Ringwood, 1968; Wyllie, 1971; Ringwood, 1975), and

(5) possibly, suites of T/P points derived from series of deep xenoliths (Boyd, 1973; Mercier and Carter, 1975; Harte, 1978).

In general, the best way to proceed is to start with what is believed to be a steady-state conductive heat flow measured at a particular depth in the crust, and to extrapolate downwards using a petrological or geochemical model (possibly based on the Birch-Lachenbruch relationship) to specify the distribution of heat-producing elements within the depth interval for which a temperature profile is required. It is also necessary to specify a thermal conductivity structure through the same interval. In general, for depths greater than 5,000 m it is necessary to consider the variation of thermal conductivity with temperature and pressure. Reliable conductivity estimates for the upper mantle are not available, partly because of uncertainties in mineralogy, grain size and texture. And, of course, extrapolations of this kind may be employed only to depths for which there are reasonable grounds to believe that equilibrium conductive heat transfer occurs.

It follows, from the acceptance of large-scale plate motions at the surface, that beyond some depth (presumably variable from place to place) material will no longer move coherently with the surface plate in the same direction at the same velocity. Beyond that depth there must be a zone in which mantle material circulates with some component of motion opposite to that of the surface plate, so that mass is conserved. Within this circulating zone, the temperature structure will be strongly influenced by the character of the motions. This circulation presumably controls the thermal condition of the lower surfaces of the plates, either by maintaining them at a constant (adiabatic) temperature, or possibly by maintaining a constant heat flux to them (constant must here be taken to mean "changing rather slowly").

It is clear, then, that "standard geotherms" like those in Figure 8 should be treated rather carefully. Geotherms in the oceanic lithosphere must be

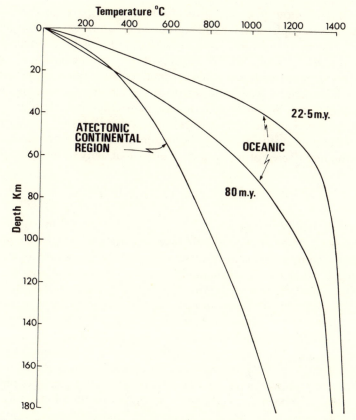

Figure 8. Continental and oceanic geotherms: The continental geotherm corresponds to an equilibrium surface heat flow of $50\ mW/m^2$ ($1.2 \times 10^{-6}\ cal/cm^2$ sec); thermal conductivity is taken as a function of temperature and pressure according to the models of Schatz & Simmons (1972) and Cull (1975), A_0 is 2.35 Wm^3 (5.6×10^{-13} cal/cm^3 sec) and h is 8.5 km; in the lower crust and mantle A is taken as 0.04 W/m^3 ($0.1 \times 10^{-13}\ cal/cm^3$ sec). The oceanic profiles are boundary layer cooling profiles calculated according to the procedure of Turcotte and Oxburgh 1969. Note that at some depth the continental and oceanic profiles must meet; with extrapolation of the profiles shown, this occurs at about 300 km, in which case that depth would be interpreted as the base of the continental lithosphere. For the calculation of the boundary layer profiles it is necessary to assume a mean mantle temperature (here 1,400°C). By assuming a lower value, say 1,200°C, the lower parts of both curves may be shifted to the left and will match with the continental profile at shallower depth, thus allowing a thinner continental lithosphere. On the other hand, unless the P fluid in the mantle is high, a temperature of 1,200°C would be inadequate to permit a partially molten oceanic low velocity zone.

changing continuously, except possibly in very old parts of the oceans. Beneath the lithosphere the thermal gradient is controlled by convection; it is not directly measurable, and any computation is model-dependent. Under different circumstances (Figures 8, 9, 10), the gradient could, variously, be close to adiabatic, have a small positive value, or be positive with local inversions. The same is true of the circulating zone under the continental plates. Within the continental plates themselves, the gradients are evidently dependent on the nature, intensity and age of the most recent tectono-thermal events affecting the region concerned, and on the concentration and degree of fractionation of radioactivity in the lithosphere.

HEAT-FLOW STUDIES AND MAGMA GENESIS

From what has been said earlier, it follows that heat-flow studies can provide only the most general guidance on questions of magma genesis. The studies cannot, for instance, resolve the important question of whether the seismic low-velocity zone under oceanic plates contains a small percentage of melt, and so serves as a possible reservoir for mid-plate oceanic volcanism. Thermal models may be readily constructed that satisfy the heat flow data and show the LVZ to be either well below its solidus or, alternatively, partially molten.

The concentration of nearly all magmatic activity around the margins of surface plates has focused attention on dynamic processes in magma genesis. Broadly, four such types of processes with some degree of physical plausibility have been invoked to explain the formation of magmas:

(1) Melting by adiabatic decompression. This process is explored in more detail below, but it depends on the relatively low thermal diffusivity of rocks. If rocks situated at depth at temperatures below their solidus for that pressure, *but above their solidus temperature for zero pressure*, are allowed to rise diapirically at rates of more than a few millimeters/year, they undergo a significant reduction in pressure but change temperature very little. In this way they may enter the field of partial melting.

(2) Shear heating or viscous dissipation. This mechanism, which is most obviously applicable to overthrust regions and subduction zones, depends upon the relationship

$$q = u\tau \tag{8}$$

where q is the amount of heat produced between two surfaces in sliding contact, u is the velocity of sliding, and τ is the shear stress between them. Again because of the relatively low thermal diffusivity of rocks, heat generated in this way diffuses away from the sliding zone rather slowly, and if heat generation continues, it results in a rapid and local elevation in temperature. This mechanism operates equally well if there is some kind of viscous coupling between the two surfaces. Various authors have presented models for island-arc volcanism along these lines (e.g., Oxburgh and Turcotte, 1968b, 1970; McKenzie and Sclater, 1968; Turcotte and Schubert, 1973). In general the main difficulty is that rather large values of shear stress (of the order of one or two kilobars) are required to produce significant effects. An interesting variant of this process is discussed by Anderson and Perkins, 1974. Using the theory of solid thermal explosions, they propose that shear in the asthenosphere could give rise to non-linear viscous heating and thermal runaway, with the generation of gravitationally unstable bodies of melt.

(3) Burial of radiogenic heat sources. This mechanism is probably confined to continental regions where large scale over-thrusting has occurred. Reference was made in a earlier section to the tendency for heat-producing elements to concentrate upwards in the continental crust; such a distribution minimizes temperatures at depth within the crust. If, however, crust with a relatively high mean value of heat production becomes deeply buried, the temperature it attains simply by reaching conductive equilibrium with its surroundings may by sufficient either to initiate melting or to set up a gravitational instability. This instability gives rise to diapiric vertical motions, which in turn result in melting by the pressure release mechanism described earlier.

The necessary burial may, in principle, be attained either by very thick sedimentation or by large-scale crustal overthrusting, such as occurs in continental suture zones (e.g., the Alps or Himalayas). This mechanism is more effective if the overlying rock mass has a high mean value of heat production, or heat production relatively uniformly distributed within it, or both (e.g. Bickle et al, 1975; Bird et al., 1975).

The depth of burial necessary for this mechanism to be effective is relatively great (i.e., of the order of 15 km) and there has to be sufficient time for the buried material to approach some kind of conductive equilibrium (e.g., 40 to 50 m.y.). Although there are no clear examples known to the writer of igneous rocks produced in this way, the mechanism is potentially capable of generating magma some time after the cessation of the main tectonic activity.

(4) Changes in fluid pressure. This mechanism of melting depends not on increasing the temperatures of the rocks but on increasing the concentration of volatiles in the system, thereby depressing the temperature of the onset of melting sufficiently that the rocks begin to meet at their ambient temperature. Volatile components are certainly present in the mantle (e.g. Green, 1972), but little is known of the factors controlling their inclusion in or expulsion from solid phases, or of any means of concentrating them in particular zones. Experience with other fluids in the upper parts of the crust suggests that, during deformation, fluids present along grain boundaries in low concentrations tend to be expelled and to migrate and concentrate in response to local pressure gradients. It is

conceivable that localized zones of ascending convective flow within the mantle might discharge volatiles upwards and thus provide a local concentration sufficient to produce melting in the overlying rocks. This mechanism, if it operates in nature, has the advantage that it is able to generate rather localized zones of melting within the lithosphere. Bailey (1970, 1974) has thoroughly explored this question.

MELTING PROCESSES AT MID-OCEAN RIDGES

We now apply some of these considerations to an analysis of melting processes at the mid-ocean ridges.

Surface observations have established that plates move apart along the crestal line of mid-ocean ridges at rates of centimeters per year, and attain full thicknesses of 80 km or more. There must in consequence be an upwelling of mantle material from greater depths to fill the void left by the separating plates. Provided that the temperature gradient within the mantle underlying the plates is superadiabatic, geotherms must be displaced upwards in the region of ascending flow, and some pattern of the kind shown in Figure 9 will be established. The exact form of this pattern is sensitive to the values adopted for the various model parameters, particularly those affecting the rheology of the mantle. Nevertheless, the general features may be regarded as well established. Along the ridge axis, where the mass-feed reaches the surface, there is maximum upward deflection of the isotherms parallel to the mass flow (Turcotte and Oxburgh, 1967). As the vertical flow approaches the surface it divides to feed the two horizontal

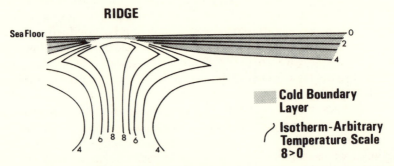

Figure 9. Pattern of isotherms expected under a mid-ocean ridge when there is superadiabatic temperature gradient in the mantle. The isotherms in the ascending zone are essentially parallel to the direction of flow. As the flow approaches the surface, the flow diverges, as do the isotherms; cooling to the upper surface begins at this stage and a cold thermal boundary layer is developed (after Oxburgh and Turcotte, 1968). The exact form and scale of the isotherm structure beneath the ridge crest depend on the detailed pattern of mantle flow, and the amount by which the mean mantle thermal gradient is super-adiabatic.

flows away from the ridge. Each horizontal flow immediately begins to cool by conduction to the ocean floor, resulting in the growth of a cold thermal boundary layer (Figure 9). This boundary layer thickens with time in proportion to $t^{1/2}$, and in as much as the mechanical properties of silicates are strongly temperature dependent, the layer may be regarded as approximately equivalent to the mechanical plate.

Provided that the melting behavior of the mantle is known, a grid of solidus isotherms may be superimposed upon the pattern of convection-displaced geotherms (Figure 10). The intersection of the two sets defines

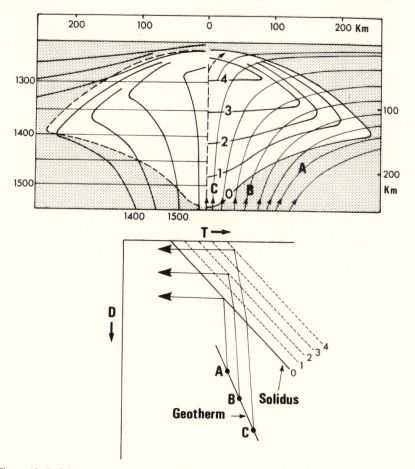

Figure 10. Left hand side: the formation of a zone of partial fusion under a mid-ocean ridge by the intersection of convection displaced isotherms (curved solid lines) with the solidus horizontal lines; temperatures in °C given for both. Right hand side: mantle material is continuously convected through the zone of partial melting, arbitrary units 1, 2, 3, and 4. Points A, B, and C originate at different depths and follow different paths through the zone of fusion, melting to different degrees.

the shape of a zone of partial melting, roughly the shape of a flattened diamond, positioned symmetrically beneath the ridge (Oxburgh and Turcotte, 1968a). The *proportions* of this diamond are particularly sensitive to the form of the mantle solidus, the assumed pattern of flow, and the deflected pattern of isotherms that the flow produces. It is unfortunately possible to construct models which violate no reasonable constraints and which, at one extreme, have partial melt zones that extend laterally to meet or perhaps form, the low velocity zone; and at the other extreme, give a small zone of localized meeting beneath the ridge. The length of the vertical axis of the diamond is particularly sensitive to the average mantle thermal gradient assumed to be disturbed by the flow, and to the depth of origin of the flow. Various models for isotherm perturbation under ridges are given by Bottinga and Allegre (1978), Forsyth and Press (1971) and Oxburgh and Turcotte (1968a, 1976). All of the models suggest some degree of isotherm inversion beneath the ridge flanks.

MECHANISM OF MELTING

Derivation of a zone of fusion in the way described above depends upon the assumption that melting occurs by adiabatic decompression. It is assumed that there is virtually no conductive heat loss from a small volume of mantle participating in the ascending flow; however, the material performs work in expanding against its surroundings and therefore experiences a drop in temperature during its ascent. Because the adiabatic gradient of temperature is less steep than the solidus, a volume of mantle that starts at a sufficiently high temperature and cools by adiabatic decompression will begin to melt when its adiabatic trajectory intersects the solidus (Figure 11). This is likely to be a realistic model for the behaviour of mantle rising under ocean ridges.

It has sometimes been suggested that isenthalpic decompression may occur in some geological situations. This process requires that decompression takes place in such a way that the expanding material does no work on its surroundings (e.g. in a rift zone?). In this case the temperature of the material should actually rise during decompression and presumably high degrees of melting would be possible. This process has been discussed by Waldbaum (1971) and Ramberg (1971); so far there has been no convincing demonstration of its importance in nature.

THE ZONE OF MELTING

Once a particle of mantle material has entered the melting zone it will begin to undergo partial melting at a rate of between .4 and .8%/km rise. An upper limit to the amount of energy available for melting, E, is given

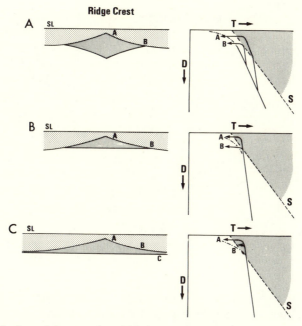

Figure 11. The form of the zone of partial fusion is very sensitive to assumptions concerning the mean temperature gradient in the mantle; compare (A), where the gradient is superadiabatic and a diamond-shaped partial melting zone is formed (cf Figure 10); melting zone: fine stipple; boundary layer: coarse stipple; S: solidus; straight solid line: superadiabatic mantle geotherm, with dot-dashed boundary-layer, near surface continuation. A and B in the T/D plot are the paths taken by material passing through points A and B in diagram to the left. Figure 11 (B) is like (A), but the initial mantle gradient is adiabatic; note flat base to the fusion zone beneath the ridge. Figure (C) is like (B) except that the mantle adiabatic gradient has a higher average temperature, and intersects the solidus whether displaced by convection or not; a flat-bottomed sub-ridge fusion zone passes laterally into the oceanic seismic low velocity zone, C.

by $E = (\beta - \alpha)C_p h$, where β and α are the solidus and adiabatic gradients respectively, C_p is the specific heat, and h is the vertical displacement of the particle after entering the fusion zone. How much melting actually occurs depends upon the uniformity of melting across the solidus/liquidus interval. Alternative suggestions illustrated by Wyllie (1971) are shown in Figure 12. Green (1971) has suggested a nearly uniform increase in the degree of melting with increasing temperature; Wyllie emphasises the consequences of the loss of phases during progressive melting, and predicts that after the loss of each successive phase there should be an interval through which temperature rises with relatively little increase in degree of melting.

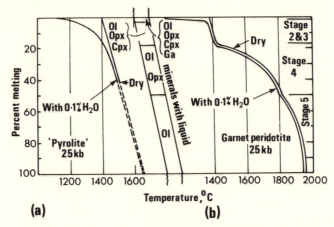

Figure 12. Alternative models for the isobaric dependence of degree of melting on temperature for ultramafic compositions both in the presence of 0.1% H_2O, and dry. Compilation of data by Wyllie (1971) from a variety of sources.

In any case the temperature throughout the melting zone is buffered by the melting behavior of the mantle. The temperature distribution around the top and sides of the melting zone will lie on the mantle solidus, which would be slightly elevated, due to the removal of the low melting fraction.

Melt will be present everywhere within the melting zone. As any individual particle passes through the zone it presumably undergoes progressive melting until the proportion of melt to residue is sufficient for the melt to form a coherent body that will rise gravitationally through the melting zone to the surface. It is not known at what degree of melting this happens, but it has been estimated to be as low as a few percent (Turcotte and Ahern, 1978). If this is correct, then the melting process must closely approximate continuous perfect fractional melting. Insofar as particles are continuously passing through the fusion zone on a variety of paths, liquid derived from all places within the zone, and thus having a wide range of compositions, accumulates near its top. These liquids of differing composition presumably mix, either during ascent through the zone or after reaching its top, in order to provide the relatively homogeneous basaltic rocks of the oceanic crust. Provided that this mixing is efficient, it provides a means of smoothing out the effects of small scale chemical inhomogeneities in the mantle. The liquid erupted at the surface at any instant will represent batches of melt derived at different stages of melting from different parts of a very large source volume. Chemical inhomogeneities in the mantle on a scale of ten kilometers or so should be smoothed out by this process while those on the scale of a hundred kilometers or so should be reflected in secular changes in magma composition.

"Degree" of Melting

If the behavior of an individual particle of mantle is followed along its path through the melting zone, it will be possible to characterize that particle as having undergone a particular degree of melting as it passes out of the melting zone. Examination of Figure 11 shows that this value will be greatest for particles rising vertically beneath the ridge axis, and least for those following paths that just graze the edge of the fusion zone. It follows that the homogeneous liquid erupted at the surface represents a mixture, the composition of which is close to that of the mean of all liquids present in the partial melting zone at any time. It is clear from Figure 11 that the contribution of low degree melts to the total is much greater than that of high degree melts: for example, although the maximum degree of partial melting might be 40%, the mean would be closer to 15%.

But even such an apparent 15% melt would not correspond to the liquid generated by the equilibrium homogeneous melting, at constant pressure, of a small volume of mantle material until that volume was 15% molten. It would differ in two main ways. First, the apparent 15% melt would represent a mixture of liquids generated over a wide P/T range and then partially re-equilibrated with the various residues that they encountered during the ascent to the surface. Secondly, because the apparent 15% melt contains a mixture of higher-degree and lower-degree melts, some phases may have contributed to the melt (say in the 40% melting zone) that would not have done so during homogeneous melting.

The observation that the thickness of the igneous layers in the oceanic crust does not vary significantly from place to place over the ocean floor and is essentially independent of spreading rate is of some importance in understanding ridge melting processes. A rapidly spreading ridge, such as parts of the East Pacific rise (half rate ~ 4.4 cm/yr), requires, for conservation of mass, an ascending mass flux about five times as great as that required by a more slowly spreading ridge, such as the Reykyanes (half rate < 1 cm/yr). The fact that the thickness of the oceanic crust is similar in both cases indicates that similar degrees of melting are achieved whether mantle material is processed rapidly or slowly through the sub-ridge melting zone. This in turn suggests three possibilities, none of which completely excludes the others:

(1) Regardless of whether the ridge is spreading rapidly or slowly, the mass flux always originates at the same depth, thus giving the same amount of adiabatic decompression and the same degree of melting in all cases.

(2) Even though faster spreading ridges might draw their mass-flux from greater depth, the thermal gradient within that part of the mantle is so close to the adiabatic gradient that mantle material always enters the zone of fusion at approximately the same temperature, regardless of the depth at which it began its rapid ascent.

(3) A chemical control operates to give a strongly nonlinear dependence between degree of melting and temperature, so that the amount of melt produced is insensitive to moderate variations in the amount of decompression.

In considering these alternatives, it should be noted that the geometry of plate motions requires that ridges migrate relatively rapidly with respect to each other and with respect to trenches, and that seismic studies have been unable to establish any anomalous velocity structure extending deeper under ridges than about two-hundred kilometers. In the case of alternative (1), this would imply that the mass flux is of shallow origin (e.g., Schubert et al., 1978), possibly associated with the main counter-flow required by the plate motion. Any deep flow associated with (2) would have virtually no associated temperature anomaly and would not be detectable by seismic observations, unless it was also characterized by a strong flow fabric, for example, with olivine c-axes parallel to the vertical flow direction.

It is the writer's present prejudice that the mass flux beneath ridges is derived from rather shallow depth (say, < 300 km) and that, although slowly spreading ridges may tap magma from slightly shallower depths than the faster spreading ones, the mantle thermal gradient is not strongly superadiabatic, and there is relatively little variation in the degree of fusion.

These considerations have an interesting application to the controversy concerning the origin of Iceland. Morgan (1972) included Iceland in his list of proposed mantle plumes and has been followed by other workers (e.g. Wilson, 1973; Burke and Wilson, 1976) in this view. Iceland lies on the northern part of the mid-Atlantic ridge, an active constructive plate margin. Various studies summarized by Bott (1974) show that, not only is the crust under Iceland between two and three times as thick as under other parts of the Mid-Atlantic ridge, but that there are also narrow elongate zones of thickened crust extending laterally from Iceland onto the two diverging plates. These zones form the Iceland-Faeroes ridge to the east and the Iceland-Greenland ridge to the west, and could be interpreted as "plume traces," as if the magmatism had been more profuse in this one localized segment of the ridge than either north or south. Morgan (1972) proposed a world wide system of relatively stable, highly localized mantle flows ascending from the core mantle boundary and able to deliver large volumes of magma to the surface through any plate, or at any plate margin that happened to lie above them.

One possible explanation for such a feature that involves no underlying plume might be compositional inhomogeneity in the mantle, such that the material feeding the flow immediately below Iceland for the last few tens of millions of years had a lower temperature of beginning of melting than the "average mantle" feeding the ridge as a whole. In consequence, given the same amount of adiabatic decompression in the ascending flow along the length of the ridge, a higher than average degree of melting would be expected under Iceland. There are some indications that the Icelandic lavas have distinctive chemical characteristics (e.g., various papers in Kristjansson, 1974; O'Hara, 1975; O'Nions and Pankhurst, 1974; Schilling, 1975); it would, however, be necessary to suppose that the lavas represented two to three times the degree of fusion that typically occurs beneath ocean ridges. It would perhaps be difficult to be certain of distinguishing an 8% melt or a 20% melt derived from different and unknown ultramafic starting compositions; however, such a possibility is not generally envisaged. Although this mechanism would require exceptional degrees of local melting, it would not require any special kind of mantle flow regime beneath Iceland.

Alternatively, it could be assumed that the degree of melting is similar everywhere along the ridge, but that mantle is "processed" more rapidly under Iceland than elsewhere. That is, mantle is fed into the fusion zone and discharged from it again more rapidly, giving locally a much higher rate of magma production, without necessarily involving abnormally high degrees of melting.

It should be emphasized that, if unacceptably high-degree melts are to be avoided, then either such a plume flow must originate at the same depth as the flows feeding the ridge elsewhere, or the mean mantle temperature gradient down to the depth of origin of the plume must be close to the adiabatic, so that a flow of deep origin produces no exceptional perturbation of the isotherms. The driving force for flows of either kind is obscure.

Morgan's (1972) proposal combined the possibility of plumes of deep mantle origin with that of chemical inhomogeneity. One possibility that represents a compromise between the two alternative approaches outlined earlier is that, in a "shallow-feed" model, a compositional inhomogeneity permits melting to begin locally at greater than average depth (and thus a higher than average degree of melting). A feed-back effect, possibly arising from reduced viscous resistance to the local body forces driving the ascending flow, might then enhance the local rate of vertical flow. There would therefore be both a higher local rate of mantle processing, and a higher degree of melting sufficient to satisfy the observed rate of surface magmatism.

CONCLUSIONS

It is evident from the foregoing that, although our understanding of thermal processes in the Earth has advanced since Bowen's time, there remain enormous areas of uncertainty. On the largest scale, we do not know whether the Earth's total surface heat flux is in equilibrium with its present internal heat generation, though this assumption is commonly made. On the other hand, we do know that plate motions are the surface manifestation of a terrestrial heat engine operating by the subtle interaction of physical and chemical processes and the transformation of thermal energy to mechanical energy. The mantle is heated by radioactivity, and cools partly because cold oceanic lithosphere sinks to depths of at least 700 m, having the same effect as ice cubes dropped into a drink on a summer day.[1] The mantle also cools in the process of plate formation by conductive heat loss to the ocean floor.

Much of the present day pattern of surface heat flow is to be seen as the transient relaxation after episodic thermal events commonly associated with, or even caused by, large scale tectonic movements. There is real difficulty in interpreting such situations on the evidence of only one instant (i.e., the present) in the relaxation history, and no unique inversion is possible. It is therefore necessary to consider families of thermal models that might satisfy the present observations, possibly limiting the range of possibilities by other geological or geophysical constraints, and then attempting to identify common features of those that remain.

Magma generation seems in general to be associated with the phases of tectonothermal activity mentioned above. There now seem to be a number of physical processes at plate margins that could plausibly lead to partial melting. They are not, however, understood in any detail. Localized zones of intense volcanism, on plate margins or elsewhere, appear to have a different explanation.

If the seismic low-velocity zone under oceans is partially molten, stress accumulation in the lithosphere and hydraulic fracturing may permit surface eruptions from this zone. The zone does not, however, appear to be generally present under continents, which may nevertheless be penetrated by mid-plate volcanism.

---

[1] The editor complained to the author that this analogy was misleading because ice cubes float; he should try whiskey.

Progress will depend on the following.

(1) A better understanding of the thermal properties of rocks, including those dependent on fabric; also their variation with temperature and pressure.

(2) A better understanding of the extent and occurrence of fluid circulation in the crust. This may in some cases partially invalidate assumptions of conductive heat transfer.

(3) A knowledge of the processes leading to the fractionation of heat producing elements in the crust.

(4) A knowledge of the pattern of circulation in the upper mantle

(5) Detailed studies of the thermal histories of particular, especially favourable, regions where good geological, geochronological and geophysical observations can be obtained to constrain the thermal history.

(6) the identification of pristine and relatively unmodified products of mantle melting and their use in constraining temperatures and depths of formation.

(7) the inversion of electrical conductivity data and seismic velocity and attenuation studies.

For the time being we must, like Bowen, continue in a frame of mind that admits of multiple working hypotheses.

REFERENCES

Anderson, D. L., 1967. Latest information from seismic observations, in *The Earth's Mantle*, T. F. Gaskell, ed., London: Academic Press.

Anderson, O. L. and Perkins, P. C., 1974. Runaway temperatures in asthenosphere resulting from viscous heating, *J. Geophys. Res. 79*, 2136–2138.

Anderson, R. N., Langseth, M. G., and Sclater, J. G., 1977. The mechanism of heat transfer through the floor of the Indian Ocean, *J. Geophys. Res. 82*, 3391–3409.

Bailey, D. K., 1970. Volatile flux, heat focusing and the generation of magma, in *Mechanisms of Igneous Intrusion*, Geol. J. Special Issue No. 2, 177–186.

Bailey, D. K., 1974. Melting in the deep crust, in *The Alkaline Rocks*, H. Sorenson, ed., New York: Wiley, 148–159.

Bickle, M. J., Hawkesworth, C. J., England, P. C., and Athey, D. R., 1975. A preliminary thermal model for regional metamorphism in the Eastern Alps, *Earth and Planetary Science Letters 26*, 13–28.

Birch, F., 1950. Flow of heat in the Front Range, Colorado, *Geol. Soc. Amer. Bull. 61*, 567–630.

Birch, F., Roy, R. F., and Decker, E. R., 1968. Heat flow and thermal history in New England and New York, in *Studies of Appalachian Geology, Northern and Maritime*, E-an Zen, W. S. White, J. B. Hadley, and J. B. Thompson, eds., New York: Interscience Publishers, 437–452.

Bird, P., Toksoz, M. N., and Sleep, N., 1975. Thermal and mechanical models of continent-continent convergence zones, *J. Geophys. Res. 80*, 4405–4416.

Bott, M. P. H., 1974. Deep structure, evolution and origin of the Icelandic transverse ridge, in *Geodynamics of Iceland and the North Atlantic Area*, L. Kristjansson, ed., Dordrecht-Holland: Riedel, 33–48.

Bottinga, Y. and Allegre, C. J., 1978. Partial melting under spreading ridges, *Phil. Trans. R. Soc. Lond. A, 288*, 501–525.

Bowen, N. L., 1928. *The Evolution of The Igneous Rocks*, Princeton: Princeton University Press.

Boyd, F. R., 1973. A pyroxene geotherm, *Geochim. Cosmochim. Acta, 37*, 2533–2536.

Bullard, E. C. and Niblett, E. R., 1951. Terrestrial heat flow in England, *Monthly Not. R. Astr. Soc. Geophys. Suppl. 6*, 222–238.

Burke, K. and Wilson, J. T., 1976. Hot spots on the Earth's surface, *Scient. Am. 235*, 46–57.

Carslaw, H. S. and Jaeger, J. C., 1959. *Conduction of Heat in Solids*, Oxford: Oxford University Press, 2nd edition.

Clark, S. P. and Jaeger, E., 1969. Denudation rate in the Alps from geochronologic and heat flow data, *Am. J. Sci. 267*, 1143–1160.

Cull, J. P., 1975. The pressure and temperature dependence of thermal conductivity within the Earth, unpub. D. Phil. thesis, Oxford.

Forsyth, D. W. and Press. F., 1971. Geophysical tests of petrological models of the spreading lithosphere, *J. Geophys. Res. 76*, 7963–7979.

Green, T. and Ringwood, A. E., 1968. Genesis of the calcalkaline igneous rock suite, *Contr. Min. & Petrol. 18*, 105–162.

Green, D. H., 1971. Composition of basaltic magmas as indicators of conditions of origin: applications to oceanic volcanism, *Phil. Trans. R. Soc. Lond. A. 268*, 707–725.

Green, H. W. II, 1972. A CO_2-charged asthenosphere, *Nature Phys. Sci. 238*, 2–5.

Hamza, V. M., 1976. Possible extension of oceanic heat flow-age relations to continental regions and the thermal structure of continental margins, *An. Acad. Bras Cienc. 48*, 121–129.

Harte, B., 1978. Kimberlite nodules, upper mantle petrology and geotherms, *Phil. Trans. R. Soc. Lond. A. 288*, 487–500.

Hawkesworth, C. J., 1974. Vertical distribution of heat production in the basement of the Eastern Alps, *Nature 249*, 435–436.

Hyndman, R. D., Lambert. I. B., Heier, K. S., Jaeger, J. C., and Ringwood, A. E., 1968. Heat flow and surface radioactivity measurements in the Precambrian shield of Western Australia, *Phys. Earth Plan. Insts. 1*, 129–135.

Jones, F. W. and Oxburgh, E. R., 1979. Two-dimensional thermal conductivity anomalies and vertical heat flow measurements, in *Terrestrial Heat Flow in Europe*, Heidelberg: Springer, 98–106.

Kristjansson, L. ed., 1974. *Geodynamics of Iceland and the North Atlantic Area*, Dordrecht-Holland: Riedel.

Lachenbruch, A. H., 1968. Preliminary geothermal model of the Sierra Nevada, *J. Geophys. Res. 73*, 6977–6990.

Lachenbruch, A. H., and Bunker, C. M., 1971. Vertical gradients of heat production in the continental crust, 2. Some estimates from borehole data, *J. Geophys. Res. 76*, 3852–3860.

Lee, W. H. K., ed., 1965. *Terrestrial Heat Flow*, Geophys. Monograph # 8, American Geophysical Union.

Lister, C. B., 1972. On the thermal balance of an ocean ridge, *Geophys. J. R. Astron. Soc. 26*, 515–535.

McKenzie, D. P., 1967. Some remarks on heat flow and gravity anomalies, *J. Geophys. Res. 72*, 6261–6273.

McKenzie, D. P. and Sclater, J. C., 1968. Heat flow inside the island arcs of the Northwestern Pacific, *J. Geophys. Res. 73*, 3173–3179.

Mercier, J. and Carter, N. L., 1975. Pyroxene Geotherms, *J. Geophys. Res. 80*, 3349–3362.

Morgan, W. J., 1972. Deep mantle convection plumes and plate motions, *Bull. Am. Ass. Petrol. Geol. 56*, 203–213.

O'Hara, M. J., 1975. Is there an Icelandic mantle plume?, *Nature 253*, 708–710.

O'Nions, R. K. and Pankhurst, R. J., 1974. Petrogenic significance of isotope and trace element variations in volcanic rocks from the Mid-Atlantic, *J. Pet. 15*, 603–634.

Oxburgh, E. R. and Parmentier, E. M., 1978. Thermal processes in the formation of continental lithosphere, *Phil. Trans. R. Soc. Lond. A. 288*, 415–428.

Oxburgh, E. R. and Turcotte, D. L., 1968a. Mid-ocean ridges and geotherm distribution during mantle convection. *J. Geophys. Res. 73*, 2643–2661.

Oxburgh, E. R. and Turcotte, D. L., 1968b. Problem of high heat flow and volcanism associated with zones of descending mantle convective flow, *Nature 218*, 1041–1043.

Oxburgh, E. R. and Turcotte, D. L., 1969. Increased estimate for heat flow at oceanic ridges, *Nature 223*, 1354–1355.

Oxburgh, E. R. and Turcotte, D. L., 1970. Thermal structure of island arcs, *Bull. Geol. Soc. Amer. 81*, 1665–1668.

Oxburgh, E. R. and Turcotte, D. L., 1976. The physico-chemical behaviour of the descending lithosphere, *Tectonophysics, 32*, 107–128.

Parsons, B., and Sclater, J. G., 1977. An analysis of the variation of ocean floor bathymetry and heat flow with age, *J. Geophys Res. 82*, 803–827.

Pollack, H. N., and Chapman, D. S., 1977. On the regional variation of heat flow, geotherms and lithospheric thickness, *Tectonophysics 38*, 279–296.

Polyak, B. G., and Smirnov, Y. B., 1968. Relationship between terrestrial heat flow and the tectonics of continents, *Geotectonics 4*, 205–213.

Ramberg, H., 1971. Temperature changes associated with adiabatic decompression in geologic processes, *Nature 539*, 234.

Rao, R. U. M., and Jessop, A. M., 1975. A comparison of the thermal characteristics of shields, *Can. J. Earth Sci., 12*, 347–360.

Richardson, S. W., 1975. Heat flow processes, in *Geodynamics Today*, p. 123–132, London: The Royal Society.

Richardson, S. W. and Oxburgh, E. R, 1978. Heat flow, radiogenic heat production and crustal temperatures in England and Wales, *J. Geol. Soc. Lond. 135*, in press.

Richardson, S. W. and Powell, R., 1976. Thermal causes of the Dalradian meta-morphism in the central highlands of Scotland, *Scot. J. Geol. 3*, 237–268.

Richter, F. M. and Parsons, B., 1975, On the interaction of two scales of convection in the mantle, *J. Geophys. Res. 80*, 2529–2541.

Ringwood, A. E., 1975. *Composition and Petrology of the Earth's Mantle*, New York: McGraw-Hill Book Company.

Roy, R. F., Blackwell, D. D., and Birch, F., 1968. Heat generation of plutonic rocks and continental heat-flow provinces, *E. P. S. L. 5*, 1–12.

Roy, R. F., Blackwell, D. D. and Decker, E. R., 1972. Continental heat flow, in *The Nature of the Solid Earth*, E. C. Robertson, ed., New York: McGraw-Hill, 506–543.

Roy, R. F., Decker, E. R., Blackwell, D. D., and Birch, F., 1968. Heat flow in the United States, *J. Geophys. Res. 73*, 5207–5221.

Sass, J. H. and Lachenbruch, A. H., 1978. Thermal regime of the Australian con-tinental crust, in *The Earth – Origin, Structure, and Evolution*, M. W. McElhinny, ed., Academic Press, in press.

Sass, J. H. and Munroe, R. J., 1974. *Basic heat flow data from the United States*, U.S. Dept. of Interior Geological Survey Open File Report 74–79.

Sass, J. H., Lachenbruch, A. H., and Munroe, R. J., 1971. Thermal conductivity of rocks from measurements on fragments and its application to heat flow determinations, *J. Geophys. Res. 76*, 3391–3401.

Schatz, J. F. and Simmons, G., 1972. Thermal conductivity of earth materials at high temperatures, *J. Geophys. Res. 77*, 6966–6983.

Schilling, J. G., 1975. Rare-earth variations across "normal segments" of the Reykjanes Ridge, 60°–53°N, Mid-Atlantic Ridge, 29°S, and East Pacific Rise 2°–19°S, and evidence on the composition of the underlying low-velocity layer, *J. Geophys. Res. 80:11*, 1459–1473.

Schubert, G., Yuen, D. A., Froidevaux, C., Fleitout, L., and Souriau, M., 1978. Mantle circulation with partial shallow return flow: effects on stresses in plates and topography of the ocean floor, *J. Geophys. Res. 83*, 745–758.

Sclater, J. G. and Francheteau, J., 1970. The implications of terrestrial heat flow observations on current tectonic and geochemical models of the crust and upper mantle of the Earth, *Geophys. J. R. Astr. Soc. 20*, 509–542.

Sclater, J. G. and Parsons, B., 1976. Reply to C. R. B. Lister and E. E. Davies, *J. Geophys. Res., 81*, 4960–4964.

Sclater, J. G., Jaupart, C., and Galson, D., 1980. The heat flow through oceans and continents, *Reviews of Geophysics and Space Physics*, in press.

Sclater, J. G., Lawver, L. A., and Parsons, B., 1975. Comparison of long wavelength residual elevation and free air gravity anomalies in the North Atlantic, *J. Geophys. Res. 80*, 1031–1052.

Sleep, N., 1975. Formation of oceanic crust: some thermal constraints, *J. Geophys. 80*, 4037–4042.

Soga, N., Klemas, G. A., and Horai, K., 1972. Thermal conductivity-density relationship for high- and low-pressure polymorphs of silicates and germanates, *J. Geophys. Res. 77*, 2610–2612.

Spetzler, H. and Anderson, D. L., 1968. The effect of temperature and partial melting on velocity and attenuation in a simple binary system, *J. Geophys. Res. 73*, 6051–6060.

Spooner, E. T. C., Beckinsale, R. D., Fyfe, W. S., and Smewing, J. D., 1974. O^{18} enriched ophiolitic metabasic rocks from E. Liguria (Italy), Pindos (Greece) and Troodos (Cyprus). *Contrib. Mineral. Petrol. 47*, 41–62.

Tozer, D. C., 1959. The electrical properties of the Earth's interior, *Phys. Chem. Earth 3*, 414–436.

Turcotte, D. L. and Ahern, J. L., 1978. A porous model for magma migration in the asthenosphere, *J. Geophys. Res. 83*, 767–772.

Turcotte, D. L. and Oxburgh, E. R., 1967. Finite amplitude convective cells and continental drift, *J. Fluid. Mech. 28*, 29–42.

Turcotte, D. L. and Schubert, G., 1973. Frictional heating of the descending lithosphere, *J. Geophys. Res. 78*, 5876–5886.

Von Herzen, R. and Maxwell, A. E., 1959. The measurement of thermal conductivity of deep-sea sediments by a needle-probe method, *J. Geophys. Res. 64*, 1557–1563.

Waldbaum, D. R., 1971. Temperature changes associated with adiabatic decompression in geological processes, *Nature 232*, 545–547.

Watanabe, T., Langseth, M. G., and Anderson, R. N., 1977. Heat flow in back-arc basins of the western Pacific, in *Island Arcs, Deep Sea Trenches and Back-Arc Basins*, Maurice Ewing Series 1, pub. American Geophysical Union, 137–161.

Williams, D. L. and Von Herzen, R. P., 1974. Heat loss from the Earth: new estimate, *Geology 2*, 327–328.

Wilson, J. T., 1963. A possible origin of the Hawaiian Islands, *Can. J. Phys. 41*, 863–870.

Wilson, J. T., 1973. Mantle Plumes and Plate Motions, *Tectonophysics 19*, 149–164.

Wyllie, P. J., 1971. *The Dynamic Earth*, Chicago:Wiley.

Chapter 6

THE FRACTURE MECHANISMS OF MAGMA TRANSPORT FROM THE MANTLE TO THE SURFACE

HERBERT R. SHAW

U.S. Geological Survey, Menlo Park

INTRODUCTION

This paper consists almost entirely of ideas about mechanics, but it is really concerned with the processes of igneous evolution. The proposed fracture mechanisms represent a viewpoint that bridges gaps in my ability to visualize a continuity of igneous processes in the earth. Although much of the discussion refers to some formal ideas about fracture mechanics, the references are made from the standpoint of a user who has some acquaintance with the history and current advances in this field, but who is far from being expert in the conceptual applications. (See Roberts (1970) for another slant on similar mechanisms of igneous intrusion.)

For these reasons, this is an odd sort of paper: it consists of a lot of review material with a few new wrinkles. It is difficult to classify neatly what is old and what is new. It may be the main point of this paper to help the reader realize that the role magma plays in the workings of the planet remains almost as elusive as it was 50 years ago when Bowen (1928) published *The Evolution of the Igneous Rocks.*

The new direction of study of this problem consists of four general ideas. (1) Fracture mechanisms play a globally intrinsic role in the transportation of magma from the melting sources to the surface. (2) In consequence of the fracture mechanisms, there is a synchronized, if complicated, seismic spectrum quantitatively related to the volume rates of magma transfer. (3) The relation of fracture mechanisms to conditions of tectonic stress and strain rates represents a rate-determining control on magma ascent paths in both oceanic and continental regions. (4) Through a process of interactive feedback relations, the evolution of magma distributions is seen to represent both a speed indicator and regulator of plate tectonic motions over the earth. The paper is devoted mainly to the first two points, but addresses the last two when describing a model for Hawaiian magmatism that is used as a reference system for the discussion.

Specifically, the paper considers the following topics, referred quantitatively only to the Hawaiian system: (1a) stress guides and magma intrusion paths, (1b) concepts of effective stress and criteria of extension fracture, (1c) tensile strengths of partly molten rock, (1d) criteria for potential depths of tensile cracks, (1e) stress conditions in the asthenosphere, (1f) stress conditions in the lithosphere and in volcanic edifices [mentioned in (1f) relative to rotation of principal stresses and in (1a) relative to local versus regional volcanic lineaments], (2a) extensional failure and seismic mechanisms, (2b) failure regimes in extension and shear in relation to a concept of failure cascades, (2c) volume changes and seismic-energy release, (2d) volumetric moments and seismic efficiency in Hawaii, (3) mechanical correlations with magma transport rates, and (4) global implications.

Conceptual Viewpoints on Fracture Mechanisms

In the geological context, an understanding of fracture mechanisms requires almost complete knowledge about the thermal, chemical, mineralogical, textural, structural, and dynamical states of a volume of rock (including mobile phases) in relation to all adjoining volumes of rock. This range extends logically to encompass all knowledge about the earth's interactions with its internal and external energy sources (e.g., Gruntfest and Shaw, 1974). Though this may be an overstatement, the fact remains that no single concept of fracture can be universally applied, except the concept of fracture as the disruption of preexisting states and the transient creation of new surfaces, or phase boundaries, where they did not exist before.

The prospect of our ever acquiring enough information to prescribe accurately the conditions of fracture in a geological context seems hopeless. It is possible, however, to obtain partial observations of how the earth has behaved and is behaving in regard to fracture processes.

The main proposition of the present paper is that igneous petrology, which involves the chronologies of intrusive and extrusive extensional fracture events of fossil and active systems, provides a data base relevant to fracture phenomena that bridges spatial and temporal gaps between observations derived from structural petrology on the one hand, and seismology on the other. The paper attempts to show how this may work, using data from Hawaii to illustrate application of the principles.

The principles discussed in the paper conform generally to ideas about geologic conditions of fracture explored by many authors, beginning with E. M. Anderson (1905, 1936, 1938). Particular emphasis is placed on the role of fluid pressure, along the lines developed by Hubbert and Rubey (1959) and Secor (1965, 1969). This approach considers only the mechanical interactions of the rock and fluid phases and does not attempt to handle questions of chemical reactions and related mechanical effects. Such chemical effects are expected to be very important in magma-rock systems, and they are accounted for empirically in this paper by considerations of fracture yield strengths versus melt fraction. Other approaches to the effects of chemical reactions on fracture mechanisms are represented by studies reviewed by Lawn and Wilshaw (1975) and discussed at length relative to magmatic systems by Anderson and Grew (1977).

The different approaches are partly matters of viewpoint, preference, and the nature of the application. The more traditional method, used here, of appealing to representations of normal stresses, effective stresses (in materials with pore fluids), and experimentally determined failure stresses, forms the basis for macroscopic criteria of fracture. The method is limited by assumptions of uniformity of stress distributions, hence it does not deal with questions of stress concentrations at crack tips, and is unsuited to treatments of crack geometry and criteria of crack propagation. It says, in effect, that when certain stress conditions are met, a crack can form, but it does not indicate anything about how the crack will propagate in time or space, except for an idealized angular relationship to the principal stress directions. Other considerations are needed to discuss more than the potentiality of fracture. This approach has the advantage, in the present context, of simplicity and the capability of outlining the general stress regimes for fracture relative to regional tectonic stress distributions.

Analytical methods using stress-intensity factors (see Lawn and Wilshaw, 1975, p. 51) seem better suited to the description of fracture criteria in the immediate vicinity of the crack tip, or on the microscopic scale. Therefore, this more focused approach can be used, together with chemical concepts on the atomic scale (involving the kinetics of chemical diffusion, the kinetics of viscous flow, and chemical reaction kinetics) to give a systematic treatment of localized fracture kinetics, or to describe crack stability conditions in greater detail. Examples of the former approach are given by the so-called "stress corrosion theory" of fracture (Anderson and Grew, 1977), and examples of stability analyses applied to igneous sheet intrusions are discussed in depth by Pollard and Muller

(1976). These methods also are limited by the fact that the stress conditions and failure criteria are idealizations requiring experimental or observational confirmation.

The formalisms used in this paper were chosen simply because they are the more familiar means of classifying geologic fracture processes. It is likely that equivalent and possibly more powerful ideas can be developed in terms of stress-intensities and concepts of stress corrosion cracking. Tests or elaboration of the present ideas along these more specific lines should be a rewarding field for future research. A suggested approach is to compare the mechanisms discussed here with other analyses of fracture that are based on the concept of effective stress (Voight, 1976, has compiled key papers), and with the more comprehensive treatment of fracture mechanisms systematically developed in Lawn and Wilshaw (1975).

NOTES ON HISTORICAL BACKGROUND

Magma Lines of Ascent

Bowen's name is associated with ideas about the chemical evolution of igneous rocks, stated in terms of "liquid lines of descent" and based on criteria of experimental phase diagrams. His thoughts on the origin of magmas were not limited to chemical concepts, however, as is indicated by reading Chapter 17, "Petrogenesis and the Physics of the Earth" in *The Evolution of the Igneous Rocks* (Bowen, 1928). Relative to the present discussion of the factors controlling magma lines of ascent is a revealing paragraph from that chapter (Bowen, 1928, p. 319):

> On the whole the production of basaltic magma by selective fusion of peridotite at a depth probably as great as 75–100 km seems to be the preferable method in spite of the great difficulties involved, but it will be plain from the foregoing discussion that the problem of the origin of magmas is far from approaching definite solution. Geophysical data help very little. All that can be said is that they are apparently consistent with the derivation of basaltic magma in the manner described. The question of the parental nature of basaltic magma can not be given a definite answer. It is becoming increasingly probable that all magmas could be derived from the basaltic by crystallization-differentiation, *but even were this a proven fact it would not follow that magmas are so derived. The decision on this point would still depend principally upon geologic considerations*; indeed it was largely such considerations rather than petrology, that led to the suggestion of the parental nature of basalt. (Italics added)

This statement could stand almost unchanged today. I happen to believe that the evolution of the igneous rocks is a dynamic process of time-dependent relations between selective fusion and crystallization-differentiation, brought about by competition among chemical transport rates of several kinds. Only Nature herself knows the path of least resistance in this composite process, and the relative proportions of the several mechanisms involved. Indeed, whatever proof exists for the operation of any specific mechanism always hinges on considerations of geologic evidence, although at the present time a wide array of new data, especially geophysical, is available to formulate that evidence.

This paper gives no proofs for any particular mechanism of magma genesis. It considers, however, that the evolution of the igneous rocks is fundamentally governed by the rate of production of mafic·magma in the mantle (basaltic magma in the broad sense), which in turn is governed by the dynamics of rheological processes there. The chemical histories of the intrusive and extrusive products are largely governed by factors that control the magma transport paths and transport times. This view applies equally to the basaltic shield volcanoes of the oceans, to the andesitic volcanoes of island arcs, and to the caldera-forming rhyolitic volcanoes of the continents.

The paper emphasizes conduit flow of various kinds, in contrast with models of diapiric magma rise, which are based on concepts of wholesale buoyant instabilities and viscous or plastic flow (e.g. Marsh and Carmichael, 1974; Kuntz and Brock, 1977). In the present sense, magma rise is generically a form of percolation process involving the evolution and interactions of transient conduit paths. This view does not rule out diapiric mechanisms, but rather contends that the history of magma rise is influenced at fundamental stages by the relationships between the internal pressure of magma, rock stresses, and the ability of extensional fracture to produce flow paths. It is shown in the discussion of Figure 8 that this is not necessarily in conflict with the simultaneous existence of wholesale plastic or viscous flow.

E. M. ANDERSON ON FAULTING, STRESS GUIDES, AND MAGMA INTRUSION PATHS: SETTING THE BALL IN MOTION

During the same period as Bowen's work the studies of E. M. Anderson (1905, 1936, 1938) in Great Britain anticipated many of the modern developments in fracture mechanics applied to geology. He very nearly (if not actually, by implication) identified the concept of *effective stress*, later defined quantitatively by Terzaghi in 1924 as *the rock stress at a point minus the fluid pressure acting at that point* (see review by Skempton, in Voight, 1976). Taking an approach analogous to but independent from

Anderson's work, Hubbert and Willis (1957) and Hubbert and Rubey (1959) used Terzaghi's concept to develop, respectively, theories for hydraulic fracture and for the mechanics of low-angle thrust faults.[1]

From the standpoint of magma intrusion paths, Anderson's studies recognized the importance of relationships between stress trajectories in the intruded rock and the internal pressure of magma. This is indicated by some short quotations from Anderson (1936), who was commenting on his earlier interpretations of the Tertiary ring structures of Mull (Bailey et al., 1924). They are given here, despite the fact that the attention of this paper is not addressed to stresses in the immediate vicinity of previously formed magma chambers, because those remarks set the stage for the principal relations of this paper.

> One may suppose that at first the magma had a specific gravity equal to that of the surrounding rocks, and that it was under a pressure just high enough to have raised it to surface level if there had been an outlet. Overlooking the fact that the rocks themselves were not quite homogeneous, one may further suppose that they were under the same horizontal and vertical pressures at every point as would obtain in a liquid with the same surface and the same specific gravity. While such was the case no fractures could arise. (Anderson, 1936, p. 135.)

Here Anderson alluded to pressure balances between equivalent rock and magma columns and pointed out that fracturing is not expected without some deviation from equal-stress conditions (he used the term "pressure" in the sense of both fluid pressure and rock stress). The former observation also anticipated the question of maximum lift of magma, as later discussed by Eaton and Murata (1960) for Hawaiian eruptions and Shaw and Swanson (1970) for Columbia basalt eruptions.

Anderson then described the formation of cone-sheets and ring-dikes as follows (The "Figure 1" of his original description is included as Figure 1 of this paper):

> "*Cone-sheets.* If, however, these conditions were modified by an increase of pressure in the magma-basin, the pressure-system in the crust would have superimposed on it a system of tensions, acting across surfaces which near the basin were roughly conical. The fine firm lines in Figure 1 . . . are intended to show the intersection of these surfaces with the plane of the diagram. A superimposed system of pressures

---

[1] In this paper I prefer the term *extensional fracture* because it includes all mechanisms of fracture resulting in the opening of space into which any mobile phase or phases may be injected. The term *hydraulic fracture* often is restricted to cases involving the injection of water or other liquids.

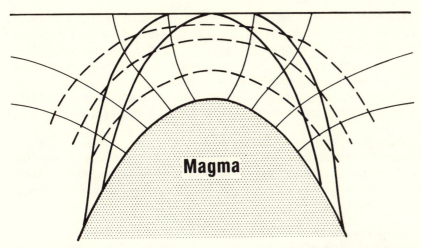

Figure 1. Stress trajectories and paths of magma injection, redrawn from Anderson (1936, Figure 5). The light full lines represent trajectories of maximum principal stresses and the dashed lines represent directions of minimum principal stresses, under conditions where the magma pressure exceeds the average lithostatic stress. Magma injection tends to follow the light full lines, normal to the directions of minimum principal stresses (forming cone sheets). For conditions of magma pressure less than the average stress, the signs of the stresses are reversed. In this case the planes normal to the directions of the minimum principal stresses (dashed lines normal to the light full lines) do not intersect the magma chamber. Injection can occur, therefore, only along planes at angles somewhat less than 90° to the minimum stress trajectories (forming ring-dikes). This condition is analogous to what is termed *extensional shear failure* in the present paper. (See text for discussion of stress fields and quantitative conditions of failure.)

would also act across surfaces which cut the former orthogonally and are indicated in section by the fine broken lines. The superimposed tensions, together with the increased pressure of the magma, might cause a series of fractures to develop, along which the magma would intrude." (Anderson, 1936, p. 135.)

"*Ring-dykes.* If the conditions were reversed and the pressure of the magma fell below that which was at first assumed, the original 'hydrostatic' pressure in the crust would be modified in a different way. Superimposed pressures would act across the surfaces whose trace is shown by the fine firm lines of Figure 1, and superimposed tensions across those which are indicated by the broken lines. It seems likely that, in this case, surfaces of fracture would originate inclined at an angle to the surfaces across which there were maximum superimposed tensions, as in the case of normal faults (Anderson, 1905)." (Anderson, 1936, p. 136.)

Here Anderson referred to the question of angular relations between directions of maximum principal stresses and planes of failure. The nature of this problem will be more evident later when failure criteria are discussed. Anticipating those relations, however, he intuitively identified a regime of what is called in this paper *extensional fracture with shear* (the ring-dikes), in contrast to *pure extensional fracture* (the cone sheets).

The condition of pure extensional fracturing is hinted at in the passage:

> *It appears, therefore, that there was no actual tension, but only a diminution of pressure, across the directions which have been followed by the cone-sheets.* These took the path of least resistance; that is, they opened fissures along planes across which the transverse pressure was least. (Anderson, 1936, p. 148; italics added.)

The italicized sentence anticipates the key principle of the present paper, the concept of *effective stresses*, already mentioned. That is, the directions of magma intrusion are governed by relations between the orientations and magnitudes of the principal stresses in the rock, as they are modified by the internal pressure of magma, to give relationships between sets of *effective stresses*. Anderson also recognized that these intrusion directions tend to be along planes approximately normal to the axis of least principal stress, and he deduced that suitable conditions for magma injection might obtain during regimes of either an expanding or contracting magma chamber.

Anderson's intent was explicitly acknowledged by Hubbert and Willis (1957, p. 162) in the following passage:

> "A phenomenon very similar to artificial formation fracturing but on a large scale is that of dike emplacement. It has been pointed out by Anderson that igneous dikes should be injected along planes perpendicular to the axis of least principal stress. This situation is entirely analogous to that for artificial formation fracturing. A striking field example of the effect of a regional stress pattern upon the orientation of igneous dikes is afforded by the Spanish Peaks igneous complex in Colorado."

Hubbert and Willis go on to gives a synopsis of the stress analysis for the Spanish Peaks by Odé (1957), followed by descriptions of hydraulic fracturing experiments quite analogous to the formation of dikes and sills emanating from a central volcanic neck.

The sequence in the growth of ideas concerning fluid-related failure mechanisms is interesting. Hubbert and Willis (1957) used the field examples of igneous intrusions to corroborate their deductions concerning "formation fracturing," and possibly even to guide their experiments. Subsequently, Hubbert and Rubey (1959) extended the ideas of effective stress in their theory of low-angle thrust faulting (or gravitational sliding of tectonic blocks), in which the formation pressure of ground water represented a controlling factor in the failure conditions. In the present paper it is similarly suggested that the formation pressure of magma in the asthenosphere plays an analogous role for plate tectonic motions. It seems obvious, in retrospect, that ideas of igneous fracturing processes have from the beginning played a major role in the development of concepts of fracturing mechanisms in the earth. These concepts can now be used, in turn, to show that igneous fracturing mechanisms play a more important part in the fracture history of the earth than is indicated by applications to isolated intrusive events.

Anderson (1936) and Hubbert and Willis (1957) gave mathematical analyses of the stress trajectories that would be generated in a rock by local pressure sources of fluid. The present paper, on the other hand, takes the approach that intrusive and extrusive traces of magma flow paths map the stress trajectories acting at the time of transport, and thus give information on relative orientations and magnitudes of the acting principal stresses. The patterns and angular relationships of these injection paths, then, form the observational data for dynamical interpretations based on other knowledge of the petrology and the material behavior, including seismic mechanisms in the case of active systems like Hawaii.

VOLCANIC FEATURES AS INDICATORS OF STRESS TRAJECTORIES

The principles introduced by Anderson (1936), and applied to patterns of radial dikes from a central volcanic neck by Odé (1957), have recently been quantified in more detail by Muller and Pollard (1977). It is clear that systematic relationships between magma sources and induced stresses have been established. As demonstrated by Pollard (1976, 1977), however, there are many details important to the interpretation of dike system evolution that have yet to be worked out. Questions exist about the en echelon character of dike systems, about dike lengths and shapes, the effect of proximity to the earth's surface, and the relations between magma pressure gradients, rock stress gradients, and conditions of dike propagation or collapse. The principles of *stress intensity factors* and *fracture toughness* offer important new insights on these questions (Pollard and Muller, 1976). Of immediate concern in this paper, however, are the more general criteria of extensional fracture occurrences and the orientations and magnitudes of tectonic stress fields.

The types of stress trajectories referred to above were supposedly induced by the action of the fluid pressure of magma in an established chamber or conduit. This process represents a limiting regime of super-position of stresses by the magma on a previously existing reference state of stress; Koide and Bhattacharji (1975) have discussed stress systems of this type in some detail. The inverse limiting regime is one in which the state of stress in the rock is primarily controlled by factors other than those produced by the magma itself, and the intrusion paths accommodate themselves to such stress states in terms of the effective stresses implied by local values of fluid pressure. All actual systems involve aspects of both regimes—which one dominates depends on whether the principal source of mechanical work on the system is imposed by magma pressure or by other forces (action of gravitational forces, tectonic compression, etc.). Subsequent examples in this section of the paper refer mainly to the latter situations.

The use of volcanic features as indicators of tectonic stress regimes was suggested by Nakamura (1969) and was invoked by Shaw and Swanson (1970) in the context of the failure criteria discussed in this paper. Fiske and Jackson (1972) experimentally demonstrated the principles of fluid injection in the context of gravitational stresses acting on volcanic edifices during the growth of composite shield volcanoes. Figure 2 illustrates some of these effects. Initially, the direction of magma injection is influenced mainly by regional tectonic stress orientations. These directions are later modified by the buttressing effects of previously formed shields as the edifice stresses evolve. In these examples, the injected fluid (experimental fluid and magma) is in the form of thin planar blades or tongues that have minimal influence on the volumetric strain of the injected material, except for the local effects of failure along the injected plane.

Jackson and Shaw (1975) extended this form of inference to stress trajectories on the scale of global plate motions. They proposed that linear volcanic chains, such as the Hawaiian-Emperor system in the Pacific, identify stress trajectories normal to the least principal stress in the plate. In a similar manner, Nakamura et al. (1977) inferred principal stress orientations in the Aleutian arc from the pattern of systematic alignments of volcanic vents. Inferred stress trajectories in the two cases appear to be reasonably consistent with plate motions. Figures 3 and 4 illustrate those conclusions.

Although many problems remain, it appears to be reasonably well established, by several workers on several different scales of observation, that systematic relationships exist between tectonic stress orientations and magma injection paths.

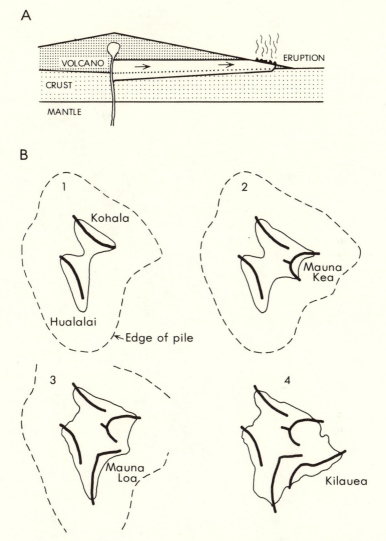

Figure 2. Magma injection paths in the evolution of rift zones during growth of the basaltic shield volcanoes of the Hawaiian Islands (modified from Fiske and Jackson, 1972, Figures 5 and 11). (A) Schematic path of magma injection from a central conduit during extensional fracture event or events governed by gravitational edifice stresses. (B) Stylized chronological sequence in the growth of rift zones (oldest to youngest stages from 1 to 4) during evolution of the shield volcanoes of the island of Hawaii. The positions of Mauna Kea, Mauna Loa and Kilauea mark the locations of the center of the next youngest shield and the formation of new rift zones. These were initially oriented according to edifice stresses induced by the existence of the older shields, and subsequently influenced by the combination of these stresses with gravitational stresses created by accumulation of the new shield.

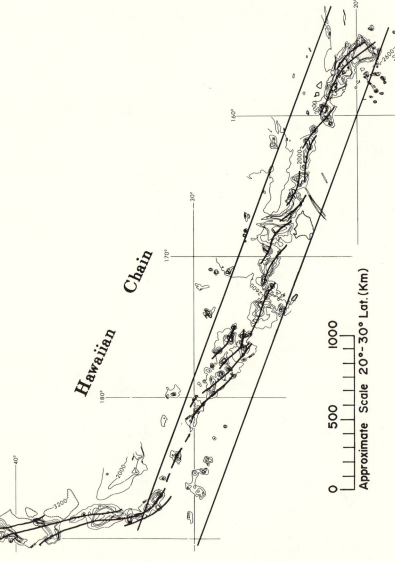

Figure 3. Volcanic loci and schematic outline of the mantle melting anomaly during growth of the Hawaiian chain (after Jackson and Shaw, 1975). The sigmoidal loci connecting successive centers of venting along the chain are considered to represent the surface traces of planes normal to the directions of the minimum principal stresses acting in the Pacific plate at the time of growth of each volcano. (See discussion in text, in Jackson and Shaw, 1975, and in Jackson et al. 1975.)

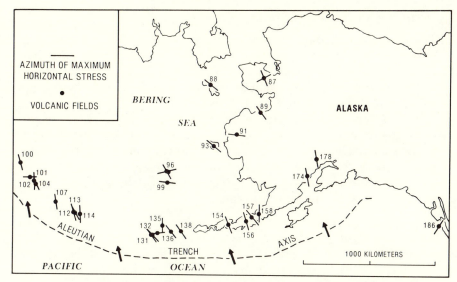

Figure 4.   Orientations of maximum horizontal stress trajectories in the Aleutians and Alaska (simplified from Nakamura et al., 1977, Figure 3; see this reference for details concerning locations of volcanoes and criteria for determination of azimuths). Arrows mark the present direction of motion of the Pacific plate relative to the North American plate after Minster et al. (1974).

FRACTURE MECHANICS APPLIED TO MAGMA ASCENT WITH SPECIAL REFERENCE TO HAWAII

The material given in this section is an attempt to complete a conceptual connection between discussions of surficial volcanic lineaments, with their implicit stress trajectories, and discussions of mantle processes. The author's viewpoints on these phenomena are expressed in other reports (Shaw, 1973; Shaw and Jackson, 1973; Jackson and Shaw, 1975; and Jackson, Shaw and Bargar, 1975). Several other models of Hawaiian volcanism exist: mine are shown here to give some pictorial continuity to conceptual frames of reference, and to serve as reference states for the discussion of types of stress regimes and tectonic domains. The correctness of any of the models is not the present issue.

This section defines *effective stress*, outlines the model for Hawaii and some of the types of control of the pressure head of magma there, and discusses several aspects of failure in partly molten rock. Symbols and definitions basically follow Hubbert and Rubey (1959) and Secor (1965); also Secor (1969) and Secor and Pollard (1975).

DEFINITION OF EFFECTIVE STRESS

The reader is referred to Hubbert and Rubey (1959, p. 139) for a thorough discussion of the decomposition of stress fields into component parts and a rigorous definition of *effective stress* following Terzaghi (1936). Other key references, with commentaries relevant to this concept, have been collected in the volume by Voight (1976).

For the present discussion, stresses across an arbitrary plane are expressed in terms of a normal component, S, and a shear component, τ. The action of an interconnected fluid pressure at any point results in a neutral stress field that produces no deformation, and a residual stress field, the normal components of which are called the *effective normal stresses*, defined for the arbitrary plane by

$$\sigma = S - p, \tag{1}$$

where p is the fluid pressure. The components of shear stress, τ, are unchanged by the fluid pressure. Following Hubbert and Rubey (1959) and Secor (1965), *compressive stress and fluid pressure are defined as positive quantities.*

A MODEL FOR HAWAII; ORIGIN OF THE FLUID PRESSURE
DISTRIBUTIONS OF MAGMA

Figure 5 is a schematic longitudinal cross section combining several features of the models for Hawaii cited above. The sketch is not drawn to scale and only identifies some of the options for stress states discussed below. Also, because stress states in this dynamic system are time-dependent, the sketch represents only one of several possible transient situations.

Figure 5 is divided into several types of tectonic regions or *stress domains*, each being associated with a characteristic, if vague, *volume domain*. The term *domain* is used in this context, or is qualified relative to even more local parts of the system. The term *magmatic system* is similarly vague, but generally it refers to sets of interacting domains on some specified scale. The *Hawaiian system* refers to the entire range of phenomena schematically portrayed in Figure 5.

The complexity of factors controlling the fluid pressure of magma can be appreciated by imagining the time-dependent changes possible in a system like that portrayed in Figure 5. Pressure magnitudes could conceivably range from negative values to values greater than the average lithostatic stress. For example, thermally induced changes in relative volumes can bring about a state of isotropic tension in liquid inclusions in crystals. Conversely, increased heating of a region containing an isolated pocket of magma that had previously reached approximate mechanical equilibrium with its surroundings could produce a local excess-pressure anomaly. This could happen without failure only if the stress differences

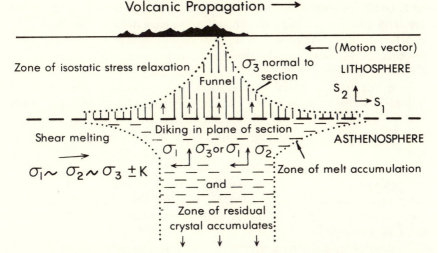

Figure 5. Schematic longitudinal section showing classification of stress domains in the Hawaiian system, based on ideas advanced in Shaw (1973), Shaw and Jackson (1973) and Jackson and Shaw (1975). The symbol S represents the total normal stress acting at a point in the rock, and σ represents the *effective normal stress* acting in rock containing an interconnected fluid phase. Subscript 1 refers to the maximum principal stress, and subscripts 2 and 3 refer, respectively, to the intermediate and minimum principal stresses.

in the rock were small compared with the tensile yield stress, as shown by criteria given below. This paper treats more typical pressure limits; however, the possibility of such widely ranging pressure conditions must be kept in mind.

Shaw and Swanson (1970, p. 287) have considered limiting factors in the control of pressures for flood-basalt eruptions. Following an event of dike formation (an extensional failure event), two simple limits might describe the pressure history of magma during injection: (1) stress adjustments in the rock matrix at the source are so rapid that a nearly constant pressure potential is maintained until all available magma is depleted, and (2) stress adjustments at the source are so slow that the magnitude of the pressure potential drops in direct proportion to the expansion represented by the injected volume.

In either case, another potential control is the rate of production of magma at the source. Consider one example (a), in which the melting rate is high. In this case, limit (1) would be analogous to a supply of magma arriving through a conduit continuous from source to surface, and fed by an infinite reservoir at constant pressure. Limit (2) would produce large, intermittent magma pulses, the volume and timing of which would depend mainly on reservoir volume and on rates of isostatic recovery of pressure at the source. These situations represent systems overdriven, so to speak, by continuous or intermittent volume sources of magma that are large relative to the instantaneous volumetric rates of injection. The exact criteria of extensional failure of the lithosphere are of secondary importance in these hypothetical cases, particularly (a-1), because the conduits stay open from source to surface during the history of flow subsequent to the initial propagation event. Criteria of extensional failure, dike propagation, and closure affect case (a-2) primarily with respect to the timing of the onset relative to the isostatic pressure cycle, and the time duration of the magma pulse before the pressure head is spent.

The limit in example (a-2) was considered by Shaw and Swanson (1970) to provide a reasonable working model for Columbia River basalt eruptions; Swanson, Wright, and Helz (1975) described the results of subsequent detailed field studies that quantify and generally confirm these deductions.

In another example (b), at the opposite extreme, the melting rate is so low that in case (b-1) the magma might never reach the surface, because there is insufficient thermal energy to overcome cooling effects during intrusion. Similarly, in (b-2), either cooling effects dominate, or extremely long times are required to produce a large enough magma reservoir to provide the potential expansion volume for a magma pulse to reach the surface during one event of extensional fracture.

Because cooling effects are dominant at low transport rates, magma pressure fluctuations tend to be more extreme than in cases (a-1) and (a-2). The situation in limit (b) may bear some resemblance to the problems involved in formation of small volume systems, such as those exemplified by lamprophyre dikes, some types of mafic volcanic necks, kimberlite pipes and diatremes. In many of these systems, of course, a gas phase plays a key role, but the present paper is restricted to discussion of two-phase magma consisting only of liquid and crystal phases. Anderson and Grew (1977) consider these analogous problems in some detail for systems with a gas phase, from the standpoint of stress-corrosion theories. They apply their concepts to the formation of kimberlite pipes.

There are many possible variants of model types combining features resembling the above four simple limits. There is one combination, example (c), that is referred to here as the quasi-steady model. In this model, isostatic adjustment of stresses at the source are rapid enough to provide sufficient pressure potential for magma intrusion to occur at roughly the rate of magma generation. In addition, the generation and intrusion rates are high enough, or have been going on long enough, that magma reaches the surface, without solidifying, at rates that also approximately balance the melting and intrusion rates (except for some fraction involved in the growth of the intrusive underpinnings of the volcanic edifice).

In model (c), there is no opportunity for accumulation of large magma reservoirs near the source, and individual magma pulses are not large enough to reach from the source to the surface during one extensional failure event. The number of events, however, is sufficiently large and frequent that a sequence of pulses moves upward in a discontinuous progression of interacting diking events. The frequency of these events is controlled by the local pressure fluctuations (up or down depending on the cooling-heating, addition-subtraction effects of magma increments) and the stress criteria of extensional failure relating to the local tectonic stress field and the effects of liquid-crystal proportions on rock strengths.

In this model, the fluid pressure of magma oscillates around a mean value approximating the average lithostatic stress. The magnitudes of pressure deviations from the mean are roughly defined by the range of yield strengths of the rock. These stress ranges, in turn, are determined in part by the local proportion of liquid magma; estimates are given later for tensile strength versus crystal-liquid ratios of basaltic magma.

If the balances were precisely steady in example (c), the result would be a continuous magma conduit with a continuous "artesian" volcanic extrusion at the surface that exactly matched the melting rate in the mantle. The steady state would also imply that fluid pressure gradients in the conduit were precisely balanced against rock stress gradients in the walls. The unlikelihood of maintaining such delicate balances is apparent in relation to even the simplest petrological models of the lithosphere and of plate tectonics, particularly when it is realized that, simultaneously, the volcanic edifice is growing, and probably propagating laterally, and that both the lithosphere and asthenosphere are in relative motion. Problems related to gradients of pressure and stress are addressed by Secor and Pollard (1975) and by Pollard (1976).

It is probably evident at this point that, even though model (c) is a hypothetical description of one combination of possible limiting factors influencing the pressure distribution, it is thought to describe some of the effects analogous to those operating in the Hawaiian system as portrayed in Figure 5.

The Mohr Diagram and Coulomb Failure Criterion

For the purpose of outlining the types of stress regime relevant to extensional fracture, the simplest graphical approach is given by the Mohr diagram of normal and shear stresses. Figure 6 shows the relationships of greatest and least principal effective stresses to normal and shear stresses, and illustrates one form of failure criterion, "Coulomb's law of failure," which is roughly applicable to some types of rocks (Hubbert

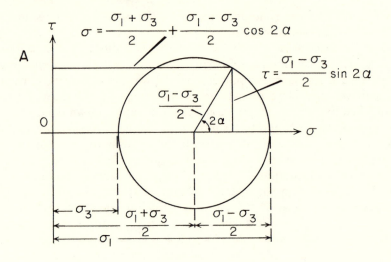

A

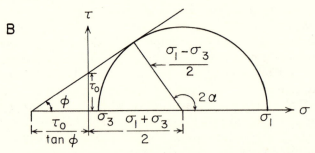

B

Figure 6. Trigonometric relations among normal stresses, shear stresses and maximum and minimum principal stresses represented by Mohr diagrams (redrawn from Hubbert and Rubey, 1959, Figures 1 and 3; see this reference for mathematical discussion of decomposition of stresses). *Diagram A* shows a complete Mohr circle and the angular relationships of points on the circle to maximum and minimum principal stresses. *Diagram B* illustrates these relations in terms of points on the circle tangent to a locus of values (σ, τ) representing conditions of shear failure; in this case the locus is a straight line, called the "failure envelope," representing Coulomb's law of failure (see Voight, 1976). The angle 2α is twice the angle between the direction of the minimum principal stress and the failure plane; α in this example is 60°. The angle ϕ is called the angle of internal friction. In this example ϕ is 30° and is equal to the angle between the failure plane and the direction of the maximum principal stress.

and Rubey, 1959, p. 124). This law is stated, in terms of *effective normal stress*, σ, in the form:

$$\tau = \tau_o + \sigma \tan \phi, \tag{2}$$

where the angle ϕ is called the *angle of internal friction* and $\tan \phi$ is called the "*coefficient of internal friction*". The intercept τ_o is in the nature of a *cohesive strength* when $\sigma = 0$; that is, when the fluid pressure, p, equals the normal stress, S. Brace (1969) has shown experimentally that the law of *effective stress* holds for the critical conditions and form of failure of a variety of rocks as long as the strain rate is not too high (e.g., less than 10^{-7} per second for Westerly granite with water as the fluid).

In Figure 6B, the intercept of the *failure envelope* with the σ axis identifies the limit of *cohesive strength* as a critical value of purely tensile stress. This condition occurs in terms of *effective stress* only if $p > S$. Hubbert and Rubey (1959, p. 142) argued that this situation is unlikely for highly jointed rock and assumed that $\tau_o = 0$; that is, the failure envelope for this case starts at the origin. Hsü (1969) took exception to that assumption and emphasized the importance of τ_o in overthrust faulting.

I believe that this issue depends essentially on the nature of the rocks and dynamics of the pore fluid conditions, and cannot be generalized. The Hubbert-Rubey analysis is one limit in a family of critical conditions for sliding. In the present discussion τ_o plays an important role because it is related to tensile strength and hence to melt fraction in partially molten rock, as will be shown. In terms of the Coulomb criterion (Figure 6B) the *tensile strength*, symbolized $-K$ following Secor (1965), is defined as

$$-K = \frac{-\tau}{\tan \phi}. \tag{3}$$

The Griffith Theory of Fracture

Many other studies have shown that the failure envelope often is not a straight line. Another form of failure criterion is given by the *Griffith theory of fracture* (Griffith, 1921, 1925), which is more applicable than the Coulomb theory to many materials. It is the criterion used by Secor (1965) to discuss *extension fracture* in crustal rocks (see this reference for fracture terminology); it is also assumed here to be more representative for partly molten rocks.

The Griffith theory actually refers to a method of analysis based on the energy balances involved in the growth of an existing crack. The fact that most rocks and minerals contain microcracks and flaws of various kinds makes this theory a logical starting point for discussion of crack propagation, particularly for rock containing a melt phase in the form of intergranular films or as a plexus of fluid-filled fractures.

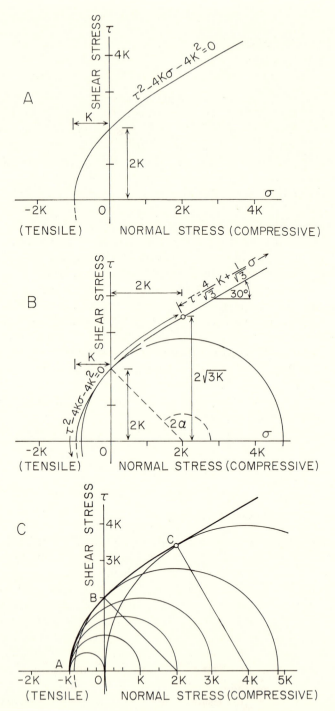

Figure 7. Mohr diagrams illustrating the Griffith and composite Griffith plus Coulomb failure laws (redrawn from Secor, 1965, figures 1, 2 and 3). *Diagram A* shows the

Many references discuss the nature of "Griffith cracks" and the application of Griffith's analysis of fracture; a detailed mathematical derivation is given by Lawn and Wilshaw (1975, p. 5 ff.), and a comparison of the relationship between the Griffith and Coulomb criteria of failure is given by Obert and Duvall (1967, p. 307–310) in terms of the Mohr representation.

The Griffith theory form of the failure envelope is defined by (Secor, 1965, p. 636):

$$\tau^2 - 4K\sigma - 4K^2 = 0. \tag{4}$$

The relationships of normal and shear stresses to this form of the failure envelope and to tensile strength, $-K$, are shown in Figure 7, taken from the detailed graphical constructions in Secor's paper. From Figure 7A it is seen that the shear stress for $\sigma = 0$ is given by

$$\tau_o = 2K, \tag{5}$$

and the tensile strength is therefore $-K = -\tau_o/2$ instead of the more negative value given by Equation (3). At values of normal stress above about $2K$ the failure envelope is assumed to be linear, in conformance with Coulomb's law. Therefore, it is referred to as a composite failure envelope. Figure 7B illustrates the Mohr relations for the composite diagram; Figure 7C identifies the corresponding regimes of failure from pure extension to pure shear failures.

Tensile Strength of Partly Molten Rock

The relationship in Figure 7, together with Equation (5), provides a simple way to estimate tensile strengths from viscometric studies of

parabolic form of the Griffith failure envelope and the relationship of intercepts on the σ and τ axes. *Diagram B* shows the composite envelope and the equations of the parabolic and linear portions. *Diagram C* shows a family of Mohr circles tangent at different points along the composite envelope: circles tangent at A theoretically give purely extensional failure along planes 90° to the σ_3 axis; circles tangent at points between A and B give failure involving both extension and shear and failure angles between those of pure extension and pure shear (theoretically, at angles between 90° and 70° to the σ_3 axis). The circle tangent at point C theoretically involves only compressive shear (with no effectively tensile components anywhere in the material) acting along a plane at a theoretical angle of 60° to σ_3. In actual materials, this limiting angle for shear failure may be closer to 70° (Secor, 1965, p. 639). The regime corresponding to points between A and C in diagram C is termed in this paper *extensional shear failure*. Strictly, this regime terminates at B relative to potential opening across the theoretical failure plane, but failure phenomena between B and C are considered to represent important transitions between the fields of dominantly compressive and dominantly tensile failures.

magma (Shaw et al., 1968; Shaw 1969). The estimates, however, are prefaced by qualifications concerning failure criteria and deformation regimes. Formation of fractures is a process involving complicated local motions and changes of state. "Simple shear," "pure tensile failure," and the various forms of failure envelopes are idealizations. Shear involves phenomena like dilatancy, and tensile failure involves local phenomena of shearing nature. These processes are usually dependent in some way on strain rates, as is shown by Brace (1969) for fracture criteria and as emphasized by Shaw et al. (1968) relative to rheological models of magma flow. Therefore, the validity of any criterion of tensile strength must be qualified by strain rate regime. At present this cannot be done quantitatively for partly molten rock, and in this paper the values of $-K$ estimated for discussion refer to incompletely specified strain-rate conditions appropriate to extensional fracture, whatever they may be. The studies of Brace (1969), however, suggest that the range of deformation rates in the asthenosphere and lithosphere represent suitable regimes for the application of the effective stress law.

The viscometry experiments of Shaw et al. (1968) in basalt magma represent conditions where the normal stress perpendicular to the direction of shear and the fluid pressure in the suspension (20 to 25 percent crystals by volume) are approximately equal; that is, departures from "hydrostatic" equilibrium were not large. Therefore, in terms of the above stress analyses, the shear resistance is analogous to a cohesive strength, τ_o. Those authors also found that the shear behavior in this case was best described by the Bingham model, representing one model of plasticoviscous materials with a "yield stress". This yield stress, designated ψ by Shaw et al. (1968), represents the limit of shear stress at vanishing shear rates and is therefore equivalent to a critical failure stress, or to a point on a failure envelope in a Mohr diagram. Combining the conditions $\psi \simeq \tau_{crit}$ with $S \simeq p$, leads to the conclusion that $\psi \simeq \tau_o$. Therefore, according to Figure 7, $\psi \simeq 2K$, and the value of tensile strength for this material is approximately $-\psi/2$.

Figure 8 was constructed partly on the basis of the above relationship. The limiting value of tensile strength for zero melt fraction was estimated from ranges of average shearing stress for the asthenosphere, with the recognition that some models infer that some amount of melt is present. Shaw (1973, Table 3) used energy balance calculations to estimate a locally acting shearing stress of 120–170 bars during melting episodes in Hawaii. Smaller *average* stresses have been estimated from models of plate motion, depending on plate speed (e.g. Schubert and Turcotte, 1972; Schubert et al., 1978), so a range from 2 to 200 bars ($K = 1$ to 100 bars) was used in Figure 8. The estimate for material roughly half molten is based on the occurrence of gash-fractures, interpreted to represent extensional fracture events, in ponded basaltic lava bodies of

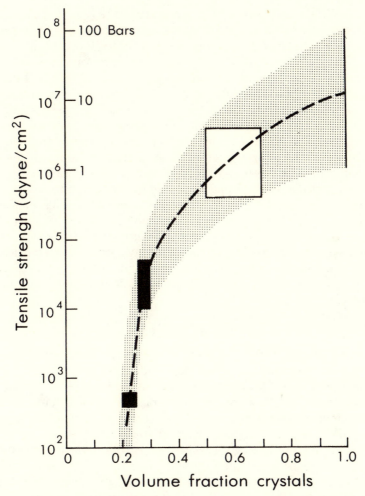

Figure 8. Approximate magnitudes of tensile strength of basaltic magma as a function of crystal-liquid ratios (see text for discussion of relationship between tensile strength and shear strength). The solid boxes are based on data for shear strength of basalt versus melt fraction from experimental data of, from left to right, Shaw et al. (1968) and Shaw (1969). The open box represents stress ranges estimated by the author for depth ranges and melt fractions related to the formation of melt-filled gash fractures in the crust of ponded lava bodies in Hawaii (e.g., Wright and Fiske, 1971, Appendix 3). The vertical bar at 100 percent solid is based on ranges of estimated shear strengths of the asthenosphere deduced from models of plate tectonics (e.g. Shaw, 1973; Schubert and others, 1978). The dashed line is a subjective reference curve for failure stresses discussed in the text. The stippled region indicates estimated ranges of uncertainty dependent on complicated factors of rock composition, texture, strain-rate regimes, and so on. The higher strength ranges tend to be more appropriate to portions of the lithosphere and the lower strength ranges to portions of the asthenosphere, though local regions in either environment may involve the entire spectrum of strength magnitudes.

Hawaii (Wright and Fiske, 1971, Appendix 3). It is estimated that the effective tensile stress for the depth of occurrence would have been in the range 4 to 4 bars for melt fractions from 30 to 50 percent by volume. The estimate of strength at a melt fraction of 70 to 75 percent was obtained from stress ranges found in the laboratory by Shaw (1969), and the estimate at 75 to 80 percent melt was based on the *in situ* measurements of Shaw et al. (1968). Whereas the estimated ranges of tensile strength in Figure 8 are each ambiguous, they are based on five different kinds of observations that form a reasonably consistent pattern. It is possible that for some conditions the estimates of strength are too high, and that the change of slope at 50 to 70 percent melt is more abrupt.

The existence of fracture phenomena in partly molten rock is not new to igneous petrologists. I have seen many examples of what appears to be simultaneously brittle and plastic phenomena in the plutons of the Sierra Nevada and New England and have been told of many similar examples elsewhere. Brittle phenomena in Hawaiian lavas are particularly notable; the transition between pahoehoe and aa lava types represents yet another regime of fracture behavior in addition to the gash-fractures mentioned.

Failure Criteria and Depth of Tensile Cracks

The implications of the foregoing examples of Mohr diagrams and failure criteria depend on stress magnitudes and orientations, and on the distribution and pressure of a fluid phase. Examples of these constraints relative to Hawaiian magmatism are outlined below, without any attempt to give a rigorous classification of deformation regimes.

In general, one of the three principal stress directions is expected to be vertical, on the average, because the earth's surface approximates a horizontal plane acted upon by zero shear stress. Extension fractures, in the strict sense, form when a Mohr circle (Figure 7) becomes tangent to the failure envelope at its intersection with the abscissa ($\tau = 0$). Therefore, such fractures will be either horizontal or vertical and are normal to the σ_3 axis. The largest Mohr circle satisfying this condition, relative to a Griffith failure envelope, is centered at $+K$ and has the values $\sigma_1 = +3K$ and $\sigma_3 = -K$ (see Figure 7C).

Extensional Failure Regimes

As pointed out by Secor (1965, p. 369), the angle of failure progressively departs from a value of 90°, measured from the σ_3 axis, as Mohr circles are increased in diameter (greater differences in maximum and minimum principal stresses). The circles tangent at, respectively, points *A*, *B* and *C* of Figure 7C involve theoretical failure angles α decreasing from 90° to about 70° relative to the σ_3 axis. The circle tangent at point *C* is centered at $+4K$ with $\sigma_1 = +8K$ and $\sigma_3 = 0$. At larger values of effective stress, all stresses are compressive. Therefore, point C marks a position on the failure envelope beyond which the opening of cracks anywhere in the material theoretically cannot occur (with qualifications relative to dilatant shear and the like). Between points A and C, failure involves a shear component plus a potential extensional component and variable failure angles (see comments by Secor, 1965, p. 638). This range of failure phenomena is referred to here as *extensional shear fracture*.

The following figures will give these ranges some quantitative perspective. If the value of K is about 10 bars (about 0 to 10 percent melt according to the dashed line of Figure 8) the regime of pure extensional fracture is represented by values of σ_1 between 10 and 30 bars for $\sigma_3 = -10$ bars, and extension fracture with shear is represented by σ_1 between 30 and 80 bars with σ_3 increasing from -10 bars to zero (that is, $S_3 = p$).

Relation of Principal Stresses to Inferred
Plate Tectonic Stresses and Gravitational Stress

The approximate total normal stress in the vertical direction is, on the average, given by

$$S_{\text{vert}} = \bar{\rho}gd \qquad (6)$$

where $\bar{\rho}$ is the mean density, g the acceleration of gravity, and d the depth. This is not necessarily true, of course, in buttressed regions or domains caught in some sort of tectonic bind, but it is the only condition that has some claim to being typical (it is presumably true in regions of isostatic compensation).

From the preceding discussion, S_{vert} has to be the minimum principal stress, S_3, if horizontal or subhorizontal extensional fractures (for example, sills) are to form; otherwise, it is either the intermediate or maximum principal stress when extensional fractures are roughly vertical. The conclusion that either the S_1 or S_2 axis is often vertical is made simply because magma reaches the surface through volcanic dikes and vents, implying that S_3 is approximately horizontal. On the other hand, it does not seem likely that S_1 can be typically vertical in the earth, because in that case both of the horizontal stresses are smaller than the mean gravitational stress, S_{vert}; that is, if this were true there would be a general tendency for relative horizontal extension in all horizontal directions contrary to abundant geologic evidence that compressional tectonics is active in many places. Therefore, a common condition seems to be that S_{vert} is the intermediate principal stress, S_2, and that S_3 and S_1 are horizontal, with S_3 normal to volcanic lineaments, which trend in the direction of the trajectory of maximum principal stress, S_1.

The above general inferences agree with the regional interpretations by Jackson and Shaw (1975) for the Hawaiian-Emperor volcanic loci and with the interpretation of Nakamura and others (1977) for trends in the Aleutian arc. Models of gravitational sliding of plates are also consistent with these interpretations (Jacoby, 1970).

As the resolution of stress fields becomes more and more focused, these generalizations become less and less appropriate. Eventually, local effects like those of Figures 1 and 2 begin to dominate. On an even more local scale, failure criteria involve analysis of stress concentrations in the vicinities of crack tips, as discussed by Pollard (1976, 1977).

Depth Limits of Extensional Fracture

Following Hubbert and Rubey, (1959, p. 142) and Secor (1965, p. 640), the ratio of fluid pressure to the vertical principal stress is symbolized by λ:

$$\lambda = \frac{p}{S_{vert}} = \frac{p}{\bar{\rho}gd}. \tag{7}$$

Therefore the vertical effective stress can be written

$$\sigma_{vert} = S_{vert} - p = \bar{\rho}gd(1 - \lambda), \tag{8}$$

and equivalently, when S_2 is vertical,

$$\sigma_2 = \bar{\rho}gd(1 - \lambda). \tag{9}$$

Defining the ratio $n = \sigma_1/\sigma_2$, Equation (9) can also be written

$$\sigma_1 = n\bar{\rho}gd(1 - \lambda). \tag{10}$$

Recalling that the largest value of maximum principal stress for pure extension fracture is $+3K$ (from Figure 7C), the maximum corresponding depth calculated from Equation (10), is found to be

$$d_{max} = \frac{3K}{n\bar{\rho}g(1 - \lambda)}. \tag{11}$$

For example, if σ_2 were exactly intermediate between σ_1 and σ_3, then, for $\sigma_1 = +3K$ and $\sigma_3 = -K$, $\sigma_2 = +K$ and $n = 3$. This gives the relation

$$d_{max} = \frac{K}{\bar{\rho}g(1 - \lambda)}. \tag{11a}$$

Using the previous example of $K = 10$ bars, and letting $\lambda = 0.99$ (fluid pressure equal to 99 percent of S_{vert}) gives $d_{max} = 3$ kilometers. But note that as $\lambda \to 1$, $d_{max} \to \infty$.

The last two sentences illustrate the extreme sensitivity of the extensional fracture depth to slight variations in pore-fluid pressure when that pressure approximates the magnitude of the gravitational load. *It is concluded that under these conditions, magma injection by extensional fracture can occur at any depth in the mantle without any specific limit on the minimum fraction of melting required.*

When σ_1 is vertical, d_{max} is three times the value in Equation (11a). On the other hand, if σ_3 is vertical, the relation becomes

$$d_{max} = \frac{-K}{\bar{\rho}g(1 - \lambda)}. \tag{12}$$

In order for d_{max} to be a positive quantity in Equation (12), λ must be greater than unity, or $p > S_{vert}$. This implies that sill-like extensional fractures represent anomalous relations of fluid pressure to gravitational loading, although several mechanisms (like buttressing, volumetric adjustments at depth, and the like) could satisfy that condition. (For further discussion, see Note at end of paper.)

The above relations are summarized in Figure 9, after Secor (1965). Increasing the fluid pressure decreases the effective stresses (Figure 9A) until eventually the Mohr circle becomes tangent to the failure envelope in the tensile region. Extensional fracture then occurs according to one of the modes of Figure 9B at some maximum depth indicated by curves like that in Figure 9C.

Stress Conditions Related to Melting in the Asthenosphere

When rock in the asthenosphere begins to melt, it changes volume according to the thermodynamics of the melting reaction. For most source rocks this change is positive and is on the order of 10 percent at constant pressure. The thermodynamics of melting in a nonhydrostatic stress field, following the Gibbs theory (Nye, 1967), indicate that for the above kind of melting relation, the first increments of melting begin to form on planes normal to the "most tensile principal stress" (Savage, 1969). This condition appears to be consistent with the extensional failure conditions discussed. *In fact, the melting reaction itself represents the first in a sequence of rock failures related to the orientation and magnitudes of the stress components.*

Considering that, according to theories of plate tectonics, the asthenosphere is in a state of shearing deformation relative to horizontal directions of transport, and that the creation of a melt phase induces a fluid pressure equal to or greater than the mean stress, then at least one of the principal stresses is reduced to zero or negative values. These observations specifically limit the conditions of failure, according to representations like those of Figures 7A, B, and C. In terms of that failure envelope, the regime of combined shear and creation of extensional openings (initially in the form of melt films, "schlieren," and the like) is bounded by values of σ_1 between $+4.8K$ and $+8K$, values of σ_3 between $-.8K$ and zero, and values of shear stress, τ, between about $2K$ and $3.3K$. If $K = 10$ bars, then $48 \leq \sigma_1 \leq 80$ bars, $-8 \leq \sigma_3 \leq 0$ bars, and $17 \leq \tau \leq 33$ bars. Larger average values of K, deduced from Figure 8 for lower average values of melt fraction in the asthenosphere, of course imply larger ranges of

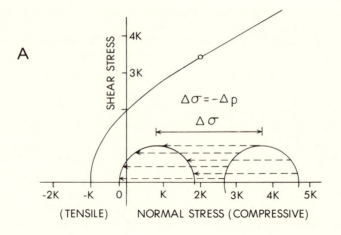

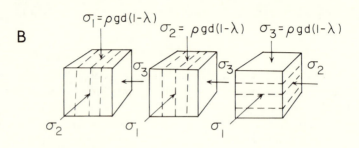

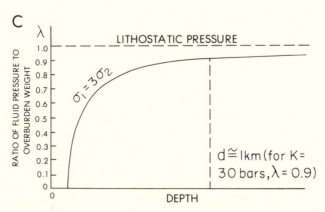

Figure 9. Resumé illustrating: (A) effect of increasing fluid pressure in reducing effective stress relative to the composite failure envelope, (B) orientations of principal stresses and potential extensional fracture planes, and (C) depths consistent with extensional fracture for conditions of the intermediate principal stress oriented vertically and equal to the gravitational load. The diagrams are redrawn and modified from relations illustrated by Secor (1965, Figures 4, 6, and 8). The depth curve in diagram C represents Equation (11) in the text; the indicated values of K and λ simply identify the characteristic depth scale of 1 km. Families of curves like diagram C converge toward unlimited depths as λ → 1.

stress. For values of $\sigma_1 < +3K$ and for $\sigma_3 = -K$, failure is purely tensile and $\tau = 0$.

The fabric of melt distribution in the asthenosphere depends on the relation of σ_2 and σ_3. If σ_2 is vertical, and the above conditions hold, then the melt segregations are planar with angular deviations of 20° or so from the vertical. If the pore pressure, p, can become slightly greater than S_{vert}, or if S_{vert} is less than the gravitational load, then extensional failure with σ_3 vertical is possible, and planar segregations of subhorizontal orientation could occur. For example, from equation 12, if d_{max} is about 100 kilometers and $K = 10$ bars, then $\lambda = 1 + 3.3 \times 10^{-4}$. Because the gravitational load at this depth is about 30 kilobars, this simply shows that the excess fluid pressure is numerically equal to the tensile strength, or 10 bars. Such small excursions from the steady state seem likely, and the resulting picture of the asthenosphere fabric is one of conjugate sets of planar melt schlieren, at relative angles of 50° to 90° from each other, created in balance with the dynamic rate of melting by the dissipation of shearing work. Shaw (1973) discussed melting rates and volumes calculated from the work rate of shear in the asthenosphere.

The above relations are key links between concepts of plate motion, considered in terms of a modified Hubbert-Rubey mechanism, and mechanisms relating the genetic conditions of melting with those of magma ascent through the lithosphere.

Stress Conditions Related to Melt Injection of the Lithosphere

The stress field orientations delineated by volcanic loci, as discussed by Jackson and Shaw (1975), are assumed to represent the long-term boundary conditions for extensional failure of the lithosphere. Stress magnitudes are indicated by the range of tensile strengths in Figure 8 applied to failure criteria in Figure 7. If tensile strengths as great as 100 bars or so exist, then maximum principal effective stresses consistent with extensional shear failures approach 1,000 bars, and shear stresses may somewhat exceed 300 bars.

Other evidence relevant to dynamic states in the lithosphere is available from seismic data and from magma transport rates, as deduced from the volcanic record of eruption rates. This evidence is explored in the next section. First, however, I will touch again on the hypothetical domain structure of the Hawaiian system, as portrayed in Figure 5.

The dynamic states in the asthenosphere, outlined in the preceding section, indicate that each time some increment of melt is generated, conditions are met at which extensional failure will take place. Therefore, each failure event implies some net migration of the melt fraction. Because the average gradients of pressure and stress are negative upward, and because strong buoyancy contrasts exist between the melt and solid phases, and because the ultimate source of mechanical work is gravitational energy release, a net upward migration of magma is virtually a necessity of the melting process. Therefore, either the melt is injected into the lithosphere at the same average rate that it is generated, or it accumulates in the asthenosphere. The latter can only occur if the minimum effective stress, σ_3, is vertical: according to the arguments advanced, this condition is transient. The extent of this storage regime, however, must relate to the dynamics of residual accumulation and "downwelling" in the mantle according to the model of Shaw and Jackson (1973), though the merits of that model are not defended here.[2]

In the discussion of effective stresses in the asthenosphere, it is plausible to consider the melt fraction as representing a pore fluid consistent with the classical definition of *effective stress* in Equation (1). Injection into the lithosphere, however, is more complicated. In some respects injection represents a condition analogous to formation fracturing, as illustrated by the experiments of Hubbert and Willis (1957) and Fiske and Jackson (1972). As magma traverses the lithosphere, however, domains are created that contain distributed melt lenses. A plexus of such injection features may also behave effectively like a porous material saturated with a pore fluid. In either case, the range of strengths is proportional to melt fraction, according to some relation like that shown in Figure 8, and the same sorts of failure criteria are assumed to apply. The principal distinction between lithosphere and asthenosphere is the wider probable range of stress magnitudes in the lithosphere related to larger failure events and to magma injection lenses of larger scale and wider spacings.

Some reflection on Figures 7, 8, and 9 should convince the reader that wide ranges in stress states and deformation rates are likely in a composite system like that portrayed in Figure 5. Comparison of Figures 9A and 7C illustrates that, when fluid pressure is a variable, one failure event may set off a series of other events, because the Mohr circle will be driven against the failure envelope over a wide range of stress differences which locally may decay in magnitude as rock stresses relax during failure. Elsewhere, however, stress differences, subsequently abbreviated $\Delta\sigma_{13}$, may increase as a result of this failure sequence. This observation, combined with domains of variable strength because of variable melt fraction (Figure 8), indicates that many paths are possible for such failure sequences. A failure sequence triggered by fluctuations in effective stresses is herein called a *failure cascade*, because it represents a composite event of gravitational energy release. Types of failure cascades will depend on balances between rock stress levels and fluid pressures governed by rates

[2] The "gravitational anchor" of Shaw and Jackson (1973) represents a situation where residual crystal accumulates in the region of melting sink as a density current into the deeper mantle. Therefore, the stress distribution above the anchor will be affected by the rate of sinking. Fast sinking rates create a "low pressure region" wherein the vertical principal stress may become the minimum principal stress.

of magma transport and stress relaxation. Systematic classification of failure sequences relative to tectonic domain structures could be a goal of future developments of magma transport theory.

Figure 10 shows a schematic model of injection of the lithosphere at Kilauea. The configuration of the region called the *source funnel* depends on the thermal history of the lithosphere path (in terms of horizontal stress gradients) and on the migration of the melting region in the asthenosphere relative to plate translation velocities. The cross section shown was influenced by the conclusions of Jackson and others (1975) on volcanic propagation rates and by Koyanagi and Endo (1971) on distributions of earthquake hypocenters beneath Kilauea volcano, Hawaii.

The inset in Figure 10 illustrates the likely effects of the volcanic edifice on rotations of stresses (Fiske and Jackson, 1972). The localization of venting depends on the interaction of thermal effects like those discussed by Jackson and Shaw (1975, p. 1863).

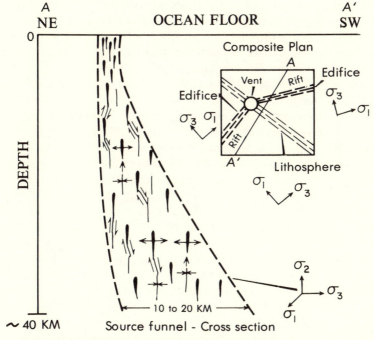

Figure 10. Schematic cross section of the lithosphere beneath Kilauea volcano representing the "source funnel" of extensional fracture events and the supply train of magma batches migrating upward from the asthenosphere. The inset shows a simplified plan of the volcanic rift systems to illustrate the contrasts in evolving stress orientations (cf. Figure 2). Migration of the source funnel is based on earthquake evidence (Figure 15) and other evidence discussed by Jackson and others (1975) and Ellsworth and Koyanagi (1977); see this last paper for seismic velocity structures relative to source depths and migration.

SEISMIC PHENOMENA RELATED TO EXTENSIONAL
FAILURE AND MAGMA ASCENT RATES

The data of seismology offer critical tests of some of the foregoing conclusions and of the possibility for direct monitoring of the degree of coupling between tectonic stress states and magma transport rates. If quantitative relationships exist, then seismic phenomena are new sources of information for monitoring of transport rates, for stress analysis, and for eruption and earthquake prediction in volcano-tectonic systems.

Among the important considerations from the standpoint of coupling of magmatic and seismic phenomena are the total amounts of mechanical energy released in dissipative processes, and the proportions of that energy that are related to fracturing processes, to magma and rock flow processes, and to radiation of seismic energy. It is evident that extensional fracture phenomena are involved in stress-release mechanisms in magmatic systems, whether or not individual events represent extensional or shear failures (e.g., according to regimes between A and C in Figure 7C). Therefore, the seismic energy budget is directly related to the volumetric strains associated with magma transport. The *rates* of stress release depend on the viscoelastic properties of rocks and on the rates of magma flow compensating extensional failure events.

A theory that attempted to take some of these factors into account for simple extensional fracture was derived by Robson et al. (1968). Their model considered principles essentially parallel to those of this paper; however, it gave a geometric model of magma flow paths that did not adequately allow for the magma flow rates and volumes that would be compatible with extensional failure events large enough to explain earthquakes of the magnitudes observed in volcanic areas. The assumptions and arguments of that paper, however, identify important limits to bear in mind concerning some of the deductions made here on the basis of empirical relations of magma transport rates and release rates of earthquake energy.

The present discussion considers some of the volumetric limitations on stress release and magma flow rates. In this context, the question of dike stability as regards lengths, shapes, and stress gradients is important (Pollard, 1976, 1977). Here it is assumed, partly from those considerations and partly from observation, that dike systems are made up of lenticular segments of various lengths: these are short compared to the lithosphere thickness, and exist in a plexus of offset lenses (Figure 11). This sort of distribution was assumed by Hill (1977) in an analysis of earthquake swarms. Hill's model was also stimulated by Pollard's (1973) discussion of concepts of dike interactions, and by discussion of an early draft of this paper. The geometry of the plexus of injection lenses is analogous in both cross section and plan view (compare Figures 10 and 11).

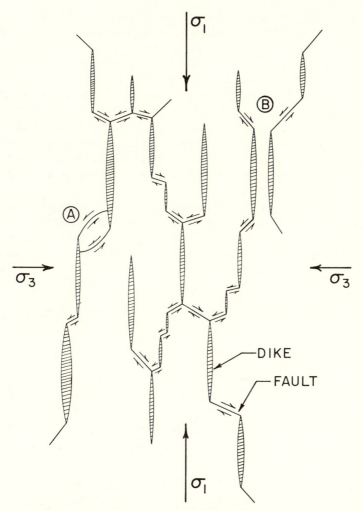

Figure 11. Schematic geometry of dike sets and interactions relative to orientations of maximum and minimum principal effective stresses (after Hill, 1977). Regions A and B show types of cracking near tips of adjacent dikes based on discussion of Pollard (1973). Relative to discussions in the present paper, this sort of geometry might apply on several different scales and represent the distributions of either horizontal or vertical magma lenses (see Figures 5 and 10). The regime of extensional shear failure (Figure 7C) is considered to have an important bearing on conditions of magma injection between lenses and possibly on the angular relations of the interconnecting fracture system.

The regime of extensional shear failure (Figure 7) is appropriate to this sort of geometry. That is, some of the shear couples shown in Figure 11 should represent transient events of extension plus shear between local domains where the stress difference between maximum and minimum principal effective stresses ($\Delta\sigma_{13}$) is suitable for pure extensional failure. In those situations the shear couple is subparallel to the σ_1 direction (deviations up to 20° or so). These would represent failures that permit magma transport from one offset segment to another, and, together with other extensional events, act as regulating valves for the gross magma flow rate. Each segment, of course, can also migrate by crack propagation at the top and closure at the bottom in a manner indicated by the analysis of Pollard (1976).

Thus, seismic failure is envisaged as a composite process in local domains of variable failure conditions (variable stress and/or melt fraction) in which there are simultaneously elements of strictly shear failure, extensional shear failure, and strictly extensional failure. The proportions of seismic energy released probably decrease in that order because of the

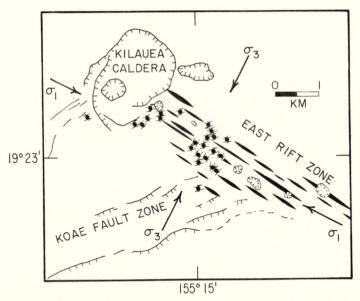

Figure 12. Examples of interrelations of postulated dike systems, stress orientations and earthquake swarms (after Hill, 1977, Figures 3, 4, and 5; see this reference for discussion of earthquake solutions). Regions represented are: (A, above) East rift zone of Kilauea volcano, Hawaii, (B, facing page) rift zone of the Reykjanes Peninsula, Iceland, and (C) system of transform faulting in the Imperial Valley, California. The "leaky transform faults" appear to represent conditions appropriate for extensional shear fracture.

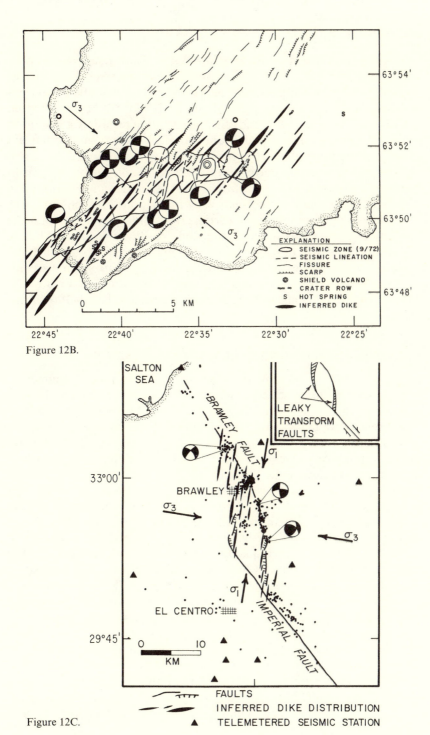

Figure 12B.

Figure 12C.

respective regimes of $\Delta\sigma_{13}$ required for failure. These could occur either as systematic failure cascades, or as intermittent processes. The range of possibilities is great, and classification will depend on coordinated analysis of seismic data (e.g., Koyanagi and Endo, 1971; Aki et al., 1977), eruption records (e.g., Swanson, 1972) and other deformation records (e.g., Swanson et al., 1976). These references are from Hawaiian studies; similar data are being generated in Iceland (Bjornsson et al., 1977).

Examples of possible correlations between diking events and earthquakes are shown in Figure 12, from Hill (1977). Additional information not discussed in that reference may exist in relations between directions of nodal planes derived from earthquake solutions and earthquake magnitudes, relative to inferred directions of dike injections. That is, higher angles relative to σ_1 and higher magnitudes may go together, decreasing to low angles and magnitudes merging with seismic tremor (e.g., Aki et al., 1977). The so-called "leaky transforms" of Figure 12C may be partly analogous to the regime of extensional shear failure of this paper.

MECHANICAL CORRELATIONS WITH MAGMA TRANSPORT RATES

Figure 13 shows a relation from Shaw (1965) for times required for laminar flow in dikes, scaled down to the viscosity of basaltic magma. It is shown that a rate of 1 kilometer per second would be predicted in a one meter dike under a pressure differential of 1,000 bars. This is the order of magnitude of the size and duration of an injection event required to produce a magnitude 5 earthquake, according to relationships given later. Although large volumetric events are theoretically possible where conduits and pressure gradients are large, Figure 13 shows that flow durations required for the production of large earthquakes are too short to be consistent with likely flow regimes. Large volumetric events also are not dimensionally consistent with the localized occurrences of the larger-magnitude earthquakes at shallow depth in Hawaii. The Kilauea deformation data of Swanson et al. (1976) and the analysis of Aki et al. (1977) describe some of the constraints on flow rates and crack propagation rates relevant to the volcanic edifice stresses. These data indicate much smaller volumetric episodes and slower flow rates.

The analysis of crack propagation by Aki et al. (1977), however, suggests that the pressure gradients in magma that compensate extensional fracture propagation may be large, because of the tendency for creation of an evaculated region inside the crack tip. If this is so, magma pressures of the order of 1,000 bars may be relevant to flow regimes (for example, 1,000 bars would be the equivalent to the mean stress at a depth of about 4 kilometers).

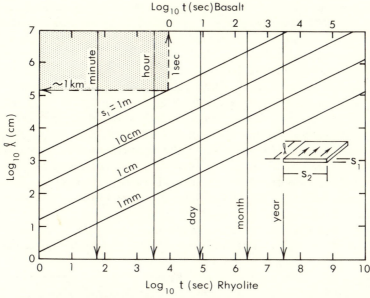

Figure 13. Calculated times for laminar magma injection in dikes (modified from Shaw, 1965). The ordinate gives the dike length (in direction of flow), and the abscissa represents integrated flow times; the scale at the bottom is for magma with a viscosity of 10^6 poises (e.g., hydrous rhyolite) driven by a pressure differential of 1,000 bars, and the scale at the top shows flow times scaled down four orders of magnitude to represent hypothetical rates for basaltic liquids (viscosity of about 10^2 poises). For purposes of discussion of dike volumes in the text it is assumed that the length S_2 is 1 kilometer; the dike width (S_1) is given on each rate curve. The shaded area highlights the hypothetical distance traveled in 1 second by basaltic magma in a 1 meter-wide conduit required to produce a magnitude 5 earthquake by the volumetric event (time lines refer to the lower scale for rhyolite). This condition is unstable relative to turbulent flow, and the curve representing a dike width of $S_1 = 1$ meter for basalt would actually be shifted downward on the graph to a position nearer the curve $S_1 = 10$ centimeters for rhyolite (but using the upper time scale for basalt). This means that, whereas a hypothetical event of laminar flow could produce a volumetric event in 1 second compatible with a magnitude 5 earthquake, a more realistic flow regime would involve either smaller flow increments (by about an order of magnitude) or longer flow durations. In the latter alternative the pulse would not be in the range of seismic periods unless it consisted of many short pulses of smaller volumes. This example refers to the energy released during an event of pure extensional fracture and does not apply to failure events also involving components of shear (see discussion in text of volumetric moments).

Relationships between the sizes of volumetric events and the corresponding earthquake magnitudes are given later. The dike dimension illustrated in Figure 13, hypothetically representing an extensional failure event of magnitude 5, might be, for example, 1 km × 1 km × 1 m; about 40 dike segments end to end would be required to traverse the lithosphere, and there would have to be about 100 events per year to account for a mean supply of .1 km^3/yr (Swanson, 1972). On the other hand, the analogous volume associated with a magnitude 2 event would be of the order 10^3 m^3 with dimensions, say, of roughly 100 m × 10 m × 1 m. A minimum of about 400 segments would be required to traverse the lithosphere, and about 10^5 events per year would be required to account for the average transport rate of .1 km^3/yr. Larger numbers of small events are more consistent with the relations of Figure 13 than are fewer large events.

These numbers simply represent limiting examples, as if the events were all extensional and only of one type. Of course, the observed distribution of earthquakes in Hawaii, cited below, involves a range of frequencies and magnitudes, typically varying from 2 to 5. There are also occasional larger events, so I expect that the frequencies of earthquakes of a given magnitude are fewer than the examples mentioned.

Shear failure is demonstrably involved in the Hawaiian earthquakes, particularly for the larger magnitudes, and the volume balances in those cases refer to the accumulated strain changes over some period of time. These precise correlations remain to be worked out, although a start on them is indicated by the data of Figures 15 and 16.

Basically this section reveals that the range of stress states, magma pressures, and path dimensions available for transport probably are consistent with a wide range of possible transient transport rates. The melt flow model derived by Mavko and Nur (1975) for stress relaxation in the asthenosphere also illustrates this point. For example, they give the stress relaxation time in the form (Mavko and Nur, 1975, Equation (11)).

$$\tau_r \simeq \eta \beta / \alpha^2 c,$$

where η is the melt viscosity, τ_r is the relaxation time, β is the effective compressibility, α is the ratio of conduit separation to path length, and c is rock porosity, which is taken to be the same as the average melt fraction in the present paper. Using the values $\eta = 100$ poises, $\beta = 10^{-12}$ cm^2/dyne, $\alpha = 1/1,000$, and $c = 10^{-5}$, gives $\tau_r = 10$ sec. The value chosen for average melt fraction is based on two lines of argument discussed later: (a) the relation of tensile strength and compressibility, and (b) the ratio of dike volume to the volume of the "source funnel" in Figure 10.

The relaxation time of 10 seconds is based on dimensional ratios and properties consistent with the present paper; however, it is derived from a flow model based on calculated permeabilities for flow in porous media.

With slight adjustments, the time is appropriate to the seismic mechanisms associated with extensional failure, and this correspondence tends to confirm the general assumptions that some part of the seismic spectrum is governed by extensional failure rates which are balanced by rates of magma flow.

VOLUME CHANGES AND SEISMIC ENERGY RELEASE

McGarr (1976) discussed the energy balances related to volume changes in the earth and gave examples of correlations between *seismic moment* (defined in terms of total volume change), earthquake magnitude, and earthquake energy. He also discussed the ratio of earthquake energy to the available mechanical work of volumetric change, and called this the *seismic efficiency* (termed S.E. in this paper). Several examples of these correlations are given in that reference. A parallel discussion is given here for Hawaii, using the earthquake data for the year 1969 (Koyanagi and Endo, 1971).

The seismic moment determined from volume changes represents an evaluation of potential energy available in the system: some part of this energy can be released as seismic radiation, depending on earthquake mechanism and efficiency. Direct correlation of earthquake magnitude with a volumetric pulse of magma assumes, on the other hand, that the earthquake mechanism is, in fact, the extensional failure event associated with flow of that volume of magma. Arguments have been given in the text suggesting that the rates at which seismic energy is released depend on combinations of mechanisms, one of which is extensional fracture. In either case, estimates of seismic moment based on volumes of magma offer a basis for comparing rates at which seismic energy is released with rates of magma transport.

Following McGarr (1976, p. 1488) the total seismic moment related to volume changes is given by

$$\Sigma M_o = K_m \mu |\Delta V|, \tag{13}$$

where ΣM_o refers to the sum of seismic moments related to a total volume change of absolute value $|\Delta V|$ (i.e., the volume change may be either positive or negative) for a material with modulus of rigidity, μ; the term K_m, not to be confused with tensile strength, is a factor with a value near unity related to the type of stress system.

The relationships between seismic moment, earthquake magnitude and earthquake energy are assumed to be approximately the same as those used by McGarr (1976, Equations (9) and (11)), as follows:

$$\log M_o = 17.7 + 1.2 M_L, \tag{14}$$

$$\log E_s = 11.8 + 1.5 M_L, \tag{15}$$

where M_L is the local magnitude and E_s is the earthquake energy. The constants in these equations vary from region to region, but these values are used for reasons of internal consistency. Several different correlation schemes relating moment, magnitude and energy are currently being discussed in the geophysical literature. Therefore, all the seismic calculations of the present paper are preliminary.

Figure 14 illustrates the relations between moment (expressed in terms of magma volume), magnitude, and energy, and Table 1 gives values of

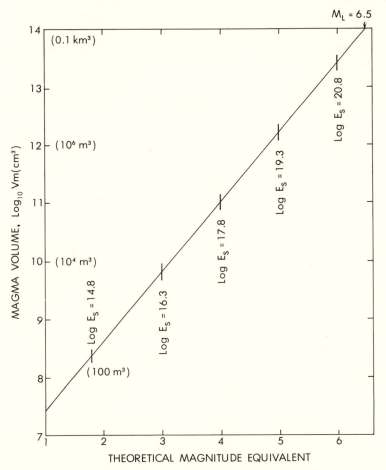

Figure 14. Theoretical relation between magma volume and local earthquake magnitude, assuming that pulses of magma transport relate to seismic moments according to Equation (13). The values E_s refer to earthquake energy calculated from Equation (15). The magnitude-moment-energy relations are according to the correlations of McGarr (1976). The upper limit of the volume scale is the average annual magma supply rate of Kilauea volcano in Hawaii, according to the estimate of Swanson (1972).

Table 1. Earthquake magnitude-depth-volumetric moment relations for Kilauea
Volcano, Hawaii, 1969 (seismic data from Koyanagi and Endo, 1971;
moments and volumes calculated from McGarr, 1976, equations 8 and 9).

Deep Source "Funnel": Depth	M_L	N	$\log M_o$ (M_o in ergs)	NM_o (erg)	$\Sigma \Delta V$ calc. ($10^6 \ m^3$)
0–4	2	~6	20.1	8×10^{20}	.003
4–7	2	~6	20.1	8×10^{20}	.003
7–9	2	12	20.1	2×10^{21}	
9–13	2–3	1	21.3	2×10^{21}	.012
9–13	2	13	20.1	2×10^{21}	
13–15	2	23	20.1	3×10^{21}	.006
	2–3	2	21.3	4×10^{21}	.023
15–17	2	21	20.1	3×10^{21}	
17–20	2	9	20.1	1×10^{21}	
	2–3	2	21.3	4×10^{21}	.12
	3–4	1	22.5	3×10^{22}	
20–23	2	15	20.1	2×10^{21}	.027
	2–3	3	21.3	6×10^{21}	
23–26	2	11	20.1	1×10^{21}	.025
	2–3	3	21.3	6×10^{21}	
26–28	2	18	20.1	2×10^{21}	.034
	2–3	4	21.3	8×10^{21}	
28–31	2	18	20.1	2×10^{21}	
	2–3	6	21.3	1×10^{22}	.15
	3–4	1	22.5	3×10^{22}	
31–34	2	17	20.1	2×10^{21}	
	2–3	8	21.3	2×10^{22}	.16
	3–4	1	22.5	3×10^{22}	
34–37	2	9	20.1	1×10^{21}	
	2–3	4	21.3	8×10^{21}	1.7
	4–5	1	23.7	5×10^{23}	
37–40	2	5	20.1	7×10^{20}	.022
	2–3	1	21.3	6×10^{21}	
Total System:	2	~600	20.1	8×10^{22}	.26
	2–3	~60	21.3	12×10^{22}	.40
	3–4	9	22.5	27×10^{22}	.90
	4–5	2	23.7	100×10^{22}	3.3
			(approx. 5% of total magma supply)*		~$5 \times 10^6 \ m^3$

* If K_m in Equation (13) is $\frac{1}{2}$, as deduced by McGarr (1976, p. 1492) for the Matsushiro
system involving extension and shear components, then the inferred volume is doubled,
or about 10 percent of the annual volume supply of magma. Calculations of average
seismic moments based on all Hawaiian earthquakes between 1961 and 1975 (F. W.
Klein, computer tabulation, Dec. 1977) gives a value of about 2×10^{25} dyne cm per
year; this value is very nearly equivalent to a volumetric moment (Equation (13))
based on the average supply rate of .1 km^3/yr (Swanson, 1972).

M_o calculated from Equation (14) for the Kilauea earthquakes recorded in 1969 (Koyanagi and Endo, 1971); Figures 15A and 15B show the distribution of hypocenters. Using the values of M_L, M_o and numbers of earthquakes, the inferred volume changes were calculated from Equation (13) and are given in the last column of Table 1. The total calculated volume in Table 1, however, is only about 5 percent of the average yearly volume supply of magma at Kilauea (.1 cubic kilometer) given by Swanson (1972). Clearly, the seismic data for the year 1969 do not adequately represent the average seismic energy release rate. Therefore, the proportionality between volumetric moment and magma transport volumes is subject to interpretation and further study.

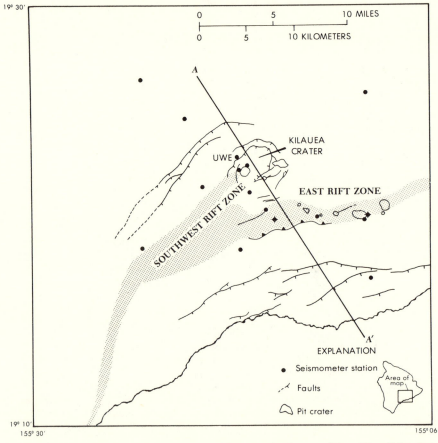

Figure 15A. Distribution of earthquake hypocenters of Kilauea volcano, Hawaii for the year 1969 (simplified from Koyanagi and Endo, 1971, Figures 3 and 5). Diagram A shows a plan view of the setting and general epicentral area of earthquake hypocenters plotted on the cross section A-A' in diagram B.

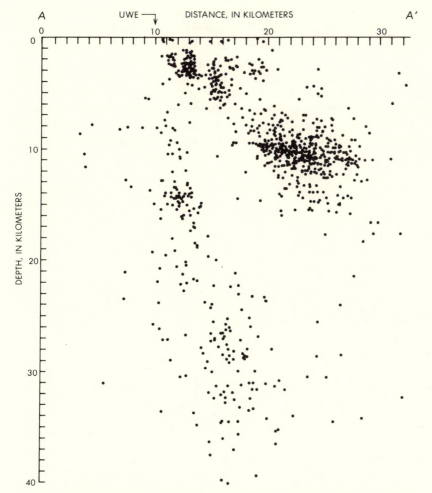

Figure 15B. Earthquake events, plotted from Koyanagi and Endo (1971), range between magnitude 2 and 5; this plot shows the general distribution without distinguishing magnitudes of events. There is no obvious correlation of magnitude and depth, but studies of earthquakes according to type, stress domain and depth are still preliminary (see above reference for details on magnitude distributions). The dense grouping of earthquakes at about 10 kilometer depth is apparently related to the volcanic rift system, and the deeper earthquakes are thought to represent the source funnel through the lithosphere as schematically outlined in Figures 5 and 10.

As a preliminary basis for interpretation, the observed seismic data of Table 1 are compared to a kinetic reference state defined by the volumetric moment (and equivalent seismic energy) that would be available if the average yearly volume of magma supply is assumed to provide the gravitational potential energy responsible for all earthquakes in Hawaii. The physical basis for this idea can be schematically visualized by assuming that magma is generated in the asthenosphere below Hawaii at a steady rate (on the year time-scale), and that yearly increments of magma buoyantly rising through the extensional fracture system in the

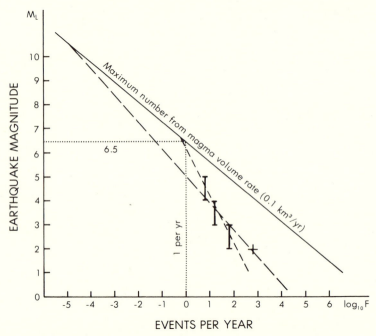

Figure 16. Number of earthquakes in each magnitude range from data for the year 1969 (data from Koyanagi and Endo, 1971). The steeper line through the data (short dashes) represents a subjective fit, assuming that the magnitude 2 earthquakes have a range similar to the other magnitude intervals; the other line (long dashes) represents a fit that passes through all data as plotted. The dashed line more nearly resembles longer term magnitude-frequency relations judging from preliminary examination of records between 1961 and 1975 (F. W. Klein, computer tabulation, Dec. 1977). The longer term data are not plotted because their evaluation is still incomplete. The solid line represents the theoretical number of earthquakes possible at each magnitude if the annual magma volume released all of its earthquake energy at that magnitude (e.g., one 6.5 magnitude earthquake per year; see Figure 14).

lithosphere (Figure 10) are compensated by equivalent downward displacements of the lithosphere, with a compensating release of gravitational potential energy. This released energy shows up in the kinetics of magma transport and in the complex spectrum of fracture mechanisms and associated earthquakes.

The maximum mechanical energy of the system is considered to be limited by the above relation. In other words, other inputs from stress-strain oscillations related to lithosphere-asthenosphere translation, strains related to secular changes in *average* thermal states, etc. are not considered in this discussion. The rationale for temporarily ignoring these factors is based on recognition of the large amount of energy potentially available from pressure-volume work associated with volume changes at asthenosphere depths. This point is referred to again in the section on seismic efficiency.

The volumes associated with magma ascent can be translated into seismic moment, magnitude, and energy relations by using Equations (13) through (15) (see Figure 14) in a manner that is the reverse of the calculations in Table 1. Because each volume increment of magma originating in the asthenosphere is of course associated with multiple events of extensional fracture before it reaches the surface, its contribution to the total seismic moment is the integral of these distributed events, only limited by the maximum available potential energy mentioned above.

These inverse calculations are combined with those of Table 1 in Figures 16 and 17. The volumetric moment associated with the average annual supply rate of .1 km^3/yr is shown in these figures as a solid line representing a limit or "lid" on the magma-related seismic moment. The line represents the locus of calculated moments that would produce the indicated number of earthquakes of given magnitude if all of the energy of magma transfer (.1 km^3/yr) were released at that magnitude (e.g. one event of magnitude 6.5 per year based on the moment-magnitude relation in Equation (14)). Because the actual energy release is distributed according to some relation between frequencies of magma transport events and their associated moments, the actual limits would be shifted downward on these plots. This could be graphically demonstrated in terms of various models of volumetric events associated with extensional fracture and magma flow; however, such modeling is not attempted in this paper.

The data in Figures 16 and 17 are based only on the earthquake data for 1969; the two dashed lines in Figure 16 represent the range of frequency-magnitude slopes that might be fitted to the data for that year. Analogous plots made by the author and based on unpublished data for the 14-year interval from 1961 to 1975 (F.W. Klein, computer tabulation, Dec. 1977) suggest that the smaller slope in Figure 16 (long dashed lines) bears the closer resemblance to long term frequency-magnitude correlations.

The plots in Figure 17 essentially show the deviations between observed numbers of earthquakes and the maximum theoretical number. The ratio of theoretical to observed frequencies is plotted in Figure 17B. According to these example correlations, the maximum theoretical earthquake magnitude based on the extrapolation of Figure 17A is about 7.6, with a theoretical recurrence interval of 50 years or so.

The limiting ratio of unity in Figure 17B, of course, represents the situation in which all of the accumulated potential energy related to magma transfer is released in a single event. Because this is not the case,

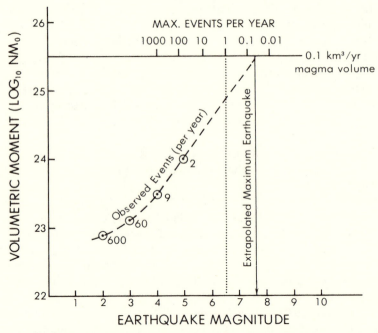

Figure 17. Relation of theoretical and observed numbers of earthquakes. *Diagram A* shows the volumetric moment-magnitude relation for the observed earthquake events of 1969 calculated from equation (14) compared with the maximum number of events calculated from equations (13) and (14) for the annual supply rate of magma. The dashed line represents the extrapolation of observed events assumed to be limited by the volumetric moment of magma transfer as a ceiling. *Diagram B* shows this extrapolation expressed as the ratio of theoretical to observed events; this figure and Figure 16 illustrate convergence of theoretical and observed relations at high magnitudes (e.g. a 7.6 earthquake about every 50 years would be the extrapolated limit if there were no smaller earthquakes).

the maximum expected magnitude is less than this limit (less than 7.6 in the above example, *based on extrapolating the data of 1969*). The general conclusion is that earthquakes in Hawaii can be expected to exceed magnitude 7 only occasionally, with recurrences ranging from tens of years to possibly a hundred years or so for the very largest events. Because such large events represent the integrated effects of multiple events of magma ascent, such earthquakes are not expected to be specifically localized by the active center of volcanic venting.

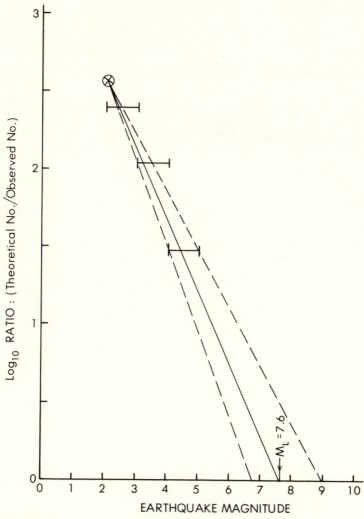

Figure 17B.

The above hypothetical examples, based on the simplest interpretation of the 1969 Kilauea data, fit reasonably well with the actual occurrences of large earthquakes in Hawaii (referring, here, generally to the Big Island). A destructive event of magnitude 7.2 occurred in 1975, and a similarly destructive event was recorded in 1868; a magnitude 6.9 event occurred in 1951, and magnitude 6.5 events occurred in 1929 and 1954 (Gutenberg and Richter, 1954; F. W. Klein, computer tabulation, Dec. 1977).

VOLUME DOMAINS IN THE HAWAIIAN SYSTEM

The extensional failure model of magma transport assumes that each increment of magma flow involves a pulse of volumetric expansion and collapse, or a sequence of such pulses. Each implies a change of seismic moment according to the McGarr relation in Equation (13) (McGarr, 1976). The validity of this assumption is subject to many additional tests related to the seismic data, such as geometric source models for earthquake sequences. The remainder of this section, however, is restricted to some observations on the total rock volumes affected by magma transport, and a relation between seismic efficiency, magnitude and magma transport volume.

Robson and others (1968) point out that, if extensional failure ultimately provides the elastic energy that is dissipated in seismic radiation, then the amount of that energy is proportional to the tensile strength, according to

$$E = \tfrac{4}{3}\beta_s K^2 V_s,$$ (16)

where β_s is the rock compressibility, K is the absolute value of tensile strength and V_s is the total volume of rock affected (Robson and others, 1968, p. 29). They also observe that the volume of melt that would compensate a change of mean stress of magnitude K is given approximately by

$$V_m = \beta_s V_s K,$$ (17)

where V_m is the magma volume appropriate to the source volume V_s. Thus, the theoretical melt fraction consistent with these balances is given by

$$f = V_m/V_s = \beta_s K.$$ (18)

Using a typical value of 10^{-12} cm²/dyne for β_s and 10 bars for K gives $f = 10^{-5}$; this fraction could be as much as an order of magnitude larger for the range of tensile strengths in Figure 8.

The value of f represents an average value for volume domains involved in extensional fracture; it could represent the volumes associated with a single event, or the integrated volumes associated with all events. If all magma lenses present in the mantle are involved in extensional fracture events during a given time period, then f would also represent the average fraction of magma in the mantle during that time.

Some comparisons based on the theoretical melt fraction are given as follows: if the cumulative seismic volume in Table 1 represents magma, then the consistent source volume is $V_s \simeq 500 \text{ km}^3$, and the volume of the 1969 funnel of hypocenters is of the order 10^3 km^3. Based on the same proportionality of $f = 10^{-5}$, the source volume consistent with $.1 \text{ km}^3$ magma per year is 10^4 km^3. This value is reasonably consistent with the total hypocentral volume of all Kilauean earthquakes. It is emphasized that these are tentative comparisons, but they suggest promise for detailed correlations between volume domains and distributions of seismic energy release. That is, according to the model, ratios of magma transport volumes to hypocenter volumes give information on volume fractions occupied by magma. Present evidence suggests that this average fraction is between 10^{-4} and 10^{-5} in the lithosphere beneath Kilauea.

SEISMIC EFFICIENCY

McCarr (1976) calculated seismic efficiences based on seismic moment considerations for the Denver earthquakes, the Matsushiro earthquakes in Japan, and for mine tremor and stope closure in South Africa (see McGarr for references to the relevant seismic data). The seismic efficiency (termed "η" by McGarr, 1976, p. 1490, and "S.E." in this paper) is obtained from the ratio of the seismic energy release to the available mechanical work of volumetric change:

$$\text{S.E.} = \Sigma E_s / \Delta W_g, \tag{19}$$

where E_s is obtained from Equation (15) and ΔW_g is estimated from the mean stress acting over the volume change determined from seismic moments. McGarr (1976) obtained values of S.E. of .3 to 3 percent for the Denver earthquakes, about 1 percent for Matsushiro earthquakes, and .24 to .4 percent for the seismic events associated with mining. The work of these volumetric events was essentially estimated from the depths, volumes, and gravitational loads.

Table 2 lists calculations of seismic energies, volumes, and seismic efficiencies for the cumulative Kilauean events of 1969 in Table 1, for a mean gravitational stress equivalent to a mean depth of 20 km. The calculated values of S.E. range from .02 percent to .2 percent for observed earthquake magnitudes 2 to 5. These are based on the calculated volumes based on seismic moments. On the other hand, using the seismic energy equivalent to the mean annual magma volume (.1 km^3, and the equivalent earthquake magnitude from Equations (13) and (14)) gives a seismic efficiency of .6 percent.

The apparent seismic efficiency from these relations increases with magnitude. Clearly, a rigorous analysis requires evaluations of individual events with their characteristic depths and magnitudes. It is evident from the following, however, that the depth-magnitude-S.E. relation is not a simple proportionality. If it were assumed that an efficiency of between about .1 and 1 percent were appropriate, as was found by McGarr for most other situations and as is found for the higher magnitude relations in Hawaii, then the appropriate mean depths for other magnitudes could be calculated. Working the problem backwards in this way, however, predicts mean depths that are too shallow compared to the hypocentral distribution in Figure 15. Thus, the exact relations between magnitude, depth, and seismic efficiency are not evident. Finding them will require a careful classification of earthquake characteristics as a function of depth. Present evidence suggests that S.E. is a function only of magnitude.

A plot of seismic efficiences from Table 2 (Figure 18) supports the volumetric relations discussed above for the following reasons: (1) The values of S.E. for $M_L = 2$ to $M_L = 5$ are based on calculated volumes and seismic energies from *observed* numbers and magnitudes of earthquakes, (2) the value of S.E. at $M_L = 6.5$ is obtained from the *calculated* earthquake magnitude obtained from the mean annual magma volume, and (3) these values, inversely derived from the same volume-moment-energy relations, fall on the same straight line.

The meaning of the slope of the line in Figure 18, and the range of S.E. values, undoubtedly relate to characteristic mechanical factors in the Kilauean system. The earlier discussion of maximum total potential energy identified the principal source as the balance of magma ascent compensated by downward displacements of superincumbent rock. If so, the net available energy is the difference between the total gravitational potential energy of the rock column from source to surface, and the energy required to lift magma from the source to the surface. This difference is about 10 percent of the total, based on typical density contrasts between magma and rock. If the seismic energy was governed, on the average, by such a proportionality, the seismic efficiency would be independent of depth. The range in Figure 18, however, would be from .2 to 6 percent instead of .02 to .6 percent. By this model, the smaller magnitudes are less efficient, because more of the energy is dissipated in the kinetics of fracture and viscous flow than in seismic radiation.

Table 2. Seismic efficiencies for Kilauea earthquakes, 1969, calculated by method of
McGarr (1976).

M_L (observed)	E_s (erg)	N	NE_s (erg)	ΣV_{calc} (cm³)	ΔW_g (erg)	S.E.
2	6.3×10^{14}	600	3.8×10^{17}	2.6×10^{11}	1.6×10^{21}	2.4×10^{-4}
3	2.0×10^{16}	60	1.2×10^{18}	4.0×10^{11}	2.4×10^{21}	5.0×10^{-4}
4	6.3×10^{17}	9	5.7×10^{18}	9.0×10^{11}	5.4×10^{21}	1.1×10^{-3}
5	2.0×10^{19}	2	4.0×10^{19}	3.3×10^{12}	2.0×10^{22}	2.0×10^{-3}
(6.5)*	3.6×10^{21}	1	3.6×10^{21}	1.0×10^{14}	6.0×10^{23}	6.0×10^{-3}
	(1.3×10^{21})**					

* Calculated magnitude of a single event equivalent to the volumetric moment of the observed magma supply rate of .1 km³/yr (Swanson, 1972), using Equations (13) and (14) and $K_m = 1$.

** Calculated by method of Robson and others (1968) for tensile strength of 10 bars, using Equations (16) and (17).

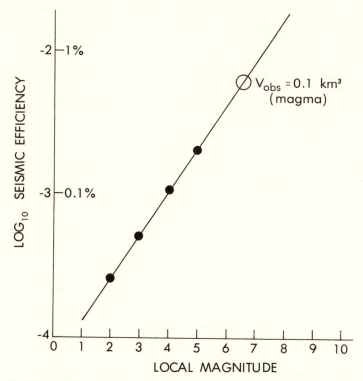

Figure 18. Seismic efficiency versus local magnitude from correlations based on Kilauean earthquakes of 1969 compared with the efficiency of a hypothetical earthquake equivalent to the annual volume of magma supply calculated from Equations (13) and (14). The efficiency is the ratio of earthquake energy from Equation (15) to the mechanical work of volume change (based on the product of volume from Equation (13) and the gravitational load stress for a mean depth of 20 kilometers). The efficiency is higher if account is taken of gravitational potential differences related to the buoyancy forces of magma ascent (see text).

GLOBAL IMPLICATIONS OF THE MECHANICS OF MAGMA RISE: PROBLEMS FOR FURTHER STUDY

The mechanical analysis proposed here suggests comments on several global tectonic factors, such as: (a) relations among plate tectonics, magma generation, and ascent rates, (b) relations between global magma generation and global seismicity, (c) implications of earthquake/extensional-fracture mechanisms for average magma ascent rates and residence times, (d) implications of magma volumes and earthquake source volumes for average seismic properties of mantle regions containing magma, (e) implications of magma paths and residence times for magma chemistry and the thermal structure of intruded rocks, and (f) seismic characteristics of active igneous systems in continental sections. Each of these categories may be a topic for major developments in the understanding of igneous processes. The implications are only touched on here.

NOTE ON GLOBAL FEEDBACK AMONG MAGMA PRODUCTION, PLATE STRESSES AND ASCENT RATES

In his comments on the history of mechanical concepts of faulting, Voight (1976, p. 99) noted the resemblance between the motions of global lithosphere plates and large-scale décollement. That analogy aptly describes my own views on the driving forces and mechanisms of plate motions. In this paper the analogy has been extended to include the role of the fluid pressure of magma in the asthenosphere as a key factor in coupling the tectonic motions with the generation of magma, and with the stress conditions permitting fracture and ascent through the lithosphere.

The key balances affecting feedback relationships are: (1) the effects of plate translation on mean shear stresses in the asthenosphere, (2) the amount of magma produced by the energy dissipation of shear in the asthenosphere, (3) the effects of the melt fraction on shear strengths, tensile strengths, and extensional failure criteria in the asthenosphere, (4) the effects of plate motions and interferences on lithosphere stresses and on the conditions of extensional failure and magma intrusion, (5) the consequent rates of tapping of magma from the asthenosphere, and (6) completion of the feedback loop in terms of resultant changes in asthenosphere properties.

I have proposed here that the concept of *effective stress* and extensional failure conditions represent the common factor that controls the extent of coupling of the various mechanisms. In its simplest form it appeals to these principles in a manner directly analogous to the arguments of Hubbert and Rubey (1959) except that qualifications analogous to the arguments of Hsü (1969) are made concerning "cohesive strength" relative to the tensile strength of partly molten rock.

Arguments concerning some of the energy balances in the above feedback processes are given in the various papers in which I have been involved, cited in the section on "A Model for Hawaii". A complete analysis of the feedback cycle is beyond the scope of this paper (cf. Shaw et al., 1971).

GLOBAL EXTRAPOLATION OF THE RELATION BETWEEN MAGMA VOLUME AND EARTHQUAKE ENERGY

From all the Hawaiian earthquakes of magnitude greater than 3.5 in the 14-year period from 1961 to 1975 (F. W. Klein, computer tabulation, Dec. 1977) I have calculated an average value of the annual seismic moment of about 2×10^{25} dyne cm (ΣNM_o/number of years) based on Equation (14). This includes the 7.2 event in 1975; with this event the average seismic moment very nearly balances the moment of 3×10^{25} calculated from Equation (13), based on magma volume for $Km = 1$. A parallel calculation based on summations of moments of all "shallow" earthquakes in the tables of Gutenberg and Richter (1954) gives an annual value of the order of 10^{28} dyne cm using the same moment-magnitude relation. The ratio $[\Sigma NM_o(\text{world})/\Sigma NM_o(\text{Hawaii})]$ is accordingly about 500.

Nakamura (1974) has estimated the global volcanic magma production rate as 6 to 8 km^3/yr. The ratio of the annual global volume to the annual volume in Hawaii (.1 km^3) is 60 to 80. Shaw (1970) and Shaw et al. (1971) estimated an annual magma generation rate in the asthenosphere of about 30 km^3. The ratio of this volume to the annual Hawaiian volume (assumed to be in approximate balance with the production rate in the asthenosphere) is 300.

Considering the possible intrusive fractions of the global production rate in continental and island-arc environments, the value of 30 km^3/yr for total magma production is reasonably consistent with Nakamura's (1974) estimate of 6 to 8 km^3/yr for the extrusive fraction.

In my opinion, the resemblance between the seismic moment ratios and magma production ratios has major significance for the sources of potential energy for global seismicity. Though the problem is regionally intricate, I suggest that global seismicity is intimately and quantitatively tied up with the earth's igneous history. The numerical ratios cited above suggest the possibility of a direct proportionality between mean annual volumes of magma ascent from melting sources in the mantle and mean annual volumetric moments of all earthquakes.

Implications of Extensional Fracture Models for Magma Ascent Rates

The balances discussed earlier concerning magma and seismic source volumes suggested a ratio of 10^{-4} to 10^{-5}. This is a minimum value representing the magma fraction involved in earthquakes. Since it is likely that most of the magma lenses in the lithosphere beneath Kilauea are eventually involved in extensional failure episodes, the value of 10^{-4} may represent a possible estimate for the *average magma fraction in the Hawaiian lithosphere*. Significant oscillations about this value, however, are likely. For example, if the time required to involve all of the affected mantle region is represented by the recurrence times of the largest earthquakes, then a fraction like 10^{-4} might represent the average over a 100-year period. If so, the maximum fraction in some volume domains might, at times, approach a value as large as 1 percent. These conclusions on magma distributions generally are consistent with the discussion of Ellsworth and Koyanagi (1977) concerning seismic velocity models for Hawaii; these authors also comment on the apparent southward migration of the mantle source.

Assuming, for the sake of discussion, that 10^{-4} is the magma fraction and 10^4 km^3 is the hypocentral volume, then the amount of magma present in the lithosphere above the hypocenter at a given time would be 1 km^3. Since the steady supply rate is $.1 \text{ km}^3/\text{yr}$, the *average* residence time of magma in the lithosphere is 10 years, according to this model.

The implication of this balance is that Hawaiian lavas represent a geologically instantaneous sample of melting in the asthenosphere. That is, the asthenosphere chemistry is sampled very directly and quickly with limited opportunities for chemical exchange on the way through the lithosphere. These chemical implications will be discussed elsewhere; they appear to be consistent with chemical models being studied by T.L. Wright (see Wright, 1971; Wright and Fiske, 1971; Wright and others, 1975).

The patterns of magma generation in relation to residence times, to ratios of intrusive and extrusive fractions, and to global seismicity suggest the possibility of a systematic global relationship. From Gutenberg and Richter's (1965) global maps of seismic regions, I have noted that most large earthquakes and the largest proportion of total global seismic moment are associated with subduction zones beneath island arcs. Nakamura's (1974) estimates of *extrusion* rates in various tectonic environments, however, show that those in island-arcs are only about $\frac{1}{10}$ of the average global rate. This suggests that on a global scale, there is a relation between the ratio of intrusive to extrusive magma volumes, magma ascent rates, and the corresponding fracture mechanisms reflected in the regional seismicity. By category, Hawaii and the oceanic spreading centers represent the lowest volume ratio of intrusive to extrusive products (and the quickest magma ascent rates), island-arc systems represent intermediate volume ratios (and intermediate residence times), and cratonic systems represent the highest volume ratios (and longest residence times). According to this interpretation, tectonic domains with the highest frequency of large earthquakes (i.e. subduction zones) are those in which the volume ratio of intrusive to extrusive igneous products is high, the volume rate of magma supply is large, and average magma ascent rates are relatively slow. Cratonic systems (e.g. Yellowstone, Wyoming; Valles, New Mexico; Long Valley, California) are thought to have yet higher intrusive/extrusive ratios, but smaller magma supply rates and the longest residence times. In these systems, stress states have had more time to relax and volume changes have taken place over longer intervals of time. Earthquake frequencies and magnitudes are therefore generally the lowest.

The above line of reasoning can provide a natural dynamic classification of magma types according to tectonic domains. According to this view, the ubiquitous occurrence of andesite in island-arc settings is a natural manifestation of that tectonic domain and its characteristic intermediate magma ascent rate (or residence time), rather than an intrinsically different original partial melt composition.

Thermal Implications of Magma Ascent Rates in Continental Sections

As remarked by Anderson (1936) the fact that a magma chamber can form in the crust suggests some period during which extensional failure conditions are not met. My analysis suggests that this would be expected only for values of $\Delta\sigma_{13}$ less than K, or for σ_3 vertical. The latter is appropriate to sill formation and the former to the development of large chambers in continental crust, wherein tectonic stress differences are diminished by stress relaxation (probably related to the thermal history of igneous intrusion).

The thermal implications of magma ascent rates and residence times in such situations are illustrated synoptically in Figure 19. This figure shows a spectrum of possible crustal thermal anomalies that might be generated by various magma injection rates over different depth intervals in crustal environments. The injection rates essentially span the range of magma production rates recorded by the evolution of the Hawaiian Archipelago (Shaw, 1973). It has been suggested by Smith and Shaw (1975) that many high-level continental magma systems of silicic affiliation are products of a magmatic history beginning with transport rates of mafic magma not

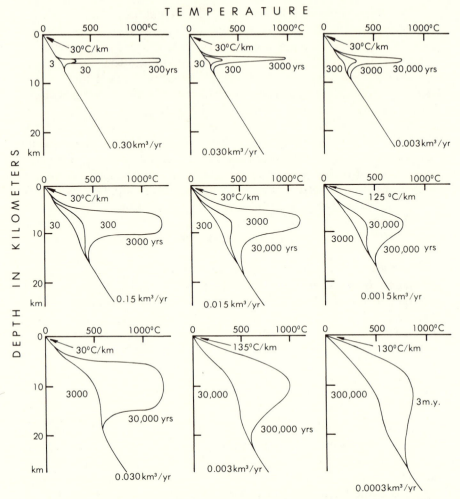

Figure 19. Diagrammatic synopses of approximate temperature profiles generated by magma injection of crustal rocks, with original gradient of 30°C/km, at volume rates approximating the total range of the Hawaiian system (see Shaw, 1973, for volume rate spectrum). The general model of thermal calculations follows the ideas discussed by Smith and Shaw (1975); the maximum temperature of 1,200°C is based on the liquidus temperature of tholeiitic basalts. The profiles represent the approximate temperatures generated if magma is injected at the given rates into a fixed volume for the times given for each curve. The left hand diagrams represent the minimum rates required to produce zones of magma of the indicated thicknesses at liquidus temperatures; if these rates persist longer, of course, the zone of liquid magma expands. The calculation is based on an available heat from magma of 10^3 cal/cm^3 injected into slab-like regions with thicknesses, respectively of 1, 5, and 10 km and with a square plan area 10 km on a side.

unlike those of the Hawaiian system, although the average rate over the lifetime of the system is lower.

Figure 19 identifies types of igneous-related thermal anomalies between two extremes: (1) sharply defined high level anomalies supported by rapid short-lived injection rates, and (2) diffusely defined regions occupying large fractions of the crustal thickness supported by lower long-lived injection rates. These conclusions on thermal regimes, however, require detailed comparisons with the dynamical concepts of tectonic domains and magma residence times as outlined above (cf. Marsh, 1978).

SEISMIC PHENOMENA IN ACTIVE MAGMA SYSTEMS OF THE CONTINENTS

The preceding outline and the discussion of Hawaiian seismicity suggest some possible tests for those earthquakes that are related to active continental magma systems. Figure 13, and the related discussion of transport rates, categorically identifies a spectrum of response times of magma flow related to magma type. Clearly, the more mafic and fluid magma types are identified with volumetric injection episodes that are potentially consistent with the periods of seismic signals.

The model of magma evolution proposed by Smith and Shaw (1975), and alluded to in the outline of thermal anomalies in Figure 19, implies that active intermediate and silicic magma systems in the crust also have an active mafic support system connecting them, umbilically as it were, to mantle sources of magma generation. If so, there is a mafic root system beneath high-level silicic magma chambers that may be associated with hierarchies of extensional fracture phenomena and earthquakes analogous to those of Hawaii. This possibility implies that earthquakes may exist *beneath* high-level silicic chambers that show no conspicuous seismicity immediately around and above the chamber because of the much lower response rates of viscous magma. In such situations, the depths of associated earthquake hypocenters do not necessarily define the minimum depths of silicic magma chambers.

The detailed analysis of seismic data in relation to models of tectonic and magmatic domains in the crust may offer a systematic basis for classifying mantle source systems that are actively supporting igneous processes in continental settings.

SUMMARY

Magma ascent in the earth is described as a process involving incremental volumetric episodes of magma injection that follow fracture pathways governed by the stress states in the rock and the locally acting fluid pressure of magma. In domains where the fluid pressure of magma approximates the average lithostatic stress, the minimum *effective stress* (minimum principal stress in the rock minus the magma pressure) is zero or negative; vertical directions of magma transport imply that the direction of the minimum stress axis is horizontal. In terms of a Mohr representation, effective stresses in such domains may enter the tensile region. The relationship between fluid pressure and rock stresses leads to two important points concerning magma transport: (1) rock in which magma represents a distributed fluid phase can fail by extensional fracture at any depth, presuming that other dynamic conditions are compatible (e.g., flow rate of melt), and (2) the tensile strengths of rock resulting from the local distributions of melt fractions influence the magnitudes of fracture events and the rates of magma transport. In relation to the second point, experimental and observational data from studies of Hawaiian lava lakes show that magma is susceptible to brittle behavior even when it contains melt fractions as high as 80 percent. The effective tensile strength of such magma is shown to range from the order of 10 bars at temperatures near the solidus, down to about 10^{-4} bars at 80 percent melt, weakening rapidly at melt fractions greater than 50 percent. These observations are not in conflict with the fact that such magma also simultaneously behaves as a plastic and viscous material under shearing deformations.

The tensile strength represents a natural scaling factor for deviations from states of uniform lithostatic stress and for average driving potentials of magma flow. The volume of a flow increment, however, also depends on the available volume of magma and the time-dependent changes of rock stress and fluid pressure. Because the kinetic spectrum is wide, the approach taken in this paper is to compare theoretical conditions of stress, fluid pressure, tensile strength, and flow rates against observations concerning compatible dike dimensions, data for average volumetric ascent rates, and rates of elastic strain release. Numerical comparisons are based on data from Kilauea Volcano, Hawaii, for the average magma supply rate and earthquake frequencies.

Mohr diagrams describing failure according to a composite Griffith-Coulomb criterion identify three general failure regimes: (a) shear failure, (b) extensional shear failure, and (c) pure extensional failure. An increase in the ratio of fluid pressure to lithostatic stress in a given rock domain can produce a failure cascade beginning in the shear mode because of progressively decreasing effective stresses; pure extensional failure is a theoretical culmination of such sequences. The advent of failure cascades, however, is governed by adjustments of stress in response to a variety of coupled magma-tectonic effects such as melt production, transport, and crystallization (thermal and mechanical effects) and effects of stress adjustments to plate and crustal motions (isostatic, plate translation, and volcanic edifice stresses). The effects of volatile and particulate transfers were not addressed in this paper though they are important to the mechanics of volcanic venting, conduit flow in diatremes, flow rates of extrusives, and the like. Because of variable distributions of stress states and magma paths, it is not possible to say whether specific extensional failure events associated with magma transport will precede, follow, or accompany events of major shear failure in the same or neighboring domains. Failure cascades of more than one type are likely in a magmatic system consisting of many different domains of stress and magma concentration.

Release of elastic strain energy in all the above failure regimes of a magmatic system relates to the total volume balances of magma ascent and compensating rock deformations. Comparison of volumetric moment-magnitude-energy relations for the earthquakes of Kilauea Volcano, Hawaii, in terms of magma transport volumes indicates that there is an approximate mechanical energy balance. It is suggested that this form of mechanical energy balance also holds on larger scales of global magma ascent and earthquakes.

Magma ascent is viewed in this paper as a form of percolation process involving intermittent directly and indirectly coupled transport in a plexus of conduits extending discontinuously from the melting source to the surface. The integral thermal and mechanical energy of a coeval magmatic system involves balances among melting rates, stress states, gravitational energy release, ascent rates, and heat transfer. In oceanic island systems like Hawaii, the episodes of extensional failure and magma transport represent the means of communication between different parts of the system. One corollary of these conclusions is that stress magnitudes and orientations change with time and location as magma increments traverse the system. Another is that the path-dependent effects of rock-magma mechanics also will be reflected in related time and spatial variations of magma chemistry.

The implications of the effective-stress law and magma-related failure mechanisms in the mantle suggest that magma production, storage, and ascent represent key factors in regulating mantle stress levels, hence plate tectonic motions. The fluid pressure of magma in the asthenosphere acts in the same sense as the hydrologic "formation pressure" in models of gravity sliding and decollement. Interaction of magma injection paths from the mantle with various types of large scale tectonic stress domains is suggested to result in a wide range of magma residence times en route from the source to the surface. Thus, consideration of the fracture mechanisms of magma ascent in relation to tectonic stress domains may identify a natural classification of evolving magma types at crustal levels corresponding with characteristic types of igneous-related thermal anomalies.

ACKNOWLEDGMENTS

The incentives for this paper on fracture mechanics have come from discussions over the years with Thomas L. Wright, who has persistently emphasized the importance of path and mechanical history on the chemistry of basaltic magma, and with the late E. Dale Jackson, who repeatedly suggested that specific mechanisms must be addressed to bridge the gap between surficial processes represented by volcanic loci and mantle rheology. I am grateful to R. B. Hargraves, T. N. Irvine, D. D. Pollard, D. P. Hill, T. L. Wright, and E. D. Jackson for reviews of the manuscript.

NOTE

The algebraic expressions relating depth, gravitational load, fluid pressure, and tensile strength require careful consideration in regard to their physical context, and the signs of the variables. For example, if it is assumed that the minimum principal stress, S_3, is vertical and equal to the gravitational load (i.e., $S_3 = \bar{\rho}gd$), then the expression in Equation (12) is obtained for the depth and is algebraically the negative of Equation (11a). In order for the depth to be physically meaningful in Equation (12), the pressure must exceed the gravitational load. Such a situation, however, represents a departure from the assumption that the vertical stress, S_{vert}, is equal to the gravitational load stress. For this balance to exist, the fluid pressure must equal the vertical stress (i.e., when the minimum principal stress is assumed to be vertical, then $p = S_{\text{vert}} = S_3$); but if this is so, then the vertical *effective stress*, σ_3, becomes zero, and there is no extensional failure unless $K = 0$. In other words, this combination of assumptions is internally inconsistent.

Two simple alternatives exist to reconcile this inconsistency, and they are physically the inverse of one another. In the first alternative, the pressure is considered to represent an anomalous pulse that transiently exceeds the vertical gravitational load and thus permits extensional fracture and sill formation (the pressure transient must be equal to or greater than the absolute value of tensile strength, K). As the pressure pulse decays, however, the injected sill presumably either collapses in thickness, is reinjected elsewhere or persists in the form of solidified traces.

In the second alternative, it is assumed that the vertical component of stress can be less than the gravitational load for tectonic reasons, such as keystone-like arching of superincumbent rocks or withdrawal of mass from below by mantle flow. In this case, letting $b = \bar{\rho}gd/S_{\text{vert}} \geq 1$, an expression equivalent to Equation (12) is obtained in the form

$$d_{\max} = \frac{-bK}{\bar{\rho}g(1 - b\lambda)} \tag{12a}$$

This expression gives the equivalent balances but avoids the implication that the fluid pressure must exceed the gravitational load for failure to occur.

Both of the above alternatives represent transient conditions. They differ in regard to the scale and duration of the anomalous stresses. The first alternative is likely to be localized and short term, whereas the second alternative may be of much larger scale and duration. In this latter instance sill-like injections might persist long enough to preserve a significant volume fraction in the liquid state within the time frame of tectonic adjustments relative to cooling and crystallization. Such may be the situation in the region of melt and residual accumulates of Figure 5. For example, if S_{vert} is about .1 percent less than the gravitational stress, and if the fluid pressure equals the gravitational stress, then sill-like injection can occur at a depth of about 40 kilometers.

REFERENCES

Aki, K., M. Fehler, and S. Das, 1977. Source mechanism of volcanic tremor; fluid-driven crack models and their application to the 1963 Kilauea eruption, *Jour. Volc. Geoth. Research 2*, 259–287.

Anderson, E. M., 1905. The dynamics of faulting, *Edinburgh Geol. Soc. Trans. 8*, pt. 3, 387–402.

Anderson, E. M., 1936. The dynamics of the formation of cone-sheets, ring-dykes, and caldron-subsidences, *Royal Soc. Edinburgh Proc. 56*, 128–163.

Anderson, E. M., 1938. The dynamics of sheet intrusion, *Royal Soc. Edinburgh Proc. 58*, 242–251.

Anderson, O. L. and P. C. Grew, 1977. Stress corrosion theory of crack propagation with applications to geophysics, *Rev. Geophys. Space Phys. 15*, 77–104.

Bailey, E. B., et al., 1924. The Tertiary and post-Tertiary geology of Mull, Loch Aline and Oban, explanation of Sheet 44, *Mem. Geol. Surv. U.K. (Scotland)*.

Björnsson, A., K. Saemundsson, P. Einarsson, E. Tryggvason, and K. Grönvold, 1977. Current rifting episode in Iceland, *Nature 266*, 318–323.

Bowen, N. L., 1928. *The Evolution of the Igneous Rocks*: Princeton: Princeton University Press.

Brace, W. F., 1969. The mechanical effects of pore pressure on fracturing of rocks, *in* A. J. Baer and D. K. Norris (eds.) *Research in Tectonics*: Canada Geol. Survey Paper 68-52, 113–124.

Eaton, J. P., and K. J. Murata, 1960. How volcanoes grow, *Science 132*, 925–938.

Ellsworth, W. L. and R. Y. Koyanagi, 1977. Three-dimensional crust and mantle structure of Kilauea Volcano, Hawaii, *Jour. Geophys. Research 82*, 5379–5394.

Fiske, R. S. and E. D. Jackson, 1972. Orientation and growth of Hawaiian volcanic rifts: The effect of regional structure and gravitational stress, *Royal Soc.* [*London*] *Proc., A. 329*, 299–326.

Griffith, A. A., 1921. The phenomena of rupture and flow in solids, *Royal Soc. London Philos. Trans., A., 221*, 163–198.

Griffith, A. A., 1925. The theory of rupture, *Internat. Cong. Appl. Mech. 1st, Delft, 1924, Proc.*, 55–63.

Gruntfest, I. J. and H. R. Shaw, 1974. Scale effects in the study of earth tides, *Trans. Soc. Rheology 18*, 287–297.

Gutenberg, B. and C. F. Richter, 1954. *Seismicity of the Earth*, Princeton: Princeton Univ. Press.

Hill, D. P., 1977. A model for earthquake swarms, *Jour. Geophys. Research, 82*, 1347–1352.

Hsü, K. J., 1969. Role of cohesive strengths in the mechanics of overthrust faulting and of landsliding, *Geol. Soc. America Bull. 80*, 927–952.

Hubbert, M. K. and W. W. Rubey, 1959. Role of fluid pressure in mechanics of overthrust faulting; I. Mechanics of fluid-filled porous solids and its application to overthrust faulting, *Geol. Soc. America Bull. 70*, 115–166.

Hubbert, M. K. and D. G. Willis, 1957. Mechanics of hydraulic fracturing, *Amer. Inst. Mining, Metall., Petrol. Engineers Trans. 210*, 153–166.

Jackson, E. D. and H. R. Shaw, 1975. Stress fields in central portions of the Pacific plate: Delineated in time by linear volcanic chains, *Jour. Geophys. Research 80*, 1861–1874.

Jackson, E. D., H. R. Shaw, and K. E. Bargar, 1975. Calculated geochronology and stress field orientations along the Hawaiian chain, *Earth and Planetary Sc. Letters 26*, 145–155.

Koide, H. and S. Bhattacharji, 1975. Formation of fractures around magmatic intrusions and their role in ore localization, *Econ. Geology 70*, 781–799.

Koyanagi, and E. T. Endo, 1971. *Hawaiian seismic events during 1969*, U.S. Geol. Survey Prof. Paper 750C, C158–C164.

Kuntz, M. A. and T. N. Brock, 1977. Structure, petrology and petrogenesis of the Treasurevault stock, Mosquito Range, Colorado, *Geol. Soc. America Bull. 88*, 465–479.

Lawn, B. R. and T. R. Wilshaw, 1975. *Fracture of Brittle Solids*, Cambridge: Cambridge University Press.

Marsh, B. D., 1978. On the cooling of ascending andesitic magma, *Phil. Trans. Roy. Soc. London, A, 288*, no. 1355, 611–625.

Marsh, B. D. and I. S. E. Carmichael, 1974. Benioff zone magmatism, *Jour. Geophys. Research 79*, 1196–1206.

Mavko G. and A. Nur, 1975. Melt squirt in the asthenosphere, *Jour. Geophys. Research 80*, 1444–1448.

McGarr, A., 1976. Seismic moments and volume changes, *Jour. Geophys. Research 81*, 1487–1494.

Minster, J. B., T. H. Jordan, P. Molnar, and E. Haines, 1974. Numerical modeling of instantaneous plate tectonics *Geophys. Jour., Roy. Astron. Soc. 36*, 541–576.

Muller O. H. and D. D. Pollard, 1977. The stress state near Spanish Peaks, Colorado determined from a dike pattern, *Pageoph. 115*, 69–86.

Nakamura, K., 1969. Arrangement of parasitic cones as a possible key to regional stress field, *Volcanol. Soc. Japan Bull. 14*, 8–20. (in Japanese with English abstract)

Nakamura, K., 1974. Preliminary estimate of global volcanic production rate, in J. L. Colp and A. S. Furumoto (eds.), *The Utilization of Volcano Energy*, Albuquerque, N. M.: Sandia Laboratories, 273–285.

Nakamura, K., K. H. Jacob, and J. N. Davies, 1977. Volcanoes as possible indicators of tectonic stress orientation—Aleutians and Alaska, Pure and Applied *Geophysics 115*, 87–112.

Nye, J. F., 1967. Theory of regelation, *Philos Mag. 16*, 1249–1266.

Obert, L., and W. I. Duvall, 1967, *Rock Mechanics and the Design of Structures in Rocks*, New York: Wiley.

Odé, H., 1957. Mechanical analysis of the dike pattern of the Spanish peaks area, Colorado, *Geol. Soc. America Bull. 68*, 567–576.

Pollard, D. D., 1973. Derivation and evolution of a mechanical model for sheet intrusions: *Tectonophysics 19*, 233–269.

Pollard, D. D., 1976. On the form and stability of open fractures in the earth's crust, *Geophys. Research Letters 3*, 513–516.

Pollard, D. D., 1977. On the mechanical interaction between a fluid-filled fracture and the earth's surface, *Jour. Geophys. Research 53*, 27–57.

Pollard, D. D. and O. H. Muller, 1976. Effect of gradients in regional stress and magma pressure on the form of sheet intrusions in cross section: *Jour. Geophys. Research 81*, 975–984.

Roberts, J. L., 1970. The intrusion of magma into brittle rocks, *in* G. Newall and N. Rast (eds.), *Mechanisms of Igneous Intrusion*, Liverpool, Gallery Press, 287–338.

Robson, G. R., K. G. Barr, and L. C. Luna, 1968. Extension failure, an earthquake mechanism: *Nature 218*, 28–32.

Savage, J. C., 1969. The mechanics of deep-focus faulting, *Tectonophysics, 8*, 115–127.

Schubert, G. and D. L. Turcotte, 1972. One-dimensional model of shallow mantle convection, *Jour. Geophys. Research 77*, 945–951.

Schubert, G., D. A. Yuen, C. Froidevaux, L. Fleitout, and M. Souriau, 1978. Mantle circulation with partial return flow: effects on stresses in oceanic plates and topography of the sea floor, *Jour. Geophys. Research 83*, 745–758.

Secor, D. T., 1965. Role of fluid pressure in jointing, *Am. Jour. Sci. 263*, 633–646.

Secor, D. T., 1969. Mechanics of natural extension fracturing at depth in the earth's crust, *in* A. J. Baer and D. K. Norris (eds.) *Research in Tectonics*, Canada Geol. Survey Paper 68-52, 3–48.

Secor, D. T. and D. D. Pollard, 1975. On the stability of open hydraulic fractures in the earth's crust, *Geophys. Research Letters 2*, 510–513.

Shaw, H. R., 1965. Comments on viscosity, crystal settling and convection in granitic magmas, *Am. Jour. Sci. 263*, 120–152.

Shaw, H. R., 1969. Rheology of basalt in the melting range, *Jour. Petrology 10*, 510–535.

Shaw, H. R., 1970. Earth tides, global heat flow, and tectonics, *Science 168*, 1084–1087.

Shaw, H. R., 1973. Mantle convection and volcanic periodicity in the Pacific; Evidence from Hawaii, *Geol. Soc. America Bull. 84*, 1505–1526.

Shaw, H. R. and E. D. Jackson, 1973. Linear island chains in the Pacific: Result of thermal plumes or gravitational anchors?, *Jour. Geophys. Research 78*, 8634–8652.

Shaw, H. R., R. W. Kistler, and J. F. Evernden, 1971. Sierra Nevada plutonic cycle: Part II, Tidal energy and a hypothesis for orogenic-epeirogenic periodicities, *Geol. Soc. America Bull. 82*, 869–896.

Shaw, H. R. and D. A. Swanson, 1970. Eruption and flow rates of flood basalts, *in* E. H. Gilmour and D. Stradling (eds.) *Columbia River Basalt Symposium*, 2nd, Proc., Cheney, Wash.: Eastern Washington State College Press, 271–299.

Shaw, H. R., T. L. Wright, D. L. Pect, and R. Okamura, 1968. The viscosity of basaltic magma: An analysis of field measurements in Makaopuhi lava lake, Hawaii, *Amer. Jour. Sci. 266*, 225–264.

Smith, R. L. and H. R. Shaw, 1975. Igneous related geothermal resources; *in* D. F. White and D. L. Williams (eds.) *Assessment of Geothermal Resources in the U.S.—1975*, U.S. Geol. Survey Circ. *726*, 58–83.

Swanson, D. A., 1972. Magma supply rate at Kilauea Volcano, 1952–1971, *Science 175*, 169–170.

Swanson, D. A., T. L. Wright, and R. T. Helz, 1975. Linear vent systems and estimated rates of magma production and eruption for the Yakima basalt on the Columbia Plateau, *Amer. Jour. Sci. 275*, 877–905.

Swanson, D. A., W. A. Duffield, and R. Fiske, 1976. *Displacement of the south flank of Kilauea volcano: the result of forceful intrusion of magma into the rift zone*, U.S. Geol. Survey Prof. Paper 963, 1–39.

Terzaghi, K., 1936. Simple tests determine hydrostatic uplift, *Eng. News-Record 116*, 872–875.

Voight, B., 1976. *Mechanics of Thrust Faults and Decollement*, Benchmark Papers in Geology, Stroudsberg, Penn.: Dowden, Hutchinson & Ross, Inc.

Wright, T. L., 1971. *Chemistry of Kilauea and Mauna Loa lava in space and time*, U.S. Geol. Survey Prof. Paper 735.

Wright, T. L. and R. S. Fiske, 1971. Origin of the differentiated and hybrid lavas of Kilauea volcano, Hawaii, *Jour. Petrology 12*, 1–65.

Wright, T. L., D. A. Swanson, and W. A. Duffield, 1975. Chemical compositions of Kilauea east-rift lava, 1968–1971, *Jour. Petrology 16*, 110–133.

Chapter 7

ASPECTS OF MAGMA TRANSPORT

FRANK J. SPERA

Department of Geological and Geophysical Sciences and Geophysical Fluid Dynamics
Laboratory, Princeton University

This paper is dedicated to the memory of Christopher Goetze.
His death is a sorrowful loss to all of those who knew him and some who did not.
His contributions to the study of rock deformation will long serve as a
benchmark for excellence. Progress in the rheology of the mantle
has been set back by his loss.

INTRODUCTION

The problem of the transport or mobilization of magma is central not only
to igneous petrogenesis but also to the evolution and dynamics of the
Earth. The generation and emplacement, for instance, of roughly 15 km^3
per year of basaltic magma along the axes of diverging plate margins is a
first-order, petro-tectonic process. A corollary of the implied mass-transfer
process is the transport of large quantities of heat from the interior of the
Earth. The heating (from 0°C to 1,200°C) and complete fusion of 15 km^3
of magma requires an amount of energy roughly equivalent to about 10%
of the annual average heat flow out of the Earth. If mean mantle temperatures were higher in the past, as some suggest (McKenzie and Weiss, 1975),
then one might suspect that magma production rates also assumed greater
values. No matter how volumetric eruption rates may have varied in the
past, the conclusion that magmatism has played a significant role in
the geological evolution of the Earth is inescapable. The almost total
restriction of high-temperature periodotitic komatiitic lavas to the
Archean ($>2.5 \times 10^9$ yrs B.P.) illustrates the importance of this view in

terms of the thermal history of the Earth.[1] Before one may use volcanic and plutonic rocks to constrain the composition and thermal evolution of the crust and mantle, however, the "petrological inverse problem" must be solved. That is, it must be shown how the final volcanic or plutonic product is influenced by the path traversed by the melt, as well as the composition and environmental conditions prevailing at the source.

The term path is used here in an extended thermodynamic sense. It refers not only to the temporal variation of intensive variables such as T, P_{total}, a_{SiO_2} and f_{O_2} but also to the gradients (spatial variation) of these variables. The temperature gradient within a mantle diapir rising through relatively cool lithosphere will, for example, be related to the rate of heat loss to the surroundings. Similarly, pressure and chemical potential gradients will influence magma flow and mass-transfer rates. It is the purpose of this paper to outline some aspects of the "petrological inverse problem" by analyzing mobilization mechanisms in terms of the transport phenomena governing the evolutionary path magmatic systems follow.

The application of transport phenomena to petrological systems consists of three main parts: momentum transport (e.g., viscous flow in conduits and diapirs, segregation and flow of melt in crystal mushes), energy transport (heat conduction, buoyancy convection, viscous dissipation) and mass transfer (diffusion and advection-aided diffusion, infiltration, nucleation and growth of crystalline phases or vapor bubbles). Natural mobilization mechanisms involve a simultaneous superposition of all three types of transport. For example, the estimation of the melting rate of a crystal is not possible unless some auxiliary assumptions about the dynamic and thermal state of the system are specified. Contrast the growth rate of a crystal in an immobile versus mobile body of magma. A primary factor governing interface-controlled crystal growth rates is the degree of supercooling (Kirkpatrick, 1974). The degree of supercooling, ΔT, may be taken as the difference between the equilibrium liquidus temperature of the crystallizing phase and the temperature at the crystal-melt interface. When a relatively large differential velocity exists between melt and crystallite, the fluid is capable of continuously sweeping away the heat produced by the crystallization process, so that the interface temperature and hence degree of supercooling is dependent on the cooling of the igneous body as a whole and not any local interface effects. At the other extreme is the case in which the buildup of crystallization heat around the growing crystal increases the local interface temperature, and therefore decreases the extent or degree of supercooling, other factors remaining the same. For ΔT's less than about 50 to 100°C, growth rates increase with an

---

[1] See Upadhyay (1978) for an account of peridotitic komatiite flows in an Ordovician ophiolite suite from Newfoundland.

increasing degree of supercooling (Kirkpatrick 1974), and so, within this temperature range, a quickly settling crystal grows faster than one moving more slowly. In semiquantitative terms, note that the ratio of heat or mass transport by advection (bulk flow) to that by heat conduction or mass diffusion is given roughly by $u \cdot l/\kappa$ or $u \cdot l/D$ respectively. In these expressions, u and l represent a characteristic velocity and length, κ the thermal diffusivity, and D the mass-diffusion coefficient. With typical values of $u = 10^{-2}$ cm s^{-1}, $\kappa = 10^{-2}$ cm^2 s^{-1}, $l = 10^2$ m and $D = 10^{-7}$ cm^2 s^{-1}, the growth (or dissolution) rate of a crystal in a magma can vary considerably depending on kinematic and thermal boundary conditions. Because of the large number of degrees of freedom associated with irreversible processes, such as mass, heat, and momentum transfer, the elucidation of mobilization mechanics promises to be a complicated business.

In the context of this paper, mobilization refers to the collection of processes that influence the final state of a magmatic system. These processes include melt generation, segregation, ascent, and emplacement or eruption. An important aspect of this paper is the realization that the genetic and transport processes associated with magmatism are best viewed as operating on a variety of concurrent temporal and spatial scales. Virtually all mobilization or transport mechanisms are evidence of mechanical or thermal instabilities. From a dynamic point of view, the eruption or emplacement of a quantity of magma is but the last in a long chain of evolutionary instabilities that characterize most chemical and mechanical systems displaced far from equilibrium (Nicolis et al., 1975; Prigogine and Lefever, 1975).

New and sometimes surprising patterns of spatial or temporal organization are exhibited by many nonequilibrium systems. An example from biology is the process of cell division. It has been suggested (Nicolis and Prigogine, 1977) that division occurs when the concentration of some metabolite reaches a threshold concentration that depends in a complicated fashion on the interplay of the nonlinear kinetics of metabolite production, the processes of mass diffusion, and the size of the cell. Cell division in this view is seen essentially as an instability or bifurcation marking a change from one organizational biochemical regime to another. An example perhaps more germane to magma transport is the onset of convection in a fluid confined between two horizontal rigid walls at different temperatures. If the upper plate is maintained at a fixed temperature and the temperature of the bottom plate is increased, a critical value of temperature difference is eventually reached, and the fluid spontaneously overturns until the temperature gradient is erased. The magnitude of the critical temperature difference depends on a number of parameters, including the thermal diffusivity, viscosity, and isothermal expansivity of the melt, the distance between the plates, and the magnitude and direction of the acceleration due to gravity vector. Below the critical value of the temperature gradient, the fluid experiences no movement and heat is simply transferred by conduction. Above this value, a new pattern of spatial organization emerges (i.e., a convection cell) that is radically different from the previous stationary state.

Another possible example of organization in non-equilibrium systems is the development of oscillatory zoning in feldspars. Here, perhaps, the balance between reaction at the interface and long-range diffusion (of mass, heat, or both) is responsible for the pattern of a spatially varying mole fraction of albite component. The interplay of competing kinetic or transport effects is very much the theme in the application of transport phenomena to petrogenesis.

The purpose of this paper is to survey some of the wide spectrum of transport processes and to give a brief and somewhat simplified (perhaps naive) account of the mechanics involved. This survey makes no pretense of being exhaustive; in recent years great effort has been applied towards an understanding of the transport and kinetic phenomena that constrain petrogenetic theories (see Hofmann et al., 1974; Yoder, 1976; and this volume for appropriate examples). The topics discussed here represent those familiar to the author that seem generally relevant to a wide range of petrologic phenomena. The aim is simply to point out the diversity of transport mechanisms and show that, in fact, these mechanisms reveal some common characteristics when they are analyzed from a dynamic point of view.

MOBILIZATION MECHANICS

In the following sections an attempt is made to give a brief introduction to the mechanics of magma mobilization and flow. The order of topics corresponds in a general way with the *inversion* of the route magma travels from source to the surface or high-level emplacement zone. Although there may be no real advantage associated with the order of topics chosen here, this author finds it easier to physically visualize processes associated with magma transport in the upper crust. This is in part due to the relative accessbility of the crust for observation as compared to, for instance, the mantle 60 or 70 km beneath Hawaii. The discussion commences, consequently, with a brief analysis of the mechanics of fluidization phenomena. This analysis finds application to situations as varied as the eruption of a basalt exsolving a volatile phase and the emplacement of diatremes within the crust. A discussion of the role played by propagating cracks in the upward movement of magma through regions undergoing penetrative deformation follows the section on fluidization. Viscous flow in conduits where pressure gradients or thermal and compositional buoyancy effects drive the flow are then discussed in relation to the conditions for turbulence. The paper ends with an account of the processes that occur at depth, an introduction to the thermodynamics of adiabatic and nonadiabatic magma ascent, and consideration of the factors that influence melt segregation at depth.

A summary section focuses on the critical role played by the rheology of both mobilized magma and the environment it passes through. Rhelogical behavior is dependent on pressure, temperature, and composition; the gradients of these; and the crystallinity and rate of deformation of both the magma and the surrounding mostly crystalline materials.

FLUIDIZATION PROCESSES

Although magma is generally thought to consist primarily of a silicate liquid phase, its eruption as a lava can be greatly facilitated by the development of a volatile phase. Estimates of the pristine H_2O and CO_2 contents of basaltic magmas (Moore, 1970; Moore et al., 1977) suggest that most if not all melts attain volatile saturation during the last few kilometers of ascent. The exsolution of a gas phase in response to a decrease in P_{total}, and an increase or decrease in T, depending on the sign of the enthalpy of the exsolution reaction, could generate volatile pressures sufficient to fluidize a system. From what is known about the chemistry and phase relations in silicic ignimbrites and ash flows, H_2O would seem to be the dominant volatile species (Ross et al., 1961; Smith, 1960; Tazieff, 1970; Pai et al., 1972). The emplacement of both kimberlites (Dawson, 1967; Hearn, 1968; Harris et al., 1970; McGetchin et al., 1973a, 1973b) and the brecciated minettes (lamprophyres) of some volcanic necks (Currie et al., 1970; Turner and Verhoogen, 1960, p. 250–256) have also been attributed to fluidization processes. In the case of the alkaline and carbonatitic rocks, a CO_2-rich gas phase is thought to have been the propellant; in some cases 10% to 20% by mass of volatiles may be retained in carbonate and hydrous phases (Harris et al., 1970).

The magnitude of the ascent velocity of the low-viscosity phase (liquid or supercritical fluid) in a fluidized system may be estimated by consideration of the physics of flow through packed beds.

Imagine a two-phase (liquid-solid or gas-solid) system, in which the less viscous phase is flowing through an immobile packed bed. In the case of a kimberlitic diatreme, for example, the packed bed would be represented by a poorly sorted mixture of xenoliths, megacrysts and brecciated country rock. It has been shown (Kay, 1968; Soo, 1967) that the pressure drop for flow through a packed bed of uniform solid particles is related to the mean fluid flow velocity, u_m, by

$$\Delta p = 3k_f \lambda \frac{1 - \varepsilon}{\varepsilon^3} \frac{L}{D} \rho u_m^2, \tag{1}$$

where k_f, λ, ε, L and D_s represent the friction coefficient, particle shape factor,[2] bed porosity, length of the fluidized zone, and nominal diameter of uniform equivalent-volume spheres. The magnitude of k_f depends on whether the flow is in laminar or turbulent motion. In the laminar region, the Blake-Kozeny equation is found empirically to be applicable. This relation gives (for uniform spherical particles)

$$u_m = \frac{\Delta p}{L} \frac{D_s^2}{150\eta} \frac{\varepsilon^3}{(1 - \varepsilon)^2}. \tag{2}$$

Equation (2) is thought to be valid for laminar flow where the relation

$$\frac{\rho D_s u_m}{\eta(1 - \varepsilon)} < 10 \tag{3}$$

holds (Bird et al., 1960). In turbulent flow through packed beds, the Burke-Plummer relation is valid when

$$\frac{\rho D_s u_m}{\eta(1 - \varepsilon)} > 1,000. \tag{4}$$

The velocity in turbulent flow may therefore be written

$$u_m^2 = \frac{4}{7} \left(\frac{\Delta p}{L}\right) \frac{D_s}{\rho} \frac{\varepsilon^3}{(1 - \varepsilon)}. \tag{5}$$

Of course, the solid particles in a diatreme or in the throat of a volcanic pipe, or the gas bubbles in a basalt, are not of uniform size. For a bed of mixed particles (either crystals, melt droplets, or gas bubbles) the pressure drop may be represented by

$$\Delta p = 3k_f \lambda \frac{1 - \varepsilon}{\varepsilon^3} \frac{L}{D_m} \exp (\sigma^2) u_m^2, \tag{6}$$

assuming a log-probability distribution of particles, where D_m is the median particle diameter as determined, for example, by sieving measurements.

As u_m increases, there is a value of the flow rate, denoted by u_{crit}, such that the bed of solid particles will be supported by the high flow rate of the

---

[2] For spherical particles, $\lambda = 1$; all other shapes have $\lambda > 1$.

fluid stream. When $u_m = u_{crit}$, the bed or tube is said to be fluidized; individual xenoliths or clasts have freedom of movement in the gas stream. With uniform fluidization, the flow is homogeneous and independent of time (i.e., the flow is steady); however, heterogeneities associated with non-uniform xenolith size and shapes probably induce channeling effects, by which the flow becomes unsteady. As the velocity of flow is increased beyond u_{crit}, entrainment of xenoliths will occur and they will be swept upwards. In some instances, geological evidence indicates that critical velocities have been exceeded, because known stratigraphic relations imply a net upwards motion for some xenoliths (McGetchin et al., 1973a, 1973b, 1975).

One may compute an approximate value of u_{crit} by noting that, at the point of fluidization, the pressure given by (1), (2), (5) or (6) must be balanced by the weight of the solid particles supported in the gas stream. This latter quantity is

$$\Delta p = (1 - \varepsilon)(\rho_s - \rho_f)gL. \tag{7}$$

If one assumes uniform-size particles, the critical mean velocity (u_{crit}) is

$$u_{crit} = \left\{ \frac{\Delta\rho\, gD_s(\varepsilon^*)^3}{3\rho_f k_f \lambda} \right\}^{1/2}. \tag{8}$$

In the laminar regime, (8) reduces to

$$u_{crit} = \frac{\Delta\rho\, gD_s^2(\varepsilon^*)^3}{150\eta(1 - \varepsilon^*)}, \tag{9}$$

whereas for turbulent flow

$$u_{crit} = \tfrac{2}{7}\{D_s(\varepsilon^*)^3\, \Delta\rho\, g\}^{1/2}. \tag{10}$$

In (8), (9) and (10), ε^* signifies the porosity of the bed at the critical velocity and $\Delta\rho$ is the density difference (taken as positive) between the solid (ρ_s) and fluid (ρ_f) phase. Although clearly a simplification, the order of u_{crit} may be calculated by the appropriate use of (9) or (10). It should be carefully noted that, while these relations give a rough idea of how u_{crit} depends on the other parameters, the precise numerical value of u_{crit} may in some cases depend on other factors not discussed here—for example, wall effects, unsteady flow, nonspherical and graded particles, fall of particles in non-newtonian melts and grain-dispersive pressures. A more complete summary of fluidization phenomena is currently in preparation by this author.

As an example of the use of these relations consider two illustrative cases: (a) gas-fluidized bed (e.g., CO_2 in a kimberlite, H_2O in an ash flow or a CO_2–H_2O mixture in a vesiculating basalt); and (b) a liquid-fluidized bed (e.g., a xenolith-packed basalt flow). For (a), the following parameters are assumed:

$$\Delta\rho = 3 \text{ gm cm}^{-3} \qquad\qquad D_s = 50 \text{ cm}$$
$$\eta = 10^{-2} \text{ gm cm}^{-1} \text{ s}^{-1} \qquad \varepsilon^* = 0.5$$

From Equation (10) one may compute $U_{crit} = 1 \text{ m s}^{-1}$. For (b) we assume

$$\Delta\rho = .5 \text{ gm cm}^{-3} \qquad\qquad D_s = 25 \text{ cm}$$
$$\eta = 10 \text{ gm cm}^{-1} \text{ s}^{-1} \qquad\qquad 0.2,$$

and compute from Equation (9)

$$u_{crit} = 0.2 \text{ cm s}^{-1}.$$

The physical parameters used to make these estimates have been taken from a number of sources (McGetchin et al., 1973a, 1975; Bird et al., 1960; Carmichael et al., 1977). For an alternative view of explosive volcanism, the reader is referred to the thought-provoking works of Bennett (1971, 1972).

In the case of gas exsolution from a melt, these simple equations are applicable if modified to account for the details of the kinetic problem concerning nucleation and growth rate of gas bubbles from the melt. The interested reader is referred to Sparks (1977) for a more complete discussion. Vapor bubbles in a viscous medium tend to increase dynamic viscosities, and there is a concurrent transfer of heat between solid and fluid. Some of these complications have been discussed with specific reference to ash-flow eruptions by Pai et al. (1972).

In terms of the applicability of fluidization phenomena to igneous petrogenesis, it should be noted that virtually all magmas contain some volatile constituents capable of exsolution at low pressure. For the more voluminous magma types (e.g., mid-ocean ridge basalts, alkali basalts of the oceanic islands, calc-alkaline island-arc volcanics and related plutonic rocks, and continental silicic volcanic rocks), however, the generation of a fluid phase by exsolution occurs at relatively shallow (crustal) depths. A basaltic melt with 1% wt. H_2O reaches water saturation at a pressure equivalent to a depth of roughly one kilometer.

More silicic melts are thought to be slightly more hydrous. Maaloe and Wyllie (1975) believe it uncommon for the water content of large granitic magma bodies to exceed about 2%. Eggler (1972) experimentally showed that the presence of orthopyroxene phenocrysts (as opposed to olivine) implies a maximum H_2O content of 2.5% in a Paricutin andesite. Precise values of the saturation depths at which fluidization is at least potentially applicable may be determined using solubility data from many sources, notably including Burnham and Davis (1971) for H_2O, and Eggler (1978) and Mysen (1977) for CO_2 and CO_2–H_2O mixtures. For another view of water and other volatile concentrations in melts, the reader should consult the works of Anderson (1974a, 1974b, 1975).

The dynamics of gas-crystal fluidization is also relevant to the eruption of kimberlitic and other diatremic rocks. Although these rocks are relatively rare, an understanding of their mechanical and thermal histories would provide insight about temperatures and the distribution of volatiles within the Earth. Complicated three-phase fluidization regimes may be applicable to the eruption of some silicic crystal-tuffs (e.g., Lipman, 1967). Studies regarding the dynamics of three-phase fluidization are not as numerous as two-phase ones, but are becoming more frequently the subject of experimental and theoretical investigations.

In conclusion, the need for careful field descriptions of the abundance, size, and distribution of the various inclusion types in rocks emplaced or presumed to have been emplaced by fluidization processes is emphasized. These data, when combined with estimates of volatile fugacities as calculated from thermodynamic expressions or phase equilibria studies (using compositions of associated phenocryst phases), are the basic information needed to estimate minimum fluidized eruption rates from Equation (6) or its more sophisticated and derivable analogues.

MAGMA TRANSPORT BY CRACK PROPAGATION

The emplacement or eruption of magma implies bulk movement of melt through the Earth's upper mantle and crust. The course rising magma follows is influenced by the magnitude and variation of regional tectonic stresses (e.g., see Jackson and Shaw, 1975; Jackson et al., 1975; Fiske et al., 1972). The presence of a low-viscosity fluid phase (e.g., a volatile phase or magmatic fluid) at a total pressure roughly equal to P_{lith} will greatly reduce the overall strength of the surroundings.[3] Migrating melts may also be thermally and chemically active. A volatile-enriched low-viscosity melt may act as a corrosive agent on the surroundings, and, with the aid of pressure forces, be injected along pre-existing or newly-formed zones of weakness. One limit to magma ascent rates is the velocity of propagation of a weakened fracture zone. The rate of crack propagation may be computed with some confidence in certain simplified cases (e.g., a perfectly elastic homogeneous substance in a homogeneous stress field); however,

---

[3] The term lithostatic pressure refers to the pressure at the bottom of a column where the density of the overburden has the average value $\bar{\rho}$. Then $p_{lith} = \bar{\rho}gz$ where z is the thickness of the overburden.

the rates of propagation of magma-filled cracks can only be roughly estimated given the present state of the art.[4]

Aki et al. (1977) proposed a model for the mechanism of magma transport in the upper crust based on migrating-fluid (magma) tensile cracks driven by an excess of magma pressure over lithostatic pressure. They claim that this mechanism can generate seismic waves by a succession of jerky crack extensions when fracture strengths of surrounding country rocks vary spatially. Their model leads to predictions regarding the spectrum of seismic-energy radiation from propagating cracks approximately in accord with the spectral features observed for volcanic tremor. It would seem then that, at least at shallow depths, excess magmatic pressure in magma-filled cracks is an important driving force for transport processes. The ultimate origin of the implied excess magma pressures is not entirely clear: if exsolution of volatiles is responsible for the pressure differential, there should be a correlation between the temperature and volatile content of a magma with the depth at which transport by the mechanism of Aki et al. (1977) becomes operative. As will be pointed out in a later section, even when no excess of magmatic over lithostatic pressure exists, fractures may open and be injected with melt, provided certain relations between the principal effective stresses hold. The important role of local and regional stress states to melt transport warrants a discussion of the factors that govern the formation, configuration and propagation rates of magma-filled cracks.

Griffith Theory of Brittle Failure

The motivation for studying and applying the Griffith theory of brittle rock failure is firmly anchored in the concept of the effective stress and the role of fluid pressure in formulating quantitative failure theories. As previously mentioned, the presence of a fluid phase (e.g., volatile or melt phase) greatly reduces the overall strength of a medium, thereby increasing the probability of the nucleation and propagation of a crack or zone of weakness within that material. An elaboration of this idea is presented in Appendix I. The sections that follow will assume basic familiarity with the ideas presented in that Appendix.

Much has been written concerning quantitative criteria for brittle failure in rocks. Perhaps one of the most successful models is that attributed to Griffith (1921, 1925), as modified and discussed by many other distinguished geologists (for example; Ode, 1957; Brace, 1960; Murrell,

---

[4] Detailed seismicity studies of earthquake swarms within the mantle in regions where other geophysical and geological evidence suggests the presence of mobilized melt may make it possible to estimate flow rates if accurate locations can be determined and if it is assumed that the propagation of melt filled cracks is accompanied by seismic signals (see Aki et al., 1977; Klein et al., 1977).

1964, 1965; McClintock and Walsh, 1962; Secor, 1965). The Griffith theory is essentially based on the supposition that all solids are filled with tiny cracks. When differences in the principal stress deviator components (see Appendix I) exceed a certain critical value, high tensile stress will occur near the ends of the Griffith cracks even when all the principal stress deviators are compressive (a compressive normal stress is taken as positive). Failure occurs by the spontaneous growth of a Griffith crack in response to sufficiently large tensile stresses developed at the tip of a suitably oriented flaw.

A particularly convenient way to represent the stress conditions for failure is by means of a stress envelope for rupture on a Mohr diagram (see Shaw, this volume). That is, on a plot of τ versus σ_{eff}, where σ_{eff} is the effective stress (Appendix I), there is a locus of (τ, σ) points, represented by some equation of the form $\tau = f(\sigma)$, that may be taken as the criterion for failure. The envelope predicted by the Griffith theory is of the form (for plane stress)

$$\tau^2 - 4K\sigma - 4K^2 = 0, \tag{11}$$

where $-K$ is the tensile strength and τ and σ represent the shear and normal components of stress on the plane of the most critically oriented cracks (Secor, 1965; Tremlett, 1978).

Natural fracturing can occur in response to either tensile or compressive principal stresses (i.e., anywhere along the failure envelope). If the magnitude of the isotropic fluid pressure or magma pressure (p_f) is equal to the ambient lithostatic pressure, fractures will open up only if the normal stress across the fracture surface is tensile. Stated another way, when $p_f = p_{lith}$, injection of magma can occur only for values of the principal stresses corresponding to the tensile (negative) part of the Mohr envelope. These stress states correspond to the formation of either tension fractures perpendicular to the minimum principal stress axes, or to conjugate shear fractures inclined to the directions of principal stresses such that the acute angle of conjugate intersection is bisected by the direction of $\sigma_{I, eff}$. It is important to note that, if $p_f = p_{lith}$, then shear fractures will open only if the stress circle at failure intersects the Mohr envelope at a point corresponding to a tensile (negative) normal stress. If, on the other hand, p_f is greater than the average of the principal stresses, shear fractures across which a *compressive* normal stress acts may be injected with magma.

The failure criteria for tension fracturing are that

$$\sigma_{III, eff} = -K, \tag{12a}$$

and that

$$\sigma_{I, eff} \leq -3K. \tag{12b}$$

Note that the effective stress is defined according to

$$\sigma_{III, \text{eff}} = \sigma_{III} - p_f.$$

In regions of the Earth where $\sigma_{I, \text{eff}} > 3K$, tension fractures could not open. If σ_I is taken as coincident with the vertical direction, the form of an igneous intrusion would be a vertical magma-filled crack (e.g., a dike). In the case when σ_I is horizontal, a sheet-like body would form. The interested reader is referred to Roberts (1970) for an account of the various configurations of magma intrusions into brittle rocks, and their dependence on the stress field.

In shear fracture, Murrell (1964) cites the criteria for open fractures to be

$$-K < \sigma_{III, \text{eff}} < 2(1 - \sqrt{2})K \tag{13}$$

$$3K < \sigma_{I, \text{eff}} < 2(\sqrt{2} + 1)K, \tag{14}$$

where $-K$ represents the tensile strength.

Values of K are rather small, varying between several tens to perhaps several hundred bars (Brace, 1960 and 1964; J. Suppe, personal communication) and are sensitive to the length of the Griffith crack (Haward, 1970). The variation of K with temperature is not well-documented, but intuitively it seems that K would decrease at elevated temperatures. The experimental demonstration of decreasing K with increasing temperature is, no doubt, complicated by the increased tendency for ductile behavior at elevated temperatures.

The conditions for failure with melt propagation require effective principal deviatoric stresses on the order of several hundred bars. This constraint does not represent an effective inhibition to failure and melt intrusion in the Earth: independent evidence, including the magnitude of geoidal undulations and stress drops accompanying earthquakes, indicates that nonlithostatic deviatoric stresses of several hundred bars may develop within the Earth. Direct measurements of $\sigma_{I, \text{ett}} - \sigma_{III, \text{ett}}$ within the upper parts of the crust support this view (Haimson, 1977).

Rates of Crack Propagation

The fracture process easiest to visualize is the catastrophic failure of a brittle substance (e.g., breaking a beer bottle) in which the velocity of crack propagation rapidly approaches $\frac{4}{10}$ of the compressional wave velocity of the material. Many materials, however, undergo an earlier phase of slower ($\sim 10^{-3}$ m s^{-1}) cracking before final catastrophic failure. The rate of subcritical or non-catastrophic crack growth is influenced by factors such as the temperature and chemistry of the crack environment (Anderson and Grew, 1977).

Fast-Cracking

Griffith (1921) was the first to apply the concept of conservation of energy to determine the conditions for spontaneous growth of a crack in terms of the tensile stresses acting at its tip. His result for the threshold tensile stress, found by minimizing the total energy of the system with respect to the length of the crack, is given by

$$\sigma_{\text{crit}} = \left\{ \frac{4\rho V_s^2 \gamma (1 + v)}{\pi a (1 - v)^2} \right\}^{1/2}, \tag{15}$$

where ρ, V_s, γ and v represent the density, shear wave velocity, surface energy and Poisson's ratio for the material, and a and σ represent the half-length of the crack and the tensile stress at the tip of the crack, respectively. For a typical *crystalline* silicate (a dunite from Twin Sisters, Mt. Washington) representative values are as follows (Birch, 1966; Swalin, 1970; Verhoogen et al., 1970):

$$\rho = 3.31 \text{ gm cm}^{-3} \qquad V_s = 4.9 \text{ km s}^{-1}$$
$$\gamma = 10^3 \text{ erg cm}^{-2} \qquad v = .24.$$

From Equation (15), the critical stress associated with a 1 cm long crack is about 50 bars. It might be intuitively expected that a melt-filled crack requires smaller stresses for propagation. Aki et al. (1978) analyzed seismograms related to the propagation of seismic waves through the partially frozen lava lake at Kilauea Iki. Their data indicate severe attenuation of shear waves traveling through the buried magma body. Their value of $V_s = .2 \text{ km s}^{-1}$ may be combined with the values

$$\rho = 2.7 \text{ gm cm}^{-3} \qquad \gamma = 10^3 \text{ erg cm}^{-2} \qquad v = .5$$

to give a critical stress of 3 bars necessary for the propagation of a 1 cm long crack through a partly consolidated magma. Although these figures should be considered as gross estimates, the calculations suggest that crack propagation may be initiated in crystal mushes at stresses at least an order of magnitude smaller than in corresponding crystalline materials.

Extensions of the Griffith theory that take account of kinetic energy terms have been proposed (Lawn and Wilshaw, 1975; Roberts et al., 1954; Anderson, 1959). The calculation of fast-crack propagation rates is possible using expressions developed by these authors. A noteworthy result is that, after the critical stress has been attained, the crack accelerates and very rapidly attains a propagation rate equal to about $\frac{2}{5}$ the P-wave velocity of the material.

The slow-cracking process, which is probably more relevant to the magma transport problem, is more realistic and consequently more difficult to analyze. As summarized by Anderson and Grew (1977),

> . . . many materials exhibit time-dependent fracture, that is, when they are subjected to a static load, they fail after some time interval that depends inversely on the magnitude of the load. Griffith's (fast-cracking) theory fails to account for this phenomenon, because it predicts that a stress only infinitesimally less than the critical value would be sustained indefinitely.

The effects of temperature and the presence of chemically reactive fluids are particularly important in determining the rate of slow-crack growth. In stress corrosion cracking,[5] the limiting crack velocity is controlled by diffusion and viscous processes in the fluid at the tip of the crack.

Slow-crack propagation rates are also dependent on the temperature and the so-called stress intensity factor K_I (Irwin, 1968; Lawn and Wilshaw, 1975), a parameter that is proportional to the strength of the material and the length of the crack. Typical slow-crack propagation values from Anderson et al. (1977, Figure 15) lie in the range 10^{-1} to 10^2 cm s^{-1} when the local temperature at the tip of the crack is $\frac{2}{3}$ of the melting temperature for the material. The high strain energy associated with large stresses in a material at the tip of a tiny elliptical flow may transiently generate high temperatures. This thermal effect will lead to increased propagation rates, perhaps in a regenerative manner. For example, in alumina (Evans, 1974) at constant K_I, increasing T from 100°C to 1,200°C ($T/T_m = .05$ to .60) increases the crack velocity by six orders of magnitude.

Stress Corrosion and Magma Transport

Although it seems impossible at the present time to calculate crack velocities with any quantitative assurance, some general comments based on analogies with ceramic systems can be made (Wilkins et al., 1976).

The primary condition for crack propagation and melt ascent is that there be tension in the solid immediately in front of the crack, and that the magma rises up and follows the propagating locus of failure (Weertman, 1968, 1971, 1972). In this sense, the crack moves no faster than that limited by viscous flow of the low-density melt phase. One expects volatile-rich, low-viscosity melts to be more mobile than highly polymerized ones. The

---

[5] Stress corrosion is a term sometimes used in the metallurgical literature to describe the behavior of loaded alloys under the influence of high temperatures and chemically active fluids. In glasses and ceramics, the same phenomena of growth of flaws in adverse environments is known as static fatigue (see Wilkins et al., 1976).

presence of a discrete fluid phase could also increase propagation rates by weakening the surrounding material by chemical reaction (Raleigh and Paterson, 1965). Wiederhorn (1967), for instance, has shown that increasing P_{H_2O} in a $Na_2O-CaO-SiO_2$ glass can increase crack velocities by several orders of magnitude, other factors remaining constant. An active fluid phase may explain the inferred or observed high emplacement velocities of the diatreme association, of some dike rocks, and perhaps of ash-flow tuffs.

Temperature also plays an active role in determining propagation rates in that the fracture strength of a material decreases at elevated temperature. The rates of chemical reactions between melt and the surroundings are also affected; high temperatures generally favor transport processes that are exponentially controlled,[6] such as mass diffusion, rates of chemical reactions, and viscous flow phenomena.

Slow-propagating cracks in the lithosphere, accompanied by magmatic activity, have been suggested as an alternative to the "hot-spot" hypothesis of volcanic ridge growth (McDougall, 1971; Turcotte et al., 1973, 1974). This idea may be difficult to test in that crack propagation at such low rates does not appear to generate seismic energy (Anderson et al., 1977). H. R. Shaw (this volume) explores in more detail a mechanism by shear failure (possibly aided by stress corrosion) for the transport of magma beneath Hawaii. It may be that careful location and magnitude-frequency studies of earthquake swarms in known volcanic regions will lead to some propagation-rate and energetic constraints on the migrating-fracture hypothesis for melt transfer. Hill (1977) has concisely summarized the qualitative features of a model for earthquake swarms in volcanic regions.

VISCOUS FLOW

At the outset it must be stated that the relationships and descriptions of flow phenomena that follow are not always *directly* applicable to magma flow. The philosophy adopted here is to consider *simple* models of flow and show with order-of-magnitude calculations where and when these models may be applicable. There is no single transport mechanism that may be universally applicable; however, the flow of magma is governed by the same balance of forces that describes flow in the oceans and atmosphere, convection in stars, and the flow of blood (a non-Newtonian fluid). Study of the fluid dynamics and heat-transfer properties of mobilized

---

[6] For instance, the temperature dependᵉnce of the viscosity of a newtonian fluid is given approximately by $\eta = \eta_0 e^{-(E/RT)}$ where η_0, E, R and T represent the viscosity at some reference T_0, the activation energy for viscous flow, the ideal gas constant and the thermodynamic temperature.

magma is a field in its infancy (perhaps prenatal?); it is to be hoped that great progress will be made in the future.

Among the simplest transport processes one may envision for the upward flow of magma is that of pipe flow. A simple model consists of a magma-filled pipe of fixed diameter extending vertically downwards. Melt may be driven upwards in response to a number of forces. In *forced convection*, pressure gradients generated by the stress deviator (Appendix I) will set up horizontal and vertical stress gradients inducing flow along the axis of the pipe. In the case of *free convection*, density differences arising from compositional and thermal differences across the pipe create buoyancy forces. The velocity field (that is, the spatial variation of the velocity) of a fluid in pipe flow depends on the rheological[7] characteristics of the fluid, and on the balance between pressure and buoyancy forces causing flow, and the viscous forces (friction) retarding it. It may be expected that, in general, flow involves the superposition of pressure as well as buoyancy forces. In the sections that follow some details related to forced, free, and mixed convection with respect to magma transport are discussed.

FORCED CONVECTION: POISEUILLE FLOW

Imagine a pipe of diameter D extending from the surface to a depth L. The conduit is filled with magma of Newtonian viscosity η. The mean flow rate (u_m) and volumetric flow rate, Q, for isothermal, laminar flow are related to the pressure gradient $\Delta p/L$ by the Hagen-Poiseuille law whereby

$$u_m = \frac{4Q}{\pi D^2} = \frac{\Delta p}{8\eta L} R^2. \tag{16}$$

In Equation (16) R represents the radius of an assumed circular pipe. For flow between vertical parallel plates the relation is not much different; the ideas discussed here can easily be extended to that case. The dependence of u_m on $\Delta p/L$, R and η is given in Table 1. From considerations of the magnitude of stress drops associated with earthquakes and the non-hydrostatic figure of the Earth, deviatoric stresses of perhaps 10^2 bars over lengths of the order of 1,000 km may be expected (Verhoogen, et al., 1970). Adoption of these figures leads to values of $\Delta p/L$ of order 10^{-1} bar km^{-1}. Note that some of the flow rates in Table 1 are of the same order as measured slowly propagating cracks in ceramic materials.

---

[7] Rheology is the science of deformation and flow. Parameters such as the crystal content, degree of polymerization, temperature, pressure and amount and nature of volatiles are all important in governing the relationship between shear stress and strain rate in magma. A simple constitutive relationship is that corresponding to Newtonian flow where the magnitude of the shear stress is directly proportional to the strain rate. The constant of proportionality is η, the newtonian viscosity.

Table 1. Dependence of u_m on $\Delta p/L$, R and
η calculated from Equation (17) for
laminar isothermal Poiseuille flow.

Conduit Radius R (m)	Pressure Gradient $\Delta p/L$ (bar km^{-1})	Mean Velocity u_m (cm s^{-1})
	$\eta = 10$ poises	
.5	10^{-2}	3.1
	10^{-1}	31.0
	1	313.0
2.0	10^{-2}	50.0
	10^{-1}	500.0
	1	5000.0
5.0	10^{-2}	313.0
	10^{-1}	3130.0
	1	31250.0
	$\eta = 10^2$ poises	
.5	10^{-2}	0.3
	10^{-1}	3.0
	1	30.0
2.0	10^{-2}	5.0
	10^{-1}	50.0
	1	500.0
5.0	10^{-2}	31.0
	10^{-1}	313.0
	1	3130.0
	$\eta = 10^3$ poises	
.5	10^{-2}	0.03
	1	3.1
2.0	10^{-2}	0.5
	1	50.0
5.0	10^{-2}	3.1
	1	313.0

A limiting case of pressure-driven flow occurs when the pressure difference originates purely from the density difference between melt and
crystalline solid. If the entire tube is filled by magma, then the pressure
difference at the base of a tube of length L is found by balancing the
lithostatic pressure against the magma pressure. For these conditions one
computes

$$\Delta p \sim \Delta \rho g L. \tag{17}$$

If $L \approx 60$ km, for example, $\Delta p \approx 2$ Kb when $\Delta \rho = .3$ gm cm^{-3}. This
differential pressure is sufficient to propel magma upwards at an average
rate of about 4 m s^{-1} (with $\eta = 300$ poises, $R = \frac{1}{2}$ m). Simple Poiseuille
flow, however, is too facile for a number of reasons. Perhaps the most
limiting assumptions are: (1) isothermal flow, and (2) the presence of a
long, deep fissure or pipe. The means whereby a simple long crack or
pipe could open (and remain open) are not clear.

FREE CONVECTION

In pure *natural* or *free convection*, the motion is caused solely by buoyancy forces. These buoyancy forces may be of compositional or thermal origin. As an example, consider an anhydrous basalt liquid with $\rho = 2.75 \text{ gm cm}^{-3}$ at $1,200°C$. The decrease in density as the temperature and mass fraction of dissolved H_2O increase is given by

$$-\Delta\rho = 5.2 \times 10^{-5}\theta + 1.0w_{H_2O} \tag{18}$$

Equation (18) is based on data in Carmichael et al. (1977) and Shaw (1974) and θ is defined as

$$\theta = T - 1,473. \tag{19}$$

Given the likely temperature gradients established across or along a narrow dike and allowing for the possibility of mass transfer of H_2O from the country rock into the magma (Shaw, 1974), local density perturbations can be significant. For instance, with $\theta = 200°K$ and $w_{H_2O} = .03$ (3 mass percent H_2O in the melt),

$$-\Delta\rho = .04. \tag{20}$$

Although this may seem to be a small difference, it could lead to significant buoyancy forces. A rough guide to the magnitude of density-induced convection velocities is given by

$$u = \left(\frac{g\,\Delta\rho\,\kappa L}{\eta}\right)^{1/2}, \tag{21}$$

where κ is the thermal diffusivity and L the length scale over which the density differences occur. With $L \sim 1.0 \text{ m}$, $\kappa = 10^{-2} \text{ cm}^2 \text{ s}^{-1}$, $\eta = 10^2$ poises and $|\Delta\rho| = .04$, velocities of the order of a few cm s^{-1} result.

MIXED CONVECTION

As indicated above, neither of the two idealized models discussed pertains very closely to the flow of magma in natural conduits. One may anticipate that most flows will be driven by *concurrent* buoyancy and pressure gradients. A large amount of experimental and theoretical effort has been invested in the study of heat and momentum transfer in mixed convection conduit flow (Goldstein, 1938; Eckert et al., 1950; Lighthill, 1953; Ostrach, 1954; Hallman, 1956; Ostrach et al., 1958; Morton, 1960; Rohsenow et al., 1962; Scheele et al., 1962; Takhar, 1968). These studies may be applied to the problems of magma flow and heat transfer to obtain some idea of the geometry and transient evolution of magmatic bodies. The flow regime exhibited in a particular case depends on the externally imposed thermal, kinematic, and dynamic boundary conditions.[8] The imposition of horizontal and vertical temperature gradients can greatly influence the flow configuration and lead to turbulence in the ascending fluid. In studies of thermal convection in tubes, it may be shown (Lighthill, 1953) that the product (R/L) Ra characterizes the kind of flow to be expected.[9] For example, at constant R and Ra, as R/L varies from 1 to $\frac{1}{100}$, flow patterns change from the thin boundary layer type (boundary layer not filling tube) to fully developed flow without a well-mixed iso-thermal core region and finally to a pattern in which a stagnant portion of fluid lies above or below the fully developed laminar boundary layer (see Figure 12 in Shaw (1965) and Figures 2, 3, and 6 in Lighthill (1953)). At very high values of Ra, turbulent flow would be expected.

ROLE OF TURBULENCE

The accurate prediction of the physical conditions necessary for turbulent magmatic flow in conduits or pluton size bodies is a difficult but important task (Appendix II). The velocity of a fluid in turbulent flow differs funda-mentally from one in laminar flow. In the former case, fluctuating velocity components are superimposed on the mean (time-averaged) velocity field. These fluctuating components are capable of redistributing heat, momen-tum, and mass (e.g., a phenocryst population or a particular chemical species) in a highly efficient manner. Rates of transfer of these quantities are therefore expected to be higher in a turbulently convecting magma body. These predictions have been shown to be accurate in numerous experimental and theoretical studies (Deissler, 1955; Eckert et al., 1950). The unequivocal petrological demonstration that a given body was in a turbulent regime, however, is difficult to prove partly because flow regimes are markedly time dependent. In a pluton cooling by free convection, for instance, temperature differences between magma and surroundings, responsible for flow in the first place, decrease with time. The efficacy of convective mixing and transport would be expected to damp out ac-cordingly, and all indications of turbulence, such as a random distribution of early crystallized solids, might be erased.

Although the conditions necessary for turbulent flow of an isothermal Newtonian fluid in a long, smooth tube are reasonably well known

---

[8] The term flow regime is meant to refer to the qualitative geometric features of a flow. A flow pattern characterized by a well-mixed isothermal core region with thin boundary layers near the walls is an example of what may be called a boundary layer regime. The pattern of flow in which temperature and flow rates vary continuously across the body is called fully developed flow.

[9] R and L refer to the radius and length of the tube. Ra is the Rayleigh number defined by Ra $= \alpha g \Delta T R^3 / \kappa v$ where α, κ, v and ΔT represent the isobaric expansivity, thermal diffusivity, kinematic viscosity and axial temperature drop respectively.

(Goldstein, 1938), the applicability of this criterion to magmatic conduit flow is of little use. The ubiquitous presence of thermal gradients in magmatic flows, coupled with the demonstration (Scheele et al., 1962), that, in non-isothermal flow, instability sets in at a much lower flow rate implies that any analysis of turbulent flow is likely to be misleading unless temperature effects are explicitly considered. Presented in Appendix II is a discussion of the magnitudes of pressure and superadiabatic temperature gradients necessary for the establishment of turbulent flow in the idealized limiting cases of forced, free, and mixed convection. The conclusion reached is that, for the magnitudes of possible vertical pressure and temperature gradients within the Earth, turbulent pipe flow may be a common phenomenon. This conclusion is especially true for low viscosity (generally basaltic) melts.

ROLE OF CONVECTION

Since numerous workers (Bartlett, 1969; Shaw, 1965; Spera, 1977) have demonstrated the likelihood of natural convection in magma chambers and ascending magmatic diapirs, little discussion is called for here.[10] The essential idea is that convection does not begin until the Rayleigh number (Ra) exceeds some critical value,[11] which depends on the imposed boundary conditions, but is usually in the neighborhood of 1,500. Ra may be thought of as the ratio of buoyancy forces favoring convection to the viscous forces retarding flow. A natural consequence of the large size of magma bodies and the physical properties of silicate melts is that magma bodies will tend to exhibit convection.

The magnitude of Ra plays an important role in the determination of the spatial and temporal evolution of convection velocities and corresponding temperatures within the convecting body. The relationship between Ra and the rates of mass and heat transport in the limiting case of boundary layer (high Ra) convection may be found in Shaw (1974), Carmichael et al. (1977) or Spera (1977). Bartlett (1969) has considered the effect of natural convection on the distribution of phenocrysts in a convecting magma chamber. He gives expressions for the variation of the particle (phenocryst) population density with height as a function of the

---

[10] It is conceivable that an upward concentration of volatiles (notably H_2O) or a downward concentration of phenocrysts in a magma chamber can more than compensate for an unstable density distribution resulting from heating from below, thereby inhibiting convection (see Elder, 1970; Hildreth, 1977; Wright et al., 1971).

[11] One should note that this criterion assumes newtonian behavior for magma. Suspended crystals in a silicate liquid result in a melt with a finite yield strength. The usual criterion for convection (Ra > Ra_{crit}) assumes Newtonian behavior and so may not be entirely appropriate.

Stokes terminal settling velocity and Ra. These calculations may be useful for low phenocryst contents, but the deviation from Newtonian behavior should be considered in highly polymerized melts or in melts with a significant crystallinity.

MAGMA ASCENT RATES: INFERENCES FROM XENOLITHS

The occurrence of lower-crustal or upper-mantle xenoliths in a basalt flow may be used to estimate the ascent velocity of the magma. A minimum ascent velocity is found by balancing frictional and buoyancy forces acting on a xenolith settling at its terminal velocity (u_n). The magnitude of u_n determines the particle Reynolds number defined as

$$\text{Re}_n = \frac{\rho_l D_n u_n}{\eta_l}, \tag{22}$$

which in turn is related to the value of the drag coefficient for a spherical nodule settling through melt of Newtonian viscosity η_l.[12] Balancing frictional and bouyancy forces and solving for u_n one finds

$$u_n = \left(\frac{8R_n \Delta \rho g}{3C_d \rho_l}\right)^{1/2}, \tag{23}$$

(Carmichael et al., 1977) where u_n, R_n, $\Delta\rho$, ρ_l and C_d represent the nodule settling velocity, radius, density difference between melt and xenolith, density of melt, and drag coefficient, respectively. C_d is related to the nodule settling velocity and hence to Re_n. For $\text{Re}_n < .1$ one has

$$C_d = \frac{24}{\text{Re}_n}, \tag{24}$$

and when $\text{Re}_n > .1$

$$C_d = 18 \, \text{Re}_n^{-3/5}. \tag{25}$$

These relations have been set out in more detail elsewhere (Carmichael et al., 1977). As an example, consider the inferred magma ascent rate based on the size of an ultramafic xenolith from the 1801 eruption of Hualalai Volcano, Hawaii. Substituting the parameters $\rho_n = 3.45$ gm cm^{-3}, $D_n = 30$ cm, $\rho_l = 2.8$ gm cm^{-3} and $\eta = 350$ poises, one computes $\text{Re}_n = 14.9$, $C_d = 3.55$ and so $u_n = 51$ cm s^{-1} (1.8 km hr^{-1}). The actual entrainment velocity must have been somewhat larger than the settling rate since the xenolith has obviously moved upwards.

[12] It is not necessary to assume spherical particles. The interested reader is referred to McNown and Malaika (1950) for a consideration of the effect of particle shape on settling velocity.

It is important to realize that this calculation assumes Newtonian behavior for the melt. Sparks et al. (1977) have pointed out, in fact, that a melt containing a few percent crystals by volume may behave rheologically like a Bingham plastic. Such a material has a rheological "equation of state" of the form

$$\sigma = \sigma_0 + \mu\dot{\varepsilon}, \tag{26}$$

where σ is the stress $\dot{\varepsilon}$ the strain rate, σ_0 is the yield strength and μ the coefficient of viscosity. A xenolith must be a minimum size if it is to sink through a Bingham plastic melt because of the finite yield strength of the magma. The minimum radius(r^*) is approximately given by (Timoshenko and Goodier, 1970; Sparks et al., 1977)

$$r^* = \frac{3K\sigma_0}{4\Delta\rho\, g}, \tag{27}$$

where K is a dimensionless constant equal to about 5.0 and σ_0 is the yield strength. There are only a few measurements of σ_0 (Shaw et al., 1968; Sparks et al., 1977); measured values are in the range of $10^2 - 10^3$ dyne cm^{-2}. There is some evidence that σ_0 depends on the volume fraction of crystals (Soo, 1967) according to

$$\sigma_0 = k_1\phi^3, \tag{28}$$

where k_1, ϕ, and σ_0 represent an empirically derived constant having the dimensions of stress, the volume fraction of suspended crystals (assumed spherical) and the yield strength respectively. Use may still be made of (24) to determine settling rates if an account of the non-Newtonian properties of the melt is made. For the non-Newtonian case, the settling velocity, u_n, becomes

$$u_n = .344 \left(\frac{\Delta\rho\, g}{\rho_l}\right)^{5/7} \left(\frac{\rho_n}{\eta}\right)^{3/7} \left(R_n - \frac{3K\sigma_0}{4\Delta\rho\, g}\right)^{8/7} \tag{29}$$

when $\mathrm{Re}_n > .1$. (The result for $\mathrm{Re}_n < .1$ can be derived in the same way.) Notice that in (29) the yield strength enters into the balance for the settling rate. Table 5 gives some numerical examples of u_n for a variety of σ_0 values. Also indicated in Table 5 are approximate ϕ values (volume fraction crystals) consistent with Equation (28) and the few measurements of ϕ versus σ_0 for basaltic magmas. Note that a 14 cm diameter xenolith would exhibit no tendency to sink in a melt with $\sigma_0 \simeq 1,200$ dyne cm^{-2}; this might correspond to a crystallinity of about 35%. These relationships point out the need for detailed observations regarding the crystallinity of nodule-bearing (generally alkalic) basaltic flows and the abundance and size distribution of enclosed xenoliths. Further measurements concerning the dependence of σ_0 on ϕ should be undertaken; Equation (28) has not been adequately tested in silicate crystal-liquid suspensions.

Table 5. Settling rates of xenoliths in basaltic magmas. The settling rate, u_n, is computed from Equation (29). These are minimum ascent rates; for xenolith entrainment the magma velocity would have to exceed the settling rates listed here. The following parameters have been assumed constant: $\Delta\rho = .65$ gm cm^{-3}, $\rho_l = 2.8$ gm cm^{-3}, $\eta = 350$ poises and $K = 5.0$. ϕ (volume fraction solids) is computed from Equation (28) with $k_1 \approx 3 \times 10^4$ dyne cm^{-2}. The effective radius is defined as $R_n - r^*$ and r^* is given by Equation (27).

Spherical Xenolith Radius R_n (cm)	Effective Radius $R_n - r^*$ (cm)	Yield Strength σ_o (dyne cm^{-2})	Volume Fraction Crystals ϕ	Nodule Settling Velocity u_n (cm s^{-1})
15.0	15.0	0	.0	51.0
15.0	13.2	300	.21	44.0
15.0	12.0	500	.26	39.2
15.0	9.1	1000	.32	28.7
15.0	3.2	2000	.41	8.8
7.0	7.0	0	.0	21.0
7.0	5.2	300	.21	15.0
7.0	4.0	500	.26	11.1
7.0	1.1	1000	.32	2.6
7.0	.0	1190	.34	0.0

ROLE OF VISCOUS DISSIPATION

Since the pioneering work of Gruntfest (1963, 1964) and Shaw (1969) the possible role of viscous dissipation in conduit flow and magma genesis has been discussed by a number of workers (Anderson et al., 1974; Fujii et al., 1974; Hardee et al., 1977). The essential idea behind viscous dissipation is as follows. In a material undergoing penetrative deformation, mechanical kinetic energy will be degraded through the action of viscous stresses to thermal energy (i.e., frictional heat). If the rate of viscous heat production exceeds the rate of heat removal by convection or conduction, the material will heat up, or melt, or both. Whether viscous dissipation is important depends, therefore, on the thermal and kinematic boundary conditions, the physical properties of the melt (notably the viscosity-temperature dependence) and the size of the body or conduit. According to Gruntfest (1963) the condition that shows whether viscous heat production can be balanced by heat conduction or not is identified by a critical value of dimensionless parameter that has since been termed the Gruntfest number, Gu. The value of Gu required for a runaway heating or melting episode depends on boundary conditions and the geometry. Gu has been defined as

$$\text{Gu} = \frac{a\sigma_o^2 l^2}{k\eta_0}, \tag{30}$$

where a is a parameter relating η to T (i.e., $\eta = \eta_0 e^{-a(T-T_0)}$ and σ, l, k, and η_0 represent the constant externally-imposed shear stress, the thickness of the layer undergoing shear, the thermal conductivity and the viscosity at T_0, a reference temperature, respectively. $Gu_{crit} = .88$ for plane shear at constant stress where the initial temperature is everywhere the same and equal to T_0. As an example, consider the possibility of viscous heating in a horizontal layer in the mantle of thickness l and initial Newtonian viscosity of 10^{21} poises. If a constant shear stress acts across the layer and $Gu > Gu_{crit}$ viscous heat generation cannot be balanced by heat conduction and thermal runaway will occur. Taking the thickness of the shear zone to be 10 km and with $a = 10^{-1}$ K^{-1}, $\sigma = 10^2$ bars and $k = 3 \times 10^5$ erg cm^{-1} K^{-1} s^{-1} (Spera, 1977; Shaw, 1968; Carmichael et al., 1977), $Gu = 3.3$. Viscous heat would be generated at a rate greater than heat could be conducted out of the plane layer. Presumably, temperature would rise in the layer until T approaches T_s, the solidus temperature, at which point partial fusion will commence. Thereafter, temperature would increase less rapidly as viscous heat production supplies the energy necessary for partial fusion. After a significant portion of melt has been generated, its upward migration would alter the thermal characteristics of the system because ascending melt would very efficiently transport heat upwards, away from the zone of deformation (Feigenson et al., 1980).

For flow of magma in a conduit, Hardee et al., 1977 showed by an approximate solution to a simplified version of the conservation of energy equation that the viscous heat-production/heat-loss balance determines a critical dike width. For widths smaller than the critical value the heat loss outstrips production and the material in the pipe freezes. For basaltic systems, computed critical dike widths are of the order of 1 meter. This is generally consistent with the observed size of basalt dikes (Fujii et al., 1974; Bodvarsson et al., 1964; Mohr et al., 1976). After flow has been established, viscous-dissipation effects could be important in maintaining or accelerating the flow.

In terms of the applicability of viscous dissipation as a mechanism for concentrating volcanic heat, mention should be made of the work by Shaw (1973). Thermal-feedback viscous-dissipation theory predicts an accelerating pattern of magma production due to the action of constant viscous stresses in the magma source region. The theory also predicts a characteristic instability time, indicative of the time necessary for the culmination of a volcanic episode. The rate of eruption of Hawaiian basaltic magmas over roughly the last 10 m.y. was found to be consistent quantitatively with the constant-stress thermal-feedback theory. Whether this is coincidence or not can only be decided by further tests, either natural or laboratory, of the theory. It may be that the observed exponential increase in the volumetric eruption rate of basaltic volcanism at Hawaii (on a 5 m.y. time scale) is not related to the magma production rate at all but instead is a manifestation of an increased availability of flow paths that ultimately lead to the surface. As noted earlier, slow crack propagation rates depend exponentially on temperature, so that, as a region of mantle is heated up by magmatic heat from below, crack propagation rates would be higher and hence the likelihood of any particular batch of magma reaching the surface would be greater. In addition to the study of rates of volcanic eruption in oceanic island or island-arc provinces, a detailed analysis of the rate of production of oceanic crust along a diverging plate margin, where adequate high resolution age data are available, might put some constraints on the melt production rate and its variation in time. Episodic sea-floor spreading would result in episodic magma production rates. Of all the environments where magma is produced it is suggested that oceanic spreading zones represent the ones where access to the surface is most available.[13]

One set of experiments described by Spera (1977, Chapter IV) illustrates that a flow system subjected to a constant strain rate may exhibit periodic thermal phenomena (i.e., episodic temperature fluctuations at a fixed point in the fluid), the frequency being orders of magnitude smaller than that of the disturbing function (i.e., the constant strain rate). For the particular geometry of his experiment [cylindrical couette flow of a viscous ($\eta_0 \approx 10^3$) fluid], the period of the temperature perturbation is approximately equal to

$$\tau = 2R_o^2 \kappa^{-1} \, \mathrm{Ra}^{-2/5}, \tag{31}$$

where R_o and κ represent the half-width of the deforming region and the thermal diffusivity ($\kappa = k/\rho C_p$), respectively. For these experiments Ra is given by the expression

$$\mathrm{Ra} = \frac{4\rho\alpha g R_o^3 R_i^2 \Omega^2}{\kappa k}, \tag{32}$$

where R_i and Ω denote the radius of the rotating shaft (source of kinetic energy) and its angular velocity (rad s^{-1}), respectively. The strain rate ($\dot{\varepsilon}$) of the fluid is related to Ω and distance r from the center of the deforming region by

$$\dot{\varepsilon} = \left(\frac{2R_i^2 R_o^2}{R_o^2 - R_i^2} \right) \left(\frac{\Omega}{r^2} \right). \tag{33}$$

[13] The presumption of accessibility is based on the inferred tensional environment as indicated by fault plane solutions of earthquakes along ridges. As shown above, in tensional environments the conditions for melt intrusion do not require $p_f > p_{\mathrm{lith}}$.

The maximum strain rate occurs at $r = R_i$, along the margins of the rotating shaft. The maximum $\dot{\varepsilon}$ is therefore given by

$$(\dot{\varepsilon})_{max} = \left(\frac{2R_o^2}{R_o^2 - R_i^2}\right)(\Omega). \tag{34}$$

When the shaft rotation rate is greater than about 1,200 rpm ($\dot{\varepsilon}_{max} = 252 \text{ s}^{-1}$), a periodic temperature fluctuation at a fixed position in space is detected by a continuously recording thermocouple located there. The amplitude and frequency of the temperature fluctuation both increase as the strain rate increases. The period of the temperature fluctuation is theoretically determined by (31) and the experimental data confirm this relationship. For example, at 1,200 rpm, the theoretical period of the temperature perturbation is 745 s versus the actually measured value of 765 s. The predicted period of the temperature oscillation should be compared to the period of the mechanical driving force (i.e., rotation period of the shaft) which is given approximately by $\dot{\varepsilon}^{-1}$ and is on the order of 10^{-3} s. The striking feature that emerges from these experiments and others (e.g. Shaw, 1967, unpublished data) is the demonstration that even a relatively simple thermomechanical system may exhibit periodic thermal phenomena in response to a mechanical source of energy which is periodic itself but on a vastly different time scale. For these experiments, the ratio of the two time scales is of the order of 10^6. Although it may be entirely coincidental, it is interesting to note that the ratio of the velocities at which lithospheric plates move to typical magma ascent rates is about the same.

DIAPIR MECHANICS

Flows involving density instabilities due to density inversions have been widely discussed in the fluid dynamic and geological literature (Chandrasekhar, 1961; Ramberg, 1963, 1967, 1968a, 1969b, 1970; Parker et al., 1955). Whitehead and Luther (1975), among others, have performed scale experiments demonstrating the physical features associated with density instability. The classical Rayleigh-Taylor instability problem is the superposition of two layers of fluid; the bottom layer, being less dense than the upper layer, is inherently unstable. By means of an isothermal linear stability analysis one can compute both the distance between successive pipes and the initial growth rate of the upwardly ascending low-density phase. Marsh et al. (1974) applied this theory to explain the observed spacing between volcanic centers in some island arc terrains. They noted that, if melt forms at depths corresponding to the top of the seismic

Benioff zone beneath the volcanic chain, then the thickness of the melt layer is on the order of 10^2 m for a characteristic volcanic center wavelength of 70 km. The experiments of Whitehead et al. (1975) indicate that a growing buoyant mass begins its migration upwards when the Stokes ascent rate for the body exceeds its rate of growth. The latter quantity is proportional to the rate of production of the low density fluid.

While there is no doubt that the Rayleigh-Taylor instability theory has applications to some problems in geotectonics and geophysical fluid dynamics, its limitation to nonreactive isothermal systems makes application to magma transport studies questionable. A brief calculation illustrates this inadequacy. Assume a mostly segregated batch of magma of viscosity 10^2 poises rising through the mantle ($\eta_m = 10^{21}$ poises). Ascent rates appropriate for carrying ultramafic xenoliths, and consistent with rates of slow cracking by stress corrosion, may be conservatively estimated to be on the order of 10^{-2} cm s^{-1}. The radius (R) of the assumed spherical diapir may be calculated as

$$R = \left(\frac{3u\eta_m}{g\,\Delta\rho}\right)^{1/2}, \tag{35}$$

where η_m is the viscosity (assumed Newtonian) of the mantle and $\Delta\rho$ represents the density difference between solid peridotite and the melt phase ($\Delta\rho \simeq .5$ gm cm^{-3}). Using these figures, one computes from Equation (35) that $R = 2,500$ km—clearly an unreasonable figure. In order to compute reasonable values of R,[14] η_m needs to be of order 10^{11} poises, many times smaller than the oft-quoted figure of 10^{21} poises for the mantle viscosity.

The calculation performed above fails because the model upon which it is based is too simplistic. Perhaps the most severely limiting factor is that, undoubtedly, heating of the country rock due to the motion of a hot diapir results in a lowering of the effective viscosity of the immediately adjacent mantle. In this view, a large blob of magma may be imagined to rise through the relatively cool lithosphere by softening a thin rind of wall rock, causing it to flow past the magma. The effective viscosity of the wall rock is significantly lowered in response to transient heating and partial melting. Quantitative phenomenological models based on country-rock/magma-diapir thermal interactions have been proposed for ascending alkali basalts (Spera, 1977; Carmichael et al., 1977) and andesite diapirs (Marsh, 1976a, 1976b, 1978; Marsh et al., 1978). These models enable one to predict temperature-depth trajectories of ascending batches

---

[14] As an example, consider the remarkable basalt flow of the 1801–1802 eruption of Hualalai Volcano on Hawaii. Its estimated flow volume of about 0.20 km^3 corresponds to an effective magma sphere of diameter 340 m.

of magma. The computed trajectories take account of most first-order processes throught to be relevant to ascent, including heat conduction and convection into country rock, partial fusion of country rock, crystallization of liquidus phases within the melt, PV work done by the expanding magma, and viscous heating. For example, a 1 km (diameter) diapir of alkali basalt, ascending at a rate of 1 cm s^{-1} from a depth of 50 km, partially fuses a rind of peridotitic country rock roughly .5 m thick to the extent of about 30%, giving effective mantle viscosities of around 10^{11} poises. The temperature loss of the diapir due to the heating effect is about 50°C. Because andesites rarely contain xenoliths, minimum ascent rates are difficult to estimate. Marsh has used ascent-rate dependent models of heat transfer between an andesitic melt and its wall rock, in conjunction with andestitic liquidus and solidus temperatures, to estimate minimum mean ascent velocities. For example, if magma is to remain molten until reaching the surface, its ascent velocity must be greater than about 3 m yr^{-1} for a body of radius 6 km and greater than 315 m yr^{-1} for a body of radius .8 km.

Although the present discussion deals primarily with the initial passage of magma through the lithosphere, repeated use of the same passageway may leave the wall rock unusually warm and therefore allow succeeding magmas to lose heat less rapidly than preceeding ones. It is very difficult to estimate the ratio of the volume melt generated at depth to the erupted volume. Presumably, in any given region, this ratio increases as time passes and the path through the lithosphere that ascending magma follows is heated up.

GENERAL FEATURES OF DIAPIR ASCENT

Of all the mechanisms proposed to account for the generation of basaltic melts within the mantle (see Yoder, 1976, Chapter 4 for a review of some) a combination of the ones originally suggested by Yoder (1952) and Verhoogen (1954, 1973) and subsequently discussed by many others (e.g., Green and Ringwood, 1967; Ramberg, 1968c; Cawthorn, 1975; Arndt, 1977) seems the most generally applicable. The essential physics of the process is based on the adiabatic decompression of an ascending solid diapir.[15] The diapir rises because it experiences a positive buoyancy force

---

[15] The assumption of adiabaticity requires that viscous heat production and heat losses by convection and conduction are negligible. Because all flows involve friction, no flow can be truly adiabatic. For a diapir to approximate adiabatic rise, a necessary (but not sufficient) condition is that its ascent time must be relatively small compared to its heat conduction characteristic time. That is, the inequality $Z/w < l^2/\kappa$ must hold. Z is the length traversed by,w the ascent rate, and l the half-width of the diapir. If a diapir of diameter 5 km ascends from a depth of 75 km, then for the flow to be considered adiabatic $w > 25$ cm yr^{-1}.

deriving from either compositional or thermal effects.[16] When the temperature-depth trajectory of the adiabatically ascending crystalline diapir intersects the peridotite solidus, partial fusion begins. The energy necessary for fusion comes from the heat content of the diapir, and so, if chemical equilibrium is maintained, the temperature-depth trajectory of the now crystal-liquid mush follows the steeper Clapeyron trajectory. Typical Clapeyron slopes for anhydrous peridotites are in the range $3-5°K \ km^{-1}$ whereas adiabatic slopes[17] lie in the range .3 to $.7°K \ km^{-1}$. After partial fusion has begun, there will be a tendency for melt to segregate. This tendency arises because the melt is more buoyant than the residual crystals and because of the filter pressing action of deviatoric stress (i.e., melt is kneaded out of the two-phase mush). Specific numerical calculations relating the depth at which partial fusion begins to the composition, temperature, and fraction of melt have been given by Cawthorn (1975), Ramberg (1968c), and Arndt (1977). These calculations are approximate, however, because considerable uncertainty remains about the size, kinetics of melting, role of frictional heating, and importance of convective heat loss within such rising diapirs. In an effort to illustrate more explicitly how these factors govern diapir ascent behavior the following macroscopic analysis is offered.

THERMODYNAMICS OF DIAPIRS

A general expression for the first law of thermodynamics (conservation of energy) as applied to a mass m in a flow system is (Lewis and Randall, 1961)

$$\Delta H + m \frac{\Delta u^2}{2} - mg \Delta Z = Q - w', \qquad (36)$$

where ΔH represents the change in enthalpy ($H_{final} - H_{initial}$) of the fluid (a function of ΔP, ΔT, and f, the mass fraction of melt formed), Δu is the change in velocity and ΔZ is the difference in fluid depth ($\Delta Z = Z_f - Z_i$). Note that $-\Delta Z$ is the increase in height above some reference level and so

---

[16] See papers by Jackson and Wright (1970) and Boyd and McCallister (1976) for a discussion of the densities of dunites, garnet websterites, garnet lherzolites and other mantle rocks.

[17] It may be shown that the adiabatic or isentropic temperature-depth relationship for a single phase is given by

$$T(z) = T_0 \exp \left[\frac{+\alpha g(Z - Z_0)}{C_p^0} \right]$$

where α and C_p^0 represent the isobaric expansivity and isobaric heat capacity of the phase and T_0 is the initial temperature (at depth Z_0).

is directly related to the potential energy increase of the mass. Q and w' on the right hand side of (36) represent the heat absorbed and energy expended in overcoming friction by the fluid, respectively. In what follows, Equation (36) is applied to the case of two-phase (e.g., crystal-liquid) flow in an attempt to evaluate the energy budget of an ascending mantle diapir undergoing partial fusion. The term fluid is used here in a completely general sense and is convenient to use when referring to the ascending diapiric material, whether it be entirely crystalline or partially fused. Note that it should be immediately obvious from simple physical considerations that the potential energy and frictional work terms are positive quantities; if the fluid moves up in the gravitational field of the earth, its potential energy increases. The expenditure of mechanical energy to overcome the retarding frictional forces will similarly always be a positive quantity. The kinetic energy will generally be positive; however, a simple order of magnitude analysis shows that the increase in kinetic energy of the fluid is generally much smaller than the other terms.[18] Because the diapir *does* move upwards to a higher potential energy state there must be compensating effects. Perhaps the most significant of these is the *decrease* in the enthalpy of the fluid. Despite the fact that partial fusion *increases* the enthalpy of the fluid (at fixed P and T), the pressure drop (ΔP) and temperature loss ($\Delta T = T_f - T_i$), incurred by the rising diapir, more than compensate for the enthalpy of melting effect. Overall, therefore, the enthalpy of the fluid decreases; it is this decrease that balances the increase in potential and kinetic energy and the production of melt.

In order to be more specific one must examine the thermal balance represented by (36) more closely. Imagine that an entirely crystalline diapir ascends without partial fusion until reaching a depth Z_i below the surface. At this depth the TZ trajectory of the diapir intersects the mantle solidus and partial fusion begins. The ambient values of P and T at the inception of partial melting are taken as P_i and T_i. Note that, in general, melt will have a tendency to segregate from the residual crystals. The energetics (not the dynamics!) of the segregation process can be modeled by allowing the melt and crystal fractions of the diapir to rise to different levels.[19] Explicitly then, the third term on the left hand side of (36) becomes

$$-mg\,\Delta Z = -m^l g(Z_f^l - Z_i) - m^{xl}g(Z_f^{xl} - Z_i). \qquad (37)$$

---

[18] Even in the case of rapidly ascending kimberlites this statement appears justified. McGetchin and Ullrich (1973b) estimate for a volatile-charged kimberlite moving upwards from a depth of 100 km in a 20 m diameter pipe that $\Delta u_m \simeq 60$ ms^{-1}. The ratio of potential to kinetic energy increase becomes $2g\Delta Z/(\Delta u)^2 \simeq 600$.

[19] Because oceanic crust is 5–7 km thick, one might suggest that the melt fraction of a ridge partial fusion episode rises at least 5 to 7 km higher than the residual crystals.

Dividing by m, noting that $m^l/m = f$ and $m^l + m^{xl} = m$, and rearranging, one obtains

$$-g\Delta Z = g(Z_i - Z_f^{xl}) + gf(Z_f^{xl} - Z_f^l), \qquad (38)$$

where Z_f^{xl} represents the depth below the surface to which the crystals rise, Z_f^l is the depth below the surface the liquid finally attains, and Z_i is the depth at which the peridotite solidus is first intersected. The change in enthalpy of the fluid accounting for both the production of melt and the segregation process is given by (Appendix III)

$$\frac{\Delta H}{m} = c^{xl}(T_f - T_i) + (1 - \alpha T_f)g(Z_f^{xl} - Z_i)$$
$$+ f[(1 - \alpha T_f)g(Z_f^l - Z_f^{xl}) + \Delta h_{\text{fusion}}]. \qquad (39)$$

If equations (38) and (39) are combined with (36), one computes for the melt fraction, f,

$$f = \frac{Q/m - w'/m - \Delta u^2/2 + \alpha T_f g(Z_f^{xl} - Z_i) + c^{xl}(T_i - T_f)}{\alpha T_f g(Z_f^{xl} - Z_f^l) + \Delta h_{\text{fusion}}}. \qquad (40)$$

In what follows some numerical estimates of f for a number of cases are given. The following parameters are assumed constant throughout:

$$\Delta h_{\text{fusion}} = 125 \text{ cal gm}^{-1} (5.2 \times 10^9 \text{ erg gm}^{-1}),$$
$$g = 981 \text{ cm s}^{-2}, \quad \alpha = 6 \times 10^{-5} \text{ K}^{-1},$$
$$T_f = 1{,}450 \text{ K}, \quad T_i = 1{,}625 \text{ K},$$
$$Z_i = 72 \text{ km}, \quad c^{xl} = .3 \text{ cal K}^{-1} \text{ gm}^{-1} (1.26 \times 10^7 \text{ erg gm}^{-1} \text{ K}^{-1}).$$

It may be seen by a straightforward order-of-magnitude calculation that $\Delta u^2/2$ is generally negligible with respect to the other terms in (40). In what follows, this term is neglected.

For the case of adiabatic $(q = 0)$, frictionless $(w' = 0)$ flow, with no melt segregation $(Z_f^{xl} = Z_f^l = 0)$, Equation (40) reduces to

$$f = \frac{\alpha T_f g(Z_f^{xl} - Z_i) + c^{xl}(T_i - T_f)}{\Delta h_{\text{fusion}}}, \qquad (41)$$

and, for example, one calculates $f = .35$ when the fluid expands from a depth of 72 km $(T_i = 1{,}350°C)$ to the surface $(T_f = 1{,}175°C)$.

A treatment of the general case of mantle diapirism that accounts for heat losses from the fluid and the internal friction is significantly more complicated. What follows here should be construed as only a guide to what typical orders of magnitude may be.

The energy dissipated by friction depends on the transport properties of the diapiric material, such as its viscosity and ascent rate, as well as the geometry of the diapir itself. In a previous section it has been shown how the uncertainty regarding the "viscosity" of the mantle can lead to unreasonable results. The problem is complicated by the fact that the effective viscosity of the rising diapir is sensitive to the fraction of melt. For the admittedly unrealistic case of Poiseuille flow of a fluid with density ρ_f and viscosity (assumed Newtonian) η_f, the friction term may be written

$$\frac{w'}{m} = \frac{8\Delta Z \eta_f u_m}{\rho_f R^2} = \frac{\Delta p}{\rho_f}. \tag{42}$$

The last equality follows from the dependence of u_m on Δp for Poiseuille flow in a circular conduit. The magnitude of w'/m is seen to depend very much on the conduit size and viscosity of the fluid. Taking a typical value of Δp to be 10^2 bars gives $w'/m \simeq 3.3 \times 10^7$ erg gm^{-1}, a figure that is much smaller than the last two terms in the numerator of Equation (40). A cylindrical diapir of 1 km radius, moving upwards from a depth of 75 km at a rate of 1 cm s^{-1}, and having an effective viscosity of 10^{11} poises, gives $w'/m = 1.8 \times 10^8$ erg gm^{-1}—a figure of comparable magnitude to $\alpha T_f g(Z_f^{xl} - Z_i)$ in (40). The latter figure is probably a generous overestimate as it is unlikely that strain rates would ever reach such large values in so large a mass. The above calculations suggest that too little is known about the rheology and configuration of mantle diapirs to neglect the energy dissipated by friction in the energy budget of rising diapirs.[20]

Finally, nonadiabatically ascending diapirs may be considered. As long as the diapir is mostly crystalline, heat losses will be governed by conduction processes. These are inherently slow compared to convection rates. The time taken for heat to travel a distance 70 km, for example, may be estimated according to $t \sim l^2/\kappa = 10^{10}/10^{-2} = 3.2 \times 10^6$ yrs. A body ascending at a rate of a 5 cm yr^{-1} will travel about 200 km during this time. However, once a significant fraction of melt has formed, heat transfer between convecting magma and wall rock could become significant. In a spherical mass of segregated melt the total heat loss per gram of liquid (i.e., $-Q/m$) is given by

$$\frac{-Q}{m} = \frac{3\bar{q}Z_f^{xl}}{\rho u_m R}, \tag{43}$$

where $\bar{q}$, Z_f^{xl}, and R represent the heat flux (erg cm^{-2} s^{-1}) out of the magma body, the segregation depth, and the radius of the assumed spherical magma sphere, respectively. The heat flux $\bar{q}$ may be related

---

[20] Consider a mantle plume of 100 km diameter with $= 10^{21}$, $u_m = 1$ m yr^{-1} and $L = 500$ km; $w'/m = 1.5 \times 10^{10}$ erg gm^{-1}, a value which indicates frictional heating could be important.

to the dimensionless Nusselt number,[21] which in turn depends on the Rayleigh number, Ra. For laminar convection

$$\vec{q} = \frac{k\,\Delta T}{R}\,\mathrm{Nu} = \frac{1}{2}\frac{k\,\Delta T}{R}\,\mathrm{Ra}^{1/4}, \tag{44}$$

where

$$\mathrm{Ra} = \frac{\alpha g\,\Delta T\,R^3}{\kappa v}. \tag{45}$$

In these expressions k, ΔT and v represent the thermal conductivity ($k = \rho C_p \kappa$) of the melt, the temperature difference between the center and margins of the convecting chamber, and the kinematic viscosity of the melt (η/ρ), respectively. If Equations (44) and (45) are substituted into (43), the result

$$\frac{-Q}{m} = \frac{3k\,\Delta T\,Z_f^{xl}\,\mathrm{Ra}^{1/4}}{2\rho u_m R^2} \tag{46}$$

is obtained. Returning to the earlier expression for the fraction of melt (ignoring the frictional work expended and kinetic energy term), one finds that

$$f = \frac{(-3k\,\Delta T\,Z_f^{xl}\,\mathrm{Ra}^{1/4}/2\rho u_m R^2) + \alpha T_f g(Z_f^{xl} - Z_i) + c^{xl}\,(T_i - T_f)}{\alpha T_f g(Z_f^{xl} - Z_f^l) + \Delta h_{\text{fusion}}}. \tag{47}$$

With Equation (47) one may estimate the fraction of melt generated for the case previously considered, this time allowing for convective heat loss. The result of this computation is $f = .22$.[22] Recall that in the earlier case, when heat transfer was ignored, $f = .35$. Obviously, heat transfer between a hot diapir and the relatively cool lithosphere it traverses can be significant.

To summarize, it has been shown that the important factors governing the degree of partial fusion are 1) the temperature and depth where partial fusion begins, 2) the depth of "segregation" of the melt, and 3)

---

[21] The Nusselt number Nu may be defined as the ratio of the total heat transported out of or into a body to the heat transported if heat conduction was the only operative heat transfer process. The numerical relationship between Nu, Ra and w has been considered by numerous workers and is summarized in the text by Rohsenow and Choi (1962). See Shaw (1974), Carmichael et al. (1977) or Spera (1977) for application of these correlations to petrological problems.

[22] In addition to the figures cited earlier, we have assumed the following: $k = 3 \times 10^5$ erg cm^{-1} k^{-1} s^{-1}, $T = 10\ K$, $R = 500$ m, $v = 10^2$ cm^2 s^{-1}, $\rho = 2.8$ gm cm^{-3}, $u_m = 10^{-2}$ cm s^{-1}.

heat loss by convection from the mostly liquid portion of the diapir to the surroundings. The work expended by the diapir on its surroundings (e.g., fracturing the country rock) and in overcoming internal friction is difficult to evaluate quantitatively unless something specific is said about the dynamics and geometry of flow. For viscous Newtonian flow in a pipe, frictional heating may or may not be important, depending on the internal viscosity of the diapir and the distribution of strain rates within it. The use of the term "segregation depth" is ambiguous; there should always be a tendency for chemical and thermal equilibration of melt with the surrounding rocks as long as chemical-potential gradients and temperature differences persist. On the other hand, there may be an effective value of the melt fraction for which the differential velocity between melt and crystal is so large as to render continued equilibration impossible. It may well be that, rather than a single pulse of magma rising upwards to the surface, a *series* of pulses is required. Each batch of crystallizing melt would act as a heat-transfer fluid; eventually the environment would be hot enough to permit relatively rapid rise either by ascent of a viscous blob or by swarms of propagating magma-filled cracks. Perhaps the cumulate-type xenoliths found in some alkali basalts represent such fossilized heat exchangers. In this context, the progression observed at Hawaii and some other volcanic centers, of early hypersthene-normative melts followed by increasingly alkalic melts, is intriguing. That segregation depths are less for tholeiitic melts than for alkalic melts seems compatible with the data from experimental phase-equilibria and thermodynamic calculations of magma-mantle equilibration conditions (Green, 1970, 1967; Yoder et al., 1962; Carmichael et al., 1977). Eggler, on the basis of experimental studies in the system H_2O-CO_2- basalt (Eggler, 1973; Eggler et al., 1978), has suggested an alternative model to explain the progression. He argues that early formed melts are *relatively* enriched in H_2O over CO_2. H_2O enrichment favors production of hypersthene-normative tholeiitic melts, whereas CO_2 charged melts are richer in alkalies. The demonstration by Mysen et al. (1975) of increasing CO_2 solubility in progressively more alkalic melts is consistent with this concept.

In the context of transport phenomena, Eggler's hypothesis and the increasing segregation depth model seem mutually consistent rather than inconsistent. Although early formed melts are relatively enriched in H_2O, the absolute concentration of H_2O remains below the saturation limit until very shallow levels are reached. The exsolution of H_2O does not contribute to the kinetic energy of the rising melt, therefore, until volatile saturation occurs at shallow depths. For most of its path, then, the magma moves upwards without the additional impetus of an exsolving gas. In contrast, a relatively CO_2-rich melt, because of the smaller solubility of CO_2 in the liquid, is more likely to become volatile-saturated at depth. As the exsolving gas expands, it performs work upon the surrounding melt, thereby increaseing the melt ascent rate. Additionally, CO_2 in distinction to H_2O shows an exothermic heat effect when exsolving from a basic melt. The energy liberated by heat of exsolution could similarly be balanced by an increase of the kinetic energy of the melt. A corollary of this reasoning is that the more rapidly ascending alkalic melts are more capable of transporting xenoliths.

These considerations highlight the fact that evolving magmatic systems are best viewed within the context of the geologic history of a region. Geochronological and eruptive volumetric rate data may be the crucial link between the chemistry of a lava and the dynamics of ascent. An important question is posed by these considerations: what fraction of magma generated within the Earth at a given time actually rises to the surface?

Porous Media Flow

On a microscopic scale, the distribution of melt in a rising diapir will depend on the grain size of the solid phases, their crystallographic orientation, and the dihedral angle between the boundaries of melt and crystals. This latter quantity is related to the magnitude of interfacial energies between the coexisting phases. For small dihedral angles, the melt phase forms a thin intergranular film which coats all grain boundaries (Adamson, 1967).

From the partial melting experiments of Mehnert et al. (1973); Busch et al., (1974); and Arndt, (1977); and from electrical conductivity arguments (Shankland et al., 1977; Waff, 1974); it appears that silicate melts tend to wet grain boundaries. Melt will initially accumulate most rapidly along polymineralic triple and quadruple junctions; however, as the fraction of melt increases, more of the grain boundaries will be coated with melt (Arndt, 1977). Analogous effects are observed in some metal systems where even a trace of a wetting liquid is sufficient to coat all crystal boundaries (Reed-Hill, 1973; Williams et al., 1968). As the polyphase assembladge is strained, the response of the suspension will depend on the fractional area of solid-solid contacts. For small degrees of melt, some solid-solid bridges will remain and the network will possess some strength. As the temperatures above the solidus increases, so will the melt fraction:[23] Mysen et al. (1977) have shown that the variation of

---

[23] Melt fraction increases at constant temperature in non-adiabatic eutectic systems. In nature, however, melting related to a continuous series of crystalline solutions might be generally expected and therefore melt fraction is a unique function of temperature (Yorder, 1976, p. 114).

the volume fraction of melt (equals ε, the porosity) with temperature is not linear. As ε increases, the area of coverage and thickness of the partial melt film increases; the fracture strength (i.e., the stress necessary to propagate liquid filled cracks) drops in the material, and the melt may be extremely mobile. On a seismic length-scale, the suspension probably behaves as a plastic material with a finite yield strength (Aki et al., 1978), preventing crystals from settling out. If pressure gradients are present, however, there will be a velocity differential between melt and crystals. The flow process could then be envisioned as a form of two-phase porous-media mixed convection. An initially heterogeneous melt distribution (perhaps resulting from an inherited metamorphric fabric) will create a spatially complicated permeability field. Highly permeable zones, favourably oriented with respect to the principal stresses, might serve as high-conductivity pathways for flow. This behavior is consistent with the ideas of Shaw (1969) regarding the role of pervasive shear strains in melt segregation, and with the recent work of Turcotte et al. (1978) pertaining to the macrophysics of melt segregation. The basic idea is that melt will flow to regions of low stress at rates greater than the surrounding deformable crystal network.

Order-of-magnitude estimates of melt flow rates may be obtained from a form of the momentum equation consistent with two-phase porous-media flow (Straus et al., 1978; Ribando et al., 1976; Turcotte et al., 1978). For low Re flow in a porous medium the Z-component (vertical) of the equation of motion is

$$-\frac{\partial p}{\partial Z} - \frac{\eta}{B} u + \Delta\rho\,\hat{g} = 0, \tag{48}$$

where B, $\Delta\rho$, η, u and p represent the permeability, density difference between melt and crystals, viscosity of the melt, vertical flow rate and the pressure, respectively. Defining

$$p = p_{\text{lith}} + p', \tag{49}$$

and noting that

$$p_{\text{lith}} = \hat{\rho} g Z, \tag{50}$$

Equation (48) may be written in the form

$$u = -\frac{B}{\eta}\frac{\partial p^*}{\partial Z}, \tag{51}$$

where $\partial p^*/\partial Z$ is defined according to $\partial p^*/\partial Z = \partial p'/\partial Z + \Delta\rho\, g$. The rate of flow is seen to be directly proportional to the pressure gradient driving flow, the density difference, and the permeability B; and inversely proportional to the dynamic viscosity.[24] Equation (48) represents a balance between the viscous forces on a moving fluid parcel, due to fluid-grain as well as fluid-fluid friction, and the pressure gradient-buoyancy forces causing flow. Pressure gradients are intimately related to the regional pattern of tectonic stresses. Consequently, one expects different segregation rates in different tectonic environments, independent of the density difference between melt and crystals.

To compute melt velocities from Equation (51), an estimate of B must be made. The dependence of the permeability (B) on the porosity (equivalent here to the volume fraction of melt) is critical, because it is the permeability that determines the differential velocity of melt and crystals. To this author's knowledge, permeabilities of silicate mushes have not directly been determined. There is, however, a wealth of information regarding flow through packed beds and consolidated media (Gonten et al., 1967; Wyllie et al., 1950; Carman, 1956; Scheidegger, 1957; Brace et al., 1968, Brace, 1977; Bear, 1972). Although it should be clearly understood that segregation of melt from a deformable crystal mush may be mechanically different from flow through a rigid porous solid, the functional form for porosity versus permeability, suggested by the studies cited above, is hesitantly adopted for the purpose of obtaining order-of-magnitude estimates of the segregation rate (i.e., the differential velocity between crystals and liquid). A first approximation for the dependence of the permeability on the volume melt fraction (porosity) is taken as

$$B = \frac{a^2\varepsilon^3}{9k(1 - \varepsilon)^2}. \tag{52}$$

The dimensionless constant k, which is related to the tortuosity, is equal to about 5.0 (Carman, 1956), and a is the radius of the solid interlocking (assumed spherical) grains. In a recent paper Brace (1977) has suggested a somewhat more general permeability-porosity relationship; however, the uncertainty regarding the size, shape and distribution of the crystal-liquid network does not warrant a more refined approximation. Table 6 is a tabulation of calculated permeabilities assuming some typical porosity and grain size values.

Combining (51) and (52), one obtains an explicit relationship between ε and the flow velocity:

$$u = \frac{a^2\varepsilon^3}{45\eta(1 - \varepsilon)^2}\left(\frac{\Delta p^*}{L}\right). \tag{53}$$

Empirically it is noted that (53) breaks down for $\varepsilon > .75$; it applies only to fully viscous flow at small values of the Reynolds number. It is also

[24] The reader should note that thermal buoyancy forces are ignored in Equation (48).

Table 6. Permeabilities and segregation velocities for melt-crystal suspensions. The permeability, B, is computed from Equation (53) and u from (54). Constant values for k (see Equation (53)) and η have been taken as 5 and 10 poises respectively. Velocities are calculated assuming $\Delta p^*/L = 100$ bars/1,000 km. Figures in parentheses were computed assuming $\Delta p^*/L = 100$ bars/10 km. To convert from volume fraction of melt (ε) to mass fraction (f), the relation $f = \rho_l \varepsilon/\rho_s + \varepsilon(\rho_l - \rho_s)$, which reduces to $f \simeq \rho_l/\rho$ at small ($\varepsilon < .20$) porosities, is useful.

Porosity ε	Permeability $B\ (cm^2)$	Flow Velocity $u\ (cm\ s^{-1})$	
	crystal diameter = .1 cm		
.01	5.7×10^{-11}	5.7×10^{-12}	(5.7×10^{-10})
.05	7.7×10^{-9}	7.7×10^{-10}	(7.7×10^{-8})
.10	6.9×10^{-8}	6.9×10^{-9}	(6.9×10^{-7})
.25	1.5×10^{-6}	1.5×10^{-7}	(1.5×10^{-5})
.35	5.6×10^{-6}	5.6×10^{-7}	(5.6×10^{-5})
.50	2.8×10^{-5}	2.8×10^{-6}	(2.8×10^{-4})
.70	2.1×10^{-4}	2.1×10^{-5}	(2.1×10^{-3})
	crystal diameter = .2 cm		
.01	2.3×10^{-10}	2.3×10^{-11}	(2.3×10^{-9})
.05	3.2×10^{-8}	3.2×10^{-9}	(3.2×10^{-7})
.10	2.7×10^{-7}	2.7×10^{-8}	(2.7×10^{-6})
.25	8.2×10^{-6}	8.2×10^{-7}	(8.2×10^{-5})
.35	3.5×10^{-5}	3.5×10^{-6}	(3.5×10^{-4})
.50	2.2×10^{-4}	2.2×10^{-5}	(2.2×10^{-3})
.70	8.5×10^{-4}	8.5×10^{-5}	(8.5×10^{-3})
	crystal diameter = .5 cm		
.01	1.4×10^{-10}	1.4×10^{-11}	(1.4×10^{-9})
.05	2.0×10^{-7}	2.0×10^{-8}	(2.0×10^{-6})
.10	1.7×10^{-6}	1.7×10^{-7}	(1.7×10^{-5})
.25	5.1×10^{-5}	5.1×10^{-6}	(5.1×10^{-4})
.35	2.2×10^{-4}	2.2×10^{-5}	(2.2×10^{-3})
.50	1.4×10^{-3}	1.4×10^{-4}	(1.4×10^{-2})
.70	5.3×10^{-3}	5.3×10^{-2}	(5.3)

important to remember that (53) applies most closely to fluid flow through a rigid porous material. Although Equation (52) is clearly an approximation, previous studies (Bear, 1972; Turcotte et al., 1978; Sleep, 1974) have shown that more complex models lead to expressions with different constants of proportionality in (52), but the same functional dependence of B on ε. Examination of Table 6 shows that permeabilities and hence segregation rates are strongly dependent on porosities (volume fraction of melt). For instance, an increase of ε by a factor of two results in flow rates twenty times larger. The porosity is unlikely to remain uniform throughout a partial melt zone, especially if inhomogeneities in the melt distribution are inherited from a subsolidus tectonite fabric. An inhomogeneous porosity field can lead to a regenerative melt segregation effect because of the non-linear dependence of permeability on porosity. In those zones where ε is initially high, B will be large. In these regions, segregation velocities are higher, and, therefore, melt will accumulate at ever-increasing rates in low-stress regions. Tabulated in Table 6 are segregation velocities for a low-viscosity melt moving under the influence of small but finite pressure gradient. The non-linear dependence of u on ε responsible for the large increase in u as ε increases is evident from the values listed in the table. These observations are consistent with experimental evidence (see Scheidegger, 1957) that supports the notion that the permeability is inversely proportional to the effective stress ($\sigma_{eff} = \sigma_{lith} - p_f$); as $\sigma_{eff} \to 0$, B increases dramatically, and, of course, so will flow rates.

CONCLUDING REMARKS: IMPORTANCE OF RHEOLOGY

It should be clear from the foregoing considerations that, of all the factors governing the rate and mechanism of magma transport, it is the rheological properties of both the melt phase and its surroundings that are most critical.

Viscosities of silicate melts can be estimated from their chemical composition, temperature, and pressure using empirical correlations (Bottinga et al., 1972; Shaw, 1972; Murase et al., 1973; Kushiro, 1976; Scarfe, 1973; Kushiro et al., 1976). The validity of these estimates has been established only at super-liquidus temperatures, where essentially Newtonian behavior is found. It has been experimentally demonstrated that the presence of gas bubbles or suspended crystals cause some melts to behave as non-Newtonian fluids[25] (Shaw et al., 1968; Sparks et al., 1977; Pinkerton et al., 1978; Murase et al., 1973). Non-Newtonian behavior is relevant to many petrologic processes including magma ascent, heat transfer by free and forced convection in magma diapirs or magma chambers, and the settling of phenocrysts and xenoliths.

There is an extensive literature concerning the rheology of hot sub-solidus silicate materials (Goetze et al., 1972; Goetze et al., 1973; Green et al., 1972; Raleigh, 1968; Stocker et al., 1973; Murrell et al., 1976). Depending on the effective stress, a variety of responses are indicated. Generally speaking, ductile behavior (plastic or viscous) is observed at

---

[25] Non-Newtonian behavior has been recognized in some single phase polymeric fluids due to effects related to the structure of the liquid itself (Spera, 1977). High-silica melts above liquidus temperatures may exhibit similar behavior.

high pressure and temperatures and small deviatoric stresses; olivine at 1,600°C, for instance, readily flows under small (~ 10 bar) deviatoric stresses (see Appendix I). Low temperatures and small effective stresses favor brittle failure. Note that, even at depths where the lithostatic pressure is large, brittle failure may ensue, provided a pressurized fluid phase is present (i.e., the effective stress is small). The effective stress concept as outlined in previous sections is extremely important in terms of magma transport, in this author's opinion. One should note that the magma-filled propagating crack and viscous diapir hypotheses are not mutually exclusive. As shown above, critical tensile stresses necessary for the propagation of cracks are smallest in materials with reduced rigidities.[26] One might expect, therefore, that cracks could most easily nucleate and grow around the partially-fused rind surrounding a viscous (mostly-liquid) blob of ascending melt. Studies of fluid-flow in non-deformable porous media indicate that fluid permeabilities increase enormously as effective stresses approaches zero. If this relationship holds true in deformable silicate mushes, then melt may be mobile even in regions with small amounts of partial melt. More likely, however, mushes behave as Bingham fluids possessing a definite yield strength. A Bingham rheology would imply that some threshold volume fraction of melt exists before melt segregation rates become large enough to preclude crystal-liquid equilibrium in the source.

Finally, in closing, it must be pointed out that many topics relevant to magma transport have been neglected in this report. Aspects such as flow differentiation (Bhattacharji, 1964; Bhattacharji et al., 1967; Komar, 1972), the formation of volcanic vents, the origin and development of compositional zoning in some magma chambers, and the density relationships between depleted and undepleted peridotites (Boyd and McCallister, 1978; Jackson et al., 1970) are just a few of the many important problems that await a clearer understanding. However, it is hoped that the techniques and approaches described in the previous sections will be of some general interest and validity to igneous petrologists and volcanologists actively pursuing the "petrological inverse problem".

SUMMARY

The main aspects of this paper are summarized below in sequential order roughly corresponding to the route magma follows from source to surface.

---

[26] The rigidity, μ, depends on density and Shear-wave velocity according to $\mu = \rho V_s^2$.

Melt Segregation

1. The rate of melt segregation from a crystal mush is governed by the rheology of the suspension and the balance among buoyancy, pressure and viscous forces.
2. Buoyancy forces may be estimated reasonably well, using density, expansivity, and compressibility data for minerals and silicate liquids. Deviatoric stress gradients can be estimated to within perhaps an order of magnitude and are dependent upon the local tectonic environment of the segregation region. Viscous forces which act to impede melt segregation depend critically upon the rheological response of the crystal-liquid system to local conditions. Viscous forces can only be poorly estimated but are probably inversely porportional to the permeability and directly proportional to the melt viscosity. Permeabilities have not (to this author's knowledge) been measured in melt-crystal systems; empirical correlations for rigid porous media suggest a non-linear, monotonically increasing dependence of permeability on volume fraction of melt. As an example, the differential velocity (segregation rate) between crystals and liquid in a mush consisting of 30% melt is 6,500 times larger than in a mush with 2% melt.

Magma Ascent Rates and Viscous Flow

1. Macroscopic energy budget calculations show that the extent of partial fusion achieved in a rising diapir depends on (1) the slope and 1 bar temperature intercept of the peridotite solidus, (2) the temperature and depth where partial fusion begins, (3) the "depth of segregation" (i.e., the depth at which the differential velocity between crystals and melt is so large as to render continued crystal-liquid equilibration impossible, (4) amount of convective heat loss between diapir and lithosphere, and (5) extent of viscous heating due to frictional processes within the diapir. Nonadiabatic heat loss effects and frictional heating cannot *a priori* be considered to be insignificant and depend strongly on the shape and size of the magmatic body.
2. Ascent rates for nodule-bearing alkalic magmas may be computed based on a buoyancy versus viscous force balance for Newtonian or Bingham-plastic melts. Typical values are of the order of a few cm s^{-1} and, interestingly, are largely consistent with both laboratory experiments and theoretical predictions of the rate of slow crack propagation by static fatigue or stress corrosion.
3. Viscous dissipation has been suggested as a mechanism operating in melt-source regions by which mechanical energy is degraded to thermal energy and provides heat needed for partial fusion. An experiment

illustrating the development of periodic temperature fluctuations at a fixed point in a continuously sheared non-Newtonian temperature-dependent viscosity fluid was described. It is concluded that even simple thermomechanical systems involving viscous dissipation may exhibit thermal periodicities of a vastly different time-scale than the mechanical periodicities driving the system.

4. Relevant factors governing the viscous flow of magmatic fluids in dikes, pipes, diapirs, cracks, and other types of magma conduits include the following: ambient pressure gradients, magma rheology, buoyancy forces due to temperature, compositional and crystal-liquid effects, and the magnitudes of horizontal and vertical temperature gradients surrounding the magma body. Melt flowing in long conduits (>1 km) is probably in turbulent flow (at liquidus temperatures) assuming likely values for temperature and pressure gradients. A pressure gradient of 1 bar/10 km can maintain magma flow rates of several cm s^{-1} ($\eta =$ 500 poises) in a 5 m diameter dike. A few percent H_2O in an otherwise anhydrous melt can generate buoyancy forces commensurate with convection velocities of a few cm s^{-1} in the melt.

Magma Ascent by Crack Propagation

1. Effective principal stress deviators of several tens to several hundreds of bars are sufficient for the initiation of brittle failure and the propagation of cracks within the crust and upper mantle. Modification of the Griffith theory of fracture in brittle solids, accounting for time-dependent and environment-dependent fracture behavior (i.e., static fatigue), indicates that the mode of failure associated with slow-cracking can occur at stresses significantly below those predicted by classical Griffith theory. Slow-crack propagation rates in ceramic materials lie in the range 10^{-1} to 10^{-2} cm s^{-1} for local crack-tip temperatures equal to about $\frac{2}{3}$ of the melting temperature of the material. Experimentally determined slow-cracking rates in ceramic materials are comparable to earthquake swarm migration rates in some volcanic regions such a Iceland and Hawaii.

Behavior of Fluidized Magma

1. Virtually all melts contain some juvenile volatiles. Estimates of the volatile contents (mostly H_2O and CO_2) of tholeiitic, alkali-basaltic, andestic and rhyolitic melts range from .5 to 4 wt %, the more siliceous melts being the most volatile enriched. If account is taken of likely magmatic temperatures, solubility data may be used to estimate depths at which the development of fluidized two-phase flow may commence due to volatile saturation. Critical fluidization velocities may be cal-

culated by balancing hydrodynamic and lithostatic pressure forces. Characteristic values lie between hundreds to tens of m s^{-1} for gas-dominated systems (e.g., kimberlites) to several cm s^{-1} for melt-xenolith systems.

ACKNOWLEDGMENTS

The constructive comments of Professor R. Hargraves and Drs. H. R. Shaw and H. S. Yoder immeasurably improved the form and content of this paper. Discussion with and editorial assistance from Professor B. D. Marsh, M. Feigenson, S. Bergman, C. J. Busby-Spera, J. Selverstone and L. Noodles is greatfully acknowledged. The reasearch reported on in this paper was supported by National Science Foundation grants EAR 74-12782 and EAR 78-03340.

APPENDIX I

SOME BASIC CONCEPTS OF STRESS THEORY

There are many excellent accounts of continuum mechanics, the theory of elasticity (Mase, 1970; Bridgman, 1964; Nadai, 1963; Jaeger, 1962) and the application of these theories to the deformation associated with magmatic intrusions (Murrell, 1965; Ode, 1957 and 1960; Roberts, 1970; Pollard et al., 1973, 1975, 1976a, 1976b). The discussion here is restricted to some general aspects of stress states in the Earth and emphasizes those topics that are more immediately applicable to the transport of magma.

Nine stress components are necessary to define the state of stress at a point in a body. The equilibrium of an arbitrary volume of a continuous material acted on by surface and body forces requires that the resulting force and moment acting on the volume be zero. The equilibrium condition implies that the stress tensor $\underline{\underline{\Sigma}}$ at a point is symmetric. The stress tensor may, therefore, be written in matrix form as

$$\underline{\underline{\Sigma}} = \begin{pmatrix} \sigma_1 & \tau_{12} & \tau_{13} \\ \tau_{12} & \sigma_2 & \tau_{23} \\ \tau_{13} & \tau_{23} & \sigma_3 \end{pmatrix}. \tag{A1}$$

In (A1) the σ_i and τ_{ij} represent the normal and shearing stresses respectively. Because $\underline{\underline{\Sigma}}$ is a symmetric tensor, it may be diagonalized by a suitable coordinate transformation. The diagonalization procedure, which can be carried out on any real symmetric tensor, determines the directions of a set of orthogonal coordinate axes such that all off-diagonal terms of the stress tensor vanish. These axes are called principal axes, and they are associated with the three mutually perpendicular principal stress planes. Across these planes only normal (compression or tensional) stresses act, and all tangential stresses vanish. The principal stresses are often denoted by the symbols σ_I, σ_{II} and σ_{III} and are called the maximum, intermediate and minimum principal normal stresses. Obviously they act in the direction of the principal axes. By convention $\sigma_I > \sigma_{II} > \sigma_{III}$. If it is assumed that the Earth's surface is a flat, stress-free horizontal plane, one of the principal stress directions is coincident with the vertical.

The magnitudes of the principal stresses may be found by solving the determinant formed by subtraction of σ from each of the diagonal terms of the stress tensor. The roots of the following cubic equation determine the magnitudes and directions of the principal stresses:

$$\sigma^3 - I_\Sigma \sigma^2 + II_\Sigma \sigma - III_\Sigma = 0, \tag{A2}$$

where

$$I_\Sigma = \sigma_1 + \sigma_2 + \sigma_3 \tag{A3}$$

$$II_\Sigma = \tfrac{1}{2}(\sigma_i \sigma_j - \tau_{ij}\,\tau_{ij}) \tag{A4}$$

$$III_\Sigma = \det \underline{\underline{\Sigma}}. \tag{A5}$$

I_Σ, II_Σ and III_Σ are known as the first, second and third stress invariants.

A convenient decomposition of the stress tensor $\underline{\underline{\Sigma}}$ is into two parts. This decomposition is effected by subtraction of the spherically symmetric mean normal stress tensor from $\underline{\underline{\Sigma}}$. The mean normal stress is defined by the relation

$$\sigma_0 = p = \frac{\sigma_1 + \sigma_2 + \sigma_3}{3} \tag{A6}$$

where σ_0 corresponds to the spherically symmetric "hydrostatic" pressure (p) of hydrodynamics. An approximation often made in petrology is that p is equal to the pressure due to the superincumbent load so that

$$p = p_{\text{lith}} = \hat{\rho}gz \tag{A7}$$

If a fluid phase such as melt is present, mechanical equilibrium requires

$$p = p_{\text{lith}} = p_f = \hat{\rho}gz, \tag{A8}$$

where $\hat{\rho}$ is the average density and z the thickness of the super-incumbent pile. Generally speaking, small deviations of p_f from p_{lith} are sufficient to transport melt at considerable rates (cm s^{-1}) and distances (1–100 km).

The remainder of the subtraction of the spherically symmetric hydrostatic pressure $\underline{\underline{P}}$ tensor denoted by

$$\underline{\underline{P}} = \begin{pmatrix} p & o & o \\ o & p & o \\ o & o & p \end{pmatrix} \tag{A9}$$

from $\underline{\underline{\Sigma}}$ is called the stress deviator, because it is a measure of the deviation of the stress state from a simple isotropic one. In matrix notation this decomposition is written as follows:

$$
\begin{pmatrix} \sigma_1 & \tau_{12} & \tau_{13} \\ \tau_{12} & \sigma_2 & \tau_{23} \\ \tau_{13} & \tau_{23} & \sigma_3 \end{pmatrix} = \begin{pmatrix} p & o & o \\ o & p & o \\ o & o & p \end{pmatrix}
$$
$$
+ \begin{pmatrix} \frac{1}{3}(2\sigma_1 - \sigma_2 - \sigma_3) & \tau_{12} & \tau_{13} \\ \tau_{12} & \frac{1}{3}(2\sigma_2 - \sigma_1 - \sigma_3) & \tau_{23} \\ \tau_{13} & \tau_{23} & \frac{1}{3}(2\sigma_3 - \sigma_1 - \sigma_2) \end{pmatrix} \quad \text{(A10)}
$$

When the stress tensor $\underline{\underline{\Sigma}}$ has been resolved by the diagonalization process into the set of principal stresses, then the decomposition into hydrostatic and deviatoric parts is easily accomplished. That is,

$$
\begin{pmatrix} \sigma_I & o & o \\ o & \sigma_{II} & o \\ o & o & \sigma_{III} \end{pmatrix} = \begin{pmatrix} p & o & o \\ o & p & o \\ o & o & p \end{pmatrix}
$$
$$
+ \begin{pmatrix} 2\sigma_I - \sigma_{II} - \sigma_{III} & o & o \\ o & 2\sigma_{II} - \sigma_I - \sigma_{III} & o \\ o & o & 2\sigma_{III} - \sigma_I - \sigma_{II} \end{pmatrix}, \quad \text{(A11)}
$$

where now

$$
p = \sigma_o = \frac{\sigma_I + \sigma_{II} + \sigma_{III}}{3}. \quad \text{(A12)}
$$

The importance of the decomposition of a stress field into its deviatoric and isotropic parts is founded on the experimental observation that deviatoric stresses alone are responsible for permanent deformation by flow or fracture in rocks. Whereas the isotropic part of the stress field does not induce any permanent deformation, it does influence the mechanism by which the material may be permanently deformed. At high temperatures and confining pressures, the release of stresses by a viscous or plastic flow process is more likely than by a brittle release. If in addition to high confining pressures, however, a fluid phase at a pressure close to the lithostatic value exists, then catastrophic failure by the propagation of cracks (i.e., brittle failure) is possible (See Shaw, this volume).

ROLE OF FLUID PRESSURE

Anderson (1951) and Hubbert and Rubey (1959) were among the first to develop and apply the concept of effective stress to failure in rocks. The essential idea can be grasped if one considers the walls of a cavity in an elastic body. If the cavity remains empty, the state of stress at a point along the boundary of the crack is that given by the principal stresses, σ_I, σ_{II} and σ_{III} at that point. If, however, the crack is filled with a fluid (e.g., a silicate liquid) at pressure p_f, then the state of stress at that point is *mechanically equivalent* to one in which the stresses $\sigma_I - p_f$, $\sigma_{II} - p_f$ and $\sigma_{III} - p_f$ act on the external surface of an *empty* crack of the same form. From (A11) one notes that the effective stress tensor at a point in a material under a fluid pressure p_f is simply the stress deviator. That is

$$\begin{pmatrix} \sigma_I - p_f & 0 & 0 \\ 0 & \sigma_{II} - p_f & 0 \\ 0 & 0 & \sigma_{III} - p_f \end{pmatrix} = \begin{pmatrix} 2\sigma_I - \sigma_{II} - \sigma_{III} & 0 & 0 \\ 0 & 2\sigma_{II} - \sigma_I - \sigma_{III} & 0 \\ 0 & 0 & 2\sigma_{III} - \sigma_I - \sigma_{II} \end{pmatrix}$$

(A13)

As pointed out in the text, it is the effective stresses that are important in establishing the criteria of failure in a material. One concludes that failure by brittle fracture is possible even at great depths if a fluid phase is present (e.g., in regions of partial fusion). *In situ* stress determinations of σ_I and σ_{III} within the crust show that non-zero deviatoric stresses are a common feature (Haimson, 1977). In the source regions of magmas, the presence of earthquakes probably indicates that differences in the principal stresses are being relieved (Klein et al., 1977).

APPENDIX II

ONSET OF TURBULENCE: ISOTHERMAL PIPE FLOW

For isothermal flow in smooth pipes, turbulence sets in when the Reynolds number, defined as

$$\text{Re} = \frac{2R u_m}{v},$$

(B1)

exceeds about 2,100 (Goldstein, 1938). In terms of Poiseuille flow, because the axial pressure gradient ($\Delta p / L$) is related to the mean velocity, the condition for turbulence may be written

$$\left(\frac{\Delta p}{L} \right) > \frac{4\rho v^2}{R^3} \, \text{Re}_{\text{crit}}.$$

(B2)

Table II presents values of $\Delta p / L$ necessary for turbulent flow as a function of v and R. The velocity of the fluid at the inception of turbulence is also given. For a melt with $\eta = 300$ poise, a 5 m wide dike would be in turbulent flow with an axial pressure gradient of .2 bar km^{-1}. The

Table 2. Critical pressure gradients and corresponding mean vertical velocities required for turbulence in circular Poiseuille magma flow under isothermal conditions. Re_{crit} is taken as 2,100 and ρ is assumed equal 3.0 gm cm^{-3}.

Radius of Pipe R (m)	Critical Pressure Gradient $\Delta p/L$ (bar km^{-1})	Critical Vertical Mean Velocity u_m^{crit} cm/s
kinematic viscosity = 10^2 cm^2 s^{-1}		
.5	202.0	2104
1.0	25.0	1041
2.0	3.1	517
5.0	0.2	208
10.0	2.5×10^{-2}	104
15.0	7.4×10^{-3}	70
kinematic viscosity = 10^3 cm^2 s^{-1}		
2.0	315.0	3.3×10^4
5.0	20.0	2100.0
10.0	2.5	100.0
15.0	.74	700.0
20.0	.3	5×10^2

order of deviatoric stress gradients in the Earth may average around 100 bars/1,000 km and so for $\eta = 300$, turbulence is expected if $R > 5$ m. At higher viscosities, pipes of larger diameter are necessary.

ONSET OF TURBULENCE: FREE AND MIXED FLOWS

Recognizing the fact that isothermal flow is highly unlikely, one may consider the conditions for turbulence in naturally convecting circular pipes. Lighthill (1953) theoretically predicted the conditions for the onset of turbulence in such systems. His predictions have been borne out for the most part by the experimental work of Lockwood and Martin (1964). The onset of fully mixed turbulent flow depends on the aspect ratio of the tube (R/L) as well as the Rayleigh number. An appropriate criterion for instability is that $[(R/L)^3 Ra]$ be greater than about 10^4.[27] The term β represents the superadiabatic temperature gradient along the vertical walls of the conduit and is equal to

$$\beta = \frac{\Delta T - (\Delta T)s}{L}, \tag{B3}$$

[27] This criterion is a general one; the precise value depends on the boundary conditions and on how one chooses to define turbulence (see discussion in Lipps. 1971).

where ΔT is the overall temperature difference between top and bottom, $(\Delta T)_s$ the corresponding adiabatic (constant entropy) temperature difference, and L the length of the magma-filled conduit. Ra is the Rayleigh number defined as

$$Ra = \frac{\alpha g \beta L^4}{\kappa \upsilon},$$ (B4)

where α, g, κ, υ represent the isobaric expansivity, acceleration due to gravity, thermal diffusivity, and kinematic viscosity of the melt, respectively. Typical values of α and κ for silicate melts are $\alpha = 5 \times 10^{-5}$ K^{-1} and $\kappa = 10^{-2}$ cm^2 s^{-1} (Carmichael et al., 1977; Shaw, 1974). Table 3 presents a tabulation of minimum superadiabatic temperature gradients needed for turbulence for a variety of dike widths and magma viscosities.

Table 3. Critical values of the superadiabatic temperature gradient for turbulence by free convection in a circular tube. Ra_{crit} is taken as 10^4. Assumed constant throughout are
$\kappa = 10^{-2}$ cm^2 s^{-1}
$\alpha = 5 \times 10^{-5}$,
$g = 10^3$ cm s^{-2} and
$L = 10$ km.

Radius of Pipe R (m)	Critical Superadiabatic Temperature Gradient (K km^{-1})
kinematic viscosity $\upsilon = 10^2$ cm^2 s^{-1}	
.5	.16
1.0	.02
$\upsilon = 10^3$ cm^2 s^{-1}	
.5	1.6
1.0	.2
3.0	7.4×10^{-3}
$\upsilon = 10^4$ cm^2 s^{-1}	
.5	16.0
1.0	2.0
2.0	7.4×10^{-2}
$\upsilon = 10^5$ cm^2 s^{-1}	
.5	160
1.0	20
3.0	7.4×10^{-1}

An examination of the approximate figures in this table indicates that turbulence could be expected for rather small temperature gradients. Again, note that for a given viscosity, increasing R tends to make the flow more prone towards instability. For example, with $v = 10^3$ cm^2 s^{-1} and $R = 1$ m, β must exceed .2°K km^{-1} for a tube 10 km long for turbulent flow to be established. If temperature gradients such as those thought to exist in the upper parts of lithospheric plates are present (of the order 10°K km^{-1}), one anticipates turbulent flow to be more the rule than the exception. Also note that increasing v tends to damp out turbulent flow. In a 1 m radius tube 10 km long, a 10°K km^{-1} superadiabatic temperature gradient is not sufficient to generate turbulence if $v > 10^{4.5}$ cm^2 s^{-1}, approximately.

The general case of "mixed" flow is probably more applicable to magma transport than either of the cases discussed above. That is, one expects that pressure as well as buoyancy forces play a significant role in magma transport. Scheele and Hanratty (1962) have proposed a criterion for the transition to unsteady (turbulent) behavior in flows where buoyancy *and* pressure forces are important. The presence of both horizontal and vertical temperature gradients enhances the possibility of turbulence. Because temperature gradients across the width of the conduit are present, there is heat transfer from the margins of the magma filled conduit to the surroundings. If heat losses from the margin of the tube are large enough, sufficiently large negative buoyancy forces can be established there. These may counteract the pressure driving forces (always acting to allow magma ascent) causing an inflexion in the velocity profile. From the experiments of Scheele and Hanratty (1962), an appropriate criterion of instability is when the parameter Ra/Pe exceeds a critical value of about 53. Pe is the Peclet number and represents the nondimensional ratio of the heat transported by convection currents to that by molecular conduction. In dimensional form, the criterion relating the magnitude of $\Delta p/L$ to the critical superadiabatic temperature gradient, β_{crit}, is

$$\beta_{\text{crit}} = \frac{1}{8\rho g \alpha R} \left(\frac{\Delta p}{L}\right) \left(\frac{\text{Ra}}{\text{Pe}}\right)_{\text{crit}} \tag{B5}$$

Collected in Table 4 are typical values of β and $\Delta p/L$ for a variety of conditions. For example, if β exceeds .44°K km^{-1} along the walls of a vertical conduit of diameter 20 m and the impressed pressure gradient is 1 bar km^{-1}, the flow will be turbulent. Large width conduits are more prone to turbulence, other factors remaining constant. If pressure gradients are are on the order of 1 bar km^{-1}, a superadiabatic temperature gradient of 5°K km^{-1} is not great enough to cause turbulent flow in a 1 m diameter conduit.

Table 4. Critical values of β the superadiabatic temperature gradient as a function of conduit size and the vertical pressure gradient. Assumed constant have been $\rho = 3.0$ gm cm^{-3} $g = 10^3$ cm s^{-2} and $\alpha = 5 \times 10^{-5}$ K^{-1}. $(\text{Ra}/\text{Pe})_{\text{crit}}$ is taken as 53.0.

Radius R (m)	Pressure Gradient $\Delta p/L$ (bar km^{-1})	Superadiabatic Temperature Gradient β_{crit} (K km^{-1})
.5	10^{-2}	.09
	10^{-1}	.9
	1	9.0
1	10^{-2}	.044
	10^{-1}	.44
	1	4.41
	10	44.1
3	10^{-2}	.015
	10^{-1}	.15
	1	1.5
	10	15.0
10	10^{-2}	4.4×10^{-3}
	10^{-1}	4.4×10^{-2}
	1	.44
	10	4.4

The general conclusion to be drawn is that, for the simplified cases discussed above, turbulence in tube flow may be expected to be common. The temperature and pressure gradients expected within the lithosphere (of order $10°$K km^{-1} and 10^{-1} bar km^{-1} respectively) indicate turbulence even for relatively narrow conduits.

APPENDIX III

The change in enthalpy is given by

$$\Delta H = H_{\text{final}} - H_{\text{initial}} \tag{C1}$$

$$\Delta H = m^{xl}h^{xl}_{T_f,P_f} + m^l h^l_{T_f,P_f^l} - mh^{xl}_{T_i,P_i}$$

$$= m^{xl}h^{xl}_{T_f,P_f^{xl}} + m^l\left[h^{xl}_{T^l,P_f^{xl}} + \int_{P_f^{xl}}^{P_f^l}\left(\frac{\partial h^{xl}}{\partial P}\right)dP + \Delta h_{\text{fusion}}\right] - mh^{xl}_{T_i,P_i} \tag{C3}$$

$$= m^{xl}h^{xl}_{T_f,P_f^{xl}} + mh^{xl}_{T_f,P_f^{xl}} + m^l(1 - \alpha T_f)v^{xl}(P_f^{xl} - P_f^{xl})$$
$$+ m^l \Delta h_{\text{fusion}} - mh^{xl}_{T_i,P_i}, \tag{C4}$$

where $h^{xl}_{T_f,P_f^{xl}}$, α, v^{xl} and Δh_{fusion} represent the specific (per unit mass) enthalpy of the crystal or crystalline assembledge at T_f and P_f^{xl} (the final pressure acting on the residual crystals[28]), the isobaric expansivity, the specific volume of the crystal (assumed incompressible) and the specific heat of fusion. Note that Δh_{fusion} depends on P and T; the values used here correspond to the average Δh_{fusion} along the PT path of the rising diapir. Thermodynamic data for evaluating the dependence of Δh_{fusion} on P and T for a number of important silicate phases are tabulated in Carmichael et al., 1977.

Rearranging (C4), using the thermodynamic identity

$$(h_{T_f,P_f^{xl}}^{xl} - h_{T_i,P_i}^{xl}) = c^{xl}(T_f - T_i) + (1 - \alpha T_f)v^{xl}(P_f^{xl} - P_i) \quad \text{(C5)}$$

and noting $(P_f^{xl} - P_i) = \hat{\rho}g(Z_f^{xl} - Z_i)$ one finally can show

$$\frac{\Delta H}{m} = c^{xl}(T_f - T_i) + (1 - \alpha T_f)g(Z_f^{xl} - Z_i)$$

$$+ f\{(1 - \alpha T_f)g(Z_f^l - Z_f^{xl}) + \Delta h_{\text{fusion}}\}$$

REFERENCES

Adamson, A. W., 1967. *Physical Chemistry of Surfaces*, 2nd ed., New York: Wiley.

Aki, K., Chouet, B., Fehler, M., Zandt, G., Koyanagi, R., Colp, J. and Hay, R. G., 1978. Seismic properties of a shallow magma reservoir in Kilauea Iki by active and passive experiments. *J. Geophys. Res. 83*, B5, 2273–2282.

Anderson, A. T., 1974a. Before eruption H$_2$O content of some high-alumina magmas. *Bull. Volcanol. 37*, 530–552.

Anderson, A. T., 1974b. Evidence for a picritic, volatile-rich magma beneath Mt. Shasta, California. *Jour. Petrol. 15*, 2, 243–267.

Anderson, A. T., 1975. Some basaltic and andesitic gases. *Rev. Geophys. Sp. Phys. 13*, 1, 37–54.

Anderson, E. M., 1951. *The dynamics of faulting and dyke formation, with Application to Britain*, Edinburgh: Oliver and Boyd.

Anderson, O. L., 1959. The Griffith criterion for glass fracture, in *Fracture*, Auerbach, B. L., Felbeck, D. K., and Thomas, D. A., eds., Wiley, New York, 331–353.

Anderson, O. L., Grew, P. C., 1977. Stress corrosion theory of crack propagation with applications to geophysics. *Rev. Geophy: Space Phy. 15*, 1, 77–104.

Anderson, O. L. and Perkins, P. C., 1974. Runaway temperatures in the asthenosphere resulting from viscous heating. *Jour. Geophy. Res. 79*, 2136–2138.

Arndt, N. T., 1977. Ultrabasic magmas and high-degree melting of the mantle. *Contrib. Mineral. Petrol. 64*, 205–221.

Bartlett. R. W., 1969. Magma convection, temperature distribution, and differentiation. *Am. Jour. Sci. 267*, 1067–1082.

Bear, J., 1972. *Dynamics of fluids in porous media*, New York: Elsevier.

---

[28] This may be calculated from Z_f^{xl}, the segregation depth.

Bennett, F. D., 1971. Vaporization waves in explosive volcanism. *Nature 234*, 538–539.

Bennett, F. D., 1974. On volcanic ash formation. *Amer. J. Sci. 274*, 648–657.

Bhattacharji, S., 1967. Mechanics of flow differentiation in ultramafic and mafic sills. *Jour. Geol. 75* 101–112.

Bhattacharji, S., and Smith, C. H., 1964. Flowage differentiation. *Science 145*, 150–153.

Bickle, M. J., 1978. Heat loss from the Earth: a constraint on Archaean tectonics from the relation between geothermal gradients and the rate of plate production. *Earth and Planet. Sci. Lttrs. 40*, 301–315.

Birch, F., 1966, Compressibility; elastic constants *in*, S. P. Clark, ed., *Handbook of Physical Constants*, Geol. Soc. Am. Memoir 97.

Bird, R. B., Stewart, W. E., and Lightfoot, E. N., 1960. *Transport Phenomena*, New York: John Wiley.

Bodvarsson, G., and Walker, G. P. L., 1964. Crustal drift in Iceland, *Geophys. J. R. Astron. Soc. 8*, 285–300.

Bottinga, Y., and Weill, D. F., 1972. The viscosity of magmatic silicate liquids: a model for calculation: *Am. Jour. Sci. 272*, 438–475.

Boyd, F. R., McCallister, R. H., 1976, Densities of fertile and sterile garnet peridotites, *Geophys. Res. Letters 3*, 9, 509–572.

Boyd, F. R. and Nixon, P. H., 1975. Origins of the ultramafic nodules from some kimberlites of northern Lesotho and the Monastery Mine, South Africa, *Phys. Chem. Earth. 9*, 431–454.

Boyd, F. R. and Nixon, P. H., 1973. Structure of the upper mantle beneath Lesotho, *Carnegie Inst. Washington Yearb. 72*, 431–445.

Brace, W. F., 1960. An extension of the Griffith Theory of fracture to rocks, *Jour. Geophy Res. 65*, 3477–3480.

Brace, W. F., 1960. Brittle fracture of rocks, *in* Judd, W. R., ed., *State of stress in the earth's crust*, New York: American Elsevier Publishing Co., Inc., 110–178.

Brace, W. F., 1977. Permeability from resistivity and pore shape, *Jour. Geophys. Res. 82*, no. 23, 3343–3349.

Brace, W. F., Walsh, J. B., and Frangos, W. T., 1968. Permeability of granite under high pressure, *Jour. Geophys. Res. 73*, no. 6, 2225–2236.

Bridgman, P. W., 1952. *Studies in large flow and fracture*, New York: McGraw-Hill.

Burnham, C. W., Davis, N. F., 1971. The role of H_2O in silicate melts I. P-V-T relations in the system $NaAlSi_3O_8$-H_2O to 10 kilobars and 1000°C. *Am. Jour. Sci. 270*, 54–79.

Busch, W.G., Schneider, G., and Mehnert, K. R., 1974. Initial melting at grain boundaries, II: Melting in rocks of granodioritic, quartz-dioritic and tonalitic composition, *Neues Jahrb. Mineral. Monatsh.*, 345–370.

Carman, P. C., 1956. *Flow of gases through porous media*, New York: Academic Press.

Carmichael, I. S. E., Nicholls, J., Spera, F. J. Wood, B. J., and Nelson, S. A., 1977. High-temperature properties of silicate liquids: applications to the equilibration and ascent of basic magma, *Phil. Trans. R. Soc. Lond. A 286*, 373–431.

Cawthorn, R. G., 1975. Degrees of melting in mantle diapirs and the origin of ultrabasic liquids, *Earth Plan. Sci. Lettrs. 27*, 113–120.

Chandrasekkar, S., 1961. *Hydrodynamics and hydromagnetic stability*, New York: Oxford University Press.

Currie, K. L., and Ferguson, J., 1970. The mechanism of intrusion of lamprophyre dikes indicated by offsetting of dikes. *Tectonophysics 9*, 525–535.

Dawson, J. B., 1967. A review of the geology of kimberlite, in *ultramafic and related rocks*, P. J. Wyllie, ed., New York: Wiley.

Deissler, R. G., 1955. *On turbulent flow*. NACA Tech. Rpt. 1210.

Eckert, E. G., and Jackson, T. W., 1950. *Turbulent convection*. Nat. Advisory Comm. Aeronaut. Tech Note 2207.

Eggler, D. H., 1972. Water-saturated and undersaturated melting relations in a paracutin andesite and an estimate of water content in the natural magma. *Contr. Mineral. Petrol. 34*, 261–271.

Eggler, D. H., 1973. Role of CO_2 in melting processes in the mantle. *Carnegie Inst. Washington Year Book 72*, 457–467.

Eggler, D. H., 1978. The effect of CO_2 upon partial melting of peridotite in the system $Na_2O–CaO–Al_2O_3–MgO–SiO_2–CO_2$ to 35 Kb, with an analysis of melting in a peridotite–$H_2O–CO_2$ system. *Am. Jour. Sci. 278*, 305–343.

Eggler, D. H., and Rosenhauer, M., 1978. Carbon dioxide in silicate melts, II: Solubilities of CO_2 and H_2O in $CaMgSi_2O_6$ (Diopside) liquids and vapors at pressures to 40 Kb.. *Am. Jour. Sci. 278*, 64–94.

Elder, J. W., 1970. Quantitative laboratory studies of dynamical models of igneous intrusions, in Newall, G. and Rast, N. *Mechanism of igneous intrustion*, Liverpool: Gallery Press, 348–362.

Evans, A. G., 1974. Fracture mechanics determinations, in *Concepts, flaws and fractography*, vol. 1, Bradt, R. C., Hasselman, P. H., Lange, F. E., eds., New York: Plenum, 221–253.

Feigenson, M. D., and Spera, F. J., 1980. Melt production by viscous dissipation: role of heat advection by magma transport. *Geophys. Res. Letters 7*, 2, 145–148.

Fiske, R. S., and Jackson, E. D., 1972. Orientation and growth of Hawaiian volcanic rifts: the effect of regional structure and gravitational stresses. *Proc. R. Soc. Lond. A 329*, 299–326.

Fujii, N., and Uyeda, S., 1974. Thermal instabilities during flow of magma in volcanic conduits. *Jour. Geophy. Res. 79*, 3367–3370.

Goetze, C. and Kohlstedt, D. L., 1973. Laboratory study of dislocation climb and diffusion in olivine. *Jour. Geoph. Res. 87*, no. 26, 5961–5969.

Goetze, C. and Brace, W. F., 1972. Laboratory observations of high temperature rheology of rocks. *Tectonophysics 13*, 583–600.

Goldstein, S., 1938. *Modern developments in fluid mechanics*, Oxford: Clarenden Press.

Gonten, D., and Whiting, R. L., 1967. Correlations of physical properties of porous media. *Soc. Petrol. Eng. Jour. 7*, no. 3, 266–272.

Green, D. H., 1970. The origin of basaltic and nephelinitic magmas. *Trans. Leicester Lit. Phil. Soc. 64*, 22–54.

Green, D. H., and Ringwood, A. E., 1967. The genesis of basaltic magmas. *Contrib. Mineral. Pet. 15*, 103–190.

Green, H. W., and Radcliffe, S. V., 1972. Deformation processes in the upper mantle, in *Flow and fracture of rocks*, AGU Geophys. Monogr. Ser., v. *16*, H. C. Heard, I. Y. Borg, W. L. Carter, and C. B. Raleigh, eds., 139–156.

Griffith, A. A., 1921. The phenomenon of rupture and flow in solids. *Phil. Trans. R. Soc. Lond. A. 221*, 163–198.

Griffith, A. A., 1925. The theory of rupture. *Internat. Cong. Appl. Mech., 1st, Delft 1924, Proc.*, 55–63.

Gruntfest, I. J., 1963. Thermal feedback in liquid flow; plane shear at constant stress. *Trans. Soc. Rheol. 7*, 195–207.

Gruntfest, I. J., Young, J. P., and Johnson, N. L., 1964. Temperatures generated by the flow of liquid in pipes. *Jour. Appl. Phys. 35*, 18–22.

Haimson, B. C., 1977. Crustal stress in the continental United States as derived from hydrofracturing tests, in *The Earth's Crust*, Heacock, J. G., ed., *Am. Geophys Union Monograph 20*, 294–307.

Hallman, T. M., 1956. Combined forced and free convection laminar transfer in vertical tubes with uniform heat generation. *Trans. ASME 78*, 1831–1840.

Hardee, H. C., and Larson, D. W., 1977. The extraction of heat from magmas based on heat transfer mechanisms. *Jour. Volcan: Geother. Res. 2*, 113–144.

Harris, P. G., and Middlemost, A. K., 1970. The evolution of kimberlites. *Lithos 3*, 77–82.

Haward, R. N., 1970. Critical stages in the fracture of an organic glass, in *Amorphous Materials*. Douglas, R. W., and Ellis, B., eds., 3rd Intl. Conf. on physics of non-crystalline solids. London: Wiley-Interscience, 431–451.

Hearn, B. C., 1968. Diatremes with kimberlitic affinities in north-central Montana. *Science 159*, 622–638.

Hess, B., Boiteux, A., Busse, H. G., and Gerisch, G., 1975, Spatio-temporal organization in chemical and cellular systems, in *Membranes, Dissipative Structures and Evolution*. Nicolis, G., Lefever, R., eds., London: Wiley Interscience.

Hildreth, W. E., 1977. "The magma chamber of the Bishop Tuff: Gradients in temperature, pressure, and composition," unpublished Ph.D. dissertation, University of California at Berkeley.

Hill, D. P., 1977. A model for earthquake swarms. *J. Geophys. Res. 82*, 8, 1347–1352.

Hoffman, A. W., Giletti, B. J., Yoder, H. S., and Yund, R. A., eds., 1974. *Geochemical transport and kinetics*, Carnegie Inst. Wash. Publ. 634.

Hubbert, M. K., and W. W. Rubey, 1959. Role of fluid pressure in mechanics of overthrust faulting. *Bull. Geol. Soc. Am. 70*, 115–166.

Irwin, G. R., 1958. Fracture, *in* Flugge, S. (ed.) *Handbuch der physik 6*, New York: Springer, 551–590.

Jackson, E. D., and Shaw, H. R., 1975. Stress fields in central portions of the Pacific plate: delineated in time by linear volcanic chains. *Jour. Geophys. Res. 80*, 14, 1861–1874.

Jackson, E. D., Shaw, H. R. and Bargar, K. E., 1975. Calculated geochronology and stress field orientations along the Hawaiian chain. *Earth Planet. Soc. Lttrs. 26*, 2, 145–155.

Jackson, E. D., and Wright, T. L. 1970. Xenoliths in the Honolulu Series, Hawaii, *J. Petrol. 11*, 405–430.

Jaeger, J. C., 1962. *Elasticity, fracture, and flow*, 2nd ed., London: Methuen and Co., Ltd.

Kay, J. M., 1968. *An introduction to fluid mechanics and heat transfer*, 2nd ed., Cambridge: At the University Press.

Kirkpatrick, R. J., 1974. Kinetics of crystal growth in the system $CaMgSi_2O_6-CaAl_2S_1O_6$. *Am. Jour. Sci. 274*, 215–242.

Klein, F. W., Einarsson, P., and Wyss, M., 1977. The Reykjunes Peninsula, Iceland, earthquake swarm of September 1972 and its tectonic significance. *J. Geophys. Res. 82*, 5, 865–888.

Komar, P., 1972. Flow differentiation in igneous dikes and sills: profiles of velocity and phenocryst concentration, *Bull. Geol. Soc. Am. 83*, 3443–3448.

Kushiro, I., 1976. Changes in viscosity and structure of melt of $NaAlSi_2O_6$ composition at high pressures. *Jour. Geoph. Res.. 81*, no. 35 6347–6350.

Lawn, B. R., and Wilshaw, T. R., 1975. *Fracture of Brittle Solids*, New York: Cambridge University Press.

Lipman, P. W., 1967. Mineral and chemical variations within an ash-flow sheet from Aso Caldera, southwestern Japan. *Contrib. Mineral Petrol. 16*, 300–327.

Lewis, G. N. and Randall, M., 1961. *Thermodynamics*, 2nd ed., New York: McGraw-Hill.

Lighthill, M. J., 1953. Theoretical considerations on free convection in tubes. *Quart. Jour. Mech. and Applied Math 6*, pt. 4, 399–439.

Lipps, F. B., 1971. Two-dimensional numerical experiments in thermal convection with vertical shear. *Jour. Atmos. Sci. 28*, no. 1, 3–19.

Lockwood, F. C., and Martin, B. W., 1964. Free convection in open thermosyphon tubes of non-circular section. *Jour. Mech. Eng. Sci. 6*, no. 4, 379–393.

Maaloe, S.; Wyllie, P. J., 1975. Water content of a granite magma deduced from the sequence of crystallization determined experimentally with water-undersaturated conditions. *Contrib. Min. Pet. 52*, 175–191.

Marsh, B. D., 1976a. Some Aleutian andesites: their nature and source. *J. Geol. 84*, 27–45.

Marsh, B. D., 1976b. Mechanics of Benioff zone magmatism, in *Geophysics of the Pacific Ocean and its margin*, G. H. Sutton, M. H. Manghani, and R.H. Moberly, eds., Am. Geophy. Union, Geophys. Monogr. *19*, 337–350.

Marsh, B. D., 1978. On the cooling of ascending andestic magma. *Phil. Trans. R. Soc. Lond. A 288*, 611–625.

Marsh, B. D., and Carmichael, I. S. E., 1974. Benioff zone magmatism. *Jour. Geophys. Res. 79*, no 8, 1196–1206.

Marsh, B. D., and Kantha, L. H., 1978. On the heat and mass transfer from an ascending magma. *Earth Planet. Sci. Lttrs. 39*, 435–443.

Mase, G. E., 1970. *Continuum mechanics*, Schaum's outline series, New York: McGraw-Hill Book Company.

McClintock, F. A. and Walsh, J. B., 1962. Friction on Griffith cracks in rocks under pressure. *Nat'l Cong. Appl. Mech., 4th, Berkeley, California, 1962, Proc.*, 1015–1021.

McDougall, I., 1971. Volcanic island chains and sea floor spreading. *Nature 231*, 141–144.

McGetchin, T. R., 1975. *Solid earth geosciences research activities at LASL, July 1–December 31, 1974.* Prog. Rep. LA-5956-PR, Los Alamos Scientific Laboratories, U.S.E.R.D.A..

McGetchin, T. R., and Nikhanj, Y. S., 1973a. Carbonatite-kimberlite relations in the Cane valley diatreme, San Juan County, Utah. *Jour. Geophy. Res. 78*, no. 11, 1854–1868.

McGetchin, T. R., and Silver, L. T., 1972. A crustal-upper mantle model for the Colorado Plateau based on observations of crystalline rock fragments in a kimberlite dike. *J. Geophys. Res. 77*, 7022–7037.

McGetchin, T. R., and Ullrich, G. W., 1973b. Xenoliths in maars and diatremes with inferences for the moon, Mars and Venus. *Jour. Geophy. Res. 78*, no 11, 1183–1853.

McKenzie, D., P., and Weiss, N., 1975. Speculations on the thermal and tectonic history of the Earth. *Geophys. J. Roy. Astron. Soc. 42*, 131–174.

McNown, J. S., Malaika, J., 1950, Effects of particle shape on settling velocity at low Reynolds numbers. *Am. Geophys. Union Trans. 31*, 74–82.

Mehnert, K. R., Busch, W., and Schneider, G., 1973. Initial melting at grain boundaries of quartz and feldspar in gneisses and granulites. *Neues. Jahrb. Mineral. Monatsh.*, 165–183.

Mohr, P. A., and Potter, E. C., 1976. The Sagatu Ridge dike swarm, Ethiopian rift margin. *Jour. Volcan. Geotherm. Res. 1*, 55–71.

Moore, J. G., 1970. Water content of basalt erupted on the ocean floor. *Contr. Mineral. Petrol. 28*, 272–279.

Moore, J. G., Batchelder, J., and Cunningham, C. G., 1977. CO_2-filled vesicles in mid-ocean basalt. *Jour. of Volcan. Geotherm. Res. 2*, 309–327.

Morton, B. R., 1960. Laminar convection in uniformly heated vertical dikes. *Jour. Fluid Mech. 8*, 227–240.

Murase, T., and McBirney, A. R., 1973. Properties of some common igneous rocks and their melts at high temperatures. *Geol. Soc. Amer. Bull. 84*, 3563–3592.

Murrell, S. A. F., 1964. The theory of the propagation of elliptical Griffith cracks under various conditions of plane strain or plane stress. *Brit. J. Appl. Phys. 15*, 1195–1229.

Murrell, S. A. F., 1965. The effect of triaxial stress systems on the strength of rocks at atmospheric temperatures. *Geoph. J.R. Astr. Soc. 10*, 231–247.

Murrell, S. A. F., and Ismail, I. A. H., 1976. The effect of temperature on the strength at high confining pressure of granodiorite containing free and chemically bound water. *Contrib. Mineral. Petrol. 55*, 317–330.

Mysen, B. O., 1977. The solubility of H_2O and CO_2 under predicted magma genesis conditions and some petrological and geophysical implications. *Rev. Geophys. Sp. Phys. 15*, 3, 351–361.

Mysen, B. O., Arculus, R. J., and Eggler, D., 1975. Solubility of carbon dioxide in melts of andesite, tholeiite and olivine nephelinite composition to 30 kbar pressure. *Contrib. Mineral. Petrol. 53*, 227–239.

Mysen, B. O., and Kushiro, I., 1977. Compositional variations of coexisting phases with degree of melting of peridotite in the upper mantle. *Am. Mineral. 62*, 843–865.

Nadai, A., 1963. *Theory of flow and fracture of solids*, vol. II, New York: McGraw-Hill Book Company.

Nicolis, G., Prigogine, I., and Glansdorff, C., 1975. On the mechanism of instabilities in nonlinear systems. *Advances in Chemical Physics 24*, 1–28.

Nicolis, G., and Prigogine, I., 1977. *Self-organization in nonequilibrium systems*, New York: Wiley.

Ode, H., 1957. Mechanical analysis of the dike pattern of the Spanish peaks area, Colorado. *Bull Geol. Soc. Am. 68*, 567 pp.

Ostrach, S., 1954. Combined natural and forced-convection flow and heat transfer of fluids with and without heat sources in channels with linearly varying wall temperatures. *NACA TN 3141*, 1–124.

Ostrach, S., and Thornton, P. R., 1958. On the stagnation of natural-convection flows in closed-end tubes. *Trans. ASME 80*, 363–366.

Pai, S. I., Hsieh, T., and O'Keefe, J. A., 1972. Lunar ash flow with heat transfer, in *Proc. of the 3rd Lunar Science Conference* (Supplement 3, *Geochimica et Cosmochimica Acta*) *3*, 2689–2711.

Parker, T. J., and McDowell, A. N., 1955. Model studies of salt-dome. *Amer. Assoc. Petrol. Geol. Bull. 39*, 2384–2470.

Pinkerton, H., and R. S. Sparks, 1978. Field measurements of the rheology of lava. *Nature 276*, 383–385.

Pollard, D. D., 1976b. On the form and stability of open hydraulic fractures in the Earth's crust. *Geophy. Res Lettrs. 3*, no. 9, 513–518.

Pollard, D. D., and Johnson, A. M., 1973. Mechanics of growth of some laccolithic intrusions in the Henry Mountains, Utah. *Tectonophysics 18*, 311–354.

Pollard, D. D., Muller, O. H., and Dockstader, D. R., 1975. The form and growth of fingered sheet intrusions. *Bull. Geol. Soc. Am. 86*, 351–363.

Pollard, D. D., and Muller, O. H., 1976a. The effects of gradients in regional stress and magma pressure on the form of sheet intrusions in cross section. *Jour. Geophys. Res. 81*, no. 5, 975–986.

Prigogine, I., and Lefever, R., 1975. Stability and self organization in open systems. *Advances in Chemical Physics 24*, 1–28.

Raleigh, C. B., and Paterson, M. S., 1965. Experimental deformation of serpentinite and its tectonic implications. *Jour Geophys. Res. 70*, 3965–3985.

Raleigh, C. B., 1968. Mechanisms of plastic deformation of olivine. *Jour. Geophy. Res. 73*, 5391–5407.

Ramberg, H., 1963. Experimental study of Gravity tectonics by means of centrifuged models. *Bull. Geol. Inst. Univ. Uppsala 42*, 1097.

Ramberg, 1967. Model experimentation of the effect of gravity on tectonic processes. *Geophy. Jour. 14*, 307–329.

Ramberg, 1968a. Instability of layered systems in a field of gravity. *Phys. Earth Planet. Int. 1*, 427–474.

Ramberg, H., 1968b. Fluid dynamics of layered systems in a field of gravity, a theoretical bases for certain global structures and isostatic adjustments. *Phys. Earth Planet. Int. 1*, 63–67.

Ramberg, H., 1968c. Mantle diapirism and its tectonic and magmagenetic consequences. *Phys. Earth Planet, Int. 5*, 45–60.

Ramberg, H., 1970. Model studies in relation to intrusion of plutonic bodies. *Geol. J. Spec.*, Issue 2, 261–286.

Reed-Hill, R. E., 1973. *Physical metallurgy principles*, New York: D. Van Nostrand Co..

Ribando, R. J., and Torrance, K. E., 1976. Natural convection in a porous medium: effects of confinement, variable permeability, and thermal boundary conditions. *J. Heat. Trans. 98.1*, 42–49.

Roberts, D. K., and Wells, A. A., 1954. The velocity of brittle failure. *Engineering 178*, 820–821.

Roberts, J. L., 1970. The intrusion of magma into brittle rocks, *in Mechanisms of Igneous Intrusion*, Newall, G. and N. Rast, eds., Liverpool: Gallery Press, 278–294.

Rohsenow, W. M., and Choi, H. Y., 1962. *Heat, mass, and momentum transfer*, New Jersey: Prentice-Hall.

Ross, C. S., and Smith, R. L., 1961. *Ash-flow tuffs; their origin, geologic relations and identification.* Prof. Pap. U.S. Geol. Survey, no. 366, 1–119.

Scarfe, C. M., 1973. Viscosity of basic magmas at varying pressure. *Nature 241*, 101–102.

Scheele, G. F., and Hanratty, T. J., 1962. Effect of natural convection on stability of flow in a vertical pipe. *Jour. Fluid. Mech. 14*, 244–256.

Scheidegger, A. E., 1957. *The physics of flow through porous media*, New York: Macmillian Co.

Secor, D. T. Jr., 1965. Role of fluid pressure in jointing. *Am. Jour. Sci. 263*, 633–646.

Shankland, T. J., and Waff, H. S., 1977. Partial melting and electrical conductivity anomalies in the upper mantle. *J. Geophys. Res. 82*, 5409–5417.

Shaw, H. R., 1965. Comments on viscosity, crystal settling and convection in granitic systems. *Amer. Jour. Sci. 272*, 120–152.

Shaw, H. R., 1969. Rheology of basalt in the melting range. *Jour. Petrol. 10*, 510–535.

Shaw, H. R., 1972. Viscosities of magmatic silicate liquids: an empirical method of predicition. *Amer. Jour. Sci. 272*, 870–893.

Shaw, H. R., 1973. Mantle convection and volcanic periodicity in the Pacific; evidence for Hawaii. *Bull. Geol. Am. 84*, 1505–1526.

Shaw, H. R., 1974. Diffusion of H_2O in granitic liquids. Part I, Experimental data, Part II, Mass transfer in magma chambers, in *Geochemical Transport and Kinetics*. A. W. Hoffman, B. J. Gilletti, H. S. Yoder and R. A. Yund, eds., Carnegie Inst. Wash., 139–170.

Shaw, H. R., Peck, D. L., Wright, T. L., and Okamura, R., 1968. The viscosity of basaltic magma: an analysis of field measurements in Makaopuhi lava lake, Hawaii. *Am. Jour. Sci. 266*, 255–264.

Sleep, N. H., 1974. Segregation of magma from a mostly crystalline mush. *Geol. Soc. Am. Bull. 85*, 1225–1232.

Smith, R. L., 1960. Ash flows. *Bull. Geol. Soc. Am. 71*, 795–842.

Soo, S. H., 1967. *Fluid Dynamics of Multiphase Systems*, Waltham, Mass.: Blaisdell Pub. Co..

Sparks, R. S. J., 1977. The dynamics of bubble formation and growth in magmas: a review and analysis. *Jour. Volcan. Geotherm. Res. 3*, 1–37.

Sparks, R. S., Pinkerton, H., and MacDonald, R., 1977. The transport of xenoliths in magmas. *Earth Planet. Sci. Letters 35*, no 2, 234–238.

Spera, F. J., 1977. "Aspects of the thermodynamic and transport behavior of basic magma," Ph.D. dissertation, University of California at Berkeley.

Stocker, R. L. and Ashby, M. F., 1973. On the reheology of the upper mantle. *Geophys. Space Phys. 11*, 391–426.

Straus, J. M., and Schubert, G., 1979. Three-dimensional convection in a cubic box of fluid-saturated porous material. *J. Fluid Mech.*, in press.

Swalin, R. A., 1962. *Thermodynamics of solids*, New York: Wiley.

Takhar, H. S., 1968. Entry-length flow in a vertical cooled pipe. *J. Fluid Mech. 34*, pt. 4, 641–650.

Tazieff, H., 1970. Mechanisms of ignimbrite eruption, in *Mechanics of igneous intrusion*, eds. Newall, G., and Rast, H., Liverpool: Gallery Press, 319–327.

Timoshenko, S., Goodier, J. N., 1951, *Theory of elasticity*, McGraw-Hill, New York.

Tremlett, C. R., 1978. A note on stress relationships determined from Mohr diagrams. *Geol. Mag. 115*, 8, 341–350.

Turcotte, D.L., 1974. Membrane tectonics. *Geophys. J. Roy. Astron. Soc. 36*, 33–42.

Turcotte, D. L. and Ahern, J. L., 1978. A porous flow model for magma migration in the asthenosphere. *Jour. Geophys. Res. 83*, no. B2, 767–772.

Turcotte, D. L., and Oxburgh, E. R., 1973. Mid-plate tectonics. *Nature 244*, 337–339.

Turner, F. J. and Verhoogen, J., 1960. *Igneous petrology*, New York: McGraw-Hill.

Upadhyay, H. D., 1978. Phanerozoic peridotitic and pyroxenitic komatiites from Newfoundland. *Science 202*, 1192–1195.

Verhoogen, J., 1954. Petrological evidence on temperature distribution in the mantle of the earth. *Trans. Am. Geophys. Union 35*, 85–92.

Verhoogen, J., 1973. Possible temperatures in the oceanic upper mantle and the formation of magma. *Geol. Soc. Am. Bull. 84*, 515–522.

Waff, H. S., 1974. Theoretical consideration of electrical conductivity in a partially molten mantle and implications for geothermometry. *J. Geophys. Res. 79*, 4003–4010.

Weertman, J., 1968. Bubble coalescence in ice as a tool for the study of its deformation history. *J. Glaciol. 7*, 155–159.

Weertman, J., 1971. Theory of water filled crevasses in glaciers applied to vertical magma transport beneath oceanic ridges. *Jour. Geophys. Res. 76*, 1171–1183.

Weertman, J., 1972. Coalescence of magma pockets into large pools in the upper mantle. *Bull. Geol. Soc. Am. 83*, 3531–3632.

Whitehead, J. A., and Luther, D. S., 1975. Dynamics of laboratory diapirs and plume models. *Jour. Geophys. Res. 80*, 705–717.

Wiederhorn, S. M., 1967. Influence of water vapor on crack propagation in soda-lime glass. *J. Amer. Ceram. Soc. 50*, 407–414.

Wilkins, B. J., and Dutton, R., 1976. Static fatigue limit with particular reference to glass. *J. Am. Ceramic Soc. 59*, 3–4, 108–112.

Williams, J. A., and Singer, A. R. E., 1968. Deformation strength and fracture above the solidus temperature. *Jour. Inst. Metals 96*, 5–12.

Wright, T. L., and Fiske, R. S., 1971. Origin of the differentiated and hybrid lavas of Kilauea Volcano, Hawaii, *Jour. Pet. 12*, 1, 1–67.

Wyllie, M. R. J., and Rose, W. D., 1950, Application of the Kozeny equation to consolidated porous media. *Nature 165*, 972–974.

Yoder, H. S., 1976. *Generation of basaltic magma*. Washington: Natl. Acad. Sci.

Yoder, H. S., and Tilley, C. E., 1962. Origin of basalt magmas: an experimental study of natural and synthetic rock systems. *J. Petrol. 3*, 342–532.

Chapter 8

MAGMATIC INFILTRATION METASOMATISM, DOUBLE-DIFFUSIVE FRACTIONAL CRYSTALLIZATION, AND ADCUMULUS GROWTH IN THE MUSKOX INTRUSION AND OTHER LAYERED INTRUSIONS

T. N. IRVINE

Geophysical Laboratory, Carnegie Institution of Washington

INTRODUCTION: TERMINOLOGY AND HISTORY OF THE PROBLEM

As background, it is useful first to review briefly the terminology of igneous cumulates. This is conveniently done by means of Figure 1 (after Wager, Brown and Wadsworth, 1960), which portrays three types of plagioclase cumulates from the Skaergaard intrusion. The rocks consist essentially of *cumulus crystals* of plagioclase that were, by some mechanism, fractionated from the main body of magma and accumulated in layers on the floor of the magma chamber. After accumulation, the crystals were cemented together by *postcumulus materials* crystallized from the interstitial or *intercumulus liquid* trapped in the pore spaces between them. The parental liquid of the Skaergaard intrusion was basaltic in composition (Wager and Deer, 1939; Wager, 1960), and in the rock in Figure 1A, all of the minerals that would be expected to crystallize from the liquid are represented in quantity among the postcumulus materials. Olivine, pyroxene, iron ore, and granophyre occur as *discrete* postcumulus constituents; sodic plagioclase is portrayed as *overgrowth* on the cumulus crystals (outlined by dashed lines). The rock appears, therefore, to have completed crystallization as a closed

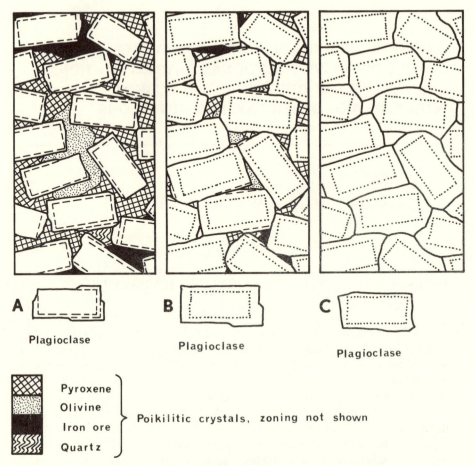

Figure 1. Diagrammatic representation of plagioclase cumulates formed from basaltic magma. After Wager, Brown, and Wadsworth (1960), with permission of Oxford University Press. For explanation, see text.

system with little or no addition or removal of liquid or other material. Such rocks are termed *orthocumulates.*[1]

In the rock in Figure 1B, however, the quantity of mafic minerals is much reduced, and the cumulus crystals are extensively overgrown by calcic plagioclase similar in composition to the original crystals (outlined by dotted lines). In the rock in Figure 1C, mafic materials are completely

---

[1] The terminology used here was largely devised by Wager, Brown and Wadsworth (1960) and Wager and Brown (1968). The term "postcumulus" was introduced by Jackson (1967).

excluded, and the plagioclase is entirely calcic and occurs in grains characterized by mutual interference boundaries. Wager and his colleagues called the rock in B a *mesocumulate*; that in C, an *adcumulate*. They believed that, in these rocks, most of the overgrowth of calcic plagioclase was in some way introduced or added during the postcumulus stage of crystallization. This process was called *adcumulus growth*.

Adcumulus growth is also encountered in other types of cumulates, especially those of olivine and/or pyroxene (e.g., Hess, 1960; Jackson, 1961), chromite (e.g., Cameron, 1975), and magnetite (e.g., Molyneux, 1970). The problems posed are as follows: How did the enlargement of the cumulus crystals occur, and perhaps more important, how were the constituents of the liquid that should have crystallized as discrete postcumulus phases removed?

Although the term adcumulus growth did not exist when Norman L. Bowen wrote *The Evolution of the Igneous Rocks*, Bowen briefly considered the problem of eliminating interstitial liquid from a mass of accumulated crystals in his discussion of the origin of monomineralic rocks (p. 167). He drew no conclusions, but suggested that the process might be analogous to that by which loose crystals of a soluble salt are converted into a solid cake on the bottom of a beaker containing a saturated solution of the salt.

The explanation of adcumulus growth that has been most widely cited in the literature was proposed by H. H. Hess, and one might guess that the idea was inspired by Bowen's remarks. Hess suggested the mechanism in 1939 in connection with his work on the Stillwater complex, and developed it more fully in his memoir on this work published in 1960. Hess postulated that, just after the fractionated crystals had accumulated, they could continue to grow from materials supplied by diffusion through the intercumulus liquid from the main body of magma above the cumulate pile. The part of the intercumulus liquid that should have crystallized as discrete postcumulus phases was thought to be eliminated upward mechanically by the continued growth of the fractionated crystals. Wager (1963) and his colleagues concurred with this mechanism, and Wager made the additional observation that the adcumulus material would have to be drawn from supercooled liquid in the main magma body. In 1972, however, G. B. Hess published a mathematical analysis of the process that indicated diffusion to be much too slow, relative to the probable rate of crystal accumulation, to supply the materials necessary for appreciable adcumulus growth. Some other explanation seemed to be required.

The original objective in this paper was to describe certain features of the layered rocks of the Muskox intrusion in the Canadian Northwest

Territories, and to outline an explanation based on infiltration metaso-matism (Korzhinsky, 1965) of cumulus minerals by migrating inter-cumulus liquid. If this explanation is correct, then the process has major importance with respect to the mechanism of adcumulus growth. In addition, during the final stages of the manuscript preparation, the author came to realize that a process called "double-diffusive convection", postulated to occur in layered intrusions by Turner and Gustafson (1978) and McBirney and Noyes (1979), almost certainly also had funda-mental significance in the history of the rocks under consideration. Given that all the necessary background information was already set out, it was convenient to include a preliminary analysis. This led to the formulation of an additional concept, to which is given the name "double-diffusive fractional crystallization."

GEOLOGY AND CRYSTALLIZATION HISTORY OF THE MUSKOX INTRUSION

The Muskox intrusion is located along the Arctic Circle in the Bear Structural Province of the Canadian Precambrian Shield, about 80 km west of Great Bear Lake (Figure 2). As exposed, it trends north-northwesterly and is about 120 km long. Over the southern half of this length it consists simply of a segmented vertical dike, 150–500 m wide, known as the "feeder dike." This unit is composed mainly of bronzite gabbro but generally also embodies either one or two internal zones of picrite parallel with its length. Some of the gabbro is chilled along the dike walls, and the picrite variously contains 15–50% granular olivine set in a gabbroic matrix. The dike is presumed to extend northward at depth like a keel beneath the main body of the intrusion (Figure 3).

The main body of the intrusion is funnelform in cross section, with lower walls dipping inward, generally at 25°–35°. The body plunges to the north at about 5°; thus, in the flat shield terrain its outcrop width gradually increases northward, reaching a maximum of about 11 km where the intrusion disappears beneath its roof rocks. The intrusion does not end at this point, however, but can be traced northwards in aeromagnetic maps for another 30 km. Over this distance there develops a major gravity anomaly that peaks some 60 km to the north and ulti-mately extends for about 230 km (Figure 2).[2] The intrusion is probably not responsible for all (or perhaps even most) of the gravity anomaly, but the indication seems clear that the exposed rocks represent only the southern extremity of a much larger mafic-ultramafic igneous complex.

---

[2] Geological Survey of Canada, Geophysics Paper 1739, 1963.

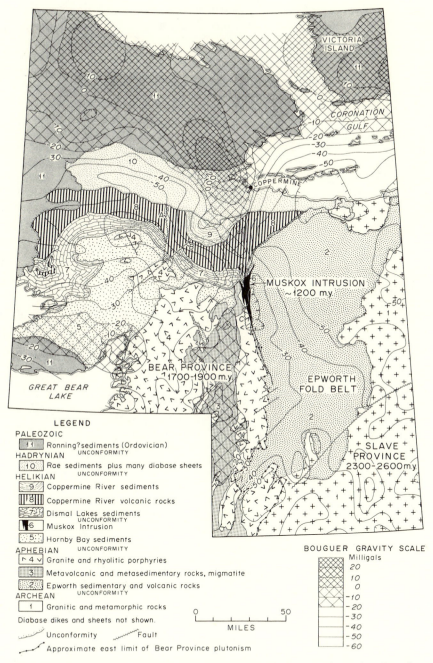

LEGEND

PALEOZOIC

▨ 11 Ronning? sediments (Ordovician)

UNCONFORMITY

HADRYNIAN

▨ 10 Rae sediments plus many diabase sheets

UNCONFORMITY

HELIKIAN

▨ 9 Coppermine River sediments

‖ 8 ‖ Coppermine River volcanic rocks

▨ 7 Dismal Lakes sediments

UNCONFORMITY

▪ 6 Muskox Intrusion

▨ 5 Hornby Bay sediments

UNCONFORMITY

APHEBIAN

∨ 4 ∨ Granite and rhyolitic porphyries

▨ 3 Metavolcanic and metasedimentary rocks, migmatite

▨ 2 Epworth sedimentary and volcanic rocks

UNCONFORMITY

ARCHEAN

☐ 1 Granitic and metamorphic rocks

Diabase dikes and sheets not shown.

⌇⌇⌇ Unconformity   〰 Fault

⌐⌐ Approximate east limit of Bear Province plutonism

0 ———————— 50
MILES

BOUGUER GRAVITY SCALE

Milligals

▨ 20
10
0
-10
-20
-30
-40
-50
-60

Figure 2. Geologic and Bouguer gravity anomaly map of the Coppermine region, Canadian Northwest Territories, including the Muskox intrusion. The intrusion plunges northward at about 5° and disappears beneath a roof of Hornby Bay sediments and older rocks at the north end of its outcrop area. The intrusion is shown here to be older than the Dismal Lakes sediments and Coppermine River volcanic rocks, but there is a strong possibility that it is more or less the same age as the volcanic rocks. The prominent gravity anomaly appears, at least in part, to reflect the northward extent of the intrusion at depth.

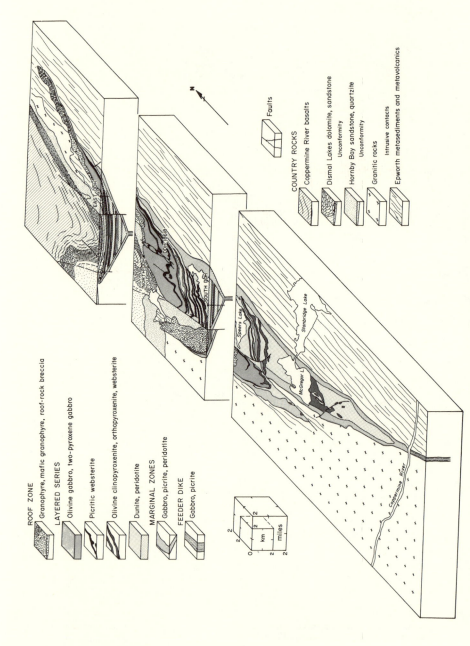

ROOF ZONE
Granophyre, mafic granophyre, roof-rock breccia
LAYERED SERIES
Olivine gabbro, two-pyroxene gabbro
Picritic websterite
Olivine clinopyroxenite, orthopyroxenite, websterite
Dunite, peridotite
MARGINAL ZONES
Gabbro, picrite, peridotite
FEEDER DIKE
Gabbro, picrite

COUNTRY ROCKS
Coppermine River basalts
Dismal Lakes dolomite, sandstone
    Unconformity
Hornby Bay sandstone, quartzite
    Unconformity
Granitic rocks
    Intrusive contacts
Epworth metasediments and metavolcanics

Faults

N

Speers Lake
Stanbridge Lake
McGregor L.
Coppermine River

EAST LOOP
NORTH LOOP
SOUTH LOOP
SOUTH DDH

km
miles

Figure 3. Exploded block diagram of the Muskox intrusion, showing its structural relations and the locations of four principal diamond drill holes. The feeder dike extends in outcrop south from the Coppermine River for about 45 km.

The main body of the intrusion is divided into four parts: two marginal zones, a layered series, and a granophyric roof zone. The marginal zones line the inward-dipping footwall contacts and are each 130–230 m thick. In general, they grade upward and inward from bronzite gabbro at the contact through picrite and feldspathic peridotite to peridotite. As in the feeder dike, the gabbro is locally chilled (especially in the south), and the chilled material appears to represent the composition of the parental liquid after partial crystallization differentiation (Irvine, 1977a). The other rocks are essentially equivalent chemically to combinations of this liquid plus gradually increasing amounts of accumulated olivine.

The layered series occupies the trough formed by the marginal zones. It is about 1,800 m thick and comprises 42 mappable layers of 18 different rock types. The layers range in thickness from 3 m to 350 m and, in general, are sharply defined and show little internal layering or structure other than planar lamination. They dip gently northward (at about 5° on the average), parallel with the plunge of the intrusion, and typically are continuous laterally from one marginal zone to the other. Some layers can be traced in outcrop for more than 25 km, and from drillhole intersections it is apparent that many of them have areal extents in excess of 250 km^2. The succession of rock types in the layered series ranges from dunite at the base through peridotite and various pyroxenites and gabbros to granophyric gabbro at the top (Figure 4). The top of the series is gradational with an irregular capping of granophyric rocks; this capping is called a "roof zone" because it is extensively charged with xenoliths of the various roof rocks.

Two features of the layered series are particularly notable: one is the dominance of olivine cumulates; the other is the repetition of rock types. In Figure 4, the olivine cumulates are termed dunite, peridotite, feldspathic peridotite, and picrite, depending on their content of discrete postcumulus minerals. In total, they make up about 1,200 m of the 1,800 m thickness of the layered series, or about 60% of the cross section of the intrusion. Olivine by itself forms more than 50% of the series by volume, and it typically is accompanied by 1–2% cumulus chromite. The amounts of olivine and chromite are grossly disproportionate with the apparent composition of the parental liquid (Smith and Kapp, 1963). The chilled margin of the intrusion is commonly equivalent to a silica-saturated tholeiite, and the current best estimate of the composition of the most primitive liquid to crystallize in the intrusion has only 19 wt % olivine in the norm and only .24 wt % Cr_2O_3 (Table 1).

The repetition of rock types is such that certain kinds of layers recur in specific sequences that evidently reflect the crystallization order of the cumulus minerals from their parental liquids (Irvine and Smith, 1967; Irvine, 1970a). The layered series has been divided into 25 of these

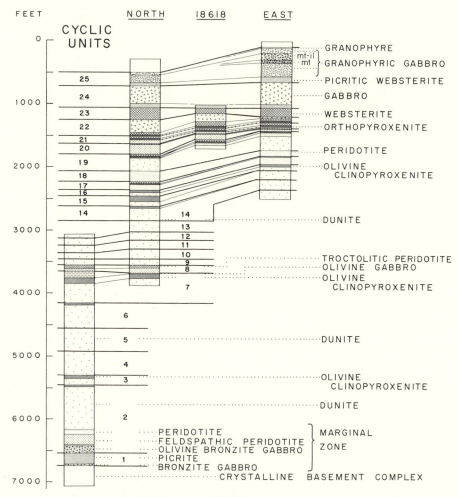

Figure 4. Drill-hole sections of the Muskox intrusion, showing the main cyclic units: mt, magnetite zone; il-mt, ilmenite-magnetite zone. Diabase dikes have been omitted for clarity. See Figure 3 for drill hole locations. Drill hole 18618 was drilled by the Canadian Nickel Co. A more detailed description of the North, South, and East drill-hole sections is given by Findlay and Smith (1965). The South drill-hole section is at left.

repetitive divisions, called *cyclic units*—a name originated by Jackson (1961) for similarly repeated stratigraphic divisions in the Stillwater ultramafic zone. In the case of the Muskox intrusion, it is evident from the sequences of rock types, the abundance of olivine, and certain other mineralogical and chemical relations described in part below, that the exposed part of the intrusion was open to at least 25 major additions

Table 1. Estimated compositions, densities and viscosities for Muskox liquids at early stages of differentiation.

	Oxides, wt %			Norms, wt %	
	1	2		1	2
SiO_2	49.20	51.08	Ol	19.07	4.23
Al_2O_3	11.98	13.65	Opx	17.94	23.18
Fe_2O_3	.48	.53	Cpx	19.35	23.05
FeO	10.02	9.72	An	24.03	27.23
MgO	14.88	9.78	Ab	13.20	15.32
CaO	9.78	11.30	Or	3.31	3.78
Na_2O	1.56	1.81	Mt	.70	.77
K_2O	.56	.64	Il	1.77	2.03
TiO_2	.93	1.07	Ap	.19	.21
P_2O_5	.08	.09	Chr	.35	.12
MnO	.18	.17	Total	99.91	99.92
NiO	.08	.03			
Cr_2O_3	.24	.08			
Total	99.97	99.95			
Density, g cm^{-3}*		2.73		2.70	
Viscosity, poise[†]		35		342	

Assumed temperature   1,325°C   1,225°C[‡]

* Calculated by method of Bottinga and Weill (1970).
[†] Calculated by method of Bottinga and Weill (1972).
[‡] From Biggar (1974).

Column Headings

1. Estimated primitive liquid, after Irvine (1977a).
2. Chilled margin composition, considered representative of liquid after about 16 wt % crystallization, from Smith and Kapp (1963).

of fresh liquid during accumulation of the layered series, with concurrent displacement and removal of the residual differentiated liquid (Irvine and Smith, 1967). The successive batches of liquid were each partly crystallized, variously progressing through some or all of the first four stages of their crystallization order before they were displaced. Some of the early batches evidently yielded as much as about 16 wt % olivine before being displaced; later batches usually precipitated 3–7%. It was in this way that the abundant olivine cumulates were produced and the cyclic units repeated. The feeder dike does not cut the layered series at the present erosion surface, however, so it must be inferred that the fresh magma was introduced by some other route. Irvine and Smith (1967) proposed that it came from the north, and flowed laterally in the space between the roof of the intrusion and the accumulating layered series (Figure 5). The fresh magma was presumably derived from a major feeder system or magma reservoir at depth in the area of the large gravity anomaly, and the displaced differentiated liquid was probably pushed on to the south and then to the surface as volcanic eruptions, the products of which have since been removed by erosion.

In the latter regard, it might be emphasized that the time period during which any given batch of fresh magma was introduced was probably very short compared with the time required for the formation of the cyclic unit that precipitated from the magma. By analogy with present-day volcanism, each new batch of magma might have been moved into the intrusion (and the old liquid displaced) in a matter of weeks or months. By contrast, the time necessary to crystallize the cumulus minerals in a typical cyclic unit is indicated by heat-transfer calculations (Irvine, 1970b) to have been of the order of hundreds of years. With a few exceptions, not pertinent here, there is no evidence that the large movements of magma directly affected the internal development of the cyclic units.

The rocks formed by crystal accumulation in the intrusion are orthocumulates in the marginal zones near the footwall contacts, and grade to mesocumulates in the lower and middle parts of the layered series (Figure 6). Parts of some pyroxenite layers are adcumulates, as is most of a layer of olivine gabbro in the middle of the layered series. The classification of the gabbroic cumulates in the upper part of the intrusion is less definite, owing to uncertainties in distinguishing postcumulus materials, but they appear to be mostly mesocumulates, grading to orthocumulates.

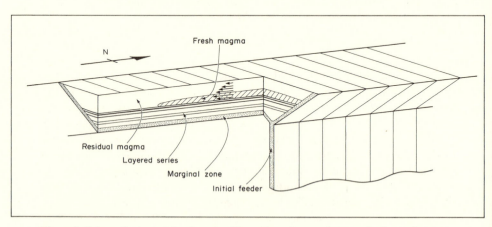

Figure 5. Schematic isometric diagram illustrating the north-to-south flow of magma through the intrusion prior to formation of a new cyclic unit.

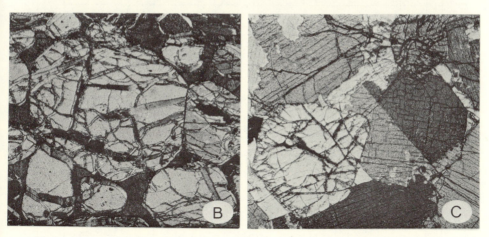

Figure 6. (A) Aerial photograph showing a 12 m layer of clinopyroxene-olivine cumulate (olivine clinopyroxenite) above a layer of serpentinized olivine-chromite cumulate (dunite), about 70 m thick. North is to the right, and the layers were intersected down dip in the North drill hole, 11 km distant, at 700–800 m below the surface. (B) Photomicrograph of an olivine-chromite mesocumulate with a small amount of postcumulus clinopyroxene. (C) Photomicrograph of a clinopyroxene-olivine mesocumulate.

THEORETICAL COMPOSITIONAL VARIATIONS IN AN IDEALIZED SEQUENCE OF CYCLIC UNITS

The concepts developed in this paper are essentially based on relations in four cyclic units in the lower part of the Muskox layered series. For background purposes, it is useful to describe the compositional variations that would be expected in such units if they had formed purely through the combination of fractional crystallization and periodic replenishment of the liquid with no complications. A hypothetical sequence of three units theoretically formed in this way is shown in Figure 7. The top unit is the most complete, comprising layers representing three stages of crystallization. Olivine is the main cumulus mineral in the first layer; it is joined by

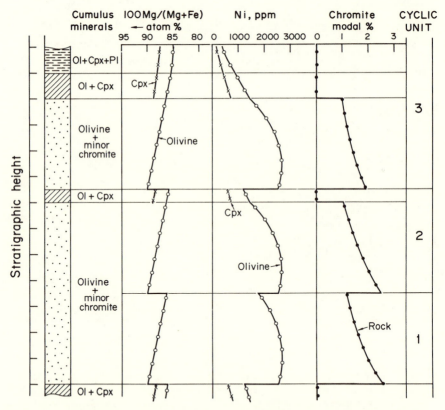

Figure 7. Hypothetical sequence of three cyclic units analogous to those in the lower part of the Muskox layered series, showing chemical and modal trends that would be expected ideally to develop through fractional crystallization and periodic replenishment of the magmatic liquid. Note that the breaks in the trends of Mg/Fe and Ni are coincident with the modal discontinuities that define the cyclic units.

clinopyroxene in the second layer and then by plagioclase in the third. This order of appearance corresponds to the experimental crystallization order of melts made from the chilled margin of the Muskox intrusion at geologically reasonable oxygen fugacities (Biggar, 1974). Chromite is shown to have precipitated in minor amounts with olivine in the first layer of the top cyclic unit in Figure 7, but is absent thereafter. Its disappearance would be expected, because a liquidus reaction relationship exists between chromite and clinopyroxene. In effect, when pyroxene begins to form, it can accommodate all the Cr available from the liquid, so chromite stops forming. This reaction relationship was suggested by Bowen (1928, pp. 279–281), documented in the Muskox intrusion by Irvine (1967), and demonstrated experimentally by Dickey, Yoder, and Schairer (1971).

The layer succession in the top cyclic unit in Figure 7 could be extended to include more stages of crystallization—for example, orthopyroxene might appear next, followed successively by magnetite, ilmenite, and apatite. More commonly in the Muskox intrusion, however, the crystallization course was interrupted by an influx of fresh liquid before even three stages were completed. Such an event is indicated in the second cyclic unit in Figure 7, in which no plagioclase was precipitated before the regression to olivine and chromite crystallization, and in the first cyclic unit, in which not even clinopyroxene was precipitated.

The stratification defined by the abrupt modal changes is described by the term *phase layering* (Hess, 1960). There may also be changes in the compositions of the cumulus minerals, a feature termed *cryptic layering* by Wager and Deer (1939); and there can be gradual changes in the modal proportions of certain cumulus minerals, here called *modal grading*. Two types of cryptic layering are illustrated in Figure 7, one defined by variation in Mg/Fe^{2+}, the other by variation in Ni or $Ni/(Mg + Fe^{2+})$. The Mg/Fe^{2+} trend was calculated by means of a Rayleigh fractionation equation (Irvine, 1977b) based on the distribution coefficient, $K_D = [(Mg/Fe^{2+})$ in liquid$]/[(Mg/Fe^{2+})$ in olivine$]$, which has an equilibrium value of about .3 (Roeder and Emslie, 1970). The trend is practically linear for olivine over the range of 16 wt % crystallization represented in the dunite layers. The Ni trends were calculated by finite difference methods, based on experimental data for the coefficient, $D = (Ni$ in olivine$)/(Ni$ in liquid$)$. These trends are convex upward with respect to the concentration of Ni because D increases strongly as the liquid becomes impoverished in olivine (Irvine, 1974a, Figure 56). At first the effect of increasing D is to maintain high Ni concentrations in the olivine even though the liquid is being depleted, but eventually the liquid becomes so impoverished that the amount of Ni entering olivine must also decline.

Modal grading is illustrated by the abundance of chromite, which decreases upward in each of the dunite layers. This type of trend, also observed in several Muskox cyclic units, develops because the olivine-chromite cotectic is curved in such a way that, as the liquid is fractionated around the curve, less and less chromite is precipitated. Such curvature has been demonstrated in the system $CaO–MgO–Al_2O_3–Cr_2O_3–SiO_2$ (Irvine, 1977c).

The key feature illustrated in Figure 7, however, is that the discontinuities in the chemical and modal trends associated with the replenishment of the liquid occur at the same stratigraphic levels as the sharp modal changes defining the regressions in crystallization path. As will be seen below, actual relations in the Muskox layered series are locally in marked contrast to this feature.

THE PROCESS OF POSTCUMULUS MAGMATIC INFILTRATION METASOMATISM

OBSERVED COMPOSITIONAL VARIATIONS IN SOME MUSKOX CYCLIC UNITS

Figures 8 and 9 show compositional data from cyclic units 4–7 of the Muskox layered series. (The cyclic units are numbered on the right-hand side of each diagram.) Each of the units embodies a dunitic layer of olivine-chromite cumulate, 100–120 m thick, exhibiting trends of variation similar to those outlined in the hypothetical example in Figure 7. The stratigraphic boundaries of the four units are located on the basis of sharp modal changes. The definition of the basal contacts of units 4 and 7 is relatively straightforward. Each corresponds to a contact of dunite over pyroxenite, marking a major regression in crystallization order. The basal contacts of units 5 and 6 are more obscure, both occurring within dunite, but in each case the unit begins with a small but distinct increase in the modal abundance of chromite, a feature that is also evident in the amount of Cr in the dunite (Figure 8). Note also that the trends defined by the analyses of olivine in units 4 and 5 are clearly reflected in the whole-rock and chromite compositions, and that these latter compositions show the same types of trends in units 6 and 7, where the olivine is completely serpentinized.

Several other minor or "subsidiary" cyclic units may also be represented in Figures 8 and 9. They are most clearly indicated in the whole-rock Mg/Fe^{2+} profiles for units 4, 6, and 7 (where they exhibit maximum iron enrichment at drill-hole depths of approximately 650, 425, and 285 m). It seemed advisable, however, to define a minimum number of units, and as will be seen below, some of the possible subsidary units may indeed differ in origin from the main units.

First consider Figure 8. It is seen that, although the Fo content (or Mg/Fe^{2+}) of olivine increases sharply near the base of cyclic unit 4 (at A)

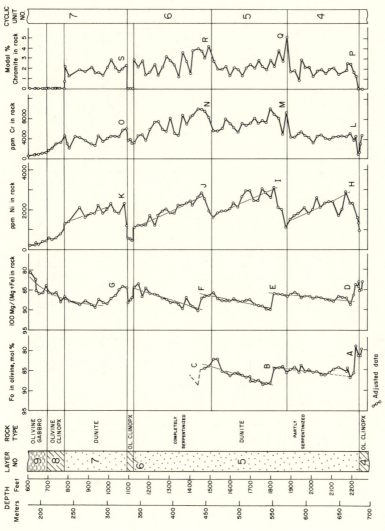

Figure 8. Mineralogical, chemical, and modal data from the rocks of Muskox cyclic units 4–7 as exposed on the South drill hole. The olivine compositions were determined by the X-ray method of Jambor and Smith (1964); Mg and Fe were determined by X-ray fluorescence, Ni and Cr by emission spectrography, and the chromite abundance by point counting. The cyclic units are defined on the basis of modal discontinuities reflecting regressions in the crystallization order of the magma. Note that the discontinuities in Mg/Fe^{2+} and Ni do not match either the modal breaks or each other.

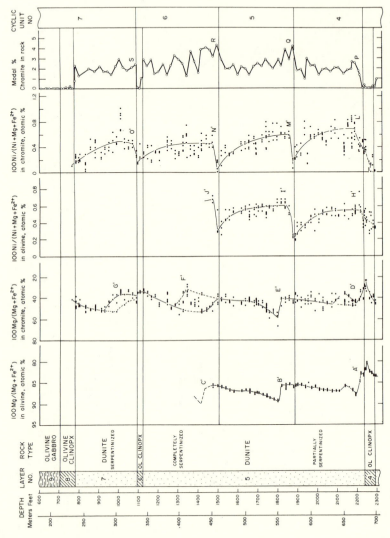

Figure 9. Electron microprobe data on Mg/Fe^{2+} and Ni in olivine and chromite in the same section as in Figure 8. The data on chromite abundance are the same as in Figure 8. As in Figure 8, the chemical breaks that would be associated with regressions in the crystallization path of the magma do not match the modal breaks that define these regressions.

and then follows a trend of gradual upward decrease like that in the hypothetical model, the increase does not occur until 6–8 m above the base of the unit. Similar relations occur at the base of unit 5 (at B), only here the sharp increase in Fo does not occur until some 20 m above the contact (see also, in Figure 9, A′ and B′). The same discrepancies are evident in the whole-rock data on Mg/Fe^{2+} (at D and E).

The olivine becomes completely serpentinized before an increase in Mg/Fe^{2+} is defined at the base of unit 6 (at C), but in the whole-rock data, the discontinuity is about 18 m above the defined contact (at F). The increase in Mg/Fe^{2+} in the lower part of unit 7 (at G) is more gradual, but a maximum is not reached until almost 50 m above the base of the unit.

With regard to the Ni data in Figure 8, the break in trend at the base of unit 4 (at H) is not closely defined, but certainly a strong increase does not occur until the third sample of dunite, about 10 m up. At the base of unit 5, the high point in chromite (at Q) or Cr (at M) is a low point in Ni (near I); and similarly at the base of unit 6, the high point in chromite (at R) and the main increase in Cr (at N) is a low point in Ni (near J). At the base of 7, the first dunite sample is relatively poor in Ni (near K), the next shows close to a maximum value.

Similar relations are revealed by the electron microprobe data in Figure 9. The Mg/Fe^{2+} trends of olivine are essentially identical to those in Figure 8, based on x-ray methods, and similar trends occur in the chromite (although with complications due to large variations in its composition, especially near the "subsidiary" cyclic units at F′ and G′, units 6 and 7). The data for Ni in olivine and chromite give a clearer indication than the whole-rock data in Figure 8 of the convex-upward profile illustrated in the hypothetical model in Figure 7. More importantly, they confirm that the first dunite samples at the base of each of units 5 and 6 is indeed poor in Ni. The Ni discontinuities, therefore, are some 3–6 m above the basal contacts.

In summary, then, the chemical breaks between the cyclic units are consistently up-section from the modal breaks, and in each case the difference for the Mg/Fe^{2+} discontinuity is 2–4 times larger than that for the Ni discontinuity.

THE INFILTRATION METASOMATISM MODEL

In searching for an explanation for the discrepancies between the chemical and modal trends described above, a possibility that first comes to mind is that the discrepancies reflect mixing of fresh and residual liquid. However, there seems to be no way, based on this concept (at least by any "first-order" theory), to produce the zones of discontinuity where olivine is rich in Ni but has low Mg/Fe^{2+}. On the other hand, the discrepancies appear

to be very satisfactorily explained by a process of infiltration metasomatism (Korzhinsky, 1965), in which the cumulus minerals are changed
in composition by reaction with the intercumulus liquid that is migrating
upward among them because of the compaction of the layered series.

The principle of the infiltration mechanism is illustrated schematically
in Figure 10. Under the assumed initial conditions in column A, equal
amounts of olivine and interstitial liquid have accumulated in two cyclic
units showing upward trends of decreasing Mg/Fe^{2+}. The phases are in
local equilibrium at all stratigraphic levels, in accord with the indicated
Mg/Fe^{2+} distribution coefficient. In B, the liquid has been removed
uniformly upward a distance ΔX_m, and the crystals and liquid are now out
of equilibrium, especially in the stratigraphic intervals immediately above
the basal contacts of the cyclic units. If crystals and liquid then react so
that local equilibrium is reestablished, the compositional profiles are
modified as in column C. Movement of the liquid another increment ΔX_m
gives the relations in D, and repeated equilibration yields the profiles in E.
The effect is that the Mg/Fe^{2+} discontinuities are shifted upward with
little change in magnitude.

The model in Figure 10 must be modified in several ways to make it more
realistic. A principal shortcoming is that the liquid was assumed (for
simplicity) to have the same content of $Mg + Fe^{2+}$ as the crystals—the

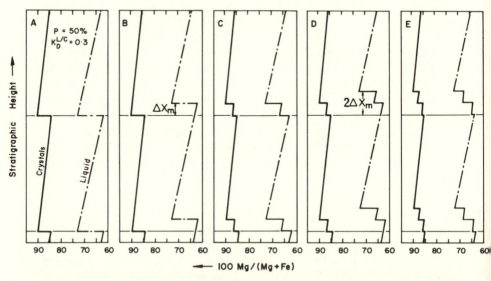

Figure 10. Schematic illustration of the effect of upward migration of intercumulus liquid)
on Mg/Fe^{2+} in cyclic units of olivine cumulates. P is porosity; K_D is the equilibrium value of $(Mg/Fe^{2+}$ in the liquid$)/(Mg/Fe^{2+}$ in the olivine$)$. For explanation,
see text.

combination being, as it were, that of an olivine melt moving through a mesh of olivine crystals. Because of this feature, the Mg/Fe^{2+} discontinuity moves very rapidly, practically at the same rate as the liquid itself. With liquid compositions poorer in $Mg + Fe^{2+}$, like those estimated for the Muskox intrusion in Table 1, the discontinuity moves much more slowly than the liquid (see below).

The shape of the discontinuity is modified when it is displaced upward, but if the size of ΔX_m is sufficiently reduced, the incremental steps are effectively smoothed and the discontinuity migrates upward with a relatively sharp, steady-state profile. In effect, a balance is developed between the sharpening of the discontinuity achieved by reducing ΔX_m, and the broadening that results from having to move the liquid a greater number of smaller increments to bring about the same displacement of the discontinuity. The units of movement are somewhat arbitrary (inasmuch as time is not necessarily involved in the calculations), but for the displacement scale of interest here, steady-state profiles are obtained at $\Delta X_m \simeq 1$ m (value established by trial calculations).

The main refinement that has been made on the scheme in Figure 10 in modeling the Muskox relations is to treat the layered series as compacting while it accumulates, rather than afterward (and with the compaction being independent of the primary cyclic variations shown by the cumulus minerals). A rigorous quantitative treatment of this process is a complicated problem in the flow of fluids through porous media, and only simple modeling has been done in the present work. Certain alternatives have been investigated, however, with results that indicate that the treatment is adequate to define general limits.

An assumption that has been made for practical reasons, but which also appears realistic, is that the porosity of the compacting zone has a steady-state profile, as illustrated in Figure 11. For simplicity it is assumed that this profile is defined by the equation $P = P_{min} + k(h_{cz})^2$, where P_{min} is a minimum final porosity, h_{cz} is distance measured from the base of the compacting zone, and k is a constant of such value that $P = P_0$ (the initial porosity) at $h_{cz} = X_{cz}$ (the total or maximum thickness of the compacting zone). The profile is closely similar to porosity profiles of water-laid sediments (muds, clays, shales, and silty rocks) compacted by burial (e.g., Baldwin, 1971, Fig. 1B). The objective in the modeling is to find a value for X_{cz} whereby the cyclic unit boundaries are infiltrated by enough liquid to cause the observed displacements of the Mg/Fe^{2+} and Ni discontinuities.

The method by which the compaction is simulated in the computer modeling is illustrated in Figure 12. Column A portrays a section of uncompacted layered series; column D shows a possible final porosity profile. Note that the final porosity is not necessarily constant; in fact, the

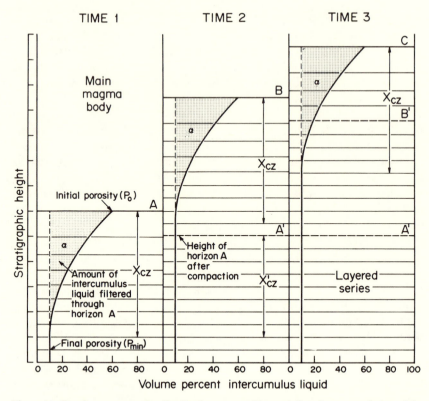

Figure 11. Steady-state porosity distribution assumed in modeling the accumulation of the Muskox layered series: X_{cz} is the current thickness of the compacting zone; X'_{cz} is the thickness of the same section of cumulates after compaction. Through the compacting zone, porosity varies according to the equation $P = P_{min} + (kh_{cz})^2$, where h_{cz} is stratigraphic height above the base of the zone, and k is a constant whereby $P = P_0$ at $h_{cz} = X_{cz}$. By the time that horizon A is compacted to A', it is infiltrated by an amount of liquid quantitatively represented by the shaded area α. Similarly for horizons B and C. The compaction is considered to go on independently of the primary cyclic variations; any of the indicated layering planes could be a boundary between cyclic units.

illustrated upward decrease is a first approximation to relations through the marginal zones into the layered series of the Muskox intrusion (cf. Figure 15, below). Column C illustrates porosity profiles when the layered series is compacted as it accumulates. The contrast between the change in the stratigraphic thickness in going from A to D and that in going from C to D is particularly notable. If the transition A–D were applicable, one would expect extensive downwarping and, perhaps, folding of the cumulates. Such deformation is not conspicuous in the Muskox layered series

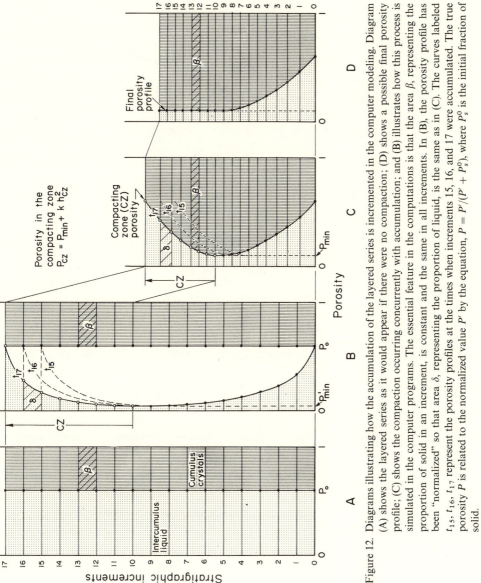

Figure 12. Diagrams illustrating how the accumulation of the layered series is incremented in the computer modeling. Diagram (A) shows the layered series as it would appear if there were no compaction; (D) shows a possible final porosity profile; (C) shows the compaction occurring concurrently with accumulation; and (B) illustrates how this process is simulated in the computer programs. The essential feature in the computations is that the area β, representing the proportion of solid in an increment, is constant and the same in all increments. In (B), the porosity profile has been "normalized" so that area δ, representing the proportion of liquid, is the same as in (C). The curves labeled t_{15}, t_{16}, t_{17} represent the porosity profiles at the times when increments 15, 16, and 17 were accumulated. The true porosity P is related to the normalized value P' by the equation, $P = P'/(P' + P_s^0)$, where P_s^0 is the initial fraction of solid.

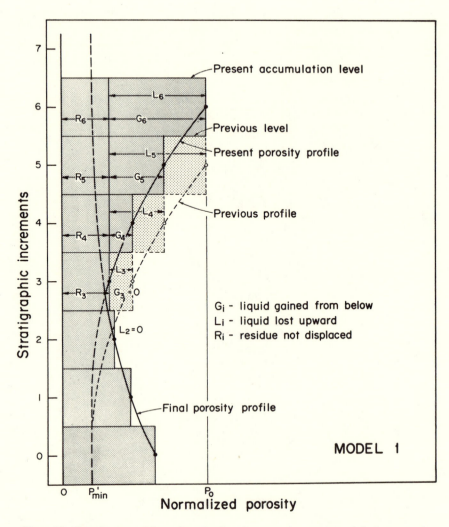

Figure 13. One of two models used to simulate the upward movement of the intercumulus liquid. (For purposes of illustration, the thicknesses of the layer increments have been greatly exaggerated in comparison with the total thickness of the compacting zone; cf. Figure 15). The porosity distribution in the compacting zone is revised every time a new layer increment is added at the top, the zone being reworked from bottom to top. The upward loss of liquid from a layer increment during a given stage of compaction must compensate for the gain from below and still yield the required reduction in porosity. The liquid gained from below is then combined with the residue, and the resulting hybrid is equilibrated with the crystals in the layer increment. Effectively, there is a maximum of mixing of liquid from different levels within the constraint that the liquid moves upward only one increment at a time. Each layer is eventually exposed to some intercumulus liquid (albeit a vanishingly small amount) from even the bottom-most part of the compacting zone as it existed when the increment was deposited; thus the migration distance is the full thickness of the compacting zone.

(the series is only slightly synformal), hence the transition C–D seems clearly to be more realistic.

In terms of the computational process, it is convenient to divide the final form of the layered series into increments having the same content of crystals. Then, only variations in the amount of liquid need be considered, as in Column B, with the porosity values being "normalized" accordingly. When accumulation is complete, the porosity values and stratigraphic thicknesses are converted to their actual values.

The physical removal of the intercumulus liquid from the compacting zone has been simulated numerically in two ways, illustrated in Figures 13 (Model 1) and 14 (Model 2). The main difference between these alternatives is that in Model 1 there is a maximum of mixing of liquid between increments within the constraint that the liquid can move only one increment at a time (an approximation of hydrodynamic dispersion), whereas Model 2 features no mixing from increment to increment. As it turns out, for a given thickness of compacting zone (X_{cz}), the models give identical displacements of a chemical discontinuity. Evidently, as long as the liquid continuously equilibrates with the crystals as it moves upward (and as long as its content of $Mg + Fe^{2+}$ does not change greatly), the displacement of a discontinuity depends only on the amount of the liquid that passes through it, not on the mixing history of the liquid. The important difference between the two models lies in the distance that the liquid moves. In Model 1, a small proportion of the liquid initially trapped at any level eventually moves the full thickness of the compacting zone; in Model 2, the migration distance of all liquid is only a fraction of this thickness. No definite statement can be made as to which of the models is the more realistic in this respect, but they would seem to represent extremes. In effect, *most* of the liquid would move at least the distance indicated by Model 2, and some might move as far as in Model 1.

A last point, in response to a question by reviewers, concerns why the intercumulus liquid should move upward. In what is undoubtedly an oversimplified explanation, the mineral grains in the cumulate pile form a self-supporting framework that is permeable to the liquid. Within the pile, the *load* per unit area supported by the framework at any given level is $(\rho_s - \rho_l)(1 - P)gh$, where ρ_s and ρ_l are the densities of the minerals (average) and liquid, g is the acceleration of gravity, h is the depth below the top of the framework (usually the depositional surface), and P is the average fractional porosity to that depth. Thus, as the pile grows, the load at any level increases, causing the framework gradually to collapse (by shifting, rotation, and deformation of the mineral grains, and by solution and redeposition effects). There is a consequent tendency for the hydraulic pressure on the intercumulus liquid to be increased, and so the liquid flows to lower pressures at higher levels. The process is, in effect, a kind of "self-

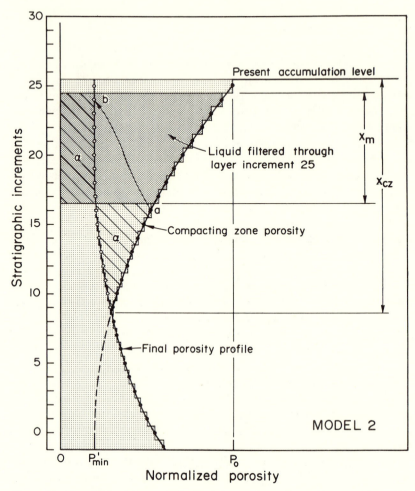

Figure 14. Alternative to the model in Figure 13. In this case, there is no mixing of liquid from one increment to the next. The distance that liquid migrates, X_m, is such that the two areas labeled α are equal.

imposed" filter pressing, and one reviewer has remarked that the chemical effects might be more appropriately termed "exfiltration metasomatism."

GENERAL COMPOSITIONAL EFFECTS COMPUTED FROM THE MODEL

To model the Muskox relations, information is needed on the composition of the liquid and the composition and modal proportions of the cumulus minerals. If these data were complete, the distribution coefficients could be calculated from them, but, as it happens, it has been necessary to use distribution coefficients from experimental studies (e.g. Roeder and Emslie, 1970) to obtain satisfactory liquid compositions (see Irvine, 1977a). The essential data that have been used are given in Table 1.

The other data that are needed for the modeling are values for the initial and final porosities of the cumulates. These have been estimated primarily on the basis of the rock analyses plotted in Figure 15. This lithologic section, also from the South drill hole, includes cyclic units 4–7 and extends downward through the marginal zone to the footwall contact.

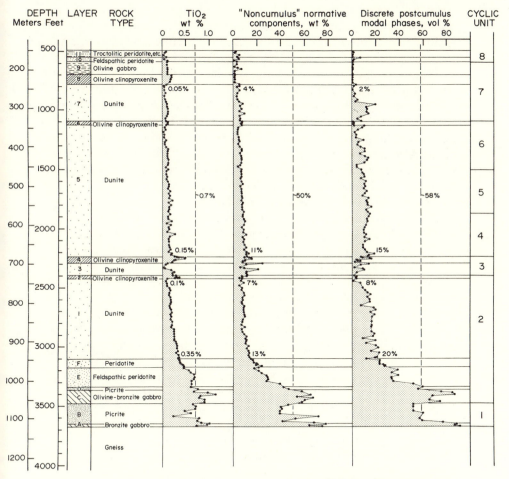

Figure 15. Compositional data reflecting the amount of pore liquid trapped in cumulates in the marginal zones and lower part of the layered series of the Muskox intrusion (including cyclic units 4–7) as sampled in the South drillhole. The dashed vertical lines indicate approximate maximum values for each compositional parameter in olivine cumulate units. For further explanation, see text.

The gabbro at the footwall is not chilled, but its composition is not greatly different from that of the best chilled material.

In principle, if there was some element in the liquid that was completely "excluded" from the cumulus minerals, then its concentrations in the bulk cumulate and in the initial liquid could be used to estimate the amount of pore liquid trapped in the cumulate (i.e., the final porosity). Unfortunately, such an element is not represented among those for which data are available. Phosphorus is the best candidate, but its concentrations are so low that the analyses would have to be reliable to .001 wt% P_2O_5 to be useful,[3] whereas those that have been obtained are accurate to only about .02 wt%. There are, however, numerous elements that are almost excluded from the fractionated minerals in the olivine-chromite cumulates—for example, Al, Ca, Na, K, Ti, Ba, Sr, and Sc. Only Ti is graphed in Figure 15, but the others show similar distributions, and the more abundant of them essentially account for the "noncumulus normative components" plotted in the next column. The latter include all normative constituents *except* those nominally equivalent to the cumulus phases (e.g., everything except Fo, Fa, and Chr in the olivine-chromite cumulates). Also graphed in Figure 15 are modal data for the discrete postcumulus phases, these comprising all modal constituents except the cumulus crystals and their overgrowth (e.g., everything except olivine and chromite in the olivine-chromite cumulates).

The marginal-zone gabbro from the South drill hole generally contains .8–1.1 wt % TiO_2—about the same as the Muskox liquid through the crystallization interval of interest (Table 1). Up-section from the gabbro in the olivine-chromite cumulates, TiO_2 drops to about .6–.7 wt % in the picrite and feldspathic peridotite, and grades off through the peridotite to about .35 wt % at the base of the layered series. It continues to decrease gradually to about .1 wt % at the top of the first dunite layer, and then, after some fluctuations through two pyroxenite layers, settles at about .15 wt % in dunite at the base of cyclic unit 4 and decreases almost continuously to about .06 wt% in the top of the dunite layer in unit 7.

The general decrease from the footwall into the layered series undoubtedly reflects a gradual decrease in the amount of trapped pore liquid. Thus, if Ti had been completely excluded from the cumulus minerals, one could infer that the amount of this liquid in the dunite in cyclic units 4–7 was reduced to about 5–10 wt % just before the liquid solidified. However, most of the TiO_2 is now situated in the olivine and chromite, so these minerals probably had at least a third of their present Ti contents

---

[3] But it then becomes a question of whether phosphorus is really excluded at this level of abundance.

when they were first precipitated. Pore material in the dunite, therefore, is probably no more than 3–7 wt % or 5–10 vol %.

A similar conclusion is reached from the data on noncumulus normative components. These components total between 4 and 11 wt % in the dunite layers, but because they almost certainly were partly contained in the original cumulus crystals, the amount of pore liquid must have been somewhat less, perhaps 4–11% by *volume*. Similarly, the postcumulus phases generally amount to 10% or less by volume, not counting overgrowth on the fractionated crystals. In the model calculations described below, the final porosity was assumed for convenience to be constant at 10 vol %.

The value used for initial porosity derives essentially from the composition of the marginal-zone picrite and feldspathic peridotite, these being the olivine-chromite cumulates with the maximum content of pore material. As indicated in Figure 15, the porosity of these rocks appears to have been about 50 wt %, based on the noncumulus normative components, or about 58 vol %, based on the discrete postcumulus modal constituents. These estimates are supported by data obtained by Campbell, Dixon, and Roeder (1977) on synthetic cumulates made with a centrifuge furnace at accelerations 500 times that of gravity. They report that an olivine cumulate made from basaltic melt under these conditions had a porosity of 63 vol %. For the present calculations, a rounded average of the above values was used, 60 vol %. The author's impression is that this value is probably too large,[4] but a large value should lead to a minimum estimate for the thickness of the compacting zone, and it is this limit that is of most interest in the present context.

Some results from the modeling are shown in Figure 16. In columns A and D, the compacting zone has zero thickness, or in effect, the pore liquid is eliminated from each increment of cumulate immediately after that increment is deposited. This possibility corresponds to the H. H. Hess (1960) mechanism of adcumulus growth. In such a case, the chemical discontinuities are not displaced from their original positions.

In the other columns, X_{cz} and X_m are respectively the thickness of the compacting zone and the distance the intercumulus liquid was migrating through the cumulate pile at the time of accumulation. Of more interest geologically, however, are X'_{cz} and X'_m, the equivalent parameters measured in terms of the compacted or *observed* thickness of the layered series. As indicated in the earlier explanation of the models, the values for X'_m from Model 2 (upper part of the diagram) are minimum migration

---

[4] Most previous estimates have been in the range 25–50 vol % (Wager and Deer, 1939; Hess, 1960; Jackson, 1961; Wager and Brown, 1978).

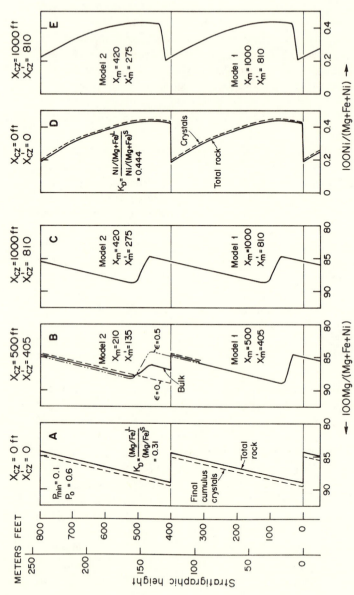

Figure 16. Compositional profiles calculated by the computer models. X_{cz} is the original thickness of the compacting zone; X'_{cz} is the thickness of the same section after compaction. X_m is the distance that the intercumulus liquid has to migrate, measured on the same basis as X_{cz}; X'_m is the migration distance measured in terms of the compacted (i.e., the observed) thickness of the layered series. In columns (A) and (D), the cumulate was compacted immediately on accumulation, hence the compositional discontinuities between cyclic units are not displaced. In the other diagrams, the discontinuities are shifted upward. The lower discontinuities are calculated by the model in Figure 13; the upper discontinuities derive from the model in Figure 14. The upper part of (B) shows a possible effect of incomplete equilibration of crystals and migrating liquid. The parameter ε is the fraction of crystals assumed to react. See text for further explanation.

distances; those from Model 1 (at the bottom) are probably maximum values.

A comparison of the computed compositional profiles with the Muskox data in Figures 8 and 9 shows that the displacements for $X'_{cz} \simeq 400$ ft (122 m) are just slightly greater than those apparent at the base of cyclic unit 4; those for $X'_{cz} \simeq 800$ ft (244 m) are closely similar to the profiles at the bottoms of units 5 and 6. It is especially notable that the contrast between displacements for Mg/Fe^{2+} and $Ni/(Mg + Fe^{2+})$ is almost identical to the contrast observed in the drillhole section. This similarity is key evidence that the modeled process is appropriate to the intrusion. The contrast arises because of the relative differences in the contents of $Mg + Fe^{2+}$ and Ni in the liquid and cumulus minerals. The liquid contains about a third as much $Mg + Fe^{2+}$ as the crystals, but only about a tenth as much Ni; thus the adjustments that must be made in the liquid's Ni content to return it to equilibrium with the crystals as it passes through a discontinuity are relatively small. The Ni content of the olivine therefore is less affected than its Mg/Fe^{2+} ratio, and accordingly, the Ni discontinuity moves more slowly. This relationship is illustrated on a theoretical basis in the section to follow.

THEORETICAL CONSIDERATIONS AND HEAT AND TEMPERATURE EFFECTS

The above modeling has been carried out in an essentially empirical way. As an aid to understanding the results and their application, it is useful to examine briefly certain theoretical aspects of the infiltration process. The following analysis of the characterisitics of discontinuities or "fronts" developed by infiltration metasomatism comes essentially from a compilation by Hofmann (1972, 1973).

If the metasomatic reaction is an exchange reaction involving only two elements or components, i and j, then their relative variations in the liquid, L, and solid, S, are conveniently dealt with in terms of atomic or molar fractions of the form[5]

$$N_i^L = c_i^L/(c_i^L + c_j^L) \equiv c_i^L/c_{sum}^L \tag{1}$$

and

$$N_i^S = c_i^S/(c_i^S + c_j^S) \equiv c_i^S/c_{sum}^S, \tag{2}$$

where the c's denote true concentrations expressed as atomic or mole fractions. If, in the infiltration process, the c_{sum}'s are constants—as they

---

[5] These fractions are usually denoted by the symbol X, but they are represented here by N because X has already been used for distance or stratigraphic thickness.

ideally should be in an exchange reaction—then the following equation (slightly modified from Hofmann, 1972, Equation 15), obtains:

$$\frac{dX}{dV}\bigg|_{N_i^L \text{ constant}} = \frac{\rho_L}{\rho_S}\left(\frac{c_{\text{sum}}^L}{c_{\text{sum}}^S}\right)\left(\frac{dN_i^L}{dN_i^S}\right), \tag{3}$$

where X is distance measured in the direction the liquid is flowing (and the front is migrating), V is the volume of liquid percolated through a *unit* cross section, and ρ_L and ρ_S are the densities of the liquid and solid. This equation describes the migration rate of a point on the composition gradient of the liquid where the ratio of i to j corresponds to the arbitrary, *fixed* value, N_i^L. Given local equilibrium, there will be a corresponding point on the composition gradient of the solid, having the same location in space and moving at the same rate, with the appropriate constant value, N_i^S. The migration rate is expressed as a function of the volume of the liquid flow per unit cross section. The actual migration velocity is

$$\frac{dX}{dt}\bigg|_{N_i^L \text{ constant}} = \frac{\rho_L}{\rho_S}\left(\frac{c_{\text{sum}}^L}{c_{\text{sum}}^S}\right)\left(\frac{dN_i^L}{dN_i^S}\right)\frac{dV}{dt} \tag{4a}$$

$$= \frac{\rho_L}{\rho_S}\left(\frac{c_{\text{sum}}^L}{c_{\text{sum}}^S}\right)\left(\frac{dN_i^L}{dN_i^S}\right)U_L P, \tag{4b}$$

where t is time, P is the fractional porosity of the solid, and U_L is the average flow velocity of the liquid (in the direction X) within the pores.

From these equations, it is evident that, other factors being equal, if dN_i^L/dN_i^S *increases* with N_i^L and N_i^S, then points with large proportions of i will migrate faster than points with smaller proportions. Thus, in the case that i is being introduced and enriched along a metasomatic front, points in the frontal zone with high proportions of i will overtake points with lower values, and so the front will be "self-sharpening." Conversely, if i is the element or component being replaced in the metasomatic exchange reaction, then the front will become more diffuse or gradational. If dN_i^L/dN_i^S *decreases* with N_i^L and N_i^S, the inverse of these relationships obtain.

To facilitate the examination of the Muskox data and modeling in the light of the above equations, let Mg be i in the case of the Mg/Fe^{2+} variations, and Ni be i for the $Ni/(Mg + Fe^{2+})$ variations (with Mg and Fe^{2+} being treated essentially as one element in the latter case). In both cases, then, i is the element *displaced* by the metasomatic exchange reaction.

Curves illustrating the relationships of N_i^L and N_i^S corresponding to the K_D values used in the modeling are shown in Figure 17. Note that in the compositional ranges appropriate to the Muskox problem, dN_i^L/dN_i^S is some 4–5 times larger for Mg/Fe^{2+} than for $Ni/(Mg + Fe^{2+})$ (i.e., 1.8 to

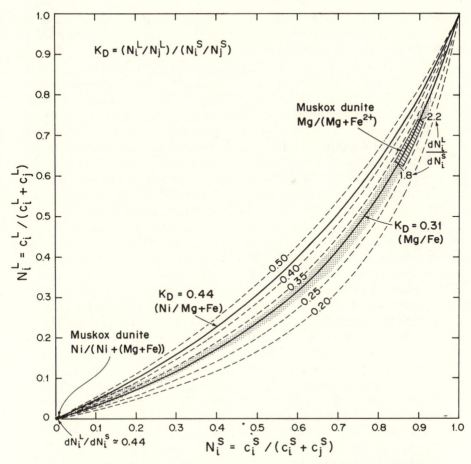

Figure 17. Plots of N_i^L versus N_i^S for different values of K_D, illustrating the relationship between $Mg/(Mg + Fe^{2+})$ and $Ni/[Ni + (Mg + Fe^{2+})]$ for the dunite (olivine cumulate) layers of Muskox cyclic units 4–7 and their parental liquid. The shaded band represents the general experimentally determined range of K_D for $Mg/(Mg + Fe^{2+})$ equilibrium at 1 atm between olivine and liquid in basic and ultrabasic magmas and analogous simple silicate systems.

2.2 versus .4 to .5). As seen from Equation (3), it is because of this feature that the Mg/Fe^{2+} discontinuities are displaced upward farther than the Ni discontinuities in the modeling—and, presumably, also in the intrusion. Along both curves in Figure 17, dN_i^L/dN_i^S increases with i—or, by the convention specified above, with Mg in the case of the one ratio of interest, and with Ni in the other. Hence, because Mg and Ni are both being displaced by the metasomatic exchange, both types of discontinuity should be diffuse or gradational. Moreover, this characteristic should be more pronounced for the Mg/Fe^{2+} discontinuities, inasmuch as they involve the larger variation in dN_i^L/dN_i^S. The model results are in accord with both these expectations.

There is, however, some discrepancy in the Muskox data in that certain of the Mg/Fe^{2+} discontinuities are relatively sharp, particularly the break at the base of cyclic unit 5 (Figure 8, B and E; Figure 9, B' and E'). If the apparent displacement of these discontinuities is indeed the result of infiltration metasomatism, then there must have been some factor that tend to sharpen them, at least in some circumstances. On the basis of the curves in Figure 17, however, it appears unlikely that this factor could have involved upward decrease in dN_i^L/dN_i^S; in fact, attempts to sharpen the discontinuities in the modeling by having K_D vary as a function of Mg/Fe^{2+} did not yield any significant improvement, even when the variation far exceeded the range that could reasonably be justified.

There are two other effects, however, that might result in sharpening, and these can be expected to be interrelated and complementary. One possibility is that c_{sum}^L/c_{sum}^S decreased *through the discontinuity* during the metasomatic process; the other is that the flow or flux of liquid (dV/dt in Equation 4a) decreased through the discontinuity. With regard to the first possibility, in the Muskox problem, c_{sum}^S essentially represents the Mg + Fe^{2+} content of the olivine in the dunite layers and so probably was almost constant (at about .66). But the Mg + Fe^{2+} content of the liquid ($= c_{sum}^L$) decreases strongly with olivine crystallization (see Table 1). Thus, if some of the liquid crystallized as it percolated through the discontinuity, c_{sum}^L/c_{sum}^S and dV/dt would both decrease. The discontinuity, therefore, would tend to be sharpened on both counts.

The reason that crystallization might occur in the liquid as it passes through the metasomatic front could relate to thermal or heat effects not accounted for in the modeling in Figure 16. In these calculations, chemical equilibrium was maintained only within the constraint that mass was conserved; no attempt was made to conserve energy. There may, however, be heat effects associated with the cation-exchange reactions that could cause crystallization, and crystallization might also occur because of local heat transfer, both between crystals and liquid as the liquid moved upward through the cumulate pile, and by conduction up or down section, depending particularly on temperature gradients within and between cyclic units. An indication of a cation-exchange effect of the appropriate sign comes from the heats of fusion of the olivine end members. According to a compilation by Yoder (1976, p. 91), the value for forsterite is about 29.3 kcal/mol, whereas that for fayalite is only 22.0 kcal/mol. Given the same qualitative relationship for olivine precipitated from the Muskox magma, then when Fo (i.e., Mg) was replaced by Fa (i.e., Fe^{2+}) and returned to the magmatic liquid in an exchange reaction, some heat would be consumed. This heat would be supplied by crystallization of a small additional amount of olivine (presumably of an Fe-enriched composition appropriate to the local equilibrium).

As for possible temperature effects, when intercumulus liquid migrates upward into the bottom of a cyclic unit, it should generally be moving into a hotter environment, at least in the period immediately after the cyclic unit began to accumulate. In such circumstances, the liquid itself might provide some of the heat necessary to bring it to thermal equilibrium by undergoing partial crystallization.[6] Also, owing to the relatively strong upward increase in temperature at the base of a cyclic unit, there would probably be a relatively strong downward conductive loss of heat from the bottom part of the unit, and this too could contribute to intercumulus crystallization within the compositional discontinuity as it migrated upward. There is a limit involved here in that the reduction of porosity— and, hence, of permeability—resulting from the crystallization might ultimately block the flow of intercumulus liquid; however, the required compositional effect for sharpening the Muskox Mg/Fe^{2+} discontinuities is not large, so the discontinuities might always have moved on before the reduction of permeability became critical.

In the course of this study, an extended attempt was made to model the heat and temperature effects just described, but a satisfactory numerical method for handling heat and mass transfer simultaneously was not established. It is believed, however, that the above discussion provides a reasonable qualitative explanation of those minor differences that exist between the shapes of the Muskox profiles in Figures 8 and 9 and the model profiles in Figure 16. Some unusual textural features in the Muskox dunite layers that might be partly related to crystallization effects like those just suggested are described in a later section of the paper.

EFFECTS OF INCOMPLETE EQUILIBRATION OF CRYSTALS AND LIQUID

A further possibility that has been considered in the modeling is that crystals and liquid do not completely equilibrate as the liquid percolates upward. This possibility was examined in several ways, but the most interesting results were obtained simply by assuming that only some of the crystals equilibrate, the remainder being completely unaffected. Results

---

[6] In some infiltration processes, temperature discontinuities migrate as relatively sharp fronts analogous to the chemical discontinuities considered here. An example has been described by Margolis (1977). Whether such discontinuities could persist over the time period involved in accumulation of a cyclic unit (say, several hundred years, on the basis of conduction models of the type used by Irvine, 1970b) is doubtful, but in the event they did, one can visualize circumstances in which interrelated chemical and thermal effects like those under consideration might variously be complementary or competing.

from a computation in which half the crystals are equilibrated are shown in the top of Figure 16B. The equilibrated crystals have the composition profile labeled $\varepsilon = .5$; those that were unaffected have the profile $\varepsilon = 0$, the same profile as in column A. The average of the two profiles, labeled "bulk," corresponds in principle to a profile defined by whole-rock analyses.

The intriguing feature of the bulk profile is that it shows a secondary discontinuity that is not unlike the previously noted "subsidiary" discontinuities in Mg/Fe^{2+} in cyclic units 4, 6, and 7 (Figure 8). As will be recalled, the latter occur at levels where the chromite has a large range of composition and, thus, clearly has not equilibrated (Figure 9).

A question is raised here whether there was sufficient time for even the olivine to equilibrate with the moving liquid. Heat-transfer modeling of the intrusion indicates that, for the stratigraphic interval of interest, the layered series accumulation rate was about .3 m (a ft) per year (Irvine, 1970b). In modeling the infiltration metasomatism, the liquid was moved at increments, $\Delta X_m = .6$ m (2 ft), or in effect, it was assumed that liquid and crystals could equilibrate in about two years. For comparison, on the basis of interdiffusion coefficients (D) for Mg with Fe^{2+} and Ni in olivine (data from Buening and Buseck, 1973; Misener, 1974) and diffusion relations for spherical bodies (from Crank, 1970, Figures 6.1 and 6.4), the homogenization of a typical olivine crystal (.06 cm in diameter) in Muskox cyclic units 4–7 should take only four to six months at the estimated initial liquidus temperature of 1,325°C ($D \simeq 3.8 \times 10^{-11}$ g cm^{-2} sec^{-1}), but could take as long as 1.5–2.0 years at 1,175°C ($D \simeq 8.4 \times 10^{-12}$ g cm^{-2} sec^{-1}). The latter temperature is probably close to the minimum value reached during liquid migration in cyclic units 4–7. Thus it would appear that enough time was generally available for olivine to equilibrate, but the limit may sometimes have been reached. Partial recrystallization of the olivine, in addition to cation exchange by diffusion, would, of course, facilitate equilibration.

Whether the secondary Mg/Fe^{2+} discontinuity generated in the model by incomplete equilibration of crystals and liquid (Figure 16, B) is really an analogue of the subsidiary Mg/Fe^{2+} discontinuities near the base of the Muskox cyclic units 4, 6, and 7 (Figure 8, near D, F, and G) is, of course, by no means certain; indeed, one may even question whether the subsidiary discontinuities in units 6 and 7 are real. On the other hand, it seems worth noting that profiles like those indicated in Figure 8 theoretically could be produced in close detail if there was a strong decrease in equilibration rates with decreasing temperature. At first, when the crystals were at high temperatures (say, about 1,300°C), there could be complete equilibration, and the Mg/Fe^{2+} discontinuity would move as a sharp front. But later, when the crystals had been substantially cooled (to below

1,200°C, for example) by the Fe-enriched liquid seeping up from below, there might be only partial equilibration, in which case a secondary discontinuity could be superimposed on the main front.

Some General Implications

The apparent success of the modeling just described suggests that, despite the idealization and approximations, the principal aspects of the natural process have essentially been covered. The important conclusions to be drawn from the results are that (1) the Muskox layered series, while accumulating, was unconsolidated to depths of possibly as much as 300 m below the depositional surface, and (2) intercumulus liquid filtered upward through the cumulate pile for distances of at least 100 m and possibly as much as 300 m. A more specific comparison of the compositional profiles for cyclic units 4–7 (Figures 8, 9) with the model calculations suggests that, in general, the liquid that infiltrated the base of a unit was mostly derived from the underlying unit. Thus, the relatively small Mg/Fe displacement apparent at the base of cyclic unit 4 could be largely effected by liquid rising from the relatively thin unit 3 (for thickness, see Figure 15); the larger displacements evident at the bottoms of units 5, 6, and 7 could respectively be caused by liquid from the thicker units, 4, 5, and 6.

These interpretations concerning distribution of pore liquid contrast strongly with views expressed by other authors about other layered intrusions. Hess (1960) and Jackson (1961), for example, argued from the thickness of slump structures that the maximum thickness of unconsolidated cumulates in the Stillwater layered series was less than 3 m. Wager and Brown (1968, p. 274) implied that the thicknesses of layered crystal mush in the Skaergaard, Rhum, and other intrusions commonly were almost negligible. The Muskox cumulates are essentially similar to those in other stratiform intrusions, however, and there seems no reason to think that the process just described should not be common to layered intrusions in general. In fact, it is the author's impression that the basin or synformal structures of the layering in bodies such as the Bushveld (e.g. Willemse, 1969), Great Dyke (Worst, 1960), Skaergaard (Wager and Deer, 1939), Sudbury (Naldrett et al., 1970), La Perouse (Rossman, 1963), and Axelgold (Irvine, 1975) intrusions can probably be at least partly attributed to preferential compaction of their more slowly cooled interior parts. The explanation of the limited thickness of slump structures may be, as Jackson (1961, p. 85) suggested, that a very small amount of postcumulus overgrowth on the cumulus minerals is sufficient to impart rigidity to the layers. It is notable in this respect that, in the Hawaiian lava lakes, the lava surface effectively becomes a "crust" when it still contains some 50 vol % liquid (Wright et al., 1968).

What is perhaps the most important implication, however, relates to the primary stage of crystallization differentiation. If it commonly happens in layered intrusions, as suggested here, that as much as 50 vol % intercumulus liquid is continuously returned to the main body of magma by compaction of the layers while they are accumulating, then this process must also be a considerable factor in the primary differentiation of the magma. In other words, it accounts for much of the separation or fractionation of crystals from liquid that has commonly been attributed to crystal settling. The process is still one of settling, but the circumstances are very different. The author is also of the view that upward concentration of intercumulus liquid in cumulate piles, owing to compaction at depth in combination with differential accumulation at the top, probably results in extensive resedimentation of cumulates—especially the less dense gabbroic types—by causing low-incidence slumping, renewed flow, and consequent redeposition of cumulus materials along the depositional surface (Irvine, 1980).

THE PROCESS OF DOUBLE-DIFFUSIVE FRACTIONAL CRYSTALLIZATION

VERTICAL ALIGNMENT OF OLIVINE IN MUSKOX CUMULATES

One of the remarkable features of the Muskox layered series is a vertical fabric in some of the dunite and peridotite layers. Examples are illustrated in Figure 18: in each case the plane of the layering is approximately horizontal. Photograph A shows relatively fresh dunite from cyclic unit 4. The common olivine grain size is about .5–.7 mm, but there are a few coarser grains, and these tend to lie parallel to the layering in the manner typical of cumulus crystals in most layered intrusions. The finer grains, however, show a faint vertical alignment and in places appear to be joined in vertical chains. Photographs B and C show coarser, serpentinized peridotite from higher in the layered series (cyclic unit 21). In B, near-vertical alignment of the olivine crystals is evident, and there is a distinct impression of continuous, relatively straight channels between them. In C, vertical grain alignment is prominent, and again, there is an apparent development of intervening parallel channels. In parts of the layered series this type of fabric is continuously developed through stratigraphic intervals of as much as 100 m (Figure 19). The vertical alignment is shown only by olivine and is almost exclusive to the olivine-chromite cumulates. A hint of vertical olivine can be seen in some of the olivine clinopyroxenite layers, but the cumulus pyroxenes in the pyroxenites and the cumulus

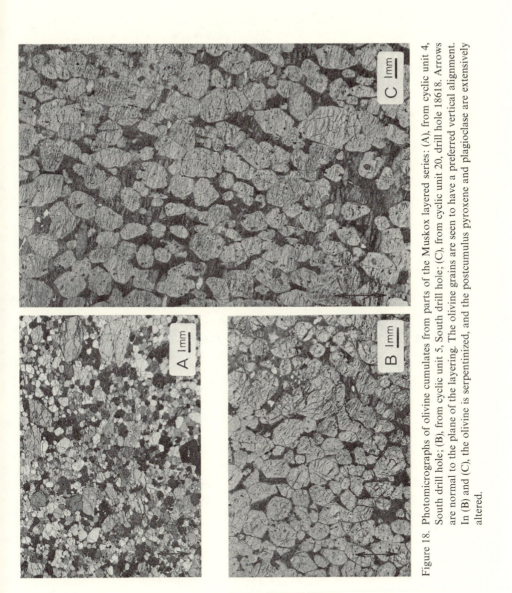

Figure 18. Photomicrographs of olivine cumulates from parts of the Muskox layered series: (A), from cyclic unit 4, South drill hole; (B), from cyclic unit 5, South drill hole; (C), from cyclic unit 20, drill hole 18618. Arrows are normal to the plane of the layering. The olivine grains are seen to have a preferred vertical alignment. In (B) and (C), the olivine is serpentinized, and the postcumulus pyroxene and plagioclase are extensively altered.

pyroxenes and plagioclase in the gabbroic layers tend invariably to lie parallel to the layering.

The vertical fabric has been seen mainly in cores from the North and South drill holes and drill hole 18618 (see Figures 3, 4); it is rare in the East drill hole and completely lacking in the marginal zones.[7] A little has been seen in outcrops along the axis of the intrusion, but its examination there is severely limited by extensive disintegration of the serpentinized olivine-rich rocks under surface weathering. None has been seen in outcrops of the fringe parts of the layered series, even though they are somewhat better preserved.

On the basis of this distribution, the vertical alignment would appear to be confined to the more compacted parts of the layered series (as judged by the distribution of excluded elements and discrete postcumulus material). The possibility is suggested, therefore, that the alignment was caused by the flow of the intercumulus liquid as it was expelled from the compacting layers. The postulate might be that the flow was sufficient to lift the cumulus grains, perhaps to a depth of a meter or so below the surface of accumulation, and to jostle them enough to rotate them into parallel orientation. Alignment of the grains could, of course, also lead to maximum compaction, even in the case that it was vertical. The absence of alignment in the fringe parts of the layered series and in the marginal zones might be explained on the basis that heat loss through the footwall contacts was sufficiently rapid in these regions to freeze more of the intercumulus liquid in place, thus deterring compaction and rotation of the mineral grains.

Another aspect of the vertical alignment that is possibly explained by the intercumulus flow model is its development relative to the boundaries of the cyclic units intersected in the lower part of the South drill hole, as illustrated in Figure 19. The relationship is not very systematic, but there is a suggestion of a rough, general upward trend in olivine orientation from horizontal to random to vertical within cyclic units, and the vertical alignment tends to be strongest near the pyroxenite layers. Thus in the upper part of unit 4 and in unit 5, the olivine-orientation sequence is horizontal to random; in unit 6, it is random to vertical with some fluctua-

---

[7] The only mineral alignment detected in the marginal zones is a local, weak planar orientation of olivine grains parallel to the footwall contacts; there are no rocks to compare with the "perpendicular feldspar" and "wavy pyroxene" rocks that formed in the Skaergaard marginal border group by *in situ* growth of crystals due to heat loss to the intrusion walls (see Wager and Deer, 1939; Wager and Brown, 1968). The impression in the Muskox intrusion is that *in situ* crystallization due to heat loss down through the footwalls, rather than promoting vertical or perpendicular mineral alignment, actually inhibited it. A somewhat different perspective is obtained, however, for the case that the heat loss was upward, as will be explained in the following section.

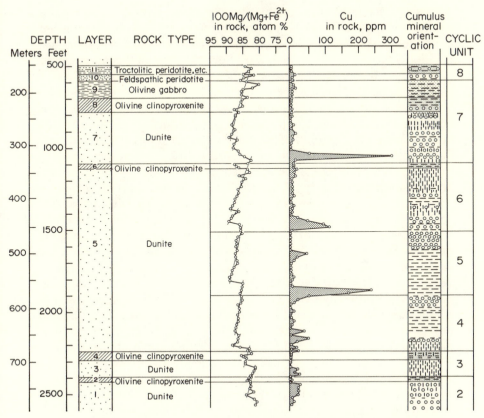

Figure 19. Data on Mg/Fe^{2+}, Cu, and olivine orientation in the Muskox cyclic units 2–8, South drill hole. The distribution of Cu essentially reflects the distribution of sulfide. In the mineral-orientation column, bars indicate horizontal and vertical alignment of olivine grains as seen in thin sections; circles indicate apparently random orientation.

tion; and, through most of unit 7, it is first alternately horizontal and random and then changes to alternately random and vertical. There are significant exceptions, however, as at the bottoms of units 4 and 7, where the orientation is vertical. (The correlation between olivine orientation and Mg/Fe^{2+} is indicated in Figure 19.)

The proposed explanation derives from viscosity estimates for the intercumulus liquid. If this liquid had the composition and temperature range suggested in Table 1, then, as indicated in the table, its viscosity probably increased upward through each of the major cyclic units by about an order of magnitude, from some 25–50 poises to perhaps 350 poises. A parallel increase in yield strength would be expected. Thus, the tendency of the intercumulus flow to lift and rotate the cumulus grains should have increased upward through each unit, a relationship that might explain the crude cyclic trend of olivine orientation just described. Moreover, with the upward flow, the effect of the more viscous liquid might be extended into the next unit up, resulting in the vertical fabric at its base, as in Muskox units 4 and 7.

A principal question that is raised, however, is whether the intercumulus flow was strong enough to lift the cumulus grains. For the steady-state porosity model in Figure 11, if the compacting zone was 300 m thick, the flow rate of liquid through the depositional surface would be about .83 times the rate of layer accumulation, or about 25 cm y^{-1} for the estimated accumulation rate of 30 cm y^{-1}. This rate is *less* by 2–3 orders of magnitude than the nominal Stokes settling velocity of the Muskox olivine grains in liquids like those described in Table 1. This deficiency might be rationalized in terms of some combination of the mechanical interaction of crystals when they are in close proximity in the cumulate pile and the yield strength characteristics of the cumulate. For example, the settling velocity of solid spheres in a 50 % suspension is only about a twentieth of their Stokes velocity (see Shaw, 1965, Table 4), and a yield strength of perhaps only 5–10 dynes cm^{-2} is probably sufficient for even an almost negligible flow of the indicated liquids to lift a typical Muskox cumulus olivine grain (see McBirney and Noyes, 1979, Figure 3). But in the end, this sort of rationalization seems to be begging the point.

The alternative explanation is that the olivine grains actually *grew* with vertical orientation along the top of the accumulating pile of layers. In such case, the rocks would qualify as "crescumulates," in the terminology of Wager and Brown (1968), and if the intercumulus flow did have a significant mechanical effect in aligning the crystals, it could have been when the crystals were still much smaller.

An apparently essential condition for this type of crystallization to occur in the axial region of the Muskox intrusion, far from the footwall contacts, is that the latent heat of crystallization be transferred *upward* into the main body of magma into liquid that had been *supercooled* in the process of convecting from cooling regions along the roof and walls down to the top of the cumulate pile (Wager, 1963; Irvine, 1970b). This possibility has been considered repeatedly since the vertical alignment was first discovered in 1965, but until recently it had always seemed somewhat unlikely for two reasons (Irvine and Baragar, 1972, p. 27–28):

(1) No examples have been found of upward-branching growths of crystals like those characteristic of the olivine crescumulates in the Rhum intrusion (Wager and Brown, 1968), or of upward-directed fan-shaped crystals like those common to the Willow Lake crescumulate layers (Taubeneck and Poldervaart, 1960). Typically, the olivine grains appear

the same at one end as the other, and although they show some tendency to form chains, it is never certain that these are due to primary growth.

(2) In the light of traditional concepts of convection in layered intrusions, with liquid sweeping down from the walls and across the floor, the impression was that crescumulate growth should probably be more prominent near the edges of the layered series than along the axis, where actually it is best developed.

These points now seem relatively unimportant, however, in the light of a new concept of the nature of convection systems in layered intrusions, described below.

DOUBLE-DIFFUSIVE CONVECTION

The suggestion has recently been made by Turner and Gustafson (1978) and McBirney and Noyes (1979) that a remarkable phenomenon called "double-diffusive convection" probably occurs in layered intrusions. This process, which owes its name to the fact that it involves both heat and mass transfer by a mechanism of coupled diffusion and convection, has been especially well illustrated experimentally by Turner and Chen (1974). It has two general manifestations, known as the "finger" and "diffusive" cases. The latter is of particular interest here, and in the present context would be expected to occur in an intrusion in which a high-temperature magmatic liquid underlies a less dense, lower temperature liquid of different composition that is being cooled at the top. In contrast to the large-scale overturn of liquid that is commonly envisaged for convection in intrusions, with double-diffusive convection the liquid column becomes gravitationally stratified on a local scale in such a way that it is subdivided into several (in some cases many) roughly horizontal layers comprised of numerous small, irregular convection cells (Figure 20). This gravity stratification, which arises in advance of crystallization, is essentially defined by a stepwise upward decrease in density from layer to layer owing to stepwise variations in composition and temperature. The density "steps" represent interfaces of finite thickness in which the liquid is relatively static and through which there is comparatively rapid heat and chemical exchange by conduction and diffusion owing to the relatively steep temperature and composition gradients. (The steps themselves may be only very small, perhaps a fraction of a degree in temperature, and fractions of a percent in composition variables.) The steps are sustained because heat diffusivity in the liquid is greater than the chemical diffusivity; consequently, as the heat moves upward, the net effect is to reduce the density of the liquid above each boundary, and to increase it below, thereby causing the gravitational instability that drives the convection within each layer. The effect of removal of heat through a boundary wall does not appear to

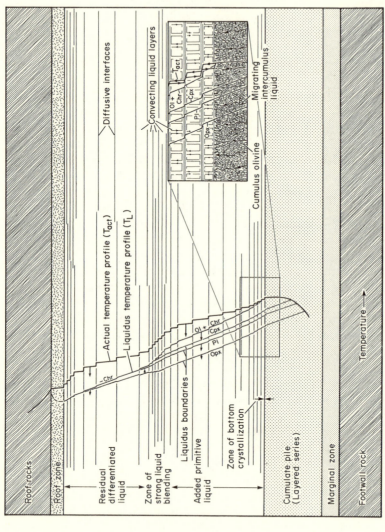

Figure 20. Schematic illustration of an intrusion undergoing double-diffusive convection and bottom crystallization in circumstances where the convection is occurring because of composition and temperature gradients established through the injection of new batches of high-temperature, primitive liquid

beneath a residue of lower-temperature, fractionated liquid. The liquid column has become stratified into convecting layers separated by diffusive interfaces through which there is heat and mass transfer by conduction and diffusion. Heat loss from the main body of liquids is essentially to the roof rocks; losses to the floor rocks result mainly in cooling of the cumulates and crystallization of intercumulus liquid (see Irvine, 1970b). Within the main body of liquids, composition effects associated with fractional crystallization at the bottom, liquid blending at intermediate levels, and contamination at the top are transmitted upward and downward (as appropriate) by the convection and diffusion. The "actual" temperature profile of the liquids consists of approximately adiabatic sections (slope about .3°C/km) interrupted by "steps" that are composition dependent. The inset enlargement portrays the temperature contrast between parts of convection cells that are rising and parts that are descending. The liquidus temperature gradient has segments that are pressure dependent (slope about 3°C/km) separated by steps that are composition dependent. Potential liquidus boundaries are also shown indicating the temperature and order of appearance of minerals at any given depth for a liquid at that depth subjected to fractional crystallization without other factors affecting its composition. Olivine and chromite crystallize at the bottom because the liquid there has the highest liquidus temperature. On the basis of relations observed in the Muskox intrusion (see Irvine, 1970b), the convection in the bottom-most liquid layer is assumed to be coupled to an upward flow of intercumulus liquid being filter pressed from the cumulate pile, and the olivine crystals are shown to be growing vertically into the diffusive interface from liquid that has become slightly supercooled by the convection. With continued heat loss from the system—and the various chemical transfer effects mentioned above—the various temperature-gradient curves would be expected to shift as indicated by the arrows on them at relative rates qualitatively indicated by the lengths of the arrows. Thus, clinopyroxene and plagioclase would be expected eventually to join olivine as crystallizing phases at the bottom; chromite should cease to form with the appearance of the pyroxene owing to their reaction relationship; and similarly, olivine should eventually stop forming with the appearance of orthopyroxene. The orthopyroxene liquidus curve is shown to be rising at the top because of blending effects at intermediate levels and contamination with siliceous material at the roof contact. In general, the convecting liquid layers would be expected to be thinnest and most numerous where the temperature and composition variations are greatest, but all relations are schematic, and the olivine grains are enormously exaggerated in size relative to other features for purposes of illustration.

have been investigated as yet, but the indications from somewhat analogous experiments (Turner and Chen, 1974, Figure 15) are that it will result in a superimposed lateral circulation within layers, with cooled liquid moving away from the wall at the bottom and hotter liquid returning at the top. The especially important characteristic of double-diffusive convection is that, not only does it transfer heat from the bottom of the liquid column to the top and sides, it also is a highly effective mechanism for mixing the contrasting initial liquids and transmitting other compositional changes.

Fractional Crystallization by Bottom Crystallization

That double-diffusive convection occurred in the Muskox intrusion is strongly suggested on at least three counts:

(1) The evidence that the magma was repeatedly stratified in composition and temperature in the manner required to initiate double-diffusive convection. This stratification was probably brought about by periodic introduction of more primitive liquid, as illustrated in Figure 5, but it could have been enhanced by roof-rock melting.

(2) The fact that the cyclic units individually are so exceptionally well differentiated. This feature implies that, as each unit evolved, its parental liquid was, in effect at least, "thoroughly homogenized." The impression is that, each time a layer increment of crystals was fractionated from the liquid, the compositional changes resulting from its removal were rapidly transmitted through the whole of the residual liquid, as would be expected if this liquid was undergoing double-diffusive convection.

(3) The fact that the layered (rock) series of the intrusion terminates against the marginal zones rather than lapping up over them (see Figure 3) would seem to be explained if lateral circulation in the convecting liquid layers acted to remove cooled liquid from the bounding walls of the magma body and transfer it in toward the axial region.

As for the vertical olivine, it is envisaged as forming in circumstances such that the bottom contact of the lowest convection layer is essentially coincident with the top of the cumulate pile and is the junction between the main body of magma and the intercumulus liquid oozing from below (Figure 20). The olivine crystals grow in the diffusive interface, together with minor chromite, on nuclei carried out of the cumulate pile by the intercumulus liquid; the constituents for their growth are drawn from the overlying liquid, which has been supercooled by the convection process. As the latter liquid is depleted in the components of the minerals, its density decreases, both because of the compositional changes and because of warming by the latent heat of crystallization; consequently, it streams upward to be mixed into the convecting layer while other, less fractionated, supercooled liquid moves down to take its place. Meanwhile, the compositional effects of bottom crystallization (most notably, the Mg/Fe decrease) are being transmitted step-by-step upward through the liquid column by the diffusion-convection process, and effects of inter-action and contamination with older magma layers and roof-rock melt (i.e., effects such as silica enrichment) are similarly being transmitted downward. These chemical transfers result in the *apparent* "thorough homogenization" of the liquid column indicated by the differentiation of the cyclic units. In reality, the liquid is *not* homogenized; a whole range of compositions is present at any given time, and the stepwise variations are continuously adjusting to the effects of the crystallization and other interactions. The result, however, is much the same as if the liquid was entirely well mixed. This process would seem appropriately called *double-diffusive fractional crystallization*.

The expulsion of intercumulus liquid by compaction certainly is not an essential condition for the occurrence of double-diffusive convection in a layered intrusion, although circumstances can be visualized in which it might cause such convection in the immediate vicinity of the cumulate floor. If, however, the two processes were coupled in the Muskox intrusion, as suggested here, then the vertical orientation of the olivine grains could be due to both mechanical alignment and upward growth, and the observed spatial association between alignment and compaction would also be explained.

This analysis is obviously in a rather speculative, preliminary form, but the author believes that, in principle, it fills a major gap that has existed in the understanding of the Muskox cyclic units since they were discovered some 15 years ago. In fact, although he is a very strong advocate of the concept that certain kinds of small-scale layering in ultramafic-gabbroic intrusions are commonly developed through current transport and de-position of crystals (cf. Irvine, 1980), the present analysis has led him also to believe that most thickly layered to massive cumulates are probably formed by growth of the cumulus minerals either *in situ* or, perhaps more commonly, within only centimeters of the cumulate floor. This is essen-tially the concept of "bottom crystallization" suggested by Jackson (1961), but it differs in that the crystallization occurs much closer to the floor than his schematic diagrams (ibid., Figures 91, 92) would suggest and is critically dependent on differences of temperature between supercooled liquid in the main body of magma and intercumulus liquid in equilibrium with the crystals in the floor.

Some Possible Implications

Although the actual role of double-diffusive convection in igneous petrology is still virtually unknown, its potential significance would appear to be enormous. The indications are not only that it is probably a principal mechanism in the fractional crystallization of magmas, but that it is also likely to prove to be highly important in transmitting effects of magma mixing and contamination through intrusions. In all these ways it could have broad significance in determining the trace-element and isotopic characteristics of igneous rocks. The author sees the process as being also a principal mechanism in the formation of magmatic ore deposits, such as the layers of chromite, magnetite, and Pt-rich sulphides that occur in stratiform intrusions, through its role in magma mixing (see Irvine, 1977c).

Perhaps the most intriguing physical feature of the process is that it could have effects in an intrusion like those just mentioned without the magma actually undergoing complete convective turnover. In an idealized case, an intrusion could become perfectly differentiated, yet some of the atoms of the original liquid might never have circulated beyond the immediate environment of the rocks in which they ultimately became fixed. By virtue of this feature also, the compositional effects of contamination might be spread throughout an intrusion, even though conspicuous signs, such as the presence of xenoliths, are confined to only very restricted regions, as along roof contacts.

On the other hand, because the compositional effects of double-diffusive fractional crystallization depend on a balance between heat and chemical transfer by the diffusion-convection system, it is possible that, even in the case of perfect fractionation of crystals and liquid at the local site of crystallization along the floor of an intrusion, if the chemical transfer to the site was slow, the overall compositional variations developed in the rocks would be small. Differences in this type of balance might explain why some layered intrusions (for example, the LaPerouse gabbro intrusion in Alaska; Rossman, 1963) show very little mineralogical variation through thousands of meters of section, whereas others (such as the Muskox intrusion) are almost perfectly differentiated through units only a few tens of meters thick.

Finally, it should be noted that these possibilities will probably also require extensive re-evaluation of quantitative models of magmatic fractionation processes. For example, if fractional crystallization does commonly occur in double-diffusive systems akin to that portrayed in Figure 20, then conventional Rayleigh fractionation equations are not strictly applicable but must be modified to accomodate the diffusion from layer to layer in the liquid column.

THE PROCESS OF ADCUMULUS GROWTH

The results of the infiltration-metasomatism modeling support the G. B. Hess (1972) conclusion that adcumulus growth does not occur at the cumulate depositional surface (although his modeling is now also open to reassessment, because it did not consider double-diffusive effects). The present results also constitute relatively direct evidence that (a) adcumulus growth occurs deep in the cumulate pile, extending at times to hundreds of meters below the depositional surface; and (b) a principal part of the process is simply compaction of the pile under its own weight, augmented by recrystallization of the cumulus minerals through continuous reaction with the migrating intercumulus liquid, and by continued growth of the crystals from the liquid in response to local thermal gradients.

In the Muskox intrusion, the extent of adcumulus growth in the marginal zones and lower edge of the layered series appears to have been limited by downward heat loss through the footwall contacts, this loss causing the residual intercumulus liquid finally to freeze in place. In the central part of the layered series, however, the amount of discrete post-cumulus material is constant at about 7–10 vol % through a stratigraphic interval of more than 1,000 m, suggesting that here the limiting factor was possibly the limit of compaction itself. By this stage, most of the rocks are still only mesocumulates; hence it is inferred that, to form extreme ad-cumulates, some additional process or processes are required. Of the possible processes, the following seem likely to be the most important:

(1) Incorporation of the "extra" components of the residual inter-cumulus liquid into solid solution in the cumulus phases.

(2) Local removal of the extra components by diffusion through the residual intercumulus liquid in response to postcumulus crystallization.

(3) Intercumulus fractionation by double-diffusive convection.

(4) Leaching of the extra components by an infiltrating vapor phase.

(5) Mechanical removal of the residual intercumulus liquid by the kneading and straining effect of tectonic deformation.

No attempt will be made to evaluate these possibilities quantitatively, but the following are some brief qualitative impressions based on observations of the Muskox intrusion and other layered intrusions, and on the remarks of other authors.

The extent to which components of the intercumulus liquid can be taken up by the cumulus minerals is obviously limited, but the effect is neverthe-less clearly important in some cases, and its potential importance increases the greater the number of cumulus phases. In the Muskox rocks illustrated in Figure 15, for example, the TiO_2 variation undoubtedly reflects the abundance of trapped intercumulus liquid, but the sole Ti-rich post-cumulus mineral, ilmenite, is found only in the gabbro and picrite. Most of the Ti in the dunite and peridotite units is situated in the cumulus chromite, some of which now contains more than 9 wt % TiO_2. Similarly, in the layered cumulates of the Duke Island ultramafic complex in south-eastern Alaska (Irvine, 1974b), substantial amounts of Al_2O_3 and Na_2O, probably derived from intercumulus liquid, have entered cumulus clinopyroxene rather than precipitating as postcumulus plagioclase.

Local elimination of certain components of the residual intercumulus liquid by diffusion through the liquid would seem necessarily to be important in the formation of adcumulates in intrusions that were relatively anhydrous and that underwent little or no syntectonic deformation. A principal point that might be made in this regard concerns the abundance of adcumulates. It is the author's impression, based on work on the Muskox, Duke Island, Skaergaard, and Axelgold intrusions and augmented by observations in the Stillwater and Bushveld complexes, that true adcumulates are generally limited or local in development. The most common cumulates by far are mesocumulates. In the Muskox intrusion, only a few of the pyroxenite layers, and one gabbro layer, consist of rocks that would qualify as adcumulates (and only marginally so), and in most instances the "missing" postcumulus phases are present in substantial quantities in other rock types only a few meters away. Wager and Brown (1968, p. 229) described the layered rocks of the Skaergaard intrusion as being mainly mesocumulates and orthocumulates, and in the outcrops that the present author has seen, adcumulates like the rock in Figure 1C are essentially confined to parts of layers less than a meter thick featuring strong modal grading. The "missing" components of the intercumulus liquid can usually be found as overgrowth on cumulus minerals only centimeters away, either up or down section, where they could easily have been concentrated by intercumulus diffusion had they originally been present in only the small amounts that might be expected to remain after compaction. Similarly, many of the rocks that would be classified as plagioclase adcumulates in the Stillwater and Bushveld complexes comprise only parts of layers that are spotted or "mottled" (poikilitic) with large postcumulus oikocrysts of pyroxene or olivine. In these cases, it is apparent that nucleation of mafic minerals in the intercumulus liquid was very slow, and so elements such as Mg and Fe^{2+} migrated by diffusion to the oikocrysts. Heat-transfer models indicate that, in the major stratiform intrusions, postcumulus crystallization must commonly have extended over thousands, if not tens of thousands, of years (Irvine, 1970b, 1974b), so there should have been ample time for intercumulus diffusion through distances of a few meters.[8]

The possibility of adcumulus fractionation by double-diffusive convection in intercumulus liquid is intriguing but as yet unexplored. Certainly it would be limited to stages in which the liquid had a low viscosity, and the porosity and permeability of the cumulate were relatively large. But given

[8] According to Hofmann and Magaritz (1977), the diffusivities of divalent cations (e.g., Ca^{2+}, Ba^{2+}, Sr^{2+} and Co^{2+}) in basalt melt range from $10^{-7} - 10^{-6}$ cm^2 sec^{-1}. The corresponding characteristic transport distances for $t = 1,000$ y are 0.6–1.8 m; for $t = 10,000$ y, they are 1.8–5.6 m.

that the adcumulus process progressed from the bottom up, and that the crystallization tended to lower the density of the liquid, then the fractionated liquid would tend to rise into the less fractionated melt above, and, accordingly, the latter might circulate down to take its place.

The importance of leaching by an aqueous vapor phase is also practically unknown. The concept stems from the experimental observation by Bowen and Tuttle (1949) that enstatite can be converted to forsterite by removing silica in H_2O vapor, and it is supported by the results of a recent study by Nakamura and Kushiro (1974) demonstrating substantial solubility of MgO and SiO_2 in H_2O vapor at magmatic temperatures. The leaching mechanism was suggested as a possible origin for adcumulates in Alaskan-type ultramafic bodies such as the Duke Island complex (Irvine, 1974b), because these bodies are typically rich in hornblende and so evidently formed from H_2O-rich magma. But criteria for specifically distinguishing vapor leaching are wanting.

The effect of tectonic deformation obviously can apply only in particular circumstances, and again, the Alaskan-type complexes are among the more likely sites, because they commonly appear to have been subjected to diapiric reemplacement after primary differentiation (Findlay, 1969; Irvine, 1974b, 1976). The author considers it significant in this regard that the largest coherent bodies of rocks that he has encountered that he would class as adcumulates are the dunite "cores" of Alaskan-type complexes at Union Bay in Alaska (Ruckmick and Noble, 1959), and at Tulameen (Findlay, 1969), Aiken Lake, and Wrede Creek (Irvine, 1974c, 1976) in British Columbia. In each case, it appears that the dunite is an early olivine cumulate that was pushed upward tectonically into overlying pyroxene-rich cumulates. The details of the mechanism by which intercumulus liquid might be removed by such deformation are quantitatively unknown, but in the type of diapiric flow that is apparently involved, the general motion is such that, because of drag along the walls of the diapir, material moves more rapidly in the interior than that at the edges until it reaches the advancing front or head of the structure. It then spreads in the head and moves to the sides to make way for the advance of more material through the interior. Thus, at any instant, the least principal stress direction must change through the structure from being parallel to the general movement direction of the diapir in its interior, to being more or less perpendicular to this direction in the head. It seems likely that, with this differential compression, interstitial liquid would be forced to the spreading head and then would escape into host rocks at its front as these rocks were extended by the advance of the diapir. Swarms of hornblende-anorthite pegmatite dikes associated with part of the Duke Island ultramafic complex (Irvine, 1974b) possibly represent intercumulus liquid released in this way.

OTHER POSSIBLE EFFECTS OF
MAGMATIC INFILTRATION METASOMATISM

SULFIDE PRECIPITATION

One last feature of Muskox cyclic units 4–7 that will be noted is the distribution of Cu illustrated in Figure 19. Distinct abundance peaks are seen in basal parts of each of units 5, 6, and 7, and slightly higher than average amounts are evident in the lower part of units 3 and 4. If a small "anomaly" in the middle of unit 5 is neglected, these peaks would seem to be a further feature of the cyclic repetition.

The abundance of the Cu essentially reflects the distribution of sulfides, and a first possibility that comes to mind with respect to the origin of the peaks is that each batch of fresh magma contained a small amount of immiscible sulfide liquid when the batch was introduced into the exposed part of the intrusion. Accumulation of the immiscible liquid with the first olivine fractionated from the magma could then result in the Cu peaks.

If the magma was saturated with sulfide liquid initially, however, then sulfide would be expected to continue to precipitate throughout the ensuing period when the rest of the cyclic unit was formed. In such case, the Ni content of the silicate liquid should have dropped rapidly because of the strong fractionation of Ni into the sulfide liquid, and accordingly, the cyclic unit should show a strong trend of upward decrease in Ni. (See results of Ni fractionation calculations by Duke and Naldrett, 1978.) In fact, however, the concentration of Ni in cyclic units 4–7 tends initially to be constant before decreasing upward (Figure 9), a trend that would be expected if only olivine and no sulfide had been fractionated (Figure 7). It would appear, therefore, that the Muskox magma was not sulfide-saturated during the primary differentiation of the cyclic units.

An alternative interpretation of the Cu or sulfide peaks is that they were produced by the infiltration process. This possibility is suggested because the main peaks occur in stratigraphic intervals where the Mg/Fe^{2+} discontinuities have apparently been displaced upward by some 10–20 m. A mechanism might be that the intercumulus liquid reached sulfide saturation as a result of crystallization during its filtration upward through the cumulate pile; then, as it crossed a chemical discontinuity from one cyclic unit into the next, the relatively large adjustment in its major element composition required to maintain local equilibrium would result in the sudden precipitation of relatively large amounts of sulfide. Sulfide solubility in basic silicate liquids is widely believed to be principally controlled by the bonding of sulfur in the melt to ferrous iron (Richardson and Fincham, 1954; Haughton, Roeder, and Skinner, 1974), hence the sharp decrease in Fe^{2+}/Mg required in the intercumulus liquid as it moved upward into the next cyclic unit may have been the controlling factor in the sulfide precipitation. Indeed, a distinct further possibility is that intercumulus liquid *resorbed* sulfide as it moved *within* a cyclic unit and its Fe^{2+}/Mg gradually increased, and then reprecipitated the sulfide when this ratio was suddenly reversed in the base of a next unit. It is intriguing to think that such a process might be an effective means for concentrating certain other economically important elements, such as the platinum-group metals (A. J. Naldrett, pers. comm., 1977).

IGNEOUS BODIES FORMED BY MAGMATIC REPLACEMENT

In addition to affecting the compositions of cumulus minerals, as described above, the process of magmatic infiltration metasomatism may cause modal changes. In fact, as originally conceived by Korzhinsky (1965, 1970), infiltration metasomatism ultimately yields sequences of monomineralic zones.

Modal effects are suggested at a few places in the Muskox layered series, but they cannot yet be demonstrated conclusively. One possible example is a downward increase in olivine in several of olivine clinopyroxenite layers overlying dunitic olivine-cumulate layers (Irvine, 1970a, Figure 8). This trend may in part be an effect of primary differentiation (as previously was suggested), but it could also reflect metasomatic introduction of olivine by way of intercumulus liquid filtering upward from the olivine-rich cumulates. Another possible example is the occurrence of olivine in an orthopyroxenite layer overlying dunite (cyclic unit 19). This olivine also might have formed by primary differentiation, but this possibility is somewhat unlikely inasmuch as olivine and orthopyroxene have not coprecipitated in other layers in the intrusion, in accord with their low-pressure liquidus reaction relationship. Alternatively, therefore, the olivine might have been introduced metasomatically.

More definite evidence of modal changes by postcumulus metasomatism can be seen in other layered intrusions. Occurrences in the Duke Island ultramafic complex and in the Skaergaard intrusion are illustrated in Figure 21. In the Duke Island examples, olivine clinopyroxenite has locally been converted to peridotite or dunite because of clinopyroxene's being replaced by olivine. The evidence of metasomatism is that the olivine- rich rock transgresses layers in the pyroxenite without displacing or distorting them, indicating replacement on an essentially volume-for-volume basis (see Irvine, 1974b, for more detail). In the example in Figure 21A, some of the displaced pyroxene appears to have reprecipitated in coarse, pegmatitic form just a short distance above (although, of course, the displaced atoms cannot specifically be identified). In the Skaergaard

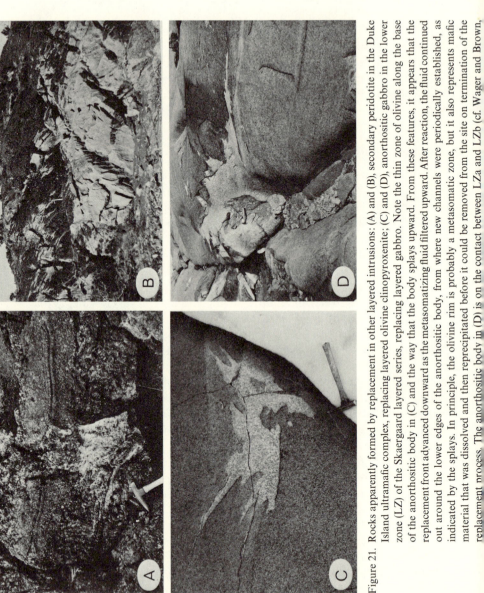

Figure 21. Rocks apparently formed by replacement in other layered intrusions: (A) and (B), secondary peridotite in the Duke Island ultramafic complex, replacing layered olivine clinopyroxenite; (C) and (D), anorthositic gabbro in the lower zone (LZ) of the Skaergaard layered series, replacing layered gabbro. Note the thin zone of olivine along the base of the anorthositic body in (C) and the way that the body splays upward. From these features, it appears that the replacement front advanced downward as the metasomatizing fluid filtered upward. After reaction, the fluid continued out around the lower edges of the anorthositic body, from where new channels were periodically established, as indicated by the splays. In principle, the olivine rim is probably a metasomatic zone, but it also represents mafic material that was dissolved and then reprecipitated before it could be removed from the site on termination of the replacement process. The anorthositic body in (D) is on the contact between LZa and LZb (cf. Wager and Brown,

examples, layered gabbroic plagioclase-olivine and plagioclase-clino-pyroxene-olivine cumulates are similarly replaced by a quasi-anorthositic rock composed of labradorite (An_{65}) and poikilitic ilmenite and augite. This type of body usually occurs within a kilometer of the intrusion con-tact around the lowest exposed (northern) edge of the Skaergaard layered series. Most of the bodies have relatively smooth lower or outer contacts highlighted by a thin (1–5 cm) zone enriched in olivine (Fo_{65}). Their upper contacts typically have an irregular, scalloped appearance, and many of the bodies splay upward (Figure 21C; see caption for discussion). Two bodies were seen to grade to mafic pegmatite, but no clear sign of the mafic silicate material displaced by the metasomatism was found.

One explanation of bodies like those in Figure 21 is that the replace-ment was effected by an aqueous vapor phase filtering through the cumulates. This possibility was suggested for Duke Island (Irvine, 1974b), and it might also be advocated for Skaergaard on the basis of the asso-ciation of the anorthosite with pegmatite. The H_2O might have come from the intercumulus magmatic liquid or, as seems more likely for Skaergaard, it might be meteoric water drawn into the intrusion by hydrothermal convection (cf. Taylor and Forester, 1973). An alternative interpretation, however, applicable in both areas, is that the metaso-matizing fluid was the intercumulus silicate liquid itself. This liquid could have moved because of the compaction of the cumulates, but rather than filtering uniformly through the pile as in the Muskox intrusion, it migrated into relatively restricted, irregular channels that are now evidenced by the replacement bodies.

An attractive feature of this explanation is that the silicate liquid would probably be a much more effective metasomatizing fluid than H_2O vapor in respect to resorbing and precipitating the various minerals involved. In view of the association with pegmatite, however, H_2O almost certainly was involved; in fact, it may even have been critical in determining the replacement reactions. In the liquidus relations of simple silicate systems, the addition of H_2O causes the diopside-forsterite cotectic to shift toward the olivine composition (Kushiro, 1969), and the diopside-anorthite eutectic to shift toward the plagioclase composition (Yoder, 1965). Given these same relations in magmas, then at Duke Island, *loss* of H_2O from intercumulus liquid that previously had been in equilibrium with both clinopyroxene and olivine might have induced the reaction whereby pyroxene was resorbed and olivine was precipitated in its place. And in the Skaergaard intrusion, H_2O loss from intercumulus liquid that had been saturated with plagioclase, augite, and olivine could have caused pyroxene to be replaced by plagioclase, or might even have led to a se-quence of reactions, with olivine replacing pyroxene and then plagioclase replacing olivine. The latter sequence is suggested by the olivine rim along the lower edges of the anorthositic bodies. Possibly the loss of H_2O occurred as the silicate liquid moved from intercumulus pore spaces into the channels mentioned above. Such loss could have been effected either by more rapid diffusion of H_2O along the channels, or by exsolution of vapor (second boiling) owing to a decrease of confining pressure in the channels.

Replacement bodies of dunite after harzburgite and orthopyroxenite, not unlike the bodies of secondary dunite at Duke Island, have been described in the Stillwater complex (Hess, 1960) and the Bushveld complex (Wagner, 1929; Cameron and Desborough, 1964), and similar bodies of dunite after harzburgite are widespread in alpine-type (ophiolitic) peridotite complexes (e.g., Burch, 1969; Irvine and Findlay, 1971; Loney, Himmelberg, and Coleman, 1971; Dick, 1977). It is widely believed that the latter bodies were formed during extraction of magma from the alpine-type harzburgite as it underwent fractional melting in the upper mantle (e.g., Dick, 1977); hence, in this case as well, the metasomatizing fluid could have been a silicate liquid. The relevant liquidus relations however, (see Kushiro, 1969), are such that the orthopyroxene to olivine replacement reaction indicated in all these examples would have to be attributed to gain, rather than loss, of H_2O by the silicate liquid.

POSSIBLE EFFECTS DURING MANTLE MELTING

A final point concerns the possibility that infiltration metasomatism like that evident in the Muskox intrusion might significantly affect the ratios of Mg, Fe^{2+}, and Ni in basic magmatic liquids as these liquids are melted and removed from their source areas, particularly in the upper mantle. In the modeling of the Muskox intrusion, the ratios of these elements in the intercumulus liquid are indicated to have been substantially changed as the liquid moved only a few meters through the compositional discontinuities between cyclic units. On the basis of studies of ultramafic xenoliths carried to the surface by basalts and kimberlites, it is known that mineral variations in the ratios of Mg, Fe^{2+}, and Ni (as well as other elements), comparable to the variations in the Muskox cyclic units, are present in the upper mantle. Thus, if a primary magmatic liquid filtered through a zone of such variation as it moved from its melting site to feeders leading to the surface, its composition could be substantially modified with respect to these elements simply by exchange reactions with little or no further fractional melting or crystallization. In the composition range of interest, variations in the minerals are considerably amplified in the liquid: for example, for a change in olivine composition from Fo_{91} to Fo_{86}, $100 \, Mg/(Mg + Fe^{2+})$ in an equilibrated coexisting liquid changes from about 75 to 65. (For comparison, the total range of this ratio in oceanic tholeiites is from about 75 to 50.) Thus, infiltration effects could go a long way in explaining certain of the nebulous chemical variations observed in basic magmas.

ACKNOWLEDGMENTS

Much of the information on the Muskox intrusion presented here stems from work done at the Geological Survey of Canada, and the first attempts at modeling the infiltration metasomatism process were also made at the Survey. I am pleased to acknowledge the field and laboratory support provided and, in particular, to note the major contributions made by C. H. Smith and D. C. Findlay toward the Muskox study. At the Geophysical Laboratory, the acquisition and processing of the electron microprobe data were facilitated through instruction and assistance by C. Hadidiacos and L. W. Finger, and Finger and F. Chayes gave helpful advice on the computer modeling. The manuscript has been improved through reviews by R. B. Hargraves, D. Rumble, W. Verwoerd, S. R. Todd, R. F. Wendlandt, and H. S. Yoder, Jr.

Finally, I would thank L. B. Gustafson, J. S. Turner, and A. R. McBirney for bringing the phenomenon of double-diffusive convection to my attention. Their realization that it probably occurs in magmas appears to be the prelude to a major advance in our understanding of igneous processes.

REFERENCES

Baldwin, B., 1971. Ways of deciphering compacted sediments, *J. Sed. Petrology 41*, 293–301.

Biggar, G. M., 1974. Phase equilibrium studies of chilled margins of some layered intrusions, *Contrib. Mineral. Petrol. 46*, 159–167.

Bottinga, Y. and D. F. Weill, 1970. Densities of liquid silicate systems calculated from partial molar volumes of oxide components, *Am. J. Sci. 269*, 169–182.

Bottinga, Y. and D. F. Weill, 1972. The viscosity of magmatic silicate liquids: a model for calculation, *Am. J. Sci. 272*, 438–475.

Bowen, N. L., 1928. *Evolution of the Igneous Rocks*, Princeton Univ. Press, Princeton, N. J.

Bowen, N. L. and O. F. Tuttle, 1949. The system $MgO-SiO_2-H_2O$, *Geol. Soc. Am. Bull. 60*, 439–460.

Buening, D. K., and P. R. Buseck, 1973. Fe–Mg lattice diffusion in olivine, *J. Geophys. Res. 78*, 6852–6862.

Burch, S. H., 1969. Tectonic emplacement of the Burro Mountain ultramafic body, Santa Lucia Range, California, *Geol. Soc. Am. Bull. 79*, 527–544.

Cameron, E. N., 1975. Postcumulus and subsolidus equilibration of chromite and coexisting silicates in the Eastern Bushveld Complex, *Geochim. Cosmochim. Acta 39*, 1021–1033.

Cameron, E. N., and G. A. Desborough, 1964. Origin of certain magnetite-bearing pegmatites in the eastern part of the Bushveld complex, South Africa, *Econ. Geol. 59*, 197–225.

Campbell, I. H., J. M. Dixon, and P. L. Roeder, 1977. Crystal buoyancy in basaltic liquids and other experiments with a centrifuge furnace, *E05 58*, 527.

Crank, J., 1970. *The Mathematics of Diffusion*, Oxford Univ. Press, London.

Dick, H. J. B., 1977. Partial melting in the Josephine peridotite. I. The effect on mineral composition and its consequence for geobarometry and geothermometry, *Am. J. Sci. 277*, 801–832.

Dickey, J. S., Jr., H. S. Yoder, Jr., and J. F. Schairer, 1971. Chromium on silicate-oxide systems, *Carnegie Inst. Washington Year Book 70* 118–122.

Duke, J. M., and A. J. Naldrett, 1978. A numerical model of the fractionation of olivine and molten sulfide from komatiite magma, *Earth Planet. Sci. Lett. 39*, 256–266.

Findlay, D. C., 1969. Origin of the Tulameen ultramafic-gabbroic complex, southern British Columbia, *Canadian J. Earth Sci. 6*, 399–425.

Findlay, D. C., and C. H. Smith, 1965. The Muskox drilling project, *Geol. Surv. Canada Paper 64–44*.

Haughton, D. H., P. L. Roeder, and B. J. Skinner, 1974. Solubility of sulfur in mafic magmas, *Econ. Geol. 69*, 451–567.

Hess, G. B., 1972. Heat and mass transport during crystallization of the Stillwater igneous complex, *Geol. Soc. Am. Mem. 132*, 503–520.

Hess, H. H., 1960. *Stillwater Igneous Complex, Montana, a Quantitative Mineralogical Study*, Geol. Soc. Am. Mem. 80.

Hofmann, A., 1972. Chromatographic theory of infiltration metasomatism and its application to feldspars, *Am. J. Sci. 272*, 69–90.

Hofmann, A., 1973. Theory of metasomatic zoning. A reply to Dr. D. S. Korzhinskii, *Am. J. Sci. 273*, 960–964.

Hofmann, A. W. and M. Magaritz, 1977. Diffusion of Ca, Sr, Ba, and Co in a basalt melt: implications for the geochemistry of the mantle, *J. Geophys. Res. 82*, 5432–5440.

Irvine, T. N., 1967. Chromian spinel as a petrogenetic indicator: Part 2. Petrologic applications, *Can. J. Earth Sci. 4*, 71–103.

Irvine, T. N., 1970a. *Crystallization Sequences in the Muskox Intrusion and Other Layered Intrusions, I. Olivine-Pyroxene-Plagioclase Relations*, Geol. Soc. S. Alfrica, Spec. Publ. 1, 441–476.

Irvine, T. N., 1970b. Heat transfer during solidification of layered intrusions. I. Sheets and sills, *Can. J. Earth Sci. 7*, 1031–1061.

Irvine, T. N., 1974a. Chromitite layers in stratiform intrusions, *Carnegie, Inst. Washington Year Book 73*, 300–316.

Irvine, T. N., 1974b. *Petrology of the Duke Island Ultramafic Complex, Southeastern Alaska*, Geol. Soc. Am. Mem. *138*.

Irvine, T. N., 1974c. *Ultramafic and Gabbroic Rocks in the Aiken Lake and McConnell Creek Map-Areas, British Columbia*, Geol. Survey Canada Paper 74-1A, 149–162.

Irvine, T. N., 1975. *Axelgold Layered Gabbro Intrusion, McConnell Creek Map-Area, British Columbia*, Geol. Surv. Can. Paper 75-1, Part B, 81–88.

Irvine, T. N., 1976. *Studies of Cordilleran Gabbroic and Ultramafic Intrusions, British Columbia: Part 2: Alaskan-type Ultramafic-Gabbroic Bodies in the Aiken Lake, McConnell Creek, and Toodoggone Map-Areas*, Geol. Survey of Canada Paper 76–1A, 76–81.

Irvine, T. N., 1977a. Definition of primitive liquid compositions for basic magmas *Carnegie Inst. Washington Year Book 76*, 454–461.

Irvine, T. N., 1977b. Relative variations of substituting chemical components in petrologic fractionation processes, *Carnegie Inst. Washington Year Book 76*, 539–541.

Irvine, T. N., 1977c. Chromite crystallization in the join Mg_2SiO_4–$CaMgSi_2O_6$–$CaAl_2Si_2O_8$–$MgCr_2O_4$–SiO_2, *Carnegie Inst. Washington Year Book 76*, 465–472.

Irvine, T. N., 1980. Magmatic density currents and cumulus processes. *Am. J. Sci., Jackson Volume*, in press.

Irvine, T. N., and W. R. A. Baragar, 1972 *Muskox Intrusion and Coppermine River Lavas. Northwest Territories, Canada*, Int. Geol. Congr. 24th, Montreal, Field Excursion A29 Guidebook.

Irvine, T. N. and D. C. Findlay, 1971. Alpine-type peridotite, with particular reference to the Bay of Islands igneous complex, in E. Irving, (ed.), *The Ancient Oceanic Lithosphere*, Canada Dept. Energy, Mines and Resources, Earth Physics Rev. Publ. 42, 97–126.

Irvine, T. N. and C. H. Smith, 1967. The ultramafic rocks of the Muskox intrusion, in *Ultramafic and Related Rocks*, P. J. Wyllie, (ed.), John Wiley and Sons, Inc., New York.

Jackson, E. D., 1961. *Primary Textures and Mineral Associations in the Ultramafic Zone of the Stillwater Complex, Montana*, U.S. Geol. Surv. Prof. Pap. 358.

Jackson, E. D., 1967. Ultramafic cumulates in the Stillwater, Great Dyke, and Bushveld intrusions, in *Ultramafic and Related Rocks*, P. J. Wyllie, (ed.), John Wiley and Sons, Inc., New York, 20–38.

Jambor, J., and C. H. Smith, 1964. Olivine composition determination with small diameter X-ray powder cameras, *Mineral Mag. 33*, 730–748.

Korzhinsky, D. S., 1965. The theory of systems with perfectly mobile components and processes of mineral formation, *Am. J. Sci. 263*, 193–205.

Korzhinsky, D. S., 1970. *Theory of Metasomatic Zoning* (translated by J. Agrell), Oxford Univ. Press, London.

Kushiro, I., 1969. The system forsterite-diopside-silica with and without water at high pressures, *Am. J. Sci., Schairer Vol. 267A*, 269–294.

Loney, R. A., G. R. Himmelberg, and R. G. Coleman, 1971. Structure and petrology of the alpine-type peridotite at Burro Mountain, California, U.S.A., *J. Petrology 12*, 245–309.

Margolis, S. B., 1977. Time evolution of a packed bed thermocline, *Bull. Am. Physical Soc. 22*, 1281.

McBirney, A. R., and R. M. Noyes, 1979. Crystallization and layering of the Skaergaard intrusion, *J. Petrology 20*, 487–554.

Misener, D. J., 1974. Cationic diffusion in olivine to 1,400°C and 35 kbar, in *Geochemical Transport and Kinetics*, A. W. Hofmann, B. J. Giletti, H. S. Yoder, Jr., and R. A. Yund, (eds.), Carnegie Inst. Washington Publ. 634, 117–129.

Molyneux, T. G., 1970. *The Geology of the Area in the Vicinity of Magnet Heights, Eastern Transvaal, with Special Reference to the Magnetic Iron ores.* Geol. Soc. S. Africa Spec. Publ. 1, 228–241.

Nakamura, Y. and I. Kushiro, 1974. Composition of the gas phase in Mg_2SiO_4–SiO_2–H_2O at 15 kbar, *Carnegie Inst. Washington Year Book 73*, 255–258.

Naldrett, A. J., J. G. Brey, E. L. Gasparrini, T. Podolsky, and J. C. Rucklidge, 1970. Cryptic variation and the petrology of the Sudbury Nickel Irruptive, *Econ. Geol. 65*, 122–155.

Richardson, R. D., and C. J. B. Fincham, 1954. Sulfur in silicate and aluminate slags, *J. Iron Steel Inst., London 178*, 4–14.

Roeder, P. L., and R. F. Emslie, 1970. Olivine-liquid equilibrium, *Contrib. Mineral. Petrol. 29*, 275–289.

Rossman, D. L., 1963. Geology and petrology of two stocks of layered gabbro in the Fairweather Range, Alaska, *U.S. Geol. Surv. Bull. 1121-F*.

Ruckmick, J. M., and J. A. Noble, 1959. Origin of the ultramafic complex at Union Bay, southeastern Alaska, *Geol. Soc. Am. Bull. 70*, 981–1018.

Shaw, H. R., 1965. Comments on viscosity, crystal settling and convection in granitic magmas, *Am. J. Sci. 263*, 120–152.

Smith, C. H., and H. E. Kapp, 1963. *The Muskox Intrusion, a Recently Discovered Layered Intrusion in the Coppermine River Area, Northwest Territories, Canada.* Mineral. Soc. Am. Spec. Pap. 1, 30–35.

Taubeneck, W. H., and A. Poldervaart, 1960. Geology of the Elkhorn Mountains Northeastern Oregon: Part 2. Willow Lake intrusion, *Bull. Geol. Soc. Amer. 71*, 1295–1322.

Taylor, H. P., Jr. and R. W. Forester, 1973. An oxygen and hydrogen isotope study of the Skaergaard intrusion and its country rocks (abstract). *Trans. Am. Geophys. Union 54*, 500.

Turner, J. S., and C. F. Chen, 1974. Two-dimensional effects in double-diffusive convection, *J. Fluid Mechanics 63*, 577–592.

Turner, J. S., and L. B. Gustafson, 1978. The flow of hot saline solutions from vents in the sea floor—some implications for exhalative massive sulfide and other ore deposits, *Econ. Geology 73*, 1082–1100.

Wager, L. R., 1960. The major-element variation of the layered series of the Skaergaard intrusion and a re-estimation of the average composition of the hidden layered series and of the successive residual magmas, *J. Petrol. 1*, 364–398.

Wager, L. R., 1963. *The Mechanism of Adcumulus Growth in the Layered Series of the Skaergaard Intrusion*, Mineral. Soc. Am. Spec. Paper 1, 1–19.

Wager, L. R., and G. M. Brown, 1968. *Layered Igneous Rocks*, Oliver and Boyd, Ltd., Edinburgh.

Wager, L. R., and W. A. Deer, 1939. Geological investigation in East Greenland, Part III, The petrology of the Skaergaard intrusion, Kangerdluqssuaq, East Greenland, *Medd. Groenl. 105*, No. 4. (reissued 1962).

Wager, L. R., G. M. Brown, and W. J. Wadsworth, 1960. Types of igneous cumulates, *J. Petrology 1*, 73–85.

Wagner, P. A., 1929. *The Platinum Deposits and Mines of South Africa*, Oliver and Boyd, Edinburgh.

Willemse, J., 1969. The geology of the Bushveld Complex, the largest repository of magmatic ore deposits in the world, *Econ. Geol. Monogr. 4*, 1–22.

Worst, B. G., 1960. The Great Dyke of southern Rhodesia, *Bull. Geol. Surv. S. Rhodesia 47*.

Wright, T. L., W. T. Kinoshita, and D. L. Peck, 1968. March 1965 eruption of Kilauea volcano and the formation of Makaopuhi lava lake, *J. Geophys. Res. 73*, 3181–3205.

Yoder, H. S., Jr., 1965. Diopside-anorthite-water at five and ten kilobars and its bearing on explosive volcanism, *Carnegie Inst. Washington Year Book 64*, 82–89.

Yoder, H. S., Jr., 1976. *Generation of Basaltic Magma*, National Acad. Sci., Washington, D.C.

Chapter 9

DIFFUSION IN NATURAL SILICATE MELTS: A CRITICAL REVIEW

ALBRECHT W. HOFMANN

Carnegie Institution of Washington, Department of Terrestrial Magnetism, Washington, DC.

INTRODUCTION

The potential importance of diffusion in igneous processes was recognized by Bowen (1921), who noted that "the growth of crystals is accomplished largely by the diffusion of material to crystalline nuclei." Bowen was, however, more intrigued by the possibility of diffusion-controlled mechanisms to explain larger features such as the formation of basic border phases in igneous bodies. In principle, a chemical gradient will be induced in an initially uniform medium if a temperature gradient is imposed on the system, and this (Soret) effect might generate chemical gradients in cooling magma bodies. Bowen conducted experiments on the diffusion in silicate melts and found that the diffusion coefficients are so small that large-scale transport phenomena in igneous bodies cannot be explained by diffusion. With some possible exceptions, this conclusion is still true today. The exceptions are in part the result of new models that combine convection with diffusion through a boundary layer in the magma body (Shaw, 1974), and in part the result of new evidence about the longevity of partially molten regions in the mantle (e.g. Shankland and Waff, 1977). Because of the time dependence of diffusion distances, this longevity might increase the effectiveness of diffusion transport to some extent. The role of diffusion in a partially molten mantle has recently been discussed by Hurley (1977), Hofmann and Magaritz (1977a,b), and Hofmann and Hart (1978).

The role of diffusion in crystal growth is the subject of an extensive literature in the material sciences. Some cases of geological interest have been discussed by Albarede and Bottinga (1972), Dowty (1977a) and Magaritz and Hofmann (1978a), all of whom were particularly concerned with the effect of diffusion control on trace element partitioning. Major-element diffusion, on the other hand, is important in the theory of nucleation and growth of crystals (see the chapter by Dowty in this book). Diffusion is also believed to be important in the formation of oscillatory zoning in feldspars (e.g. Bottinga et al., 1966; Sibley et al., 1976). Finally, because of the extreme slowness of diffusion in crystalline silicates, diffusion in the melt phase appears to be the only mechanism capable of producing local equilibrium during partial melting (Hofmann and Hart, 1978).

Since Bowen's work on diffusion in melts was published in 1921, no additional measurements appear to have been made for about forty years. Recently, however, a considerable amount of new experimental work has been done on diffusion in natural melts and glasses. It is the purpose of this paper to review all the recent experimental results that have come to the author's attention. An attempt is made to categorize and evaluate these results and to give the reader enough background to assess the data himself. The various geological applications are not discussed, and the reader is referred to the references given above.

THEORY

This section and the one following are intended to be a selective rather than comprehensive review of theory and experiment. For a more detailed treatment the reader is referred to textbooks such as that by Crank (1975). A very useful and thorough review of theory, experiment, and results of diffusion in oxide glasses has been given by Frischat (1975).

Diffusion processes may be classified in the following categories: self-diffusion, tracer diffusion, and chemical diffusion. In self-diffusion, no chemical concentration gradients are present, and the diffusion process is observed by labelling some of the atoms with radioactive or stable-isotope tracers. In tracer diffusion, a labelled tracer is introduced in the form of an isotope (usually radioactive) of a chemical element that is not otherwise present in the substance; the tracer concentration is so small that no other appreciable concentration gradients are introduced. In chemical diffusion, bulk chemical concentration gradients are present in the specimen. Because of the impurities present in almost all natural minerals and in all natural melts and glasses, there is likely to be no practical difference between self-diffusion and tracer diffusion. However, a fundamental difference between these and chemical diffusion exists. This difference is quantitatively most significant in the case of major elements.

A phenomenological description of self diffusion and tracer diffusion, in one dimension is by Fick's first law

$$J = -D \frac{\partial c}{\partial x}, \tag{1}$$

where J is the flux of tracer $[g \ cm^{-2} \ s^{-1}]$, D is the diffusion coefficient $[cm^2 \ s^{-1}]$, c is the tracer concentration $[g \ cm^{-3}]$, and x is the distance $[cm]$ in the direction of diffusion. Equation (1) may be applied to a volume element of material together with the requirement of conservation of mass. This leads to the continuity equation known as Fick's second law

$$\frac{\partial c}{\partial t} = D \frac{\partial^2 c}{\partial x^2}. \tag{2}$$

This equation treats D as a constant and is valid for self-diffusion and tracer diffusion, that is, in a system where the material properties are uniform. Equation (1) also applies to the case of binary chemical diffusion, for example, interdiffusion of KCl and NaCl. Because of electroneutrality, the exchange of K and Na is coupled, so that the overall process can be described by a single coefficient. However, the magnitude of this coefficient will in general vary with concentration (in the above example, KCl relative to NaCl). In this case, Fick's second law becomes

$$\frac{\partial c}{\partial t} = \frac{\partial}{\partial x}\left(D \frac{\partial c}{\partial x}\right). \tag{2'}$$

Chemical diffusion in multicomponent systems is described by Onsager's (1945) extension of Fick's laws

$$J_i = -\sum_{j=1}^{n} D_{ij} \frac{\partial c_j}{\partial x} \tag{3}$$

where n is the number of independently variable components. Again, the value of D_{ij} will in general vary with composition. However, even with constant D_{ij}, the flux of each component will depend on the gradients of all components in the system. This can easily lead to locally reversed concentration gradients. This may be seen by comparing Equations (1) and (3). In binary diffusion, described by (1), the flux is always "downhill" relative to the concentration gradient, because of the negative sign. In multicomponent diffusion, described by (3), this restriction is removed, and the flux of a component i may be locally "uphill" (against its own concentration gradients), because the sum of other terms in Equation (3) may be greater than and opposite in sign to the ith term. The D_{ij} coefficients can be measured by determining concentration profiles of all the components in the system. In principle, this presents no particular difficulty, because the gradients can be analyzed using the electron microprobe. Gupta and Cooper (1971) have shown that the set of coupled equations (3) can be uncoupled by diagonalizing the (D) matrix for the vector equations equivalent to (3). The uncoupled equations then have the same form as Fick's laws [Equations (1) and (2)] and can be solved in the same way. In practice, this requires analysis of the chemical paths of all components of the system; to my knowledge no system of more than three components has been analyzed in this manner. Also, this approach is not valid if the coefficients are concentration-dependent.

A simplified approach in describing multi-component diffusion is to regard the system as pseudobinary. The flux of one component is then simply viewed as being balanced by a counter flux in which all the other species are combined. Cooper (1968) has developed the formal theory of this method. He showed that an "effective binary diffusion coefficient" (EBDC) can be used with Fick's laws under certain conditions. In this context, it is useful to distinguish between real space and composition space; the term *compositional path* refers to the progression of concentrations in composition space. Cooper's conditions are as follows: (1) all the concentration gradients must be parallel to the same direction in real space and (2) the system must either be in a steady state or else have no finite length in the direction of diffusion. Condition (2) permits use of EBDCs in infinite and semi-infinite systems, but not in finite systems, because in finite systems the compositional path changes with time.

Cooper (1968) pointed out that the convenience gained by using an EBDC for the description of an experiment is at the expense of generality, because the EBDC is dependent not only on composition itself but also on the direction of the compositional path (that is, the direction in composition space). This loss in generality becomes especially serious when "uphill" diffusion is encountered; this causes the EBDC to undergo extreme variations through discontinuities, infinite values, and negative values.

The temperature dependence of the diffusion coefficients is usually described by the so-called Arrhenius relation

$$D = D_0 \exp\left[-E/RT\right] \tag{4}$$

and a plot of log D versus T^{-1} usually gives a straight line over a significant range of temperatures. On this plot, the preexponential factor D_0 (sometimes called "frequency factor") is derived from the intercept, and the "activation energy" (E) is derived from the slope of the line. R is the universal gas constant, and T is the absolute temperature. The concepts of activation energy and frequency factor are based on theoretical and experimental evidence on diffusion in crystalline materials. D_0 is related to the jump frequency, jump distance, defect structure, and geometry of the crystal lattice. E is related to the energy of formation of imperfections and to the energy of motion required to cause an atom to jump from one site to another (see for example Manning, 1968). It is hard to see how this physical interpretation can be carried over to the description of liquids. Nevertheless, as will be seen, many diffusion data for both silicate glasses and liquids fit straight Arrhenius lines quite well. However, diffusion data for rare gases and molecular gases in silica glass show distinct curvature on the conventional Arrhenius plot. These results do yield straight lines when one plots D/T versus T^{-1}. This is equivalent to replacing D_0 by $D_0{}'T$ in Equation (4) (Doremus, 1973, p. 131–134).

The pressure dependence of the diffusion coefficient is given by an expression analogous to the Arrhenius equation.

$$D = D_0 \exp\left[-PV/RT\right] \tag{5}$$

where V is the activation volume (see for example Lazarus and Nachtrieb, 1963). Again, the applicability to liquids of the physical concept of an activation volume is questionable.

The concept of a driving force in diffusion is frequently applied to diffusion processes. This has been discussed in detail by Manning (1968, pp. 13–19 and throughout the book). For the purpose of the present paper it is sufficient to warn the reader that the term "driving force" can easily cause confusion. This is especially true in the case of the notion that a concentration gradient is the "driving force of diffusion." Diffusive movement is primarily random (as a result of kinetic energy of the particles) and the presence of a concentration gradient does not necessarily make it easier for a given particle to move in a given direction (Manning, 1968). *Physical* driving forces affect the direction of an atomic jump and may be caused by an electrical, gravitational, or stress field. Generalized *thermodynamic* driving forces are not necessarily actual forces that act on the individual atoms. "Instead they arise from the tendency of a system which has fluctuated away from equilibrium to return to equilibrium conditions" (Manning, 1968, p. 213). The tracer diffusion experiments described in this paper were conducted in the absence of significant driving forces. The results of the chemical-diffusion experiments, on the other hand, are probably affected by driving forces connected with concentration gradients. However, none of the experiments described here were designed for an assessment of these forces.

The frame of reference used in measuring the flux of atoms is another possible cause of confusion. This subject has been discussed in detail by Brady (1975). It is sufficient to note here that, in tracer experiments, when there is no bulk flow of material, the various possible frames of reference are not significantly different from one another. However, in chemical diffusion experiments the choice of reference frame is not trivial, because there is a net flux for every component. Thus, depending on the frame of reference chosen, different diffusion coefficients can be derived from a given experiment. Consequently, it is in theory not valid to compare directly the diffusion coefficients of water in melts measured by weight loss on the one hand, and by the width of a hydration front on the other (see Figure 6). However, I believe that the magnitude of this error is smaller than the experimental uncertainty of the particular measurements discussed.

EXPERIMENTAL METHODS

In principle, diffusion coefficients can be determined from the bulk gain or loss of a component or tracer in a substance if the initial and boundary conditions are known. Many convenient solutions to Equation (2) are given by Crank (1975), and more complex cases can be evaluated by finite-difference methods. The main objection to this approach is that it yields the least detailed information about the actual nature of the transport processes that occur in the specimen. As a result, the measured "diffusion" coefficient will be spurious if the bulk transport is dominated by some process other than volume diffusion. Examples of this are common in the literature on argon diffusion in minerals, as has been extensively discussed by Mussett (1969) and by Giletti (1974). Competing transport processes in glasses may include migration of the tracer along the surface or internal fractures, convective flow of the melt, or deformation of the glass specimen. In liquids, the most difficult experimental problem is convection.

An apparently successful application of the bulk-transport method for measuring diffusivity in melts has been described by Winchell and Norman (1969). Spheres of melt, doped with a volatile tracer, were held suspended by platinum wire, and the diffusion coefficients were determined by measuring the bulk loss of tracer from the melts. This loss was caused by the combined processes of diffusion and surface evaporation. Another example is given by Muehlenbachs and Schaeffer (1977). These authors measured self-diffusion of oxygen in silica glass by suspending silica fibers in a stream of CO_2. The oxygen-isotope composition of the CO_2 and the initial SiO_2 differed by a known amount, and the diffusion coefficient was determined by measuring the bulk change in the isotopic composition of the silica fibers.

The bulk transport method is particularly convenient when the transport distances are so short that it is difficult or impossible to measure them accurately. However, because of the above mentioned pitfalls associated with this method it is very important to make sure that diffusion is the actual mechanism of transport. This can be done in several ways: (1) cross checking the results with one or more different experimental techniques; (2) varying the size and geometry of the specimen; (3) varying the duration of the experiments. Test (1) is desirable in any experiment and needs no further elaboration, but it is often not practically feasible. Test (2) is the best available substitute for measuring the actual concentration gradient in the sample, and test (3) should be an essential part of any kinetic experiment. A particularly illuminating application of test (3) may be somewhat imprecisely called the zero-time experiment. This consists simply of terminating the diffusion experiment as soon as the experimental run conditions of temperature (or pressure, PO_2, or other externally imposed conditions) have been reached. In this way, the amount of diffusive and non-diffusive transport that occurred during the combined run-up and quench times is determined. Hofmann and Giletti (1970) found (to their surprise and dismay) that this form of (short-duration) transport dominated the total transport in experiments run for as long as six weeks. It should also be stressed that the time test is useful only if the variation in run durations is sufficiently large. For example, in many experiments a change in duration by a factor of two changes the transport distance or amounts by a factor of $\sqrt{2}$, a value frequently close to or below the experimental reproducibility. Therefore, depending on this reproducibility, it is usually desirable to vary the duration by at least a factor of ten.

The preferred method for measuring diffusivity is the determination of actual transport distances and concentration gradients. Only by measuring both the concentration profile and its dependence on run time is it possible to ascertain unequivocally whether the transport process is fully described by the diffusion equation and the prescribed initial and boundary conditions. Because of the relatively large diffusion coefficients in silicate glasses and melts, this is generally not difficult. For convenience, the experiments are usually designed for one-dimensional diffusion in the direction normal to a prepared surface and parallel to the container walls. Examples of some experimental designs are shown in Figure 1. The thin-source method is widely used on glasses and melts. The couple method

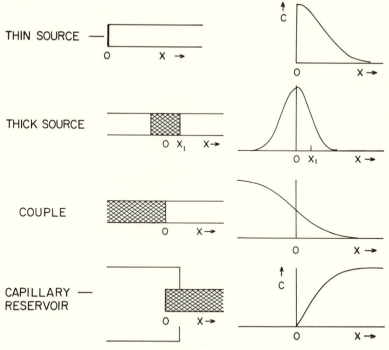

Figure 1. Schematic illustration of several experimental methods for measuring one-dimensional diffusion. The initial distribution of the tracer source is indicated by a heavy vertical line at $X = 0$ for the thin-source experiment, and by cross-hatching for the extended-source experiments. The corresponding concentration profiles, C versus distance X, after the diffusion experiment, are shown on the right.

has been used in several studies on melts (Medford, 1973; Hofmann, 1974; Smith, 1974). The thick-source method is useful when a radioactive tracer is introduced into the sample by irradiation of a portion of the sample with neutrons (Varshneya et al., 1973) or high-energy particles (e.g. deuterons; see Hofmann and Brown, 1976). The capillary-reservoir method was designed specifically for liquids and has been used, for example, by Towers and Chipman (1957), Koros and King (1962) and by Henderson et al. (1961) to measure diffusion of Ca, Si, O, and Al in $CaO-SiO_2-Al_2O_3$ melts.

The concentration profile is measured after the diffusion experiment by a variety of techniques, including the use of the electron microprobe (in studies of chemical diffusion), the ion microprobe (Jambon and Semet, 1978), autoradiography (for tracer diffusion using beta emitters, see Towers et al., 1953; Hofmann and Magaritz, 1977a), and sectioning (by chemical etching or mechanical grinding) in conjunction with gamma emitters. In all cases, it is convenient to design the experiment (by the choice of specimen geometry and run duration) in such a way that the specimen can be treated as a one-dimensional, semi-infinite medium. It may be noted that, in many thin-source experiments, the distribution of tracer on the surface of the specimen is intentionally non-uniform and the diffusion process is therefore clearly three-dimensional. For example, the tracer may be deposited only in the central portion of the surface of a specimen cylinder in order to avoid surface migration of tracer around the corner and along the sides of the cylinder. Nevertheless, it has been proved by Tannhauser (1956) that the ordinary one-dimensional solution of the diffusion equation is applicable to this case without error if the specimen is sectioned parallel to the surface, and if the total tracer concentration is determined for each section.

The following solutions of equation (2) are useful for the calculation of diffusion coefficients from the measured concentration of one-dimensional diffusion experiments.

SEMI-INFINITE SOURCES (DIFFUSION COUPLE):

$$C = \tfrac{1}{2}C_1 \operatorname{erfc} y, \tag{6}$$

where $y = x/2(Dt)^{1/2}$, $\operatorname{erfc} y = 1 - \operatorname{erf} y$, $\operatorname{erf} =$ error function, $C =$ concentration at distance x from the interface between the semi-infinite source which has an initial concentration C_1 at $x \le 0$ and the semi-infinite medium, $x > 0$, which has an initial concentration $C = 0$. $D =$ diffusion coefficient, $t =$ duration of the experiment.

THIN SOURCE AT ORIGIN OF SEMI-INFINITE MEDIUM:

$$C = C_0 \exp(-y)^2, \tag{7}$$

where C_0 is the concentration at the origin ($x = 0$) and time t.

EXTENDED SOURCE WITHIN INFINITE MEDIUM:

$$C = \tfrac{1}{2}C_0\{\operatorname{erf}(y' - y) + \operatorname{erf}(y' + y)\}, \tag{8}$$

where $y' = h/2(Dt)^{1/2}$, $y = x/2(Dt)^{1/2}$, and $h =$ half-thickness of the source, which is symmetrical about the origin $x = 0$.

The diffusion coefficient may be found from Equations (6) through (8) for any concentration value C at its distance x from the origin. To utilize the entire concentration profile (rather than a single pair of C and x), it is convenient to linearize the profiles. This is done easily for Equation (7) by taking the logarithm of both sides, and for Equation (6) by first converting the measured concentrations into values of y (using Equation. (6)) and then plotting this parameter y versus the distance x. These linearized profiles can then be evaluated by standard methods of linear regression.

If the concentration (or radioactivity) profile is measured by sectioning, each value of C is the average of the segment removed, and the overall accuracy depends in part on the thickness of the sectioning steps and the curvature of the concentration profile. (Similarly, if a concentration profile is analyzed by electron microprobe, the accuracy depends in part on the effective beam width.) Because of this, it may be preferable to use the residual-activity method of Frischat and Oel (1966). Instead of the sectioned material itself being counted, the activity of the specimen that remains after each cut (residual activity) is measured. Equation (6) and (8) must accordingly be modified by integration. For example, the solution of the thin-source case [Equation (6)] becomes

$$A_{res} = A_{tot} \text{ erfc } y, \tag{9}$$

where A_{res} is the residual ($=$ after sectioning) and A_{tot} the total activity of the specimen. Equation (9) may then be linearized in the same way as Equation (6).

Other solutions, for example for the case of sectioning experiments in conjunction with beta emittors, where absorption must be taken into account, have been given by Frischat and Oel (1966) and Frischat (1975).

DIFFUSION IN LIQUID SILICATES

CHEMICAL DIFFUSION

Bowen (1921) analyzed interdiffusion between plagioclase and diopside melts at 1,500°C by measuring the refractive index of the quenched glass as a function of distance, calculating concentration values from the refractive indices, and fitting the concentration-distance curve to an error-function solution somewhat similar to Equation (8) but appropriate to the case where the overall system has a finite length. The fit between theoretical and observed profiles was generally not good, and Bowen ascribed this to a variation in the diffusion coefficient as a function of composition.

It is instructive to consider Bowen's evaluation of his results from a formal point of view. The molten plagioclase-diopside couple is not a binary diffusion system, but may be described as part of the compositional system $Na_2O-CaO-MgO-Al_2O_3-SiO_2$. Because Bowen attempted to determine a binary composition profile (in real space) by measuring the refractive index of the quenched glass, his evaluation of the diffusion coefficient resembles that of an effective binary diffusion coefficient. However, the experiments were made on specimens of finite length, and the use of an EBDC is not valid in such an experiment (see section on diffusion theory). In addition, because the diffusion path in multi-dimensional composition space is generally curved (that is, non-binary) and dependent on the end-member compositions of the diffusion couple, Bowen's calibration of composition versus refractive index (made by preparing a series of binary diopside-plagioclase mixtures) is in principle not applicable to the actual compositional path. It seems probable that these flaws in Bowen's evaluation (detected with the benefit of over 50 years of hindsight) would not change the order of magnitude of the diffusion coefficients he obtained ($D = 10^{-6}$ to 10^{-7} cm^2 s^{-1}). These values are, in fact, similar to more recent results, and Bowen's petrological conclusions, especially the dismissal of the Soret effect with regard to any important petrological consequences, are still valid.

More recent studies of chemical diffusion in natural melts were made by Medford (1973), Yoder (1973) and Smith (1974). All three authors used the diffusion-couple method and measured the composition gradients by electron microprobe. Medford used a mugearite basalt with an increased CaO concentration (10.0% versus 7.5%) in one half of the diffusion couple. Medford argued that the concentration differences in the other components were too small to be measured and that therefore the CaO gradient alone was sufficient to determine a diffusion coefficient for calcium. However, in such an experiment there can be no net flux of Ca without a counter flux of one or more other components. Therefore, Medford's method of evaluating only the Ca gradient actually yields EBDCs (effective binary diffusion coefficients), even though Medford believed them to be the "principal" coefficients, D_{ii}.

Because the measured calcium profiles did not yield good fits to theoretical profiles (assuming a single, constant diffusion coefficient), Medford also determined D values for concentration-dependent diffusivity, using Boltzmann's (1894) method (see also Crank, 1975, p. 105–106). In this way, he also obtained a well-fitted Arrhenius line (see Figure 4). However, the EBDCs obtained in this way are not directly comparable to the tracer-diffusion results to be discussed later.

Yoder (1973) measured diffusion gradients in a basalt-rhyolite couple from an experiment conducted at 1 kb water pressure and 1,300°C. Although Yoder did not attempt to calculate diffusion coefficients from his measured microprobe traverses, EBDC values of 1 to 5 $\times$ 10^{-9} cm^{2}s^{-1} can be estimated from his published figure of the major-element gradients. The similarity in length of all the gradients implies strong coupling between the major components.

Smith (1974) studied alkali interdiffusion in couples of basalt, andesite, and rhyolite melts. He enriched one side of each couple with one alkali metal, e.g. Na, and the other side with another alkali metal, e.g. K, and then calculated the EBDCs from the observed concentration gradients of the alkalis.

The EBDC values were found to depend on composition, as expected, and generally increased with increasing proportion of the heavier alkali metal present. At any given composition, however, the lighter alkali had a greater EBDC than did the heavier alkali (except in obsidian, where they were nearly the same). This indicated clearly that, for example, in a couple consisting of Na-basalt and Rb-basalt, a significant portion of the Na flux was balanced by other cations (in addition to Rb), and Smith identified

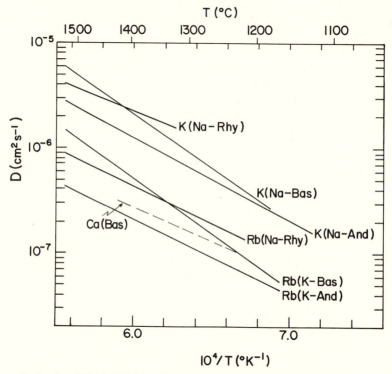

Figure 2. Results of Smith (1974) for alkali interdiffusion in basalt (Bas), andesite (And), and rhyolite (Rhy). The two interdiffusing alkalis are indicated on each line along with the matrix composition. For example, the notation K(Na − Rhy) indicates the EBDC (effective binary diffusion coefficient, see text) of K in a couple consisting of Na-rhyolite and K-rhyolite (with equimolar concentrations of Na and K), measured at the point in the diffusion profile where the alkali ratio K/(K + Na) = .5. Not shown are the EBDCs for the corresponding other member of each couple, e.g., Na(K − Rhy). For any given couple, the EBDCs decrease in the order Na ≥ K ≥ Rb, with a maximum difference of a factor of five. The dashed line labeled Ca(Bas) gives Medford's (1973) results for Ca diffusion in a mugearite basalt.

Ca as the main balancing species. The Ca-concentration profile has a minimum and a maximum as a result of uphill diffusion induced by the alkali gradients.

Some of Smith's results are shown (with permission) in Figure 2. The EBDCs for alkali interdiffusion are mostly in the range of $D = 10^{-7}$ to 10^{-5} cm^2 s^{-1}. Smith also experimented with basalt-rhyolite interdiffusion at one atmosphere and confirmed Yoder's results that indicated an effective D value for the bulk homogenization of the two melts of about 10^{-9} cm^2 s^{-1}. Evidently, the presence of water in Yoder's experiment (saturated at 1 kb) did not significantly enhance the rate of overall diffusional homogenization.

TRACER DIFFUSION

Tracer diffusion, like self-diffusion, is comparatively easy to measure and interpret, because the more complex aspects of multicomponent diffusion—the dependence of the coefficient on composition and compositional gradients—are eliminated. Also, the results are probably directly applicable to problems of isotopic and trace element migration in a magma. On the other hand, these results are not directly applicable to interdiffusion of major elements in magmas of different bulk composition (e.g., basalt and rhyolite).

Included in this review are several diffusion studies on the compositional system $CaO-Al_2O_3-SiO_2$, because data on Si and Al diffusion are available only for this system. The results of these studies, all done by the capillary-reservoir method, are compiled in Figure 3 (Towers et al., 1953; Towers and Chipman, 1957; Koros and King, 1962; Henderson et al., 1961; Oishi et al., 1975). In this compositional system, the ionic radius is positively correlated with the magnitude of the diffusion coefficient. Henderson et al., 1961, gave a qualitative explanation for the relative behavior of Ca, Al, and Si in terms of dominant size of anionic complexes formed by these cations. However, they were unable to explain the high diffusivity of oxygen. The discrepancy between the oxygen diffusivities of Koros and King on the one hand, and Oishi et al. on the other, remains unresolved. It should be noted that the bulk composition was similar but not exactly the same in all these studies (about 40% SiO$_2$, 40% CaO, 20% Al$_2$O$_3$, by weight). Also shown on Figure 3 are the results of Winchell and Norman (1969) for diffusion of Sb in the system 41% SiO$_2$, 22% Al$_2$O$_3$, 37% CaO.

A closer analogue of natural melts is one of the compositions used by Winchell and Norman (1969). This is, in weight per cent, 48 SiO$_2$, 13 Al$_2$O$_3$, 20 CaO, 2 Na$_2$O, 11 Fe$_2$O$_3$, 4 MgO, 2 K$_2$O. Their results for

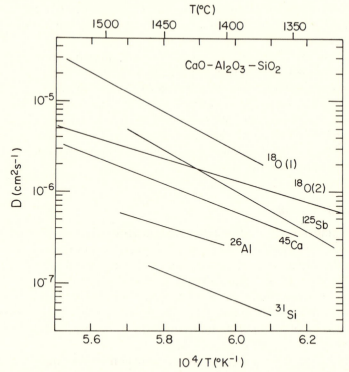

Figure 3. Tracer diffusion in the system CaO–Al$_2$O$_3$–SiO$_2$. The results are taken from the following sources: ^{18}O (1): Koros and King (1962); ^{18}O (2): Oishi, et al. (1975); ^{125}Sb: Winchell and Norman (1969); ^{45}Ca: Towers and Chipman (1957); ^{26}Al: Henderson, et al. (1961); ^{31}Si: Towers and Chipman (1957).

Cs diffusion are shown in Figure 4 together with data on natural basalt by Muehlenbachs and Kushiro (1974) for oxygen, by Hofmann and Magaritz (1977a) for Ca, Sr, Ba, and Co, by Hofmann and Brown (1976) for Na, Mn, Co, and V, and by Magaritz and Hofmann (1978a) for Eu and Gd.

The method used by Muehlenbachs and Kushiro involved measuring the bulk exchange of oxygen between beads of melt and an O$_2$ or CO$_2$ atmosphere. A portion of their experiments were done by determining the time t_0 required for the sample of radius r_0 to reach isotopic equilibrium with its environment. The diffusion coefficient was then calculated from the relation $D = r_0^2/4t_0$. Strictly speaking, this equation is an exact solution of the diffusion equation if t_0 is the time when 95% of the net mass transfer (that would be needed for complete equilibrium) has been attained (see Crank, 1975, Equation (6.22) or Figure 6.4). Because the approach to true equilibrium is an asymptotic process, the magnification

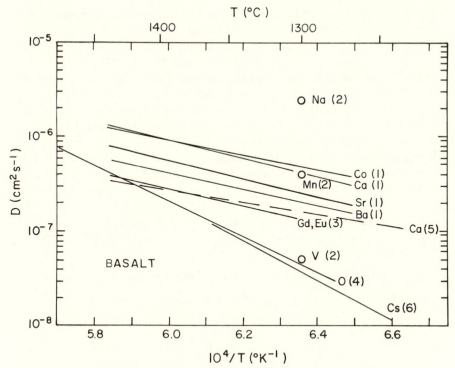

Figure 4. Tracer diffusion in basalt. Arrhenius lines and single-temperature points are labeled by the appropriate chemical element and a number in parentheses. The numbers refer to the following references: (1) Hofmann and Margaritz (1977a); (2) Hofmann and Brown (1976); (3) Magaritz and Hofmann (1978b); (4) Muehlenbachs and Kushiro (1974); (5) Medford (1973); (6) Winchell and Norman (1969). The dashed line for Ca(5) actually pertains to effective binary diffusion coefficients rather than to tracer diffusion coefficients (see text).

of analytical errors in the calculation of diffusion coefficients becomes unnecessarily large when the degree of equilibration is as high as is implied by $D = r_0^2/4t_0$. However, Muehlenbachs and Kushiro did a portion of their experiments at smaller degrees of equilibration (with correspondingly smaller error magnification) and the results of the two methods appear to be in reasonable agreement.

The experiments by Winchell and Norman (1969), Hofmann and Magaritz (1977a) and Magaritz and Hofmann (1978b) were done by the thin-source method. The main drawback of this method is the apparently inevitable convective contribution to the total transport. This effect can be verified by time studies and minimized by allowing the diffusion gradient to become at least 1 cm long (in the experimental configuration used by Hofmann and Magaritz (1977a). The time studies show that con-vection occurs primarily during the initial heating and melting of the charge and affects the region in the vicinity of the tracer source only. The thick-source experiments (Hofmann and Brown, 1976), where the tracer is introduced by high-energy deuterons within the sample, show a very much smaller convective contribution to the diffusion transport.

The effect of pressure on diffusion in silicate melts has been studied by Watson (1978) in the compositional system 29 Na_2O, 5 CaO, 10 Al_2O_3, 56 SiO_2 (in weight per cent) between 1 and 30 kb and 1,100° to 1,400°C. Watson developed an elegant modification of the thin-source method that allowed him to seal the end of the platinum capillary tube (which contained the starting glass) after a thin source of ^{45}Ca tracer had been deposited. By sealing the capsules he also minimized convection. This was documented by beta-activity profiles (using autoradiography) and time studies. Figure 5 shows his results for 1 kb and 20 kb. An expression for the full range of his results is

$$D_{ca} = [0.0025 \exp(-23,107/RT)] \exp[P(0.7297T - 1261.32)/RT],$$

where P is in kb, and T in deg K, and R in cal deg^{-1} mole^{-1}.

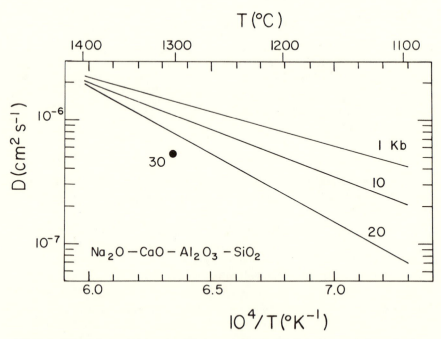

Figure 5. Effect of pressure on the diffusion coefficients of ^{45}Ca in a soda-lime-alumina-silica system. The results are taken, with permission, from Watson (1979).

DIFFUSION IN SILICATE GLASSES

CHEMICAL DIFFUSION

Diffusion of water in obsidian has been studied by Shaw (1974) at 750 to 850°C and by Friedman and Long (1976) at 95 to 245°C and by Jambon et al. (1978) at 750 to 900°C. Also of interest is Arzi's (1978) work on water diffusion into a specimen of Westerly Granite during the melting process itself. These results are shown on Figure 6. Shaw measured the water uptake by the weight increase of the bulk specimen, and found that the time dependence of the sorption curve shows a rather abrupt bend when the specimen approaches saturation. Consequently, hydration takes place in a relatively narrow region ("sharp hydration front"). Such a front will form if the diffusion coefficient increases with H_2O concentration. Jambon et al. (1978) obtained H_2O diffusion data as a byproduct

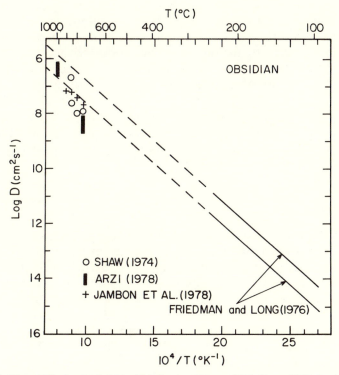

Figure 6. Water diffusion in obsidian (Friedman and Long, 1976; Shaw, 1974; Jambon et al., 1978) and in partially molten granite (Arzi, 1978). The two solid lines are envelopes for all the results of Friedman and Long on twelve different obsidian samples, and the dashed lines are extrapolations of these envelopes to higher temperatures.

of Cs tracer diffusion measurements in (previously) hydrated obsidian under 3 kb argon pressure. They calculated D_{H_2O} from the weight loss of the obsidian specimens. Their results are similar to those given by Shaw (1974), but their experiments do not include time studies and have no direct bearing on the problem of the concentration dependence of the diffusion coefficient. The measurements of Friedman and Long were made by observing the advance of the hydration front under the microscope. Arzi obtained H_2O diffusion data from the rate of advance of the melt front (water saturated) in a granite specimen. He concluded from a comparison of his results with Shaw's (1974) data that the diffusion coefficient is rather insensitive to water concentration. This conclusion is based on the similarity of results obtained for obsidian specimens with different initial water content (ranging from .2 to $>3\%$ H_2O), but Arzi ignored the shape of Shaw's sorption curves, which were the main basis for Shaw's conclusion that the diffusion coefficient does increase with water concentration. Both arguments are open to question. The effect due to the variation of initial water content was not studied systematically by either author. Aside from this, Arzi's experimental method is inherently incapable of detecting any concentration dependence because the diffusivity is measured by the rate of advance of a melt front. This yields a parabolic rate law, $X = Kt^{1/2}$, irrespective of any concentration dependence of the diffusion coefficient (Boltzmann, 1894). Most of Shaw's sorption curves have non-zero intercepts when extrapolated to time zero, and it is not entirely clear whether the difference in shape (from a curve for constant diffusivity) is significant when all possible experimental errors are considered. The problem will probably not be resolved until actual water-concentration profiles are determined.

Tracer Diffusion

The following discussion is confined to diffusion in glasses of granitic (= obsidian) and basaltic composition. For an extensive review of diffusion in simpler, synthetic glasses the reader is referred to Frischat (1975). Most of the published results are for alkalis and alkaline earths, but a few preliminary values have been determined for trivalent cations. All the measurements compiled in Figure 7 were made using the thin-source technique. Sippel (1963) irradiated the sample with low-energy (2 MeV) deuterons to generate a near-surface layer (~ 20 μm thick) of ^{24}Na ions by the reaction $^{23}Na(d, p)$ ^{24}Na. After the diffusion experiment the specimen was sectioned (by grinding) to determine the activity profile. Carron (1968), using essentially the same techniques and different obsidian specimen, extended the measurements to higher temperatures

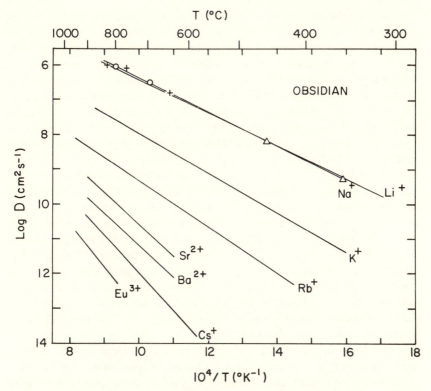

Figure 7. Tracer diffusion in obsidian. Data sources are: Li—Jambon and Semet (1978);
Na—Sippel (1963, triangles), Carron (1968, crosses), Margaritz and Hofmann
(1978a, circles); K, Rb, Cs—Jambon and Carron (1973); Sr, Ba—Magaritz and
Hofmann (1978a); Eu—Magaritz and Hofmann (1978b). Individual data points
are shown only for Na to indicate agreement between results of three different
authors.

(830°C), and Magaritz and Hofmann (1978a) confirmed these results
using a thin source of ^{22}NaCl evaporated from aqueous solution on the
surface of the specimen. The agreement between all three sets of data is
excellent (Figure 7). Diffusion of K, Rb, and Cs was measured by Jambon
and Carron (1973) using tracer deposition from chloride solution and
sectioning by chemical (HF) etching and weighing; Li diffusion was measured by Jambon and Semet (1978) using the stable isotope ^{7}Li as tracer
(deposited from LiNO$_3$ solution). The tracer profile was then determined
using an ion microprobe and scanning a surface parallel to the diffusion
direction. It is surprising that, despite the difference in ionic radius, Li
and Na diffusion are found to be very nearly identical over the entire
temperature range. Perhaps Li does not migrate by self-diffusion in this

type of experiment but by chemical diffusion with unexpectedly strong coupling to Na. An alternative interpretation, suggested by Jambon and Semet (1978), is that a critical ionic radius exists below which the mobility is not affected by size. Dowty (1977b) advanced similar arguments for diffusion in crystalline silicates. Sr, Ba, and Eu diffusion in obsidian was measured by Magaritz and Hofmann (1978a,b) using radioactive tracers and mechanical-grinding techniques. Except for the above-noted Li–Na similarity, all the diffusivities in obsidian appear to decrease with increasing ionic charge and radius. This correlation is at least qualitatively similar to the behavior observed in liquid basalt.

A first attempt to measure the effect of water content on cation diffusion was made by Jambon et al. (1978). In these experiments (which were mentioned above in connection with water diffusion), [134]Cs tracer was deposited on the surface of obsidian previously hydrated at 800°C under 2 kb water pressure. The sample was then held in an argon atmosphere at 3 kb pressure and at temperatures between 750 and 900°C. Cs diffusion was enhanced by up to four orders of magnitude in hydrous obsidian compared with dry obsidian. Jambon et al. made an attempt to evaluate the dependence of D_{Cs} on the water concentration. However, these calculations were made with the assumption that the water concentration was quasi-constant with time and varied only as a function of depth of penetration. Under the experimental conditions this assumption can only be approximately correct, and these measurements are therefore preliminary and need confirmation by further experiments conducted under conditions of constant water concentration.

The first diffusion measurements in basalt glass were reported by Rálková (1965). She used the thin-source method with ^{137}Cs and ^{90}Sr but did not determine the activity profile after the experiment. Instead, she measured the activity remaining on the surface as a function of time. This technique depends critically on the accuracy of the absorption coefficient used in evaluating the results. Rálková estimated absorption in the sample from absorption curves in aluminum, corrected to the density of basalt glass. Thus it appears that the effect of the mean atomic number on the absorption coefficient was neglected. Additional complications resulted from the presence of the daughter isotope ^{90}Y in the Sr diffusion experiments, and this required substantial corrections to the measured count rates. If Rálková's technique can be checked and verified by direct measurements of the activity profiles, it should be particularly useful and easy to apply because it is non-destructive.

Rálková's results are shown together with more recent results of Jambon and Carron (1978) on Figure 8. It is important to note that Rálková's basalt glass recrystallized at temperatures above 750°C, whereas the glass used by Jambon and Carron showed only traces of augite

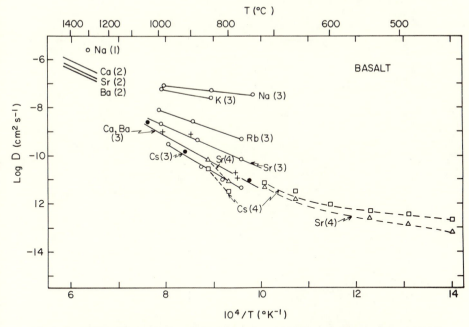

Figure 8. Tracer diffusion in basalt liquid (at $T > 1,250°C$) and glass (at $T < 1,000°C$). Data sources are given by numbers in parentheses: (1) Hofmann and Brown (1976); (2) Hofmann and Magaritz (1977); (3) Jambon and Carron (1977); (4) Rálková (1965). The results for Ca (crosses) and Ba (solid dots) are represented by a single line. Note, however, that one of the crosses lies close to the line for Sr. Rálková ascribed the discontinuity in her results for Sr and Cs (between 700 and 800°C) to the onset of devitrification.

crystallized during annealing times that exceeded the diffusion times by a factor of ten. Jambon and Carron (1978) noted a transition point at 730°C, with a sharp discontinuity in the diffusion coefficients. However, they did not report results below this temperature. The two sets of experiments are therefore not strictly comparable. Rálková's results do not yield a straight Arrhenius line except at temperatures below 600°C. Perhaps the curvature of the Arrhenius lines is also related to a glass transition. Discontinuous or curved Arrhenius lines have not been observed in obsidian but are relatively common in synthetic binary and ternary glasses (see Frischat, 1975).

The results of Jambon and Carron (1978) on the diffusion of Na, K, Rb, Cs, Ca, Sr, and Ba in basalt glass are compared with the respective diffusivities in obsidian on Figure 9. The similarity of the results is remarkable. Differences exceeding a factor of two are found only for Na and for a single (possibly anomalous) value of Ba. In this respect, the

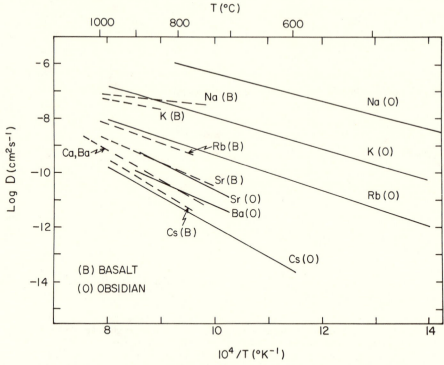

Figure 9. Comparison of tracer diffusion in glasses of basalt, dashed lines labeled (B), and obsidian, solid lines labeled (O). These Arrhenius lines are taken from Figures 7 and 8, omitting the results of Rálková (1965). Note the close agreement of the diffusivities in the two glasses. Exceptions are the diffusivities of Na and K. There are no data for Ca diffusion in obsidian.

behavior of natural melts appears to be very different from that of simpler synthetic glasses in that the latter commonly show a strong dependence of diffusivities on composition.

RELATION TO VISCOSITY

Bowen (1921) related his diffusion results to the viscosity using a theory developed by Einstein and now generally known as the Stokes-Einstein relation. In this theory the diffusing atoms are treated as noninteracting, neutral, hard spheres of radius r which are suspended in a viscous medium of viscosity η at (absolute) temperature T (see Doremus, 1973, p. 156):

$$D\eta = \frac{k_B T}{6\pi r} \tag{10}$$

where k_B is the Boltzmann constant. This relation and its more sophisticated variants have been quite popular in the literature on crystal growth. One obvious deficiency of the theory in ionic liquids is the assumption of neutral spheres. Nevertheless, Donaldson (1975) found that the Stokes-Einstein relation predicts diffusion coefficients in high-temperature melts within about a factor of 20. A sufficient body of diffusion data is now available to assess the validity of this relationship in natural silicate melts and glasses. Figure 10 shows two comparisons of measured versus calculated diffusivities, one in basalt at 1,300°C, and one in obsidian at 800°C. In basalt the disagreement is one to two orders of magnitude, and in obsidian the disagreement is as high as ten orders of magnitude. The viscosities used in this calculation were 10^2 poises for

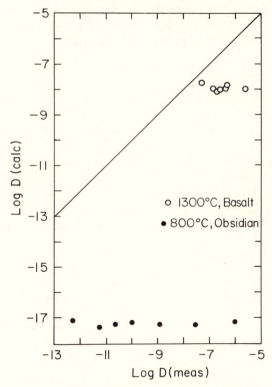

Figure 10. Comparison of measured diffusivities, D(meas), and diffusivities calculated from the Stokes-Einstein relation, Equation (10), D(calc). The two temperatures chosen for the comparison are 1,300°C for basalt and 800°C for obsidian, because data are available for the largest variety of elements. The diffusivities are taken from Figures 4 and 7. The solid line is the locus of agreement between observed and calculated values. The misfit demonstrates the inadequacy of the Stokes-Einstein equation.

basalt and 10^{11} poises for obsidian (see Murase and McBirney, 1973). The diffusivities were for tracer diffusion of alkalis, alkaline earths, and rare earths. Other considerations lead to the same conclusion. As noted by Doremus (1973), the activation energies of diffusion and viscosity are generally different. The comparison of chemical diffusion (effective binary diffusion coefficients) in basalt, andesite, and rhyolite by Smith (1974) shows that the diffusivities are rather similar in all three compositions (see Figure 2), but the viscosities at, for example, 1,300°C, differ by 4 to 5 orders of magnitude. Finally, the positive correlation of diffusivity with ionic radius in the system $CaO-Al_2O_3-SiO_2$ (Figure 3) is not compatible with Equation (10). If one compares tracer diffusivities of cations with similar ionic radius but different charge (e.g., K^+ and Ba^{2+}, or Na^+ and Eu^{3+}), one invariably finds substantially smaller diffusivities for the more highly charged ion. This demonstrates quite clearly that any theory, such as the Stokes-Einstein theory, that does not take ionic charge into account is both qualitatively and quantitatively inadequate.

COMPENSATION

Winchell (1969) and Winchell and Norman (1969) showed that in silicates there is a positive correlation between the experimental activation energy, E, and the preexponential factor D_0 in the Arrhenius equation. The scatter in this correlation is quite large when diffusion in crystalline silicates is included but is reduced considerably when only melts and glasses are considered.

Winchell (1969) gives the equation for all silicates

$$\log_{10} D_0 = 1.45 \times 10^{-4} E - 5.94 \qquad (11)$$

(E in cal/deg mole) and Winchell and Norman (1969) obtained for diffusion of alkalis and several Period V elements in $CaO-Al_2O_3-SiO_2$ glasses the relation

$$\log_{10} D_0 = 1.23 \times 10^{-4} E - 5.79 \qquad (12)$$

They also noted that correlations of this type are to be expected from Eyring's absolute reaction rate theory.

Figure 11 shows $\log D_0$ versus E for tracer diffusion in natural melts and glasses (compiled from Figures 4, 7, and 8). The linear regression yields the equation

$$\log_{10} D_0 = 1.08 \times 10^{-4} E - 5.50 \qquad (13)$$

The correlation is considerably better for basalts than for obsidians,

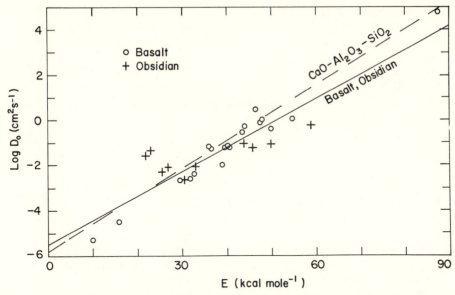

Figure 11. Correlation between $\log_{10} D_0$ and E [see Equation (4)] for tracer diffusion of alkalis and Period V elements in CaO–Al$_2$O$_3$–SiO$_2$ melts, and for tracer diffusion in basalt and obsidian melts and glasses. The dashed line corresponds to Equation (12) and is taken from Winchell and Norman (1969). The solid line corresponds to Equation (13) and is calculated from the results shown in Figures 4, 7, and 8.

but there is a rough, overall agreement between all these data and those of Winchell and Norman (1969) for Ca–Al$_2$O$_3$–SiO$_2$ glasses. Equations (11), (12), and (13) are all of the form

$$\log_{10} D_0 = aE + b \qquad (14)$$

which may be combined with (4) to yield

$$\log_{10} D = \left(a - \frac{1}{2.303RT}\right) E + b \qquad (15)$$

Winchell (1969) called this correlation "compensation law" because D_0 and E, as a result of the correlation, compensate each other in such a way that, at some critical temperature, all the Arrhenius lines intersect at a single value for the diffusion coefficient. It can be seen by setting the first term on the right-hand side in (15) equal to zero that a critical temperature,

$$T_c = \frac{1}{2.303Ra} \qquad (16)$$

exists where

$$\log_{10} D_c = b, \tag{17}$$

irrespective of the value of the activation energy E. In other words, if all the diffusion data conformed perfectly to (14), then all the Arrhenius lines (4) would intersect at a single point given by (16) and (17). The critical temperatures and diffusivities for (11), (12), and (13), respectively, are

$$\log_{10} D_c = -5.94, -5.79, -5.50$$
$$T_c = 1,234, 1,504, 1,770 \; (°C).$$

The three pairs of D_c and T_c values show relatively good agreement in $\log D_c$ but considerable differences in T_c. This merely reflects the general property of the Arrhenius equation that the diffusion coefficient becomes increasingly insensitive to absolute temperature differences as the temperature increases. For example, most of the cation diffusivities shown in Figure 4 change by less than one order of magnitude within a 200°C temperature interval (1,250 to 1,450°C).

The correlation equation has practical value in that it predicts similar diffusion coefficients (approximately $D = 10^{-6} \; cm^2 \; s^{-1}$) for all cations at high temperatures that exist in the earth's upper mantle. On the other hand, in crustal regions and temperatures below 1,000°C the diffusivities diverge by many orders of magnitude. To use the correlation equation for more specific predictions would require an independent knowledge or prediction of either E or D_0.

PREDICTION AND MECHANISM OF DIFFUSIVITY

The diffusion coefficient of an ionic species can be predicted from the electrical conductivity in special cases only, namely when the electrical current is carried exclusively (or predominantly) by this species. The diffusivity can then be predicted from the Einstein equation (see, for example, Doremus, 1973, p. 152–154)

$$\sigma = \frac{Z^2 F^2 Dc}{RT}, \tag{18}$$

where σ is the electrical conductivity, Z is the ionic charge, F is the Faraday, D is the tracer diffusion coefficient, and c the concentration of the current-carrying ion. The assumptions hold approximately true for Na diffusion in obsidian, and Carron (1968) found good agreement between the conductivity of obsidian and the diffusivity of Na.

Except for the special case covered by the Einstein equation, no quantitative prediction of diffusion coefficients in silicates is possible on any purely theoretical basis. A survey of the results of both alkalis and alkaline earths indicate that the diffusivities are in some way related to ionic radius. In general, at any given temperature and matrix composition, and for a given ionic charge, the diffusivity decreases with increasing ionic radius, (exceptions to this rule are found in the diffusivities of Ca in basalt glass and Li in obsidian; see Figures 7 and 8). Intuitively interpreted, this means that mobility is inhibited by a large size of the diffusing particle. A second qualitative observation is that diffusivity also decreases with increasing ionic charge. This effect appears to be at least in part responsible for the failure of the Stokes-Einstein theory. An example is the diffusivity of V^{+5} in basalt, which at $1,300°C$ is two orders of magnitude smaller than the diffusivity of Na^+, even though V^{+5} has the smaller ionic radius. A straightforward intuitive explanation is that the more highly charged ions are subject to greater local binding forces, which inhibit mobility. When these observations are related to current models of the structure of silicate melts (e.g., Hess, this volume), the diffusion behavior may be qualitatively explained by the following model: the cations, depending on their size and charge (or some related property such as ionization potential or ionic field strength), form complexes (with network formers such as SiO_2) of variable stability. A cation is mobile only during the time it is in a dissociated state, so that the net mobility depends on the stability of the complexes formed as well as on the size of the dissociated ion. One difficulty with this model is that the ionic radius is not a well-defined quantity, because it depends on the coordination number (see Whittaker and Muntus (1970), and because the coordination number is not necessarily a well-known or even a well-defined quantity. Uncertainties in the charges of some species and covalent bonding produce further difficulties.

In the absence of a quantitative theory it is desirable to have empirical relationships that can perhaps be used for the purpose of predicting diffusion coefficients. Winchell and Norman (1969), for example, gave empirical relationships between D_0 and ionic radius, and between D_0 and a "modified cationic field function (ionization potential divided by the Pauling radius). They obtained an apparently excellent linear plot of $\log D_0$ versus r^{-1} for monovalent species and $(1.59r)^{-1}$ for multivalent species where r is the ionic radius (in Å) of the appropriate oxidation state. The plot included the species Na, K, Rb, Cs, Mo, In, Sn, Sb, Te, I in a $CaO-Al_2O_3-SiO_2$ melt. Unfortunately, the present author was unable to verify their plot with any "reasonable" set of ionic radii, and the validity of this correlation is uncertain.

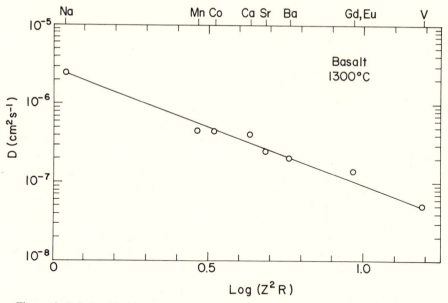

Figure 12. Relationship between ionic charge, Z, ionic radius, R, and diffusivity for tracer diffusion in basalt at 1,300°C. The diffusivities are taken from Figure 4. The ionic radii are taken from Whittaker and Muntus (1970), assuming octahedral coordination of all the cations concerned.

An attempt to find an empirical relation between diffusivity, ionic radius and ionic charge resulted in Figure 12, which is applicable only to basalt at 1,300°C. The linear regression analysis yields

$$\log_{10} D_{1300} = -1.47 \log (Z^2 r) - 5.56, \qquad (18)$$

where Z is the ionic valence and r the ionic radius (in Ångstroms). Unfortunately, the above approach failed in attempts to describe diffusion in obsidian by a similar correlation. The ionic radii used were those of Whittaker and Muntus (1970) for octahedral coordination, assuming a $+5$ oxidation state for V. Choosing higher coordination numbers for the larger ionic species does not result in drastic changes of the plot. Nevertheless, considerable caution is in order in using equation (18) to predict diffusion coefficients of other species, at least until the relationship has been tested with a greater variety of cations. On the other hand, the remarkable similarity of diffusivities of alkalis and alkaline earths in basalt glass compared to obsidian glass may reasonably be used to infer that glasses of intermediate composition will also have similar diffusivities.

The relation (18) does illustrate the general qualitative effects of ionic size and charge on the mobility of the cations, both in basalt and obsidian.

Beyond this, however, the mobility of the cations depends also on the structure of the melt in a manner that is not well understood. For example, the Arrhenius line for Eu and Gd in basalt has a slope of zero between the temperatures of about 1,320 and 1,210°C (Magaritz and Hofmann, 1978b). This feature was omitted from Figure 4 for sake of clarity and because this temperature range extends below the liquidus (at about 1,170°C). Because diffusivities normally decrease with decreasing temperature, this constant diffusion coefficient over a temperature range of about 100°C is extremely unusual, and the explanation must somehow lie in the structure of the melt. On the other hand, the similarity of some diffusion coefficients in glasses of very different structure and compostion (see Figure 9) indicates that some cation mobilities are very insensitive to melt structure. Magaritz and Hofmann (1978b) noted that those cations whose mobilities appear to be sensitive to melt structure (e.g., the rare earths) are also strongly partitioned between two immiscible melts (Watson, 1976; Ryerson and Hess, 1978). It remains to be seen whether this qualitative correlation between diffusivity and equilibrium partitioning is a general property of silicate melts.

A final feature worth noting is that none of the Arrhenius lines for cation diffusion in obsidian (Figure 7) show any detectable change in slope. This is in marked contrast with the behavior in the binary and ternary glasses reviewed by Frischat (1975), nearly all of which show some evidence of a glass transition in their diffusivities (i.e., a change of slope of the Arrhenius line).

In conclusion, it must be admitted that the mechanism of diffusion in silicate melts and the relation of diffusion to melt structure is not well understood, at least on a quantitative level. On the other hand, many regularities, such as the almost universally straight Arrhenius lines and the negative correlation of D with ionic size and charge are quite apparent, and these regularities may be expected ultimately to contribute to the understanding of silicate melts. Meanwhile, the body of diffusion data is already quite important to our understanding of igneous processes (although a discussion of this topic has not been included in this paper). Large gaps remain to be filled in diffusion measurements on network-forming ions (Si, Al), covalently bonded species, and in determining the effect of volatiles on the mobilities.

ACKNOWLEDGMENTS

The quality and completeness of this review has been improved substantially by the reviews and criticisms of Albert Jambon, John Brady, and Eric Dowty. I am grateful for their help. Harry D. Smith kindly gave permission to use results from his Ph.D. dissertation for this review. I hope that a more complete account of his work will be published separately. A substantial portion of the work reviewed here was supported by NSF grant No. EAR 77–13498.

REFERENCES

Albarède, F. and Y. Bottinga, 1972. Kinetic disequilibrium in trace-element partitioning between phenocrysts and host lava. *Geochim. Cosmochim. Acta 36*, 141–156.

Arzi, A. A., 1978. Fusion kinetics, water pressure, water diffusion and electrical conductivity in melting rock, interrelated. *J. Petrology 19*, 153–169.

Boltzmann, L., 1894. Zur Integration der Diffusionsgleichung bei variabeln Diffusionscoefficienten. *Annalen d. Physik* (N. F.) *53*, 959–964.

Bottinga, Y., A. Kudo and D. Weill, 1966. Some observations on oscillatory zoning and crystallization of magmatic plagioclase. *Amer. Mineralogist 51*, 792–806. 792–806.

Bowen, N. L., 1921. Diffusion in silicate melts. *J. Geology 29*, 295–317.

Brady, J. B., 1975. Reference frames and diffusion coefficients. *Am. Jour. Sci.*, *275*, 954–983.

Carron, J-P., 1968. Auto diffusion du sodium et conductivite electrique dans les obsidiennes granitiques. *Comptes Rendus Acad. Sci.* Paris, Série D, *266*, 854–856.

Cooper, A. R., 1968. The use and limitiation of the concept of an effective binary diffusion coefficient for multi-component diffusion. In: *Mass Transport in Oxides*, edited by J. B. Wachtman, Jr. and A. D. Franklin. Natl. Bur. Stds. Spec. Publ. 296, 79–84.

Crank, J., 1975. *The Mathematics of Diffusion*. Second edition. Oxford University Press.

Donaldson, C. H., 1975. Calculated diffusion coefficients and the growth rate of olivine in a basalt magma. *Lithos 8*, 163–174.

Doremus, R. H., 1973. *Glass Science*. New York: John Wiley & Sons.

Dowty, E., 1977a. The importance of adsorption in igneous partitioning of trace elements. *Geochim. Cosmochim. Acta 41*, 1643–1646.

Dowty, E., 1977b. Relative mobility of cations and anions in minerals. *Geol. Soc. America, Abstr. w. Programs 9*, 955–956.

Friedman, I. and W. Long, 1976. Hydration rate of obsidian. *Science 191*, 347–352.

Frischat, G. H., 1975. Ionic diffusion in oxide glasses. Bay Village, Ohio: Transtech Publications.

Frischat, G. H. und H. J. Oel, 1966. Eine Restaktivierungsmethode zur Bestimmung von Selbstdiffusions-Koeffizienten in Festkörpern. *Z. Angew. Phys. 20*, 195–201.

Giletti, B. J., 1974. Diffusion related to geochronology in *Geochemical Transport and Kinetics*, A. W. Hofmann, B. J. Giletti, H. S. Yoder, Jr. and R. A. Yund, (eds.), Carnegie Inst. Wash. Publ. *634*, 61–76.

Gupta, P. K. and A. R. Cooper, 1971. The [D] matrix for multicomponent diffusion. *Physica 54*, 39–59.

Henderson, J., L. Yang, and G. Derge, 1961. Self-diffusion of aluminum in CaO–SiO_2–Al_2O_3 melts. *Transactions, Metallurgical Soc.* AIME *221*, 56–60.

Hofmann, A. W., 1964. Strontium diffusion in a basalt melt and implications for Sr isotope geochemistry and geochronology. *Carnegie Inst. Wash. Yearbook 73*, 935–941.

Hofmann, A. W. and L. Brown, 1976. Diffusion measurements using fast deuterons for *in situ* production of radioactive tracers. *Carnegie Inst. Year Book 75*, 259–262.

Hofmann, A. W. and B. J. Giletti, 1970. Diffusion of geochronologically important nuclides in minerals under hydrothermal conditions. *Eclogae geol. Helv. 63*, 141–150.

Hofmann, A. W. and S. R. Hart, 1978. An assessment of local and regional isotopic equilibrium in the mantle. *Earth Planet. Sci. Lett.*, *38*, 44–62.

Hofmann, A. W. and M. Magaritz, 1977a. Diffusion of Ca, Sr, Ba, and Co in a basalt melt: Implications for the geochemistry of the mantle. *Jour. Geophys. Res. 82*, 5432–5440.

Hofmann, A. W. and M. Magaritz, 1977b. Equilibrium and mixing in a partially molten mantle. In *Magma genesis*, H. J. B. Dick, (ed.), Bull. 96, Oregon Dept. of Geology and Mineral Industries, 37–41.

Hurley, P. M., 1977. Estimates of possible diffusional effects in trace element separation. *Earth Planet. Sci. Lett. 34*, 225–230.

Jambon, A. and J.-P. Carron, 1973. Etude expérimentale de la diffusion des éléments alcalins K, Rb, Cs dans une obsidienne granitique. *Comptes rendus Acad. Sci.* Paris, 276, Serie D, 3069–3072.

Jambon, A. and J.-P. Carron, 1978. Etude expérimentale de la diffusion cationique dans un verre basaltique: alcalins et alcalinoterreux. *Bull. Soc. Fr. Min. Crist. 101*, 22–26.

Jambon, A. and M. P. Semet, 1978. Lithium diffusion in silicate glasses of albite, orthoclase, and obsidian composition: an ion microprobe determination. *Earth Planet. Sci. Lett. 37*, 445–450.

Jambon, A., J.-P. Carron, and F. Delbove, 1978. Données preliminaíres sur la diffusion dans les magmas hydratés: le césium dans un liquide granitique à 3 kb. *Comptes rendus Acad. Sci. Paris 287*, Ser. D, 403–406.

Koros, P. J. and T. B. King, 1962. The self-diffusion of oxygen in a lime-silica-alumina slag. *Trans. AIME 224*, 299–306.

Lazarus, D. and N. H. Nachtrieb, 1963. Effect of high pressure on diffusion. In: *Solids under Pressure*, W. Paul and D. M. Warschauer, (eds.), McGraw Hill Book Co., Inc.

Magaritz, M. and A. W. Hofmann, 1978a. Diffusion of Sr, Ba, and Na in obsidian. *Geochim Cosmochim Acta 42*, 595–605.

Magaritz, M. and A. W. Hofmann, 1978b. Diffusion of Eu and Gd in basalt and obsidian. *Geochim Cosmochim. Acta 42*, 847–858.

Manning, J. R., 1968. *Diffusion Kinetics for Atoms in Crystals*. Princeton: D. Van Nostrand Co., Inc..

Medford, G. A., 1973 Calcium diffusion in a mugearite melt. *Can. J. Earth Sci. 10*, 394–402.

Muehlenbachs, K. and I. Kushiro, 1974. Oxygen isotope exchange and equilibrium of silicates with CO_2 or O_2. *Carnegie Inst. Wash. Year Book 73*, 232–236.

Muehlenbachs, K. and H. A. Schaeffer, 1977. Oxygen diffusion in vitreous silica-utilization of natural isotopic abundances. *Canadian Mineralogist 15*, 179–184.

Murase, T. and A. R. McBirney, 1973. Properties of some common igneous rocks and their melts at high temperatures. *Geol. Soc. America. Bull. 84*, 3563–3592.

Mussett, A. E., 1969. Diffusion measurements and the potassium-argon method of dating, *Geophys. J. Roy. Astron. Soc., 18*, 257–303.

Oishi, Y., R. Terai and H. Ueda, 1975. Oxygen diffusion in liquid silicates and relation to their viscosity In: *Mass Transport Phenomena in Ceramics*, A. R. Cooper and A. H. Heuer, (eds.), New York: Plenum Press. p. 297–310.

Onsager, L. 1945. Theories and problems of liquid diffusion. *Ann. N.Y. Acad. Sci. 46*, 241–265.

Rálková, J., 1965. Diffusion of radioisotopes in glass and melted basalt. *Glass Technology 6*, 40–45.

Ryerson, F. J. and P. C. Hess, 1978. Implications of liquid-liquid distribution coefficients to mineral-liquid partitioning. *Geochim. Cosmochim. Acta, 42*, 921–932.

Shankland, T. J. and H. S. Waff, 1977. Partial melting and electrical conductivity anomalies in the upper mantle. *Jour. Geophys. Res. 82*, 5409–5417.

Shaw, H. R., 1974. Diffusion of H_2O in granitic liquids. Part I. Experimental Data; Part II. Mass transfer in magma chambers. In: *Geochemical Transport and Kinetics* A. W. Hofmann, B. J. Giletti, H. S. Yoder, Jr., R. A. Yund, (eds.), Carnegie Inst. Wash. Publication *634*.

Sibley, D. F., T. A. Vogel, B. M. Walker, and G. Byerly, 1976. The origin of oscillatory zoning in plagioclase: a diffusion and growth-controlled model. *Am. Jour. Sci., 276*, 275–284.

Sippel, R. F., 1963. Sodium self diffusion in natural minerals. *Geochim. Cosmochim Acta 27*, 107–120.

Smith, H. D., 1974. An experimental study of the diffusion of Na, K, and Rb in magmatic silicate liquids. Ph.D. dissertation, University of Oregon.

Tannhauser, D. S., 1956. Concerning a systematic error in measuring diffusion constants. *J. Appl. Phys. 27*, 662.

Towers, H. and J. Chipman, 1957. Diffusion of calcium and silicon in a lime-alumina silica slag. *Trans. AIME 209*, 769–773.

Towers, H., M. Paris and J. Chipman, 1953. Diffusion of calcium ion in liquid slag. *Trans. AIME 197*, 1455–1458.

Varshneya, A. K., A. R. Cooper, R. B. Diegle, and S. A. Chin (1973). Measurement of self-diffusion in solids by neutron activation of masked samples. *J. Am. Ceram. Soc. 56*, 245–247.

Watson, E. B., 1976. Two-liquid partition coefficients: Experimental data and geochemical implications. *Contrib. Mineral. Petrol. 56*, 119–134.

Watson, E. B., 1979. Calcium diffusion in a simple silicate melt to 30 kbar. *Geochim. Cosmochim. Acta, 43*, 313–322.

Whittaker, E. J. W. and R. Muntus, 1970. Ironic radii for use in geochemistry. *Geochim. Cosmochim. Acta 34*, 945–956.

Winchell, P., 1969. The compensation law for diffusion in silicates. *High Temperature Science 1*, 200–215.

Winchell, P. and J. H. Norman, 1969. A study of the diffusion of radioactive nuclides in molten silicates at high temperatures, in *High Temperature Technology*, Third International Symposium Asilomar 1967, 479–492.

Yoder, H. S., Jr., 1973. Contemporaneous basaltic and rhyolitic magmas, *Amer. Mineral. 58*, 153–171.

Chapter 10

CRYSTAL GROWTH AND NUCLEATION THEORY AND THE NUMERICAL SIMULATION OF IGNEOUS CRYSTALLIZATION

ERIC DOWTY

Department of Geological and Geophysical Sciences, Princeton University

INTRODUCTION

During the first half of the twentieth century, while Bowen and his collaborators were elucidating the phase relations of igneous rocks, the theories of crystal growth and nucleation were evolving only rather slowly. It was not until after the Second World War, with the development of solid-state electronics, that experimental and theoretical effort became intense. For this reason, and because it is necessary to know the equilibrium phase relations in order to model the driving forces of kinetic processes, the understanding of the role of kinetics has lagged far behind that of phase equilibria.

Despite the recent progress in kinetic theory, it seems fair to say that crystal-growth and nucleation theory cannot generally provide quantitative predictions of rates even in well-known simple systems. Furthermore, although empirical knowledge of phase relations is voluminous, thanks in large measure to the work of Bowen, we still lack the precise values of thermochemical quantities necessary for workable kinetic theory. Properties such as heat of melting and surface energy are only qualitatively known, even for some of the most common minerals. Nevertheless, the qualitative or semi-quantitative application of crystal-growth and nucleation theory can provide valuable insight into many geologic

© 1980 by Princeton University Press
Physics of Magmatic Processes
0-691-08259-6/80/0419-67$03.35/0 (cloth)
0-691-08261-8/80/0419-67$03.35/0 (paper)
For copying information, see copyright page

processes. A sound understanding of the basic principles of this theory is necessary for the interpretation of rock textures.

The aims of this paper are twofold: first, to review the basic theory of crystal nucleation and growth and the experimental data pertaining to actual rates, and second, to show how these may be applied to igneous crystallization by numerical modeling. The review of theory is intended to cover all important subjects, but complete detail cannot be given for each of them. For more complete coverage of the basic aspects, the reader is referred to any of several texts and review articles (e.g. Turnbull and Cohen, 1960; Chalmers, 1964; Christian, 1965; Jackson, 1967; Strickland-Constable, 1968; Laudise, 1970; Uhlmann, 1972; Brice, 1973; Ohara and Reid, 1973; Woodruff, 1973; Flemings, 1974; Kirkpatrick, 1975). In particular, the classical equations for nucleation and growth (Equations (5), (16), (21), (24), and (29) of this paper) are developed in most of these sources. No systematic attempt has been made to identify original references; for historical surveys, see Buckley (1951) and Van Hook (1961). Emphasis in this paper will be on the application of theory to structurally complex minerals and multicomponent igneous liquids, and on the ways in which nucleation and growth phenomena influence shape, composition, and size of crystals in igneous rocks.

SYMBOLS USED

K	all-purpose constant	ΔH_f	enthalpy of fusion
k	Planck's constant	$\Delta \bar{H}_m$	partial molar enthalpy of mixing
R	gas constant	ΔS_f	entropy of fusion
σ	surface energy	$\Delta \bar{S}_m$	partial molar entropy of mixing
D	diffusivity	ΔS_t	$= \Delta S_f + \Delta \bar{S}_m$
η	viscosity	τ	incubation period of nucleation
ν	vibrational frequency	α	roughness parameter of Jackson
t	time	ξ	anisotropy parameter of Jackson
d	lattice or interplanar spacing	R_n	homogeneous nucleation rate
V_m	molar volume	R_g	growth rate
T	actual temperature (Kelvin)	R_{cont}	growth rate, continuous mechanism
T_f	fusion temperature		
T_l	liquidus temperature	R_{sc}	growth rate, surface nucleation
ΔT	supercooling (T_l or $T_f - T$)		mechanism small-crystal case
ΔG_a	free energy of activation	R_{lc}	growth rate, surface nucleation
ΔG_c	free energy difference crystal-liquid		mechanism large-crystal case
ΔG_{tr}	free energy of activation of transport (diffusion)	R_{sd}	growth rate, screw dislocation mechanism
ΔH_{tr}	activation enthalpy of transport	R_{lat}	rate of spreading of layers

Notes—The constant capital K has different values in different equations

—The symbol R with a subscript refers to a rate; without subscript it is the gas constant.

—Symbols not in this list may have different meanings in different equations.

NUCLEATION THEORY

Nucleation is classified as either homogeneous, if it arises entirely through random thermal fluctuations in the liquid, or heterogeneous, if the presence of another phase in contact with the liquid facilitates the process.

Homogeneous Nucleation

Of the many factors affecting the rate of nucleation, the most important (that is, those which are most variable with supercooling or supersaturation) are the work, or Gibbs free energy, of forming a nucleus, and the rate of transport of atoms from the liquid onto incipient nuclei. These are both exponentially dependent on supercooling and/or temperature. Other influences are lumped into the pre-exponential factor, which is not constant with temperature or supercooling, but is much less sensitive to them than are the exponential factors.

Free Energy of a Nucleus

According to classical chemical kinetic theory (e.g. Benson, 1960; Laidler, 1965; Pannetier and Souchay, 1967), the rates of many processes are proportional to the exponential of a free energy of activation ΔG_a:

$$R_0 = K \exp\left[\frac{-\Delta G_a}{RT}\right], \tag{1}$$

where the constant K includes various geometric and population factors, as well as a vibrational frequency factor that is dependent on temperature. The free energy of activation is in effect an energy barrier that must be surmounted as the system goes from its initial to final state, even though the free energy of the final state may be considerably lower than that of the initial state. Such an energy barrier opposes the beginning of the precipitation of one phase within another, because the energy of the interface between the two phases contributes a term to the total free energy of the system. The free energy change on crystallization is thus

$$\Delta G = V \Delta G_c / V_m + \Sigma a\sigma, \tag{2}$$

where V is the volume of the crystallized material, ΔG_c is the bulk free energy change per mole, V_m is the molar volume, a represents the area of each of the various surfaces of the crystals, and σ is the specific surface energy, or surface tension, of each. Energies of edges and corners will also be significant if the new crystals or nuclei are small. For a given negative (favorable) ΔG_c, the total free-energy change ΔG_a is a function of the size of the nuclei (Figure 1). The peak of the curve in Figure 1 defines the

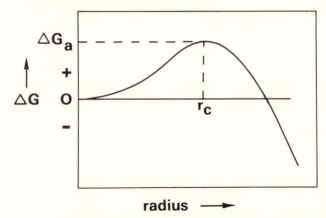

Figure 1. Gibbs free energy of spherical nuclei versus size (Equation (5)). Nuclei above the critical size, r_c, should grow spontaneously, as the free energy of the system is hereby reduced. The free energy of the critical size nucleus, G_a, is the free energy of activation for nucleation (neglecting transport processes).

contribution from the surface energy of a spherical nucleus to the activation energy, and a nucleus of the radius at which this peak occurs is called the critical nucleus: any nuclei larger than this will grow spontaneously, because the total free energy is reduced by their growth, but any smaller nuclei are unstable and prone to dissolution. The fact that a few nuclei may reach the critical size and grow to macroscopic crystals is due to random thermal fluctuations in the liquid.

The surface energy of a crystal is of course dependent on shape, since different faces or other surfaces have different values of σ. Critical-size nuclei are expected to have the equilibrium shape (e.g. Hartman, 1973a), which is simply the shape that minimizes the Gibbs free energy (Equation (2)) at constant volume. This shape can be predicted from the crystal structure by the methods of Hartman and Perdok (1955; Hartman, 1973b) or Dowty (1976a). However, there is little to be gained by this, because the absolute values of surface, edge, and corner energies cannot easily be predicted or measured. Therefore, for simplicity, nuclei will be assumed to be spherical, with the value of σ representing an average for all parts of the surface. Equation (2) then becomes

$$\Delta G = \tfrac{4}{3}\pi r^3 \, \Delta G_c/V_m + 4\pi r^2\sigma, \qquad (3)$$

where r is the radius. Differentiating this with respect to r yields the critical radius, r_c:

$$r_c = \frac{-2\sigma V_m}{\Delta G_c}, \qquad (4)$$

which, substituted back into Equation (3), gives

$$\Delta G_a = \frac{16\pi\sigma^3 V_m^2}{3\Delta G_c^2}. \tag{5}$$

This is the maximum free energy of a nucleus, and therefore it is the activation energy appearing in Equation (1) and Figure 1.

For crystallization from melts, the bulk free energy of crystallization ΔG_c is conventionally related to supercooling, $\Delta T = T_l - T$, where T_l is the equilibrium liquidus or fusion temperature and T is the actual temperature. A derivation of this relation for one-component systems will be found in many texts, but here the general relation for multicomponent systems will be given. For the free energy change of the system ΔG_c^l with crystallization of any component at the equilibrium liquidus temperature T_l, we have

$$\Delta G_c^l = \Delta H_f + \Delta \bar{H}_m - T_l(\Delta S_f + \Delta \bar{S}_m) = 0, \tag{6}$$

where ΔH_f and ΔS_f are the molar enthalpy and entropy of crystallization of the pure component, and $\Delta \bar{H}_m$ and $\Delta \bar{S}_m$ are the partial molar enthalpy and entropy of mixing at the liquidus composition. This equation is essentially equivalent to Equation (26) of Weill et al. (this volume), but includes the assumption that the enthalpy of crystallization of the pure component is independent of temperature. This assumption is not strictly true, and if the heat capacities of the crystal and its own liquid are known for the temperature interval of interest, the variation of ΔH_f with temperature can be taken into account in evaluating the integral in Equation (26) of Weill et al. However, the values of the heats of fusion and the heat capacities are seldom known with sufficient precision to warrant this correction. Defining ΔS_t as the total entropy change on crystallization, we have

$$\Delta S_t = \Delta S_f + \Delta \bar{S}_m = \frac{\Delta H_f + \Delta \bar{H}_m}{T_l}. \tag{7}$$

Assuming now that not only ΔH_f but also $\Delta \bar{H}_m$ is insensitive to changes in temperature, the free energy change at the actual temperature R of nucleation is

$$\Delta G_c^T = \Delta H_f + \Delta \bar{H}_m - T\frac{\Delta H_f + \Delta \bar{H}_m}{T_l}. \tag{8}$$

If we also substitute Equation (7) into Equation (6), we obtain

$$\Delta G_c^l = \Delta H_f + \Delta \bar{H}_m - T_l\frac{\Delta H_f + \Delta \bar{H}_m}{T_l} = 0, \tag{9}$$

which, subtracted from Equation (8), yields

$$\Delta G_c^T = \Delta T \frac{\Delta H_f + \Delta \bar{H}_m}{T_l} = \Delta T \Delta S_t. \tag{10}$$

For a one-component system, the relation simplifies to

$$\Delta G_c^T = \Delta T \frac{\Delta H_f}{T_f}, \tag{11}$$

where T_f is the fusion temperature of the pure component. Enthalpies of mixing in multi-component systems can be obtained from calorimetric measurements (e.g. Weill et al., this volume). Depending on the magnitude of ΔH_f of the crystallizing component, the enthalpy of mixing may or may not be important: for example, the low ΔH_f of silica means that small heats of mixing will have a large relative effect on ΔG_c, whereas the same $\Delta \bar{H}_m$ would be much less significant for pyroxene or olivine, which have much higher ΔS_f.

Transport Factor

The overall rate is also proportional to the rate of arrival of atoms on the crystal surface, and there is another energy barrier associated with atomic movement across the interface and/or through the liquid. The activation energy of this process, ΔG_{tr}, can be added to that of critical nucleus formation, or separated into a factor which has the form of diffusivity D.

$$D = K \exp\left[\frac{-\Delta G_{tr}}{RT}\right], \tag{12}$$

and since $\Delta G_{tr} = \Delta H_{tr} - T \Delta S_{tr}$,

$$D = K \exp\left[\frac{\Delta S_{tr}}{R}\right] \exp\left[\frac{-\Delta H_{tr}}{RT}\right] = K' \exp\left[\frac{-\Delta H_{tr}}{RT}\right]. \tag{13}$$

For crystallization of simple compounds from their own liquids, D can be predicted to a first approximation (within one or two orders of magnitude) by the Stokes-Einstein relation (Ree and Eyring, 1958; Donaldson, 1975):

$$D = \frac{kT}{3\pi r_0 \eta} \tag{14}$$

where k is Boltzman's constant, η is viscosity and r_0 is the radius of the diffusing particle. In silicates, the nature of the growth unit or "diffusing particle" is usually unknown, so that r_0 cannot be specified, but this relation is normally used only to obtain relative values of the transport factor. A more serious problem in silicate liquids is that the viscosity is generally determined by aluminum-oxygen and silicon-oxygen linkages, whereas other cations seem to be free to move about in the interstices. The measured diffusivities of univalent and divalent ions are many orders of magnitude greater in high-silica liquids than is predicted from the Stokes-Einstein relation (Hofmann, this volume). The application of this relation in general to crystallization in multicomponent silicate systems is therefore evidently invalid.

Pre-exponential Factor and Overall Expression

The above derivation of the critical nucleus factor is simplified in that it neglects the rate of the reverse reaction, and the fact that a state of equilibrium does not exist with respect to incipient nuclei while macroscopic crystals are being formed. The average number of atomic clusters n_i per unit volume containing any specific number of atoms is given by the Boltzmann distribution (e.g. Turnbull, 1956):

$$n_i = N \exp\left[\frac{-\Delta G_i}{RT}\right], \tag{15}$$

where N is the number of atoms per unit volume and ΔG_i is the free energy required to form each cluster. A statistical-mechanical derivation of the rate would be based on the probability of individual clusters within such populations either increasing or decreasing their size. However, Equation (15) itself describes a state of equilibrium, and is no longer valid when clusters larger than the critical size are continually growing into macroscopic crystals.

Such considerations do not affect the exponential part of the rate equation, but do find expression in the pre-exponential part, along with the other factors mentioned above as included in K in Equation (1). A complete expression for nucleation rate includes Equation (1), with the value of ΔG_a from Equation (5), multiplied by the transport factor (Equation (13)) and the pre-exponential factor. The value of ΔG_c in terms of ΔT from Equation (10) can also be inserted. A widely-quoted equation is that of Turnbull and Fischer (1949; see also Turnbull, 1952; 1956), which is based on the theory of absolute reaction rates:

$$R_n = n^* N \frac{kT}{h}\left(\frac{\sigma}{kT}\right)^{1/2}\left(\frac{2V_a}{9\pi}\right)^{1/3}\exp\left[\frac{-16\pi\sigma^3}{3\Delta G_c^2 RT}\right]\exp\left[\frac{-\Delta H_a}{RT}\right], \tag{16}$$

where n^* is the number of atoms on the surface of a critical-size nucleus, N is the number of atoms of the precipitating substance per unit volume of liquid, k and h are Boltzmann's and Planck's constants, and V_a is the volume per atom of the precipitating substance. Nuclei are assumed to be spherical in this version of the equation. The proportionality of n^*

to σ and ΔG_c is obtained from Equation (4); thus the pre-exponential factor is dependent on $\sigma^{5/2}$, $T^{1/2}$, and ΔG_c^{-2} or $(\Delta S_t \Delta T)^{-2}$.

The application of Equation (16) to high degrees of supercooling is doubtful, because it was derived on the explicit assumption that $1/n \ll 1$, where n is the number of atoms in a critical nucleus; in reality, the size of the critical nucleus decreases with increasing supercooling. However, no other applicable expressions seem to be much better in this respect, and Equation (16) does show the qualitative features that are found in the laboratory, beginning with the experiments of Tamman (1898), and Tamman and Mehl (1925). These features include the following: an initial interval of ΔT over which the nucleation rate is small or negligible, called hereafter the critical ΔT of nucleation; a rapid rise of nucleation rate to a peak with increasing ΔT; and a usually slower decrease thereafter, as the exponential factor due to the critical nucleus approaches unity and the transport factor begins to control the rate. Increasing the value of surface energy increases the critical ΔT of nucleation and shifts the peak of the curve to higher ΔT; increasing ΔS_t has the opposite effect.

It is generally conceded that the essentially thermodynamic approach outlined above is not likely to lead to accurate expressions for nucleation rate, because macroscopic thermodynamic quantities like surface energy are not practicable when applied to the very small atomic or molecular units involved in nucleation (e.g. Dunning, 1969). Future advances in quantitative nucleation theory will probably depend on approaches through statistical mechanics. Although much has already been done in this line (e.g. Lothe and Pound, 1969), progress is slow, even for structurally simple substances. Application of statistical mechanics to silicate crystallization will require elucidation of the actual atomic structures of the melt and of crystal surfaces, and the dependence of these structures on temperature and pressure.

Heterogeneous Nucleation

The formation of nuclei on preexisting solid phases can greatly reduce the critical ΔT of nucleation because the surface energy associated with the nucleus is reduced: solid-solid surface energy is generally less than the solid-liquid surface energy, especially if there is epitaxy (e.g. Gebhardt, 1973; Matthews, 1975). Much has been written on heterogeneous nucleation (e.g. Chakraverty, 1973), but most of this literature pertains to growth from the vapor, in part because direct observation of the substrate is easier than in melt growth. No general theory of heterogeneous nucleation can be offered, since the reduction in surface energy is dependent not only on the identity of the substrate material, but also on its physical state (surface roughness, grain size, presence of dislocations, and the like).

The factors affecting the rate are the same as for homogeneous nucleation, however, and a gross picture of the difference between homogeneous and heterogeneous nucleation may be obtained simply by reducing the value of σ in Equation (16).

A complete absence of material that could serve as a substrate for heterogeneous nucleation in a melt is in practice unattainable, and it is generally considered that there is no such thing as completely homogeneous nucleation. It is especially difficult to minimize heterogeneous nucleation in the laboratory, primarily because of the large ratio of container surface area to sample volume. A liquid-vapor interface seems to be much less likely to promote heterogeneous nucleation than a liquid-solid interface, presumably because of much larger surface energies. For example, Murase and McBirney (1973) measured surface energies of 250–350 ergs cm^{-2} for typical silicate melts in air, whereas crystal-liquid surface energies for silicates tend to be in the neighborhood of 50 ergs cm^{-2} or less (see below). For nucleation on the liquid-gas interface, it is the crystal-gas surface energy which is pertinent, but there is no reason to believe that this would be appreciably less than the liquid-gas surface energy. The technique of suspending liquid-silicate beads in platinum-wire loops (Lofgren, this volume) is usually more successful in minimizing heterogeneous nucleation than are experiments in which the silicate is crystallized in capsules or crucibles. Presumably this is because the liquid-solid interface is largely eliminated in favor of a liquid-gas interface (see also Turnbull, 1952). The frequent failure of vapor bubbles to act as nucleation sites (e.g. Swanson, 1977) also testifies to the high energy of the gas interface.

In natural magmas, external surfaces are always much less important because of the greater volume of the liquid, and heterogeneous nucleation is not necessarily expected to be as completely dominant as it is in most laboratory experiments. Nevertheless, various "foreign" particles may be present, and later-forming crystals can always nucleate on early ones. Textural and other evidence suggest a considerable range in the degree of heterogeneity of nucleation in igneous rocks (see below and Lofgren, this volume).

INCUBATION TIME

The statistical Boltzmann distribution of sizes of atomic clusters in the liquid (Equation (15)) depends on temperature, with larger clusters being favored by a lower temperature. If a melt is equilibrated at a temperature above the liquidus and then quenched to an appreciable ΔT, it takes time for the new distribution of cluster sizes to become established. The new distribution will be steady-state rather than equilibrium if nuclei are being formed, but in any case the nucleation rate is depressed while this state is being attained. A number of different theoretical expressions for the incubation or induction period τ necessary for the nucleation rate to reach its steady-state value have been offered (e.g. Turnbull, 1948; Dunning, 1969; Walton, 1969), but most have the form

$$\tau = K \frac{n^2}{D}, \tag{17}$$

where n is the number of atoms in the critical nucleus and D is the transport factor (Equation (13)). The most important aspect of incubation time is its significance with respect to correlation of laboratory results with natural igneous rocks, since sudden temperature changes are much more common in the former. Lofgren (this volume) discusses the abundant evidence of the effects of incubation time in laboratory experiments.

CAVITATION

In rapidly crystallizing systems it is sometimes observed that introduction of a single seed into the supercooled liquid causes abundant nucleation on many centers not connected with the location of the original seed. This has been attributed to the phenomenon of cavitation, the most reasonable explanation for which is that the collapse of cavities produced by volume change on rapid crystallization generates a shock wave (e.g. Chalmers, 1964; Hunt and Jackson, 1966). The shock wave locally raises the pressure and the liquidus temperature, thereby increasing ΔT and causing nucleation. In some systems it is also possible to initiate nucleation by stirring, tapping the container or other mechanical agitation, which may have a similar effect.

CRYSTAL GROWTH THEORY

There are three principal processes in crystal growth, any of which may be the controlling factor in the overall growth rate; (1) interface kinetics—the movement of material across the interface and its attachment on the crystal surface; (2) material transport—the movement of material through the liquid; and (3) heat transfer—the removal of latent heat of crystallization from the interface. Which of these processes is the dominant factor depends on the relative diffusion and heat-transfer rates, the particular growth mechanism, and the supercooling. Generally, interface kinetics is dominant at low supercooling, while either material transport or heat transfer becomes more important at high supercooling. According to

Tiller (1963; 1977), the total supercooling or supersaturation can in principle be subdivided into contributions from each process, one of which is normally the controlling factor. The fundamental rate equation of Jackson (1969) is based on tracing the free-energy state of the system through all these processes. Only by considering *all* processes can the growth rate be predicted quantitatively and *a priori*; however, by identifying the dominant process or processes and making suitable approximations, one can obtain semi-empirical relationships that may be useful in describing the dependence of growth rate on ΔT. All the growth-rate equations developed below are in fact based on such simplifications.

INTERFACE TYPE AND GROWTH MECHANISM

The mechanisms of attachment of atoms onto the crystal surface are divided into two main types, continuous growth and layer growth. In continuous growth, atoms are able to attach themselves wherever they land; this type of growth is normally associated with a "rough" or nonplanar interface, which has a high concentration of unsatisfied bonds and therefore high surface energy. In layer or lateral growth, normally associated with a "smooth", planar, or singular interface, with a low concentration of unsatisfied bonds and low surface energy, atoms are not strongly localized on the surface, although once they land there, there may be small probability of their leaving again. Atoms may move rather freely over the surface (surface diffusion), and are usually able to attach themselves permanently only at the edge of layers or steps, where they are in effect bonded to two surfaces. In layer growth, there is usually some difficulty in starting a new layer, but once initiated, the layers can spread rapidly across the surface. Crystals with macroscopically flat, crystallographically rational faces have presumably grown by the layer mechanism, whereas smoothly rounded crystals more probably have grown by the continuous mechanism, but it should be realized that there is not necessarily a sharp distinction between the two mechanisms; insofar as it is possible to observe or model the actual microscopic surface structures of crystals, they appear to be neither perfectly smooth nor ideally rough.

The nature of the interface in simple compounds can be predicted by the theoretical model of Jackson (1958a), who proposed that the surface roughness is determined by the value of the parameter

$$\alpha = \frac{\Delta S_f}{R} \xi, \tag{18}$$

where ξ, the anisotropy factor, is the fraction of bond energy in a monolayer that attaches the atoms of the layer to each other, rather than to the preexisting crystal surface. Values of α less than about 2 should correspond to a rough interface, while values greater than 4 should correspond to a smooth interface. It would appear intuitively that this should also predict the growth mechanism, that is, continuous versus layer growth, and this seems to be borne out in practice (e.g. Kirkpatrick, 1975), but Jackson et al. (1967) warn that the connection is not justified theoretically.

Although Jackson's theory has been very successful for simple substances, its application to structurally complex silicates is not straightforward. In metals or simple molecular crystals, the entropy of melting per mole presumably corresponds to the entropy per actual growth particle, but the true growth units in igneous melts are unknown. There is certainly no justification for using the value of ΔS_f for the usual gramformula units, which are only mathematical conventions. Entropies of melting *per atom* for oxides and silicates are mostly in the range .4(SiO_2) to 3.6(MgO) cal mol^{-1} degree^{-1}. The anisotropy factor is not defined for growth layers more than one atom thick, since the model itself only considers one atomic layer, but when growth layers are greater than one atom thick, ξ is presumably at least qualitatively correlated with the attachment energy of Hartman (1973b) or the template fraction of Dowty (1976a). Both of these quantities are measures of the bond energy per layer, holding that layer to the crystal surface.

In multicomponent systems, ΔS_f should presumably be replaced by ΔS_t (Equation (7)), and therefore the probability of having a smooth interface is usually increased over that in the pure system. Small weight percentages of water can have a large effect on liquidus temperature and should be rather important in this respect. For example, cristobalite is not faceted and grows by a continuous mechanism from a pure anhydrous SiO_2 melt (Wagstaff, 1969), but quartz in rhyolites and especially waterrich "porphyries" is often euhedral. Quartz grown in the laboratory from water-rich granite and granodiorite melts is also faceted (Swanson, 1977).

The interface model of Temkin (1966; Bennema and Gilmer, 1973; Woodruff, 1973) is somewhat more comprehensive than that of Jackson, in that it considers more than one layer, but it generally seems to lead to similar predictions about surface structure and/or mechanism. This and other aspects of interface structure are reviewed by Woodruff (1973).

CONTINUOUS GROWTH

The only energy barrier, or activation energy, encountered when atoms are able to attach directly to the surface is that of transport across the interface, similar to ΔG_{tr} for nucleation (Equation (12)), although Turnbull (1950) points out that transport is not precisely the same for growth and nucleation. The rate of attachment, R_{att}, is then

$$R_{att} = K \exp\left[\frac{-\Delta G_{tr}}{RT}\right].\qquad(19)$$

Atoms detaching themselves from the surface must also overcome the bulk free energy difference between crystal and liquid, so the rate of detachment R_{det} is

$$R_{det} = K' \exp\left[\frac{-\Delta G_{tr}}{RT}\right]\exp\left[\frac{-\Delta G_c}{RT}\right].\qquad(20)$$

The net growth rate R_{cont} is the difference between (19) and (20):

$$R_{cont} = R_{att} - R_{det} = Fd_0 v \exp\left[\frac{-\Delta G_{tr}}{RT}\right]\left(1 - \exp\left[\frac{-\Delta G_c}{RT}\right]\right),\qquad(21)$$

here showing explicitly the factors in the pre-exponent, which include F, the volume fraction of the precipitating substance in the liquid; f, the fraction of sites available on the surface; d_0, the thickness of a single atomic layer; and v, the atomic vibrational frequency in the liquid, which is proportional to temperature. When the crystal consists of more than one type of atom, the values of f and d_0 would need to be averaged over the crystal structure. The value of f should approach unity for very rough interfaces.

LAYER GROWTH

Surface Nucleation

If each layer is initiated by a clustering of atoms at some random location on the surface, with the cluster forming a surface or two-dimensional nucleus, the energy barrier that must be overcome is similar to that in three-dimensional nucleation (Equation (2)). Layer growth normally takes place on rational crystallographic planes that have relatively low surface energy. If the surface nucleus is a plate parallel to the face, with height equal to the crystallographic lattice or d-spacing for this plane, the only contribution it makes to the total surface free energy is that of the side faces, because the upper surface is atomically identical to the area covered by the plate. There is no reason for the nucleus to have height greater than d, since this would only increase the surface energy, but if there are planes of low surface energy at spacings less than d, the height could be less. The latter is especially likely if there is pseudosymmetry, that is, pseudo-d-spacings at half or other fractions of the true d-spacing.

The derivation of the rate of formation of two-dimensional nuclei follows the same procedure as for three-dimensional nuclei. The first step is to find the Gibbs free energy of a critical nucleus, which is the activation energy ΔG_a. Assuming that the nucleus has the shape of a disc with height d and radius r, its contribution to the total free energy of the system is

$$\Delta G = d\pi r^2 \, \Delta G_c / V_m + d2\pi r\sigma, \tag{22}$$

neglecting edge and corner energies. Differentiating ΔG, with respect to r, we find the critical radius to be $-\sigma V_m/\Delta G_c$, which substituted back into Equation (22) gives

$$\Delta G_a = \frac{-d\pi\sigma^2 V_m}{\Delta G_c}. \tag{23}$$

The rate of formation of two-dimensional nuclei per unit area of the face in question is proportional to the exponent of this activation energy. The overall rate of growth depends on the size of the face in question relative to the undercooling. If surface nuclei form slowly, and if the face is not large, each layer spreads over the surface before a new one is initiated. This is known as the *small-crystal case*, and the overall growth rate R_{sc} is proportional to the rate of formation of two-dimensional nuclei, the area A of the face, a transport factor as in three dimensional nucleation, and the thickness d of the layer (Hillig, 1966; Flemings, 1974):

$$R_{sc} = KdA \exp\left[\frac{d\pi\sigma^2 V_m}{\Delta G_c RT}\right] \exp\left[\frac{-\Delta H_{tr}}{RT}\right]. \tag{24}$$

Here the pre-exponential factor K would be more like the pre-exponential in the equation for three-dimensional nucleation (16) than the pre-exponential in the equation for continuous growth (21); as written, K includes the entropy part of the transport factor (Equation (13)). The first exponent will have a negative sign because ΔG_c is negative.

As the face becomes larger or the rate of nucleus formation increases, the overall rate approaches that for the *large-crystal case*, which is more realistic for observable crystals (Calvert and Uhlmann, 1972). Here the number of nuclei or layers existing at any one time on the face is at a steady state, so that the area disappears from the expression, but the rate of lateral propagation of the layers, R_{lat}, becomes important. Calvert and Uhlmann (1972) give for the overall growth rate R_{lc} in the large-crystal case

$$R_{lc} = K(R_{sc}R_{lat}^2)^{1/3}. \tag{25}$$

The rate of lateral spreading R_{lat} is dependent on the rate of arrival of atoms from the liquid to the surface, the rate of their motion along the surface (surface diffusion), and the attachment kinetics on the edge of the layer itself. Atoms arriving at the edge of the layer may be rapidly attached, and the expression for the rate of this process should be similar to that for the rate of continuous growth (21). If we assume that the rates of transport through the liquid, across the interface, and parallel to the surface can be represented by one factor with activation enthalpy ΔH_{tr}, the lateral transport rate is

$$R_{lat} = Kd\left(1 - \exp\left[\frac{-\Delta G_c}{RT}\right]\right)\exp\left[\frac{-\Delta H_{tr}}{RT}\right], \qquad (26)$$

and the overall rate for the large-crystal case becomes

$$R_{lc} = Kd\exp\left[\frac{d\pi\sigma^2 V_m}{\Delta G_c RT}\right]^{1/3}\left(1 - \exp\left[\frac{-\Delta G_c}{RT}\right]\right)^{2/3}\exp\left[\frac{-\Delta H_{tr}}{RT}\right]. \qquad (27)$$

By this time, the pre-exponential factor K has become very complicated and contains elements from the expressions for continuous growth and three-dimensional nucleation. In this simplified expression the transport factor $\exp(-\Delta H_{tr}/RT)$ should be considered to represent the process (interface transport, surface diffusion, or volume diffusion in the liquid) that is the slowest, and therefore controls the rate.

The expressions for growth by surface nucleation predict a critical interval of ΔT over which the rate is insignificant, as in the expression for three dimensional nucleation. This interval is much smaller for two-dimensional than three-dimensional nucleation, a few degrees at most. Nevertheless, experiments usually show some growth in this interval. Cahn (1960; Cahn et al., 1964) suggested that a diffuseness of the interface could account for the observed non-zero growth rates at very small ΔT, and derived an equivalent for Equation (24) that includes an empirical diffuseness factor. Cahn's theory predicts that there should always be a transition from layer growth to continuous growth as supercooling is increased. Jackson et al. (1967) criticised this theory on the grounds (1) that it assumes that solidification is a first-order phase transition, whereas according to Jackson et al. it is second-order; (2) that actual measurements of growth rates as a function of ΔT do not show the expected change of mechanism; and (3) that there seems to be no way even in principle to evaluate or predict the diffuseness parameter independently of crystal growth measurements. On the other hand, Glicksman and Schaeffer (1967) reported a change in the morphology of crystalline phosphorus from faceted to non-faceted with increasing ΔT: this is predicted by the theory of Cahn but not by that of Jackson (1958a). Also, Nagel and Bergeron (1972) reported data on the growth of SrB_4O_7 that seemed to show the expected change of mechanism, although this result appears only after a rather dubious temperature correction (see below). Evidently Cahn's theory is flawed, but whether diffuseness is a useful concept remains to be decided. For some time, especially after the discovery of the screw dislocation mechanism (see below), many theorists tended to disbelieve in the existence of the surface nucleation mechanism because of the large discrepancies between predicted and observed rates at small ΔT. However, the overall rate data and observed morphologies discussed below demonstrate that surface nucleation is in fact a common mechanism (see also Flemings, 1974, p. 311). Some modification to the theory, such as Cahn's, may be necessary if the anomalously small critical ΔT's are actually observed for growth by this mechanism.

Twins

Surface nucleation becomes unnecessary, or is facilitated, if the crystal face abuts against another crystal with a favorable surface. A new layer can be easily initiated at the intersection if atoms bond to both the substrate and this second surface. An optimal situation of this type is a reentrant interfacial angle due to twinning, if the angle is acute enough. However, the atomic structure of the twin boundary must not be too unfavorable energetically. Twinned crystals are very often observed to grow faster, or to be larger than untwinned ones (e.g. feldspars: Becke, 1911; Smith, 1974, section 17.3). In some cases, such as a spinel twin in cubic crystals, the increased growth rate of the reentrant faces causes them to grow themselves out of existence, and more than one twin plane may be necessary for a self-perpetuating arrangement (Flemings, 1974, p. 318).

Screw Dislocations

The necessity for surface nucleation is removed if screw dislocations with axes non-parallel to the face are present (Frank, 1949; Burton et al., 1951). This leads to a self-perpetuating continuous growth layer in the form of a spiral. The growth rate is then inversely proportional to the spacing of steps in any radial direction, that is, the tightness of the spiral, and directly proportional to the lateral spreading rate R_{lat} (Equation (26)). The tightness of the spiral is limited by the fact that its innermost turn has a configuration very similar to that of a surface nucleus, and is likewise subject to instability (dissolution) if it becomes smaller than the critical size. According to Burton et al. (1951) the idealized spiral is Archimedean, with the equation in polar coordinates r (radius) and θ (angle)

$$r = 2\theta r_c,\tag{28}$$

where r_c is the radius of the critical two-dimensional nucleus. Real spirals are seen to have shapes reflecting the anisotropy of the growth directions of the crystal. Since the critical radius is $-\sigma V_m/\Delta G_c$ (from Equation (22)), and the average spacing of layers is equal to this value, the expression for overall growth rate R_{sd} has the form

$$R_{sd} = Kd \frac{-\Delta G_c}{\sigma \Delta V_m} R_{lat}.\tag{29}$$

The growth rate is not usually strongly dependent on the density of dislocations (Burton et al., 1951; Flemings, 1974).

DETERMINATION OF MECHANISM

The growth rates for each of the mechanisms have characteristic dependences on T and ΔT, so that in principle it is possible to determine the mechanism by measuring the growth rate at small ΔT where volume diffusion in the liquid is unimportant. For simple crystals forming from pure melts, for which the Stokes-Einstein relation approximately describes the transport factor, such methods can be extended to larger ΔT by computing the reduced growth rate R_r:

$$R_r = R_m \eta \left(1 - \exp\left[\frac{-\Delta H_f \Delta T}{RT T_f}\right] \right)^{-1}, \tag{30}$$

where R_m is the measured growth rate (e.g. Klein and Uhlmann, 1974; Kirkpatrick, 1975). The reduced growth rate is considered to be a measure of the fraction of available sites on the crystal surface (f in Equation (21)). A plot with R_r vertical and T horizontal should give a horizontal line for continuous growth, a straight line of positive slope passing through the origin for screw dislocation growth, and a curve of positive, increasing slope for surface nucleation growth. Also, for surface nucleation, a plot of log R_r versus $1/(T\Delta T)$ should give a straight line of negative slope.

This method of identifying the growth mechanism has been extensively used recently (see Kirkpatrick, 1975 for further review) and seems to give reasonable results, but the plots are sometimes ambiguous. For example, for sodium disilicate (Meiling and Uhlmann, 1967), the R_r versus ΔT plot is not a perfectly straight line. Often the log R_r versus $1/(T\Delta T)$ plot gives curved rather than straight lines (Klein and Uhlmann, 1974), even when surface nucleation is otherwise indicated. The latter behavior has been reproduced in numerical models of crystallization (Gilmer and Bennema, 1972) and has been attributed to a change in surface energy with ΔT. This again suggests inadequacies in the equations for surface-nucleation growth.

DIFFUSION AND HEAT FLOW

Diffusion

Several types of material transport in crystal growth can be described by the usual equations for diffusion (e.g. Fick's laws), with diffusion coefficients such as those in Equations (12) and (13). These include transport within the liquid (volume diffusion); across the interface (interface diffusion); in layer growth, parallel to the interface (surface diffusion), and from the surface onto the edge of a layer. A rigorous solution to the problem of the influence of diffusion involves setting up differential equations to describe the flux across the various surfaces or other boundaries in the system. Theoretical solutions have been derived for certain limiting conditions, especially for growth by screw dislocation (Burton et al., 1951; Bennema and Gilmer, 1973; Ohara and Reid, 1973), but again the requisite diffusion parameters for interface and surface diffusion are difficult if not impossible to obtain independently of crystal growth measurements.

At small supercooling, the growth rate in the continuous mechanism is limited by the detachment rate. In the layer mechanism, growth at small supercooling may be limited by formation of two-dimensional nuclei or by surface diffusion. At larger supercooling in one-component systems, the rate-controlling factor in either continuous or layer growth becomes the transport of material across the crystal-liquid interface. At high supercooling in multicomponent systems, there is usually a transition to control of the overall rate by volume diffusion in the liquid: this has important effects on many growth properties. The transition incidentally does not necessarily occur at the peak of the growth rate curve; effects due to volume diffusion may often be seen at lower supercooling.

When volume diffusion is slow relative to the rate of advance of the interface, concentration gradients are established in the liquid. In growth of a planar or smoothly curved interface, this should cause a reduction in growth rate with time, because the liquid in the vicinity of the interface is continuously being depleted in the crystallizing component. The growth rate R_g at a constant ΔT (Christian, 1965) is

$$R_g = K(D/t)^{1/2}, \tag{31}$$

where D is the diffusivity of the slowest-moving component in the liquid and t is time. However, both theory and experiment demonstrate that a single interface may be unstable under these circumstances. Because the concentration of the crystallizing component increases away from the interface, as shown in Figure 2, the supercooling also increases, producing a condition known as constitutional supercooling. In the absence of constitutional supercooling, crystals form with normal or compact morphologies, that is, without re-entrants or gaps in their outermost surfaces. The breakdown of such solid interfaces is called morphological instability (Chalmers, 1964; Cahn, 1967; Sekerka, 1973; Woodruff, 1973), and may result from constitutional supercooling in the following way: any protrusion of part of the interface ahead of the rest places it in a region of the liquid with greater supercooling and allows it to grow faster. Planar interfaces are thus kinetically unstable with respect to various non-compact morphologies such as cellular interfaces, hopper and skeletal crystals, dendrites and spherulites (for illustrations of these various morphologies, see the references above and also Flemings, 1974, and Lofgren, this volume). The various types of protrusion in these morphologies tend to bypass and isolate areas of the liquid that are depleted in the crystallizing component, and thus the rate of advance of the outermost extremities attains a steady state, rather than decreasing with time. Since any concentration gradient produces relative instability of a planar interface, formation of compact morphology in multicomponent systems evidently demonstrates that the rate in these cases is at least partly determined by interface processes rather than volume diffusion. In other words, surface attachment kinetics stabilizes a planar interface (see references above or Kirkpatrick, 1975).

Heat Removal

In pure metals, interface attachment kinetics is so rapid that growth rates are often limited by removal of latent heat from the interface. The effects of latent-heat buildup at the interface are very similar to those of volume diffusion: the gradient of decreasing temperature away from the interface may cause time-dependent growth rates or morphological instability. Because liquid silicates have much higher viscosities than liquid metals, it has generally been assumed that thermal diffusion in liquid silicates is much faster than chemical diffusion, so that heat transfer plays no significant role. In some recent experiments, the temperature of the growth interface was actually measured as it advanced over a thermocouple. For sodium disilicate (Hopper and Uhlmann, 1973) and anorthite (Klein and Uhlmann, 1974), the temperature increase over the bulk liquid was insignificant. However, for PbB_4O_7 (Laird and Bergeron, 1968), a temperature increase of $23°$ was measured at peak growth rate. These experiments were in pure or nearly pure systems in which the role of volume diffusion was at a minimum; in multicomponent systems the role of diffusion could only be greater, and any temperature difference at the interface would presumably be decreased. Nevertheless, it would not be wise to rule out effects of heat transfer for rapid crystallization of minerals with large ΔH_f (e.g. olivine).

ADSORPTION

Atoms can adhere to crystal surfaces without necessarily being permanently bonded to structural sites: this phenomenon is generally known as adsorption. Because the structure on the interface may be considerably different from that in the interior of the crystal, foreign atoms or even large molecules may be adsorbed on the surface and may interfere with the uptake of the crystal constituents, thus slowing or poisoning growth (Buckley, 1951; CNRS, 1965). It may be that only a very low concentration of such poisons in the liquid is necessary to slow growth significantly, because some of the poisons are very selectively adsorbed onto the surfaces. The effects of growth poisons and other adsorption phenomena are well documented for growth of minerals from aqueous solution, an especially important example being the effect of magnesium on calcite growth (Berner, 1975; Folk, 1974). Compositional effects such as sector-zoning, discussed below (also Lofgren, this volume), demonstrates that adsorption also occurs in igneous rocks, but its effects on growth rates in these rocks are entirely unknown.

SHAPES OF CRYSTALS

The factors determining shapes of crystals are very different for the low-supercooling regime, in which interface processes control the growth rate, as opposed to the high-supercooling regime, in which volume diffusion (or heat transfer) in the liquid controls the rate.

INTERFACE-CONTROLLED GROWTH

Growth by the continuous mechanism is expected to be reasonably isotropic and crystals are generally rounded. Growth by any of the layer mechanisms results in faceted crystals, and the growth rates on the different faces may be very disparate. If the values of the anisotropy factor ξ in Jackson's model are very different for different directions, a crystal may grow partly by the continuous mechanism and partly by layer mechanisms.

For faceted crystals, the crystal structure is an important determinant of morphology, i.e., relative development of faces. The factors which have been considered to be important are the lattice geometry and other translational symmetry elements, and the surface energies of the individual faces. According to the "Law of Bravais," as extended by Donnay and Harker (1937), the morphological prominence of a face (area on a given crystal or frequency of occurrence on many crystals) should be directly dependent on the lattice or d-spacing perpendicular to it. The area of a face is inversely related to its growth velocity. Donnay and Donnay (1961) showed that in some cases one must consider the pseudo-symmetry of the crystal, which can result in reduction of effective d-spacings. The influence of crystal geometry alone was extended yet further by Schneer (1970, 1978) and McLachlan (1974), who proposed that the d-spacing should be weighted by the X-ray structure factor of each face, which is a measure of the extent to which the atoms are arranged in well-defined planes parallel to the face in question. On the other hand, Hartman and Perdok (1955; Hartman, 1973b) and Dowty (1976a) proposed weighting the *d*-spacings by the inverse of the surface energy of the particular face. The two weighting schemes are not incompatible; if the atoms are arranged in well-defined planes, it is to be expected that bond energy will be concentrated within planes, rather than across them, and the face in question should have low surface energy.

Each of the equations for growth rate by layer mechanisms (24, 27, 29) shows a different dependence on *d*-spacing and surface energy. In the equations for surface nucleation (24, 27), the layer or interplanar *d*-spacing appears in both the exponent and pre-exponent. Assuming reasonable values of the parameters, i.e., $\sigma = 20 \text{ erg cm}^{-2}$, $\Delta S = 20 \text{ cal mol}^{-1} \text{ deg}^{-1}$, $V_m = 100 \text{ cm}^3 \text{ mol}^{-1}$, and $T = 1,000°\text{K}$, and neglecting the variation of area of the face, Equation (24) becomes

$$R_{sc} = Kd \exp\left[\frac{-10.88d}{\Delta T}\right] \tag{32}$$

for *d* in angstroms. The equation for large-crystal growth (27) is similar, but *K* for large-crystal growth is different and the factor in the exponent is divided by three. Thus, if ΔT is relatively small, that is, much less than about 3 to 10 degrees for the above parameters, the exponential factor is very sensitive to differences in *d*, with a large *d* giving a small growth rate. This sensitivity decreases as ΔT increases, and the occurrence of *d* in the pre-exponent, which gives variation in the opposite direction, becomes more important. This shows the basis for the Bravais-Donnay-Harker law, and also how this law may break down at large supercoolings.

The surface energy σ that appears in Equations (24) and (27) is not that of the face on which growth is taking place, but that of the side faces of a plate nucleus. A relatively low surface energy of the face itself implies that most of the strong bonds in the structure are parallel rather than perpendicular to the surface, and that the surface energies of faces nearly perpendicular to the one on which growth is occurring are high. Thus the postulate of Hartman and Perdok (1955) and Dowty (1976a) about the role of surface energy is qualitatively consistent with the growth rate equations, but the actual dependence is seen to be more complex than the simple linear function assumed by these authors.

For screw-dislocation growth, (Equation (29)), the surface energy and lattice spacing appear only in the pre-exponent, and the predicted variation of growth rate with *d* is opposite to that given by the Bravais-Donnay-Harker theory, although the dependence on σ is as predicted by Hartman and Perdok (as pointed out previously by Bennema and Gilmer, 1973).

In principle, then, the growth mechanism can be determined from the morphology, at least for growth at low ΔT. The lattice spacings are well known, and relative (not absolute) surface energies can be predicted by the methods of Hartman and Perdok (1955) or Dowty (1976a). For example, the morphology of olivine is predicted very well by the Bravais-Donnay-Harker theory, and still more closely approximated when surface energies and stereochemistry are taken into account (Fleet, 1975; 'T Hart, 1978). This implies that olivine grows by the surface-nucleation mechanism. Conversely, if the mechanism is known to be surface nucleation, information about ΔT of growth may be obtained. Of course, predictions of morphology from the crystal structure alone cannot always be accurate, because the nature of the growth medium also has an effect, through pre-structuring of growth units and face-selective poisoning. The mechanism may also be different for different faces, since the anisotropy factor and the probability of suitably oriented screw dislocations vary with direction.

The growth rates for the different faces determine the relative perpendicular distances from each face to the center of the crystal. According to the theorem of Wulff (1901; see also Herring, 1951), if these central distances are proportional to surface energies, the crystal has the equilibrium shape. For the surface-nucleation mechanism, it is obvious that there is no such simple dependence of growth rate on surface energy, and thus, even for very slow growth, the crystal should not have the equilibrium shape. However, for the screw-dislocation mechanism, there is a simple direct dependence of rate on surface energy, if it is assumed that the surface energy on the main face is inversely related to that on the side faces of layers. Equation (29) also contains d, the layer height, but it appears that the rate of lateral spreading, which also appears in the equation, should tend to be inversely proportional to layer height. Thus crystals grown entirely by the screw-dislocation mechanism may have at least approximately the equilibrium shape.

DIFFUSION- OR HEAT-CONTROLLED GROWTH

As noted above, when the overall growth rate is controlled either by volume diffusion in the liquid or removal of latent heat from the interface, a planar interface becomes unstable and various non-compact morphologies develop. Any protuberances are able to grow faster than surrounding areas. In effect, the edges and corners of a crystal growing by the layer mechanism form natural protuberances; since a point on the edge or corner receives material (or radiates heat away) over a larger solid angle than a point on a face, the contours of concentration or temperature are closer to the interface. Thus, as ΔT is increased and diffusion or heat transfer begins to become important, the first unstable morphologies developed are generally various types of hopper or skeletal crystals. As dendritic morphologies develop on increasing supercooling, the dendrites often originate from corners or edges. Crystals growing by the continuous mechanism generally do not have sharp corners or edges; they often develop, as the first sign of instability, a cellular morphology (Tiller, 1963), which is a sort of inverse honeycomb structure. Large planar faces may also develop this morphology. With greater supercooling, the cells may become dendrites. Dendrites sometimes retain a constant crystallographic orientation, but misorientations due to twinning, dislocations, and other factors will become favored if they result in a fast-growing direction being oriented toward the non-depleted or cooler liquid. Volume changes on crystallization may also cause pressure gradients that could lead to misorientations (Tiller, 1977). When such misorientation is extreme, the crystallites form a radiating aggregate, or spherulite (Keith and Padden 1963). In some cases, for example in chalcedony or fibrous quartz grown from aqueous solution (Frondel, 1978), the fibers are elongated in the direction of the axes of screw dislocations, which are favored growth directions. If the crystal normally grows by a layer machanism, the dendrites also show planar faces (for example Cahn, 1967), but the faces developed are not necessarily those present on a compact crystal, because there may be changes in relative growth rate with increasing ΔT (above).

Many igneous minerals show a sequence of non-compact morphologies with increasing ΔT in the approximate order discussed above, i.e., compact to skeletal to dendritic to spherulitic (e.g. Lofgren, 1974; Donaldson, 1976; Fenn, 1977), although not every morphological type is observed for each mineral. These morphologies can be very useful in estimating ΔT and/or cooling rate of igneous rocks, and the experimental data are discussed by Lofgren (this volume).

The spacing of instabilities such as dendrite arms is a function of diffusion and growth rates, and can be used quantitatively if the relation between ΔT and growth rate is known. Flemings (1974, p. 148) gives the following general empirical expression for S, the spacing of the instabilities;

$$S = a(R_g)^{-b} \tag{33}$$

where R_g is the rate of growth, a is a constant depending primarily on diffusion which must be determined for each system, and b is normally in the range $\frac{1}{3}$ to $\frac{1}{2}$ but must likewise be determined for each system. The parameters a and b are dependent on the composition and temperature, and extrapolation from one magma to another, or even from one stage of crystallization to another, is unwarranted. Some attempts have been made to determine spacings of instabilities theoretically (e.g. Kotler and

Tiller, 1968; Hardy and Coriell, 1969), but while the results are qualitatively consistent with observations, no useful quantitative predictions can be made for the general case.

COMPOSITIONS OF CRYSTALS

Even under conditions in which there is no significant reaction between liquid and already-formed crystals (perfect fractionation), the composition of crystals may be influenced to varying degrees by kinetic factors. Consider isothermal crystallization in the plagioclase system. At some temperature T below the liquidus (Figure 3A) the composition C_1 is unstable. To attain equilibrium, it would break down to liquid composition C_2 and solid composition C_3. If volume diffusion in the liquid is slow relative to the rate of growth, the equilibrium liquid composition C_2 can in fact be attained or nearly attained at the interface, while there is a gradient in liquid composition from C_2 to C_1 away from the interface (Figure 2). For this case, in which surface equilibrium is nearly maintained, the composition of the crystal C_3 is near that of equilibrium for the temperature T, and is more albitic than the composition of true equilibrium for the liquid C_1.

There is no reason, however, to assume *a priori* that a state of surface equilibrium exists; in fact, there must be some disequilibrium or supercooling across the interface, or there would be no growth. Figure 3B shows

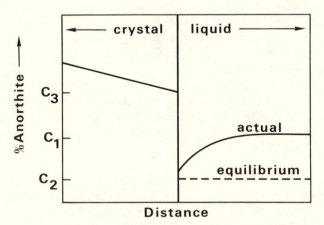

Figure 2. Composition in the plagioclase system near the crystal-liquid interface. The bulk liquid composition C_1, and the equilibrium liquid and solid compositions C_2 and C_3, are the same as in Figure 3. As shown, interface equilibrium is nearly attained, that is, the liquid composition at the interface is close to C_2, in which case the solid composition must be close to C_3.

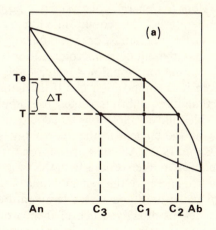

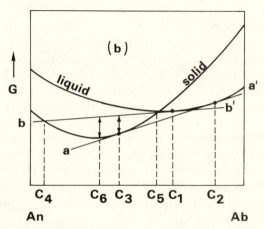

Figure 3. (a) Temperature-composition diagram for the plagioclase system. C_1 is the actual liquid composition, T_l is the equilibrium liquidus temperature and C_2 and C_3 are the equilibrium liquid and solid compositions at the actual temperature T. (b) Gibbs-free-energy/composition diagram for the plagioclase system at temperature T. The tangent to both the solid and liquid curves, a-a′, defines the equilibrium compositions C_2 and C_3. The tangent b-b′ to the liquid curve at the actual composition C_1 defines the solid compositions whose formation could reduce the free energy of the system: such compositions lie between the two intersections C_4 and C_5 of the tangent b-b′ with the curve for the solid. The lengths of the vertical lines with arrows between the curve for the solid and the tangent give the free energy per mole gained when solids of these compositions form from the liquid. This energy is larger for the solid composition C_6 than the equilibrium composition C_3.

schematically the Gibbs free energies of crystal and liquid as a function of composition at temperature T. The equilibrium liquid and solid compositions are found on such a diagram by drawing the tangent to both curves. Thus, if the liquid has the equilibrium composition C_2, the only solid composition whose formation would not increase the free energy of the system is the equilibrium composition C_3. If the liquid at the interface is more anorthitic than its equilibrium value, for example C_1, the thermodynamically allowable solid compositions that could form from this liquid are found by drawing the tangent to the free energy curve of the liquid at C_1 (Baker and Cahn, 1971). All compositions on the free energy curve of the solid between its two intersections with this tangent, that is, C_4 and C_5, can reduce the free energy of the system. None of these compositions are in equilibrium with the liquid C_1 at temperature T, and thermodynamics does not tell us directly which of them would actually form. However, the free energy gained per mole on formation of any of the solids is given by the vertical distance between the free energy curve of the solid and the tangent to C_1. If the topology of the free-energy-composition diagram is as shown in Figure 3B, the most favored composition on the basis of free energy gain is definitely not the equilibrium solid composition C_3, but is more anorthitic, C_6. The experimental results of Lofgren (1974), who crystallized plagioclase at substantial ΔT and found compositions always more anorthitic than the equilibrium value for that temperature, are consistent with this analysis and the assumed topology. A possible alternate explanation and/or contributing factor to the experimental results is that the heat of crystallization raised the temperature at the interface, causing more anorthitic plagioclase to form. However, Klein and Uhlmann (1974) actually measured the temperature at the interface during the crystallization of pure anorthite and found no significant increase over the bulk temperature.

Jackson (1958b), using a different approach, predicted that in an ideal system (no heat of mixing) the partition coefficients of elements between crystals and liquid at the interface should remain fairly close to the values for equilibrium. Thus liquid composition C_1 should crystallize compositions close to the equilibrium solid composition for T_e, whatever the actual temperature. This also is consistent with Lofgren's results, if the actual conditions were in between surface equilibrium and complete disequilibrium (homogeneous liquid). Jackson's analysis only applies to ideal systems, however, and the magnitude of possible deviations in non-ideal systems was not predicted.

A rigorous determination of the composition that should crystallize from an unstable liquid is more complex than simply determining the maximum bulk free-energy gain, since differential volume-diffusion rates and other kinetic factors must also play a role. The techniques of non-equilibrium or irreversible thermodynamics (e.g. de Groot and Mazur, 1962; Denbigh, 1958; Prigogine, 1961) are in principle applicable to this problem, but rather arbitrary assumptions about the coupling between rates and thermodynamic driving forces are usually necessary. For example, rates are usually assumed to be linear with ΔG or other thermodynamic parameters, but not all the growth-rate equations are linear in ΔG_c (although some approach linearity at low ΔT). Baker and Cahn (1971) reviewed several treatments of the problem based on irreversible thermodynamics and found them all inadequate for reasons such as this nonlinearity, and also because some of the theoretical results simply disagreed with experiment.

In principle, the free energy difference ΔG_c appearing in the nucleation and growth-rate equations should be that given by the vertical distance between the free energy of the actual solid crystallizing and the tangent to the liquid composition (Carmichael et al., 1974); however, because the free-energy/composition relations are in fact unknown, this is not a practicable method.

For minor and trace elements, it is conventional to consider crystal/liquid partition coefficients to be constant at a given temperature. This is probably valid if surface equilibrium is maintained, but the establishment of diffusion gradients in the liquid causes the composition at the interface to differ from that in the bulk, with effects corresponding to those on major elements described above (Burton et al., 1953; Albarede and Bottinga, 1972). That is, if the element in question is partitioned into the liquid, its concentration in the crystal is greater than would be predicted from the bulk liquid composition, and if it is partitioned into the crystal, its concentration is less. If interface equilibrium does not hold, then Jackson's (1958b) conclusion that partition coefficients remain close to the values for the equilibrium temperature implies that the coefficients at a given actual temperature of crystallization are a function of composition.

The occurrence of adsorption adds further complications. In addition to slowing the growth in some cases, adsorbed foreign atoms may be incorporated into the crystal. The occurrence of sector zoning in igneous minerals (e.g. Hollister and Gancarz, 1971; Nakamura, 1973; Dowty, 1976b; Kitamura and Sunagawa, 1977), seems to demonstrate that adsorption has some effect, and at least the majority of sector-zoning models assume the importance of adsorption. If surface equilibrium can be assumed, incorporation of adsorbed elements can be modeled by postulating that the partitioning of elements between the crystal interior and the crystal surface, and between the crystal surface and the liquid, are described by separate coefficients (e.g. Kröger, 1973, p. 11; Brice, 1973,

p. 125). Of course both these coefficients must be determined empirically, and the diffusion properties of the liquid must be known. Some minerals, such as pyroxene, seem to be susceptible to adsorption effects, while others, such as feldspar, are less so (Dowty, 1978).

OSCILLATORY ZONING

Most hypotheses to explain oscillatory zoning in plagioclase and other minerals have involved either cyclic changes in external conditions, such as convection, degassing, and the like, or an interaction between crystal growth and diffusion (Smith, 1974). Many variations on the latter idea have been offered since the paper by Harloff (1927), but all seem either to be inconsistent with crystal-growth theory and observations, or are too vague for meaningful tests. A recent example is the paper by Bottinga et al. (1966), which proposes that the growth mechanism varies with supercooling, as predicted for crystals in general by Cahn (1960). According to this hypothesis, the rapid growth rate of a crystal by the continuous mechanism at relatively high supercooling causes a diffusion gradient to become established in the liquid. This reduces the supercooling at the interface, so that the mechanism changes to surface nucleation. At the slow growth rate of this mechanism, diffusion in the liquid is able to eradicate the concentration gradient, causing the supercooling to increase again. The mechanism changes back to continuous growth and a new cycle begins.

Unfortunately there is considerable uncertainty about Cahn's theory, and in any case there is no evidence either from measured growth rates as a function of ΔT or observed morphology that plagioclase ever grows by the continuous mechanism (see below). As mentioned above, the theory of growth by the surface nucleation mechanism at small supercooling may require modification. Perhaps it will be profitable to take up the question of the interaction of growth and diffusion again when such modifications have been made.

THE DATA ON RATES AND OTHER PARAMETERS

THEORETICAL PREDICTIONS AND INDEPENDENT MEASUREMENT OF PARAMETERS

The parameters that most affect the absolute rates of nucleation and growth and the shape of the curves of rate versus supercooling include the entropy of melting, the surface energy, and diffusion rates.

The most striking aspect of the value of surface energy σ is its effect on the critical interval of supercooling over which there is negligible nucleation; in fact, most estimates of surface energies have come from observation of this interval in homogeneous nucleation experiments (Woodruff, 1973). Turnbull (1950) found that the ratio of surface energy per atom measured in this way to the heat of fusion per atom was almost constant—about .3 to .5 for a large number of metals. The range of surface energies measured was large, from 10 to about 300 ergs cm^{-2}. However, all these metals have similar, very simple structures, and some more complicated materials deviated from the relationship. Turnbull himself suggested that the relation between heat of fusion and surface energy should be dependent on structural complexity, so there is little security in extrapolating his numerical results to silicates.

Kotze and Kuhlmann-Wilsdorf (1966) proposed a theoretical relation for surface energy, based on a dislocation model of melting. In its simplest form, this relation is

$$\sigma = .0258Gb, \tag{34}$$

where G is the modulus of rigidity at the melting point and b is the Burgers vector of a dislocation, assumed in this case to be the interatomic distance in the melt. This purely theoretical equation predicted the values measured by Turnbull (1950) remarkably well and deserves testing on more complex material. Unfortunately, the requisite data on minerals, in this case elastic constants near the melting point, are again almost entirely lacking. Such relations probably only apply in any case to the interface between a crystal and its own melt; evidence discussed below suggests that surface energy can be a sensitive function of liquid composition.

The heats of fusion of oxides and silicates are only poorly known. For example, there are no thoroughly reliable values for anorthite, forsterite or iron-rich pyroxenes. In standard compilations of thermodynamic data (Kubachewski et al., 1967; JANAF, 1971, 1974, 1975; Barin and Knacke, 1973; Barin et al., 1977; Robie et al., 1978) the heat of fusion entries for simple oxides are often estimates based on structural or chemical similarity with other compounds. In assessing such reliable measurements as do exist, I have found that there is a strong negative correlation between the entropy of fusion of simple oxides of metals with a valence of four or less, per mole of oxygen atoms, and $z^2 s^{-1} O^{-1}$, the square of the formal charge of the cation divided by the cation-oxygen distance, per oxygen atom. It also appears that the entropies of fusion of complex oxides and silicates can be approximated by a sum of the entropies of fusion of the component oxides, estimated from the z^2/s relationship. A correction for entropy of mixing seems not to be warranted for most compounds. Least-squares analysis of about 30 compounds for which reliable data are available led to the predictive equation for the entropy of fusion ΔS per oxygen atom in cal deg^{-1} mol^{-1} of oxygen atoms:

$$\Delta S/O = 8.74 - 1.57(z^2 \ s^{-1} \ O^{-1}), \tag{35}$$

where $z^2 \ s^{-1} \ O^{-1}$ is an average value for all cations with s in angstroms. Standard error of the estimate is .6. Estimation of entropies of fusion by this method is probably sufficiently accurate for purposes of computing ΔG_c in the nucleation and growth-rate equations, considering the other uncertainties involved (for example, in σ).

Although the rates of interface transport and surface diffusion can be obtained only from crystal-growth experiments, the rates of volume diffusion in the liquid can be measured independently, and the available data are reviewed by Hofmann (this volume). Unfortunately the diffusivities of silicon and aluminum, which are expected to be the slowest-moving ions (Dowty, 1977; Hofmann, this volume), and probably control the overall transport process in many cases, have not yet been measured in any igneous liquid.

Crystal Growth and Nucleation Data for Minerals

The available data on rates of nucleation and growth of igneous minerals are reviewed in this section, not only for their own importance, but for the information they give on parameters such as surface energies and transport factors, and on nucleation and growth mechanisms.

The pioneering work of Tammann (1898) on organic and other systems having low melting points inspired corresponding measurements on silicates. These early works are summarized by Kittl (1912) and Eitel (1954). Much of this work was qualitative, but Kittl (1912) used a microscope heating stage to measure growth rates as a function of ΔT for a number of minerals (Table 1a). Most of the samples were natural minerals, although diopside and nepheline were synthetic. The samples were enclosed in silica glass, so that even the synthetic minerals probably crystallized from multicomponent liquids, as is suggested by the reduced melting points of diopside and nepheline. In all cases the crystals appeared to be euhedral, although very often acicular. Schumoff-Deleano and Dittler (1911) determined relative nucleation densities for several minerals as a function of ΔT (Table 1b). Again, the minerals were natural specimens and the systems must be considered to have been multi-component.

Winkler (1947) made measurements at six temperatures on the nucleation and growth of nepheline from a melt which contained over 10% LiF as a flux (Tables 1a, 1b). Unless this composition happens to be a eutectic (in which case LiF would have precipitated), there must have been some variation in liquidus temperature as crystallization proceeded,

Table 1a. Measured Growth Rates of Minerals and Related Substances.

Crystal	Melt Composition* wt %	Liquidus Temp. K	ΔT at Peak	Rate at Peak, $cm\ sec^{-1}$	Reference
Cristobalite	—	1996	50	2×10^{-6}	(a)
Anorthite	—	1830	300	1.5×10^{-2}	(b)
Anorthite	—	1830	>200	1.5×10^{-2}	(c)
Diopside	—	1664	>100	2.2×10^{-2}	(c)
$Na_2Si_2O_5$	—	1147	60	1×10^{-3}	(d)
GeO_2	—	1389	100	1×10^{-5}	(e)
SrB_4O_7	—	1270	100	1.6×10^{-2}	(f)
PbB_4O_7	—	1048	120	2×10^{-4}	(g)
Olivine	—**	1653	20	2.5×10^{-5}	
Wollastonite	—	1593	20	3×10^{-5}	
Bronzite	—	1663	30	1.2×10^{-4}	
Hypersthene	—	1523	20	1.2×10^{-4}	(h)
Diopside	—	1553	30–50	9×10^{-5}	
Spodumene	—	1618	20	2×10^{-5}	
Augite	—	1583	10–20	4×10^{-5}	
Leucite	—	1603	20	1.5×10^{-5}	
Nepheline	—	1150	30	1×10^{-5}	
Nepheline	Neph + 10% LiF	1110	20	3.6×10^{-4}	(i)
Plagioclase	$An_{66}Ab_{34}$	1760	60	1×10^{-5}	
Plagioclase	$An_{48}Ab_{52}$	1710	50	1×10^{-5}	(j)
Plagioclase	$An_{40}Ab_{60}$	1690	40	1×10^{-5}	
Diopside[#]	$Di_{60}Ab_{40}$	1578	95	6.7×10^{-4}	
Diopside[#]	$Di_{50}Ab_{50}$	1558	115	5.2×10^{-4}	
Diopside[#]	$Di_{40}Ab_{60}$	1533	111	1.5×10^{-4}	
Diopside[#]	$Di_{52}An_{28}Ab_{20}$	1553	60	3.0×10^{-4}	
Diopside[#]	$Di_{60}An_{40}$	1553	100	1.4×10^{-4}	(k)
Plagioclase[#]	$Di_{30}An_{50}Ab_{20}$	1633	140	3.7×10^{-4}	
Plagioclase[#]	$Di_{30}An_{40}Ab_{20}$	1593	130	2.0×10^{-4}	
Plagioclase[#]	$Di_{40}An_{40}Ab_{20}$	1573	150	$.9 \times 10^{-4}$	
Alkali Fs	$Ab_{90}Or_{10}$ + 2.7% H_2O	1218	200	5.0×10^{-6}	
Alkali Fs	$Ab_{90}Or_{10}$ + 5.4% H_2O	1113	150	5.2×10^{-6}	
Alkali Fs	$Ab_{90}Or_{10}$ + 9.0% H_2O	1048	75	2.5×10^{-6}	
Alkali Fs	$Ab_{70}Or_{30}$ + 1.7% H_2O	1243	160	3.2×10^{-6}	(l)
Alkali Fs	$Ab_{70}Or_{30}$ + 4.3% H_2O	1113	130	3.2×10^{-6}	
Alkali Fs	$Ab_{70}Or_{30}$ + 9.5% H_2O	1038	75	2.4×10^{-6}	
Alkali Fs	$Ab_{50}Or_{50}$ + 2.8% H_2O	1228	220	6.3×10^{-6}	
Alkali Fs	$Ab_{50}Or_{50}$ + 6.2% H_2O	1088	110	5.7×10^{-6}	

Table 1a. Measured Growth Rates of Minerals and Related Substances.

Crystal	Melt Composition* wt %	Liquidus Temp. K	ΔT at Peak	Rate at Peak, cm sec^{-1}	Reference
Quartz	Gr$^\infty$ + 3.5% H$_2$O	1143	150	10^{-8}	
Quartz	Gd$^\infty$ + 6.5% H$_2$O	963	100	10^{-7}	
Quartz	Gd + 12% H$_2$O	943	200	10^{-8}	
Alkali Fs	Gr + 3.5% H$_2$O	1163	200	10^{-7}	
Alkali Fs	Gd + 6.5% H$_2$O	953	100	10^{-7}	(m)
Alkali Fs	Gd + 12% H$_2$O	923	200	10^{-8}	
Plagioclase	Gr + 3.5% H$_2$O	1233	150	10^{-6}	
Plagioclase	Gd + 6.5% H$_2$O	1213	180	10^{-6}	
Plagioclase	Gd + 12% H$_2$O	1173	300	10^{-8}	

* Pure liquid if not specified.

** Compositions of crystals in Kittl's experiments were not usually specified, except for synthetic diopside CaMgSi$_2$O$_6$ and nepheline NaAlSiO$_4$. Most experiments were done in silica-glass containers, and some admixture of SiO$_2$ in the liquid is suggested by reduced melting temperatures.

Only the crystal which is first on the liquidus in Leonteva's (1948) experiments is given here.

∞ Gr = granite, Gd = granodiorite. See Swanson (1977) for compositions.

References
  (a) Wagstaff (1969)
  (b) Klein and Uhlmann (1974)
  (c) Kirkpatrick et al. (1976)
  (d) Meiling and Uhlmann (1967)
  (e) Vergano and Uhlmann (1970a)
  (f) Nagel and Bergeron (1972)
  (g) DeLuca et al. (1969)
  (h) Kittl (1912)
  (i) Winkler (1947)
  (j) Leonteva (1949)
  (k) Leonteva (1948)
  (l) Fenn (1977)
  (m) Swanson (1977)

but such effects will be neglected in the following discussion. The rates Winkler obtained are shown in Figure 4A and some theoretical approximations to these results using Equations (16) and (27) are shown in Figure 4B. The enthalpy of fusion of nepheline has not been measured, so it was estimated as 21 kcal mol^{-1} at the melting point by the method described above. The best-fit value of 22 erg cm^{-2} for σ is essentially based on the position of the nucleation rate peak. The high value of activation energy ΔH_{tr} in the transport factor, 150 kcal mol,$^{-1}$ is required by the rapid drop-off in rates of both nucleation and growth with increasing ΔT. Although the uncertainty in this value is very large, such a high activation energy suggests that the limiting transport factor in this case may be silicon diffusion in the multicomponent liquid.

Leonteva (1948, 1949) measured growth rates in the system Di–An–Ab and also the pure plagioclase system (Table 1a). The peak growth rates in these multi-component systems, from 10^{-5} to 10^{-4} cm/sec, are considerably less than those of the same minerals crystallizing from pure liquids (see above).

Table 1b. Measured Nucleation Rates or Densities of Minerals.

Crystal	Melt Composition* wt %	Liquidus Temp. K	ΔT at Peak	Rate or Density at Peak[§]	Reference
Nepheline	Neph + 10% LiF	1110	58	1.1×10^{-3} nuc cm^{-2} sec^{-1}	(i)
Alkali Fs	Ab$_{90}$Or$_{10}$ + 2.7% H$_2$O	1218	>400	>6 × 10^3 nuc cm^{-3}	
Alkali Fs	Ab$_{90}$Or$_{10}$ + 5.4% H$_2$O	1113	200	2.5 × 10^3 nuc cm^{-3}	
Alkali Fs	Ab$_{90}$Or$_{10}$ + 9.0% H$_2$O	1048	130	2.2 × 10^3 nuc cm^{-3}	
Alkali Fs	Ab$_{70}$Or$_{30}$ + 1.7% H$_2$O	1243	>350	>6 × 10^3 nuc cm^{-3}	(l)
Alkali Fs	Ab$_{70}$Or$_{30}$ + 4.3% H$_2$O	1113	200	4.2 × 10^3 nuc cm^{-3}	
Alkali Fs	Ab$_{70}$Or$_{30}$ + 9.5% H$_2$O	1038	110	1.2 × 10^3 nuc cm^{-3}	
Alkali Fs	Ab$_{50}$Or$_{50}$ + 2.8% H$_2$O	1228	>400	>6 × 10^3 nuc cm^{-3}	
Alkali Fs	Ab$_{50}$Or$_{50}$ + 6.2% H$_2$O	1088	165	.8 × 10^3 nuc cm^{-3}	
Quartz	Gr$^{\infty}$ + 3.5% H$_2$O	1143	220	10^7 nuc cm^{-3}	
Quartz	Gd$^{\infty}$ + 6.5% H$_2$O	963	>400	> 10^8 nuc cm^{-3}	
Quartz	Gd + 12% H$_2$O	943	300	10^8 nuc cm^{-3}	
Alkali Fs	Gr + 3.5% H$_2$O	1163	250	10^7 nuc cm^{-3}	
Alkali Fs	Gd + 6.5% H$_2$O	953	>200	> 10^8 nuc cm^{-3}	(m)
Alkali Fs	Gd + 12% H$_2$O	923	100	10^8 nuc cm^{-3}	
Plagioclase	Gr + 3.5% H$_2$O	1233	250	10^7 nuc cm^{-3}	
Plagioclase	Gd + 6.5% H$_2$O	1213	>200	> 10^8 nuc cm^{-3}	
Plagioclase	Gd + 12% H$_2$O	1173	100	10^7 nuc cm^{-3}	
spinel	—	1633	>200	48 nuc cm^{-2}	
gehlenite	—	1573	185	100 nuc cm^{-2}	
melilite	—	1453	90	72 nuc cm^{-2}	(n)
hedenbergite	—	1433	70	44 nuc cm^{-2}	
aegirine	—	1293	60	20 nuc cm^{-2}	

* Pure liquid is not specified.

$^{\infty}$ Gr = granite, Gd = granodiorite. See Swanson (1977) for compositions.

[§] Where nucleation densities only are given, insufficient information is available to determine absolute rates; however, the observed densities can be used to compare minerals measured by the same investigator.

References
(i)   Winkler (1947)
(l)   Fenn (1977)
(m) Swanson (1977)
(n) Schumoff-Deleano and Dittler (1911)

The growth rate of anorthite crystallizing from its own melt was measured by Klein and Uhlmann (1974) and Kirkpatrick et al. (1976) (Table 1a). The rates measured in the two studies are in good agreement, although those of Kirkpatrick et al. suggest a somewhat lower ΔT at the peak of growth rate. The dependence of reduced growth rate on ΔT in both studies indicated a surface nucleation mechanism. Although the faces present were not reported, the crystals were elongated in the c direction

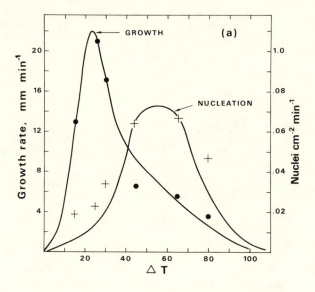

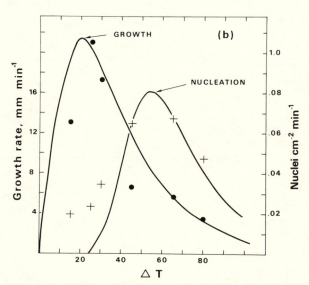

Figure 4. Nucleation (crosses) and growth rates (dots) of nepheline from Winkler (1947): 4(a) shows curves as drawn by Winkler (1947); 4(b) show curves drawn according to Equations (16) and (27) (this chapter), with $\Delta H = 21$ kcal mol^{-1}, $\sigma = 22$ erg cm^{-2} and $\Delta H_{tr} = 150$ kcal mol^{-1}. Vertical scale is arbitrary for the theoretical curves. According to Winkler, the measured nucleation rates which fall above the curves at small ΔT are due to heterogeneous nucleation.

in both studies, which is normal in natural anorthite (Smith, 1974). Kirkpatrick et al. (1976) also measured the growth rate of diopside from its own melt. Crystals were elongated parallel to c and the plot of reduced growth rate versus ΔT indicated a screw dislocation mechanism. The growth rate of cristobalite from pure SiO_2 was measured by Wagstaff (1969). The crystals were not faceted, and the plot of reduced growth rate versus ΔT indicated continuous growth.

Fenn (1977) measured growth rate and nucleation density in alkali feldspars ($Or_{10}Ab_{90}$, $Or_{30}Ab_{70}$ and $Or_{50}Ab_{50}$) growing from hydrous melts. Again, because of the presence of more than one component, liquidus temperatures must have varied with crystallization, but such complications should not be important for small degrees of crystallization and will be ignored. Fenn reported his nucleation results in terms of nucleation density rather than rate because of severe problems with incubation time. The results show pronounced trends with changing water content of the melt. Examples are shown in Figures 5A and 5B. At low water content, the onset of appreciable nucleation is at very large ΔT, as expected from the relatively low ΔS_f of alkali feldspar. Some nucleation occurs at smaller ΔT, but this may be attributable to heterogeneous nucleation from impurity particles within the liquid. With increase of water content, nucleation occurs at much smaller ΔT and peak nucleation density decreases. Growth rates show similar but less pronounced trends.

A decrease in the critical ΔT of nucleation could be ascribed either to increase in ΔS_t or decrease of σ. The decrease in liquidus temperatures at higher water contents, to which a small negative heat of mixing (Burnham and Davis, 1974) may contribute, shows that ΔS_t is increased, but this is not nearly sufficient to cause the observed effect. However, the main features of the trends can be reproduced qualitatively by a large reduction in surface energy, from about 40 to about 15 ergs cm^{-2}, as shown in Figures 5B and 5C, with Equations (16) and (27) again being used. The only aspect of the observations not accounted for by this change in surface energy is the reduction in nucleation density. This could be a result of a greater incubation time at smaller ΔT predicted by Equation (17); since the nucleation curves for high water contents peak at smaller ΔT, they should be more severely affected.

Some change of surface energy is probably to be expected on addition of water. In a melt with 9% water by weight, as many as 30% of the oxygen atoms could be hydroxyls if all the water dissociated. Although complete dissociation apparently does not occur at these high water contents in the system albite-water (Burnham and Davis, 1974), the dissociation that does occur should reduce the electrostatic interaction between cations

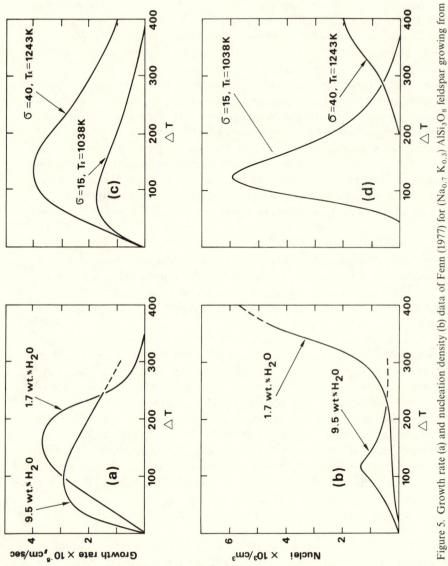

Figure 5. Growth rate (a) and nucleation density (b) data of Fenn (1977) for $(Na_{0.7} K_{0.3})$ $AlSi_3O_8$ feldspar growing from hydrous melts. Theoretical curves in (c) and (d) are according to Equations (16) and (27), with $\Delta H_{tr} = 10$ kcal mol^{-1} and $\sigma = 40$ erg cm^{-2} at 1.7% H_2O and $\sigma = 15$ erg cm^{-2} at 9.5% H_2O.

and oxygen across the interface, because the effective charge of the oxygens would be halved.

The measured growth rates peak at large supercooling and drop off only rather slowly thereafter with increasing ΔT. This behavior is consistent with a low activation energy of transport: a value of 10 kcal mol^{-1} was used for the calculated curves shown in Figure 5. Because the morphologies developed at high ΔT were of the non-compact type, that is, dendrites and spherulites, it may apparently be concluded that the limiting transport factor was volume diffusion in the liquid, rather than interface transport or surface diffusion. This apparently low activation energy of 10 kcal mol^{-1} is certainly not suggestive of silicon-aluminum interdiffusion, for which an activation energy of 75–100 kcal mole^{-1} has been inferred (Sipling and Yund, 1974) in crystalline feldspar. Activation energies for alkali and water diffusion in melts are all within the range 10–30 kcal mol^{-1} (Hofmann, this volume). In the systems studied by Fenn, the compositions of the crystals do not differ from that of the liquid in Al/Si ratio, but only in Na/K ratio and water content. The relatively small decrease in peak growth rates between experiments at different water contents, despite drastic decreases in liquidus temperature, is also consistent with small activation energy, but it must be noted that viscosity probably decreases also with increasing water content.

Swanson (1977) measured nucleation densities and growth rates for quartz, plagioclase, and alkali feldspars in granodiorite and granite melts with various amounts of water. His results for growth rates (Table 1a) are generally similar to those of Fenn (1977), and the decrease in growth rates with increasing water content is even more striking in the granodiorite system. It is difficult to compare nucleation rates, since only nucleation densities were given and the times of the experiments were not reported in detail by either Fenn or Swanson, but the critical ΔT's of nucleation seem to be generally smaller in Swanson's systems. These smaller critical ΔT's, if not an effect of heterogeneous nucleation, again suggest rather small effective surface energies, in the range 10–30 erg cm^{-2}. Both Fenn and Swanson reported only "internal" nucleation, that is, those crystals which did not obviously form on the container walls.

In Swanson's results, both the nucleation and growth-rate curves tend to peak at fairly high ΔT, and the growth rates drop off only rather slowly with increasing ΔT. This is somewhat surprising in this case, because the crystals do not have the same Si/Al ratio as the melt, and some diffusion of these elements through the liquid is necessary.

Growth and diffusion rates in the systems studied by Fenn and Swanson are generally too small to permit meaningful determination of the growth mechanism by rate measurement at small ΔT, and the use of viscosity to derive reduced growth rates at higher ΔT is definitely unwarranted in these multicomponent systems. However, in all cases the crystals were faceted. Feldspars were elongated parallel to a, with prominent $\{010\}$ and $\{001\}$ forms. The Donnay-Harker method predicts that $\{110\}$ and $\{1\bar{1}0\}$ should also be important, and this is usually the case in natural crystals. The habit of the synthetic feldspars of Fenn and Swanson also differs from that found in laboratory growth of pure anorthite, which is elongated parallel to c, and for which the surface-nucleation mechanism has been inferred (below). Thus the screw dislocation mechanism should not be ruled out for the feldspars grown by Fenn and Swanson. Quartz occurred in Swanson's experiments in flat $\{0001\}$ plates, rather than the more normal elongated prismatic shape with prominent rhombohedra which is predicted by the Donnay-Harker method. The dominant growth mechanism for quartz in this case may be by the screw dislocations parallel to $[11 * 0]$ discussed by Frondel (1978).

Some other materials related to silicate minerals whose growth rates have been measured include $CaMgSi_2O_6$–$CaAl_2SiO_6$ pyroxenoids (Kirkpatrick, 1974), for which the surface nucleation mechanism was inferred; sodium trisilicate (Scherer, 1974) and sodium disilicate (Meiling and Uhlmann, 1967), which were both inferred to grow by the screw dislocation mechanism; and GeO_2 (Vergano and Uhlmann, 1970a), which like SiO_2 grows by the continuous mechanism (Table 1a). The raw growth rate data (Table 1a) for PbB_4O_7 (DeLuca et al., 1969) and SrB_4O_7 (Nagel and Bergeron, 1972) appear to be consistent with the surface nucleation mechanism. If a correction for increase in temperature at the interface is applied to the SrB_4O_7 data, they show behavior as predicted by Cahn (1960), that is, a change from layer growth to continuous growth with increasing ΔT. However, the interface temperature was not actually measured, and the correction involves a rather dubious extrapolation of the temperature difference in another system (PbB_4O_7).

It should be noted that measurements are normally made on the fastest-growing direction in crystals. This enhances the probability of finding the mechanism to be continuous growth rather than layer growth, and screw-dislocation rather than surface nucleation, since the expected rates decrease in the order mentioned, at least for small ΔT. There is no reason that the growth mechanism must be the same for all directions, particularly in such anisotropic structures as chain silicates. Crystals grown in the laboratory are often less perfect than natural crystals, and this also would tend to increase the likelihood of growth by the screw dislocation mechanism.

Some experiments and measurements have been made which do not directly furnish growth or nucleation rates as a function of ΔT, but do nevertheless give valuable information about parameters or mechanisms.

Gibb (1974) measured the critical ΔT for nucleation of plagioclase in a natural basalt. Problems with incubation time were encountered, but the critical ΔT is apparently small, certainly less than 50°, indicating effective surface energy of less than about 30 erg cm^{-2}. Kirkpatrick et al. (1978) have determined plagioclase nucleation and growth rates during cooling at constant rates in plagioclase melts of composition $An_{50}Ab_{50}$. This type of experiment does not give nucleation rate as a function of ΔT, but may be more directly applicable to the crystallization of igneous rocks; the rate as a function of ΔT may be indirectly obtained anyway from numerical modeling similar to that described below.

Kirkpatrick (1977) determined growth and nucleation rates of plagioclase in basalt sampled in different parts of Hawaiian lava lakes. The clustering of plagioclase and other crystals (olivine and augite), and some ophitic textures, indicated heterogeneous nucleation, which evidently continued over a considerable time interval. Nucleation rates ranged from 7×10^{-3} nuclei cm^{-3} sec^{-1} to 2 nuclei cm^{-3} sec^{-1}, the more rapid rates being observed near the surface. Supercooling could not be directly measured, but the very low growth rates, 2 to 11×10^{-10} cm sec^{-1}, suggested small ΔT. Both nucleation and growth rates depended on the fraction of basalt crystallized, as well as the cooling rate. Donaldson (1975) derived instantaneous growth rates of olivine in quenched basalt by measuring the composition gradients in the liquid adjacent to the interface and using diffusion coefficients from the Stokes-Einstein relation to find the rate of advance of the interface. The rate of about 10^{-7} cm sec^{-1} is reasonable, but in addition to the uncertainty of applying the Stokes-Einstein relation to complex silicate liquids, it is not clear at what stage during the growth and quenching processes the diffusion gradients became established.

The laboratory data show very clearly the influence of increase in complexity of the system in decreasing peak growth rates. This is due in part simply to lower liquidus temperatures, but more importantly to the effects of multi-component diffusion. From values of about 10^{-2} to 10^{-3} for pure anorthite and diopside, the rates drop to 10^{-5} to 10^{-3} for plagioclase and pyroxene in the Di–An–Ab system. Even in the pure plagioclase system, rates drop by 10^3 from pure anorthite, evidently reflecting the necessity for Si/Al interdiffusion. The rates measured in the Di–An–Ab system are somewhat higher than those inferred (below) for laboratory crystallization of lunar basalts, as is expected on the basis of the more complex composition of the latter.

A similar decrease of growth rate with chemical complexity of the melt is evident in granitic systems. The rates for feldspars and quartz measured by Swanson (1977) in granite and granodiorite are much lower than those for pure alkali feldspars, plagioclase, or SiO_2 (cristobalite). Swanson's experiments are probably directly applicable to natural granite and granodiorite melts, although whether natural magmas very commonly have such high water contents is perhaps questionable.

The fact that multicomponent diffusion strongly influences *peak* growth rates does not mean that it is the major controlling factor in all geologic crystallization. The measurements of Kirkpatrick (1977) on Hawaiian basalts and the numerical modeling described below show that most crystallization, even in moderately rapidly-cooled basalts, takes place at relatively small ΔT, well below that of peak growth rate. At such small ΔT, interface processes are probably the rate-controlling factor, as is confirmed by the frequency of compact morphologies in non-glassy basalts. Even in glassy basalts, observed diffusion gradients are not large (Anderson, 1967; Bottinga et al., 1966; Donaldson, 1975) and may have been established during a quench, rather than during growth of phenocrysts.

Although the available measurements appear to provide growth rates as a function of ΔT for typical basaltic and granitic compositions at least within a couple of orders of magnitude, the situation with respect to nucleation rates is much less satisfactory. Major experimental problems include incubation time, avoidance of heterogeneous nucleation (if it is desired to measure homogeneous nucleation), and the fact that true nuclei are submicroscopic, and measurement of their number usually requires a period of growth at small ΔT, subsequent to the incubation period. These problems are by no means insurmountable, and experimental methods will no doubt improve greatly with time. However, application of laboratory nucleation rates to natural magmas will always be less secure than application of growth rates, because of the difficulty of ascertaining the extent to which nucleation is heterogeneous.

THE SIZE OF CRYSTALS—MODELING IGNEOUS CRYSTALLIZATION

A great deal of information is potentially obtainable from the textures of rocks, and petrologists have always attempted to extract this information, despite the general lack of experimental data and sound theoretical background. Several aspects of textures can be useful, including crystal shape, as discussed above and by Lofgren (this volume). The remainder of this paper will be addressed to one important aspect, the sizes of crystals.

I am currently developing techniques for numerical simulation of igneous crystallization, with the principle objectives being the prediction of grain sizes and mineral zoning trends. The development is still in a very early stage, with many of the techniques to be worked out, and because of the lack of fundamental thermochemical data describing the equilibrium phase relations as well as kinetics, the modeling at this stage can only be qualitative at best. Nevertheless, a brief account of the methods and some preliminary results are offered here, in order to show the principles involved, the potential usefulness of this approach, and some of the problems which are inherent in the application of laboratory experiments to natural crystallization. Specifically, modeling of the crystallization of a single component under various conditions and modeling of basalt crystallization using the simplified system diopside-anorthite-albite (Di–An–Ab) will be described.

The modeling described here covers only part of the range of processes involved in igneous crystallization. On a larger scale, Kirkpatrick (1976) and Walker et al. (1976), for example, have attempted to model the inter-action of heat flow and crystallization in magma bodies. Towards the other end of the scale, Loomis (1980) has modeled some aspects of the interaction of diffusion, heat flow and crystallization at the interface between individual plagioclase crystals and liquid, and Monte-Carlo computer models of actual atomic behavior on a crystallization interface have been developed (Gilmer and Bennema, 1972). Eventually these models on different scales should be integrated to provide a comprehensive picture of crystallization.

Grain Size Distributions

Measurements of the frequency of crystal sizes provide a record of the nucleation and crystal growth history of an igneous rock. Assuming that no dissolution takes place, the size of a crystal is dependent on its time of nucleation and subsequent growth rate. Although the growth rate may vary, such variation affects all crystals of the same mineral simultaneously, so that the order of time of nucleation is always the same as the order of size if the liquid remains homogeneous with respect to composition and temperature. Therefore, in a size-distribution plot of a single mineral, such as is shown in Figure 6, the size axis can be interpreted as representing time of nucleation. Of course, the relation between time and size on the horizontal scale is variable and depends on the growth rate, which is a function of supercooling. If nucleation was continuous (that is, if crystals of every possible size exist), and if only one nucleation mechanism was operative, the supercooling at any given time could in principle be obtained from the number of crystals of that size, by using measured or theoretical nucleation rates.

A quantitative solution or inversion of the size-frequency measurements into a time-supercooling plot would require complete knowledge of the nucleation and growth rates as a function of supercooling, as well as some rather complex mathematical techniques. Qualitative information can be

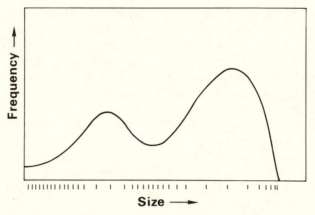

Figure 6. Hypothetical crystal-size/frequency plot. The short vertical lines below the horizontal axis are intended to delimit equal time intervals. Assuming that the nucleation and growth mechanisms do not change, the vertical axis (number of crystals)
is correlated with supercooling, as the nucleation rate is an increasing function of
supercooling. When supercooling is high, as during the generation of the two
peaks, growth rate is also high, so that the size increment between successively
nucleated crystals is large, hence the greater intervals between time points on
the size axis.

obtained without these requirements, however. For example, a peak in the
size-distribution plot indicates a burst of nucleation, which is most likely
caused by an interval of relatively high ΔT. Such a burst of nucleation is
expected if nucleation is homogeneous, as the supercooling must reach a
critical value for nucleation to take place; however, once crystals are
formed, their rapid growth and release of latent heat, and/or change in
liquid composition, causes the supercooling to decrease to a steady-state
value. Such crystallization histories have been obtained in the numerical
models described below. On the other hand, relatively flat distribution of
crystal size (about the same number of crystals of every size) suggests
continuous nucleation, as might be expected if nucleation is heterogeneous
and occurs at low supercooling (as observed by Kirkpatrick, 1977).

It is unfortunate that complete size distributions of igneous rocks have
never been considered worthy of publication (an exception: Kraus, 1962),
despite the fact that the information for them is routinely obtained in
some methods of modal analysis, and that average grain sizes are sometimes reported. Different cooling histories and different types of nucla-
tion/growth relationships can lead to distinctly different grain size
distributions. As sedimentary petrologists have long been aware, the mean
alone does not provide a satisfactory description of grain size.

Grain size distributions measured in thin section are biased toward
smaller sizes, and must be corrected analytically. If the grains are spherical,

the correction is straightforward, (e.g. Krumbein and Pettijohn, 1938): it is carried out by using the moments of the distribution, which incidentally offer a compact means of describing the results of size analysis. The problem becomes much more difficult if the grain shape is more complex (Kendall and Moran, 1963), but some solutions have been offered for common grain shapes (e.g. Grey, 1970), and the statistical manipulations present no great difficulties with modern computer techniques.

Basic Algorithm for Modeling Crystallization

Figure 7 shows a flow diagram of the computer procedure. In this particular case, heat is removed at a constant rate, which is the simplest possible model for natural igneous rocks. More complex functions of heat removal versus time could easily be programmed. For modeling laboratory experiments on dynamic crystallization (Lofgren, this volume), the temperature can also be reduced at a constant rate, ignoring heat removal; and crystallization can also be made to occur at a constant temperature.

The crystal growth and nucleation equations used are those given above, specifically Equations (16) and (27). The equation for large-crystal surface nucleation has been used in the work described below, although some minerals evidently crystallize by the screw dislocation mechanism. Differences between the shapes of the curves for the two mechanism are not large. The absolute growth and nucleation rates are scaled independently, either by using experimental results, or by trial-and-error adjustment to obtain the best fit to observed grain-size distributions. A single heat-capacity value was obtained for multicomponent systems by taking the weighted average of all the end-member crystals. No attempt has been made to deal directly with heat or diffusion gradients in the crystallizing system. This is undoubtedly an oversimplification, but should not lead to large errors, because the effects of diffusion are usually included in the estimated or measured growth rates. For example, in the multicomponent systems studied by Winkler (1947), and Swanson (1977), the observed growth rates must have been influenced to a large extent by diffusion in the liquid; such influence in the system of Fenn (1977) was probably less.

If the time interval used in each cycle is too large, simulated crystallization becomes cyclic, because the release of latent heat increases the temperature above the liquidus. It has so far always been possible to eliminate such cycling by reducing the time interval. The computation procedure is normally terminated at about 90% crystallization, both because interference effects would become important at this stage in a real rock, and because the calculations become increasingly unstable (i.e., spurious cycling begins) as the amount of liquid decreases and the bulk solidification rate (which is proportional to the surface area of crystals) increases.

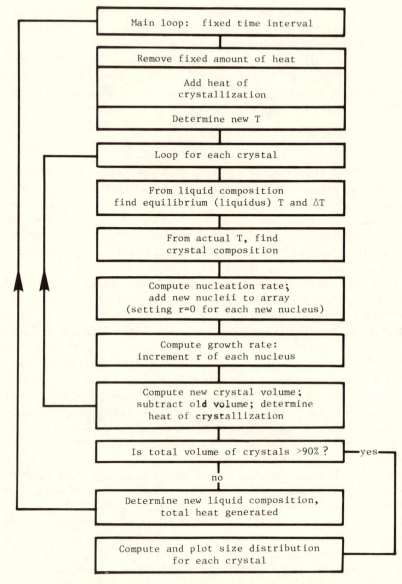

Figure 7. Flow diagram of the computer algorithm for numerical modeling of crystallization.

ANALYTICAL EXPRESSIONS FOR PHASE RELATIONS

Although phase diagrams are available for important systems, the analytical representation of phase relations presents serious difficulties. Empirical expressions using arbitrary mathematical functions are still the most accurate method in simple, well-known systems (e.g. Weill et al., this volume), but the applicability of this method is very limited. In the long run, a sound thermodynamic description is much to be desired, as components could then be added as necessary. A thermodynamic model is also required for extrapolation of the experimental results into metastable regions: this extrapolation is vitally necessary in kinetic modeling.

The state of the art of thermodynamic representation of igneous phase relations is discussed by Weill et al. (this volume), Hess (this volume), and Wood and Fraser (1978). In the absence of a satisfactory comprehensive theory and the requisite empirical data (such as heats of fusion and mixing), I have provisionally used a simplified "ideal" model to simulate crystallization. In view of the limitations of the kinetic data, great accuracy is not required. In this model, heats of mixing of components have been ignored, and the mixing entropy of the system, or the activities of the crystallizing components, have been represented by the Temkin (1945) scheme, that is, all cations are assumed to be mixed randomly. This scheme has been used with success in several silicate systems (Wood and Fraser, 1978). The liquidus temperature of diopside, for example, is

$$T_l = \frac{T_f}{1 - \left(\dfrac{R}{\Delta S_f} \ln \dfrac{a_{Di}}{a_{Di}^0} \right)}, \tag{36}$$

where a_{Di} is the activity of diopside in the liquid in question, and a_{Di}^0 is the activity of pure diopside. The fusion entropy ΔS_f is for a unit of diopside in which the cations total one, that is, $Ca_{.25}Mg_{.25}Si_{.5}O_{1.5}$. The activity of diopside is

$$a_{Di} = (X_{Ca})^{.25}(X_{Mg})^{.25}(X_{Si})^{.5}, \tag{37}$$

where X_{Ca}, X_{Mg}, and X_{Si} are the amounts of Ca, Mg, and Si in the melt as mole fractions of all cations. For pure diopside a_{Di}^0 is thus .354.

The values of ΔS_f used in such liquidus expressions were arbitrarily adjusted to provide reasonable approximations (shown in Figure 8) to the true phase equilibria in the binary systems, thus compensating for real heats of mixing and inadequacies of the entropy model. The measured melting entropies recommended by Weill et al. (this volume), and not these adjusted values, were used in the growth and nucleation rate equations, and to obtain the latent heat generated during crystallization.

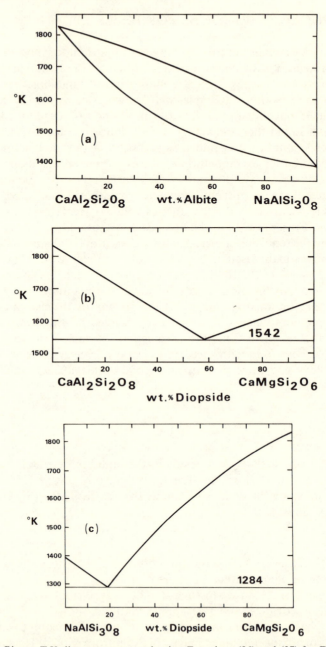

Figure 8. Binary *T-X* diagrams computed using Equations (36) and (37) for Di–An and Di–Ab, and Equations (38) and (39) for plagioclase, with "adjusted" heats of melting of 16.7 kcal mol^{-1} for albite, 25.6 kcal mol^{-1} for anorthite, and 33.0 kcal mol^{-1} for diopside.

The simple Temkin (1945) model provides a convenient definition of an "ideal" silicate melt, and presumably gives the *maximum* entropy of mixing. It may seem unreasonable to expect silicon and cations like Ca and Mg to mix over the same "sites", and this presumably led Weill et al. (this volume) to prefer a two-site model. However, considered in terms of mixing of neutral or nearly neutral oxide "molecules" such as SiO_2, MgO, CaO, and the like, the model adopted here is seen to be reasonably consistent with conventional notions of melt structure (for example, Hess, this volume). In a pure silica melt, each SiO_2 "molecule" associates closely with two others, forming an essentially continuous framework. In a melt with composition $2MgO \cdot SiO_2$ (for example), each SiO_2 molecule is still associated with two other oxide molecules, but there is a high probability that these other molecules will be "MgO". The latter presumably do not attach strongly to other groups, so that the polymerization of the melt is much reduced. Oxides with several metal atoms in the neutral unit, such as Na_2O, are considered to "dissociate" into non-neutral units. The proven dissociation of H_2O in silicate melts (e.g. Burnham and Davis, 1974) provides some justification for this. The numerical values of mixing entropy computed by this model in the system Di–An–Ab are only slightly larger for most compositions than those calculated by the two-site model of Weill et al. (this volume). Of course the actual heats of mixing measured by Weill et al., and other considerations such as liquid immiscibility discussed by Hess (this volume), suggest that mixing according to this model is *not* completely random. In particular, the frequencies of different polymeric arrangements of silicon and oxygen are evidently not what would be predicted from random attachment of molecules: liquid immiscibility implies that the favored configurations for silicon/aluminum tetrahedra in some systems are either (1) sharing no oxygens with other tetrahedra, or (2) sharing all four oxygens. However, this is not the place to discuss such complications, or the potential general applicability of this mixing entropy model; the representations of phase relations which it provides should be sufficient for the qualitative purposes of numerical simulation of crystallization.

Determination of Supercooling and Crystal Compositions

Crystal-liquid kinetics are determined by the supercooling or superheating of each solid phase. This ΔT is calculated with respect to the liquidus temperature for that phase, whether the temperature represents stable or only metastable equilibrium (Baker and Cahn, 1971; Flemings, 1974). For example, in the normal course of crystallization in a eutectic system, shown in Figure 9A, the liquid, after initially crystallizing component A, is expected to become saturated with component B at a composition

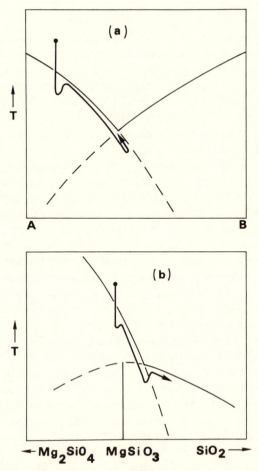

Figure 9. Schematic temperature/composition diagrams for (a) a eutectic system and (b) a peritectic system (specifically forsterite-silica, showing only the region near the peritectic). The thin solid lines are the stable liquidus curves, and the dashed lines are their metastable extensions. The eutectic and peritectic points are at the intersections of the liquidi; horizontal tie lines are omitted. The heavier lines with arrows show typical paths of actual liquids during real crystallization. In the eutectic (a) the liquid supercools to some critical ΔT of nucleation for component A, then, after crystals of A form, it reverts to smaller supercooling and closely parallels the liquidus curve. Although the liquid becomes saturated with B when it crosses the metastable extension of the B liquidus, its composition continues to be controlled by crystallization of A until the critical ΔT of nucleation for B is reached, determined with respect to the liquidus curve for B. Thereafter it tends to revert to a steady-state position just below the eutectic. In the peritectic (b), the liquid first nucleates forsterite, then follows the forsterite liquidus with some small ΔT until the critical ΔT of nucleation for enstatite is exceeded. Here enstatite would crystallize, but olivine would also continue to crystallize until its meta-stable liquidus is crossed. The liquid would parallel the enstatite liquidus curve until the enstatite-silica eutectic is reached (not shown in this diagram).

poorer in B, and at a temperature lower than the true eutectic. In a peritectic system, for example forsterite-silica (Figure 9B), the liquid, after initially crystallizing olivine, becomes saturated with pyroxene (i.e., it crosses the dashed line) at a composition richer in pyroxene, and at a temperature slightly greater than the true peritectic. It may nucleate pyroxene whenever it reaches the necessary supercooling with respect to the pyroxene liquidus, but it does not stop crystallizing olivine until it has crossed the metastable extension of the olivine liquidus. Olivine could then begin to react with the liquid, or dissolve. Melting and dissolution have not been studied as throughly as crystallization, either theoretically or experimentally, but rates should also be dependent on ΔT (for example, Meiling and Uhlmann, 1967; Wagstaff, 1969; Vergano and Uhlmann, 1970b).

The general effect of such metastable phase relations is to extend the range of liquid compositions from which a crystal can form. Also, metastable phase relations, combined with nucleation kinetics, can lead to changes in the order of appearance of phases. For example, in Figure 9B, if the liquid composition is slightly to the right of the peritectic, and if the critical ΔT of nucleation of olivine is less than that of pyroxene, olivine could nucleate metastably before pyroxene. In the eutectic system (Figure 9A), if the liquid is slightly enriched in component A with respect to the eutectic, but crystal B has a smaller critical ΔT of nucleation than crystal A, B could nucleate before A. At least some cases of changes in order of appearance or length of crystallization interval in natural or synthetic rocks, caused by changes in cooling history, may be due to such effects. It was predicted by Gibb (1974) from his experimental results that the nucleation of plagioclase might often be delayed, and a rather late nucleation of plagioclase with respect to the equilibrium sequence has been observed in olivine-normative lunar basalts (Walker et al., 1976). Such an effect is also predicted theoretically for homogeneous nucleation, because the melting entropy per unit volume of plagioclase is less than that of the other major early phases, olivine, pyroxene, and spinel.

Changes in the order of appearance of phases and variations in compositions can also arise from non-equilibrium thermodynamic effects. As discussed above, lack of equilibrium between liquid and crystals can cause it to be thermodynamically possible to form a range of solid compositions from a given liquid, rather than a single equilibrium composition. Under certain conditions, lack of equilibrium may make it thermodynamically disadvantageous or even impossible to form the true stable phase before a metastable phase crystallizes (Baker and Cahn, 1971). Either this type of effect or the effect of varying critical ΔT's of nucleation probably accounts for most specific instances in which crystallization follows Ostwald's rule, according to which a transition from one phase to a more stable one proceeds through any other phases which are intermediate in stability.

In dealing with the problems of numerical simulation of crystallization, it is necessary to specify the composition of crystals forming solid solutions: in the modeling performed to date, this applies only to plagioclase. The free-energy/composition relations are not precisely known for most igneous minerals, let alone liquids, so it is not possible to determine ΔG_c from these, or to find the solid composition which gives the greatest free energy change on crystallization. The composition of the plagioclase solid solution was therefore simply taken to be the equilibrium composition for the temperature in question, and ΔG_c was determined from the supercooling (Equation (10)). The procedure is as follows. First, the ratio of An to Ab in the melt is applied to the pure plagioclase system to determine the equilibrium liquidus temperature and equilibrium crystal composition in that system. The analytical expressions for this are as follows (Bowen, 1913):

$$\frac{x}{x'} = \exp\left[\frac{\Delta H_f^{Ab}}{R}\left(\frac{1}{T_f^{Ab}} - \frac{1}{T_l}\right)\right], \tag{38}$$

$$\frac{1-x}{1-x'} = \exp\left[\frac{\Delta H_f^{An}}{R}\left(\frac{1}{T_f^{An}} - \frac{1}{T_l}\right)\right], \tag{39}$$

where T_l is the equilibrium liquidus temperature, x is the mole fraction of Ab in the liquid, and x' is the mole fraction of Ab in the solid. There appears to be no analytical solution to these equations for T_l, so they are solved by the Newton-Raphson iterative method. Then, knowing the liquidus temperature and the liquid composition x, the equations can be solved analytically for the equilibrium solid compostion x'. A weighted average value of ΔS_f is obtained from the solid composition, and this is used in Equation (36) with the activity of plagioclase of this composition in the liquid to determine the ternary liquidus temperature. The actual temperature is then subtracted from the ternary liquidus temperature to determine the supercooling. This supercooling is then subtracted from the equilibrium temperature originally found in the pure plagioclase system, and the resulting temperature is used with Equations (38) and (39) to find the actual composition of the crystallizing plagioclase. This procedure is not incompatible with the establishment of a diffusion gradient in the liquid; in fact, the assumption that the crystallizing solid has the equilibrium composition implies surface equilibrium, which can only exist in the presence of a diffusion gradient. On the other hand, under such conditions, the ΔT at the interface would be overestimated by using the bulk composition of the liquid. Again, we rely on the fact that growth rates are either adjusted by trial and error or are derived from measurements in similar systems, and can be considered to be dependent on the bulk super-cooling regardless of the actual ΔT or composition at the interface.

Crystallization of a Single Component System

To show how grain-size distributions are dependent on cooling history, some simple models of crystallization in a one-component system have been carried out, using the data from Winkler's (1947) study of the crystallization of nepheline. The theoretical curves for homogeneous nucleation and growth derived above (Figure 4) were used, scaled so that peak rates were the same as those found by Winkler.

First, to model Winkler's own results, isothermal crystallization was simulated. The average grain sizes obtained are compared with Winkler's in Figure 10. Any discrepancy here between the two curves is presumably a result of inadequacy or inapplicability of the theoretical expressions, and/or changes in liquidus temperature during Winkler's experiments because of the presence of the LiF flux. Winkler observed a maximum in the grain-size/supercooling plot at about 20° ΔT. This result is incompatible with homogeneous nucleation theory, and was not observed in the numerical simulation, because the theoretical nucleation rate decreases much faster than the growth rate as ΔT goes to zero. Instead, grain size increased with decreasing ΔT, until nucleation effectively ceased. The finite nucleation rate observed by Winkler at small ΔT evidently represents heterogeneous nucleation, the presence of which he recognized himself. It should be noted, however, that the shape of Winkler's nucleation-rate curve (Figure 4A) is similar to those drawn by Fenn (1977) for crystallization of alkali feldspar, in that there is a gently-sloping region of finite nucleation rate at low ΔT. Also, Tammann and Mehl (1925) emphasized

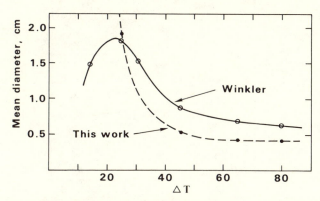

Figure 10. Grain size in isothermal crystallization of nepheline: actual results of Winkler (1947), and numerical simulation using growth and nucleation rate curves of Figure 4.

that experiments usually show measureable nucleation rates at small ΔT, which is contrary to the predictions of homogeneous nucleation theory. This is probably not a refutation of that theory, but a testimony to the presence of small quantities of foreign bodies in the liquids, and the inevitability of a certain amount of heterogeneuous nucleation. The extent of heterogeneous nucleation may be limited by the abundance of such foreign bodies. In his analyses of grain sizes in natural dikes, Winkler (1948, 1949) apparently assumed that the observed peak at low super-cooling is a general phenomenon. As pointed out by Shaw (1965), and confirmed by the results given here, this is not consistent with homo-geneous nucleation theory. However, if heterogeneous nucleation is as important in natural magmas as it was in Winkler's and Fenn's experi-ments, then small grain size at very slow cooling rates is to be expected.

Since nucleation occurs continuously, the grain size obtained for crystallization at a single temperature is an essentially flat line, as shown in Figure 11A, at least if interference effects and diffusion are ignored.

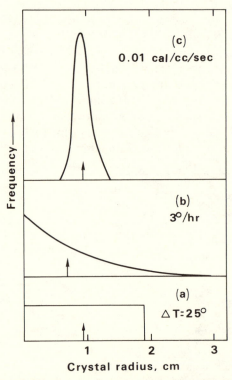

Figure 11. Frequency distribution of grain sizes obtained numerically with three types of cooling history: (a) isothermal; (b) constant rate of temperature and (c) constant rate of heat loss. The arrows indicate means.

The average grain sizes obtained when the temperature is reduced at a constant rate are shown in Figure 12 and a typical size-frequency plot is shown in Figure 11B. This is presumably the type of thermal history produced in laboratory dynamic crystallization experiments (Lofgren, this volume), assuming that the thermal mass of the furnace as a whole swamps the effect of latent heat generation. Also shown in Figure 12 is a relationship derived theoretically by Winkler from his isothermal experiments. The discrepancy between the two curves is larger than can be accounted for by poor fit of the equations to the rate data, and throws doubt on Winkler's derivation. The size-frequency distribution shown in Figure 11B is characteristic of this type of cooling history, if the growth rate peaks at lower ΔT than the nucleation rate. Under conditions of constantly decreasing temperature, nucleation in this case does not begin until growth rate is already decreasing, which leads to ever-increasing numbers of smaller crystals.

The size distribution obtained for crystallization at constant rate of heat removal, which is the most realistic geologically of the simple models, is shown in Figure 11C. In this model, there is no nucleation until the liquid supercools to the critical ΔT of nucleation, but once a certain number of crystals have nucleated, the temperature goes back to near the

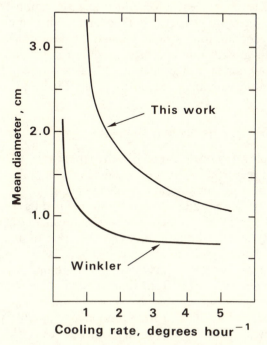

Figure 12. Mean grain size of nepheline versus rate of temperature decrease; theoretical curve of Winkler (1947) and results of simulated crystallization.

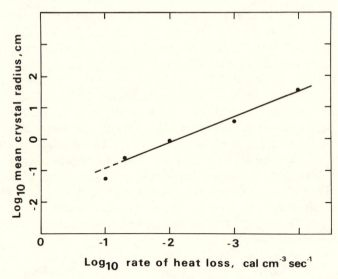

Figure 13. Mean grain size versus rate of heat loss, simulated crystallization of nepheline. Dotted line signifies incomplete crystallization.

melting point as latent heat is generated (as, for example, in Figure 9). Nucleation is therefore restricted in time, and a fairly uniform grain size results. If the cooling rate is rapid enough, however, ΔT increases continuously until growth rate becomes insignificant, producing glass. The log-log plot of grain size as a function of heat loss rate in the numerical models shows a reasonably linear relationship (Figure 13).

CRYSTALLIZATION IN THE SYSTEM Di-An-Ab

As an example of models more directly applicable to natural rocks (although still very simplified), some results in the haplobasalt system are presented. For the initial attempt at modeling basalt textures, the pyroxene-phyric or quartz-normative Apollo 15 lunar basalts were chosen. These basalts, together with their Apollo 12 equivalents, are possibly the most intensely studied basaltic rocks in the solar system. Their most outstanding characteristic is the presence of skeletal phenocrysts of clinopyroxene, with extreme magnesium-iron zoning and some sector zoning. The phenocrysts vary from about a tenth of a millimeter to over a centimeter in length, but seem to be rather uniform in size in any one thin section; the size of crystals in the groundmass—mainly pyroxene and plagioclase, with minor ilmenite and other accessories,—varies sympathetically with the phenocryst size. There was considerable debate about the origin of this texture, whether it was produced in a single stage of crystallization on extrusion, or by crystallization of phenocrysts at depth and groundmass on extrusion (see Dowty et al., 1974; Lofgren et al., 1974; and Lofgren, this volume, for references). The textures and mineral zoning patterns were reproduced by single-stage cooling in the laboratory (Lofgren et al., 1974; Grove and Walker, 1977), but the match between laboratory and natural textures is not exact, and there is still some uncertainty about the details of the cooling history.

As an approximation to these basalts, liquid composition of $Di_{60}An_{30}$ Ab_{10} was used. Equations (16) and (27) were again used for nucleation and growth rates as a function of ΔT, with the ΔH_f values adopted by Weill et al. (this volume), an arbitrary value of 20 erg cm^{-2} for surface energy of both phases, and an activation energy of 50 kcal mol^{-1}.

The initial modeling was an attempt to reproduce the dynamic cooling experiments of Lofgren et al. (1974), using constantly decreasing temperatures. The mean grain sizes of pyroxene and feldspar were estimated from Lofgren et al.'s figures, and the model nucleation and growth rates were scaled until reasonable agreement of mean grain sizes was obtained (Table 2). In addition to the observed grain sizes, the experimentally

Table 2. Grain size in laboratory and numerical analogs of pyroxene-phyric basalt

Cooling Rate	Mean Radius, mm		% of Feldspar Crystallized
	Pyroxene*	Feldspar	
Lofgren et al. (1974) constant temperature decrease**			
30 deg hr^{-1}	.13	—	<100
10 deg hr^{-1}	.21	.02	<100
2.5 deg hr^{-1}	.84	.05	100
1.2 deg hr^{-1}	—#	.10	100
Numerical model, constant temperature decrease			
30 deg hr^{-1}	.18	(.0007)	<2
10 deg hr^{-1}	.29	.02	100
2.5 deg hr^{-1}	.52	.03	100
1.0 deg hr^{-1}	.86	.04	100
Numerical model, constant rate of heat loss			
.01 cal cm^{-3} sec^{-1}	.24	(.0007)	<2
.005 cal cm^{-3} sec^{-1}	.32	(.002)	10
.002 cal cm^{-3} sec^{-1}	.54	.04	100
.001 cal cm^{-3} sec^{-1}	.60	.04	100
.0005 cal cm^{-3} sec^{-1}	.93	.09	100

* Phenocrysts only.

** Average volume was estimated, then the radius of an equivalent sphere was derived to obtain mean radii.

Too large to estimate.

determined cooling rate at which the groundmass is largely glass is a valuable datum. The effect on grain size of increasing the model growth rate is in some ways the same as decreasing the nucleation rate, but antipathetic variation of these parameters is limited by the necessity to match the cooling rate for the transition between crystalline and glassy groundmass.

Having derived the supposedly appropriate nucleation and growth rates by this method, namely peak nucleation rates of 10 nuclei $cm^{-3} sec^{-1}$ and 7,500 nuclei $cm^{-3} sec^{-1}$ for diopside and plagioclase respectively, and peak growth rates of 5×10^{-5} cm sec^{-1} for both, crystallization at various constant rates of heat loss was simulated. Results of this are also shown in Table 2. It is the grain-size distributions from this type of model that are expected to be close to those of the natural samples, and from which actual rates of heat loss could be obtained by comparison. However, because of obvious oversimplification of the basalt composition and crystallization sequence as well as other approximations, such quantitative comparisons are unwarranted at this time. The correspondence shown in Table 2 is only approximate. These models do shed some light qualitatively on several aspects of the crystallization of the basalts.

Again there is a distinct difference between constant heat loss and constant temperature decrease, as shown in Figure 14. In the constant heat loss models (Figure 14B), at least five different time-temperature regions may be identified: (1) a rapid temperature drop until pyroxene is nucleated; (2) a very rapid rise back to within 2 degrees of the liquidus; (3) a long interval of pyroxene crystallization with moderately decreasing temperature; (4) after the delayed nucleation of plagioclase, a very rapid rise back to near the cotectic: this rise also involves a rapid change in liquid composition (see also Figure 15B); (5) an interval of near-cotectic crystallization with only slowly decreasing temperature. The constant-temperature-decrease model (Figure 14A) shows similar trends in ΔT, but there is a notable difference in behavior after plagioclase begins to crystallize. Since the temperature is not allowed to rise, all the plagioclase crystallization takes place at a very high ΔT, and the liquid composition does not revert to the cotectic (Figure 15A). The differences between the two modes of crystallization are undoubtedly exaggerated in these simple models as opposed to a more realistic system. Adding more components and allowing solid solution in pyroxene would tend to even out the slopes of different parts of the liquidus, and some heterogeneous nucleation could reduce the dips and subsequent rebounds caused by delayed nucleation. However, some differences will persist, and it will clearly not be reasonable to expect an extremely exact reproduction of the textures of natural rocks by laboratory crystallization at constantly decreasing temperature. In the absence of experiments carried out at constant rate of heat loss, which

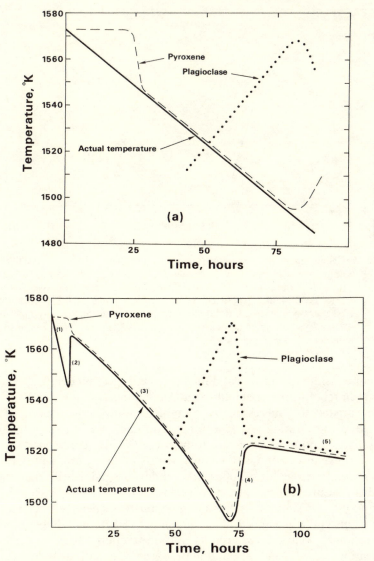

Figure 14. Temperature versus time plots for simulated crystallization of model pyroxene-phyric basalt ($Di_{60}An_{30}Ab_{10}$): (a) constant temperature decrease, 1 deg hr^{-1}; (b) constant heat-loss, .001 cal sec^{-1}. The numbers in (b) denote time-temperature regimes discussed in the text. The solid lines show the actual temperature, and the dashed and dotted lines show the liquidus temperatures for pyroxene and plagioclase. The point at which the dashed and dotted lines cross indicates the time at which the liquid reached the cotectic composition; the plagioclase liquidus is metastable to the left of this point, and the pyroxene liquidus is metastable to the right, though in the constant-heat-loss model, the curves converge to near the cotectic (stage 5).

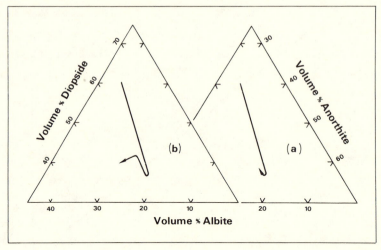

Figure 15. Change in composition of the liquid with time during simulated crystallization of pyroxene phyric basalt: (a) constant temperature decrease, 1 deg hr^{-1}; (b) constant heat loss, .001 cal sec^{-1}. In both cases, the liquid overshoots the cotectic, crystallizing pyroxene alone, until the critical ΔT of nucleation for plagioclase is exceeded. In (a) all plagioclase crystallizes at a very high super-cooling and the liquid never attains the cotectic. In (b), the liquid reverts to near the cotectic (see Figure 14).

present considerable technical difficulties, numerical modeling should be helpful in relating experimental to natural textures.

The models are successful in reproducing only some aspects of the textures of the natural rocks. The initial short interval of high supercooling of pyroxene and rapid reversion to near the liquidus corresponds to the limited range in size of pyroxene phenocrysts in the natural and synthetic rocks. Combined with the apparently random spatial distribution, this is highly suggestive of homogeneous or quasi-homogeneous nucleation. The phenocrysts could have nucleated on previously-formed spinel crystals or iron droplets (Grove and Walker, 1977), but if they did, almost all such nucleation sites must have been used up in the initial burst of pyroxene nucleation. Solid solution in the pyroxene was not included in the modeling, but extreme zoning would obviously occur over the long interval of crystallization, as observed in the actual rocks. The long-delayed nucleation of plagioclase in the models corresponds with in-ferences drawn from the increase in the Al contents of the natural pyroxene phenocrysts up to very high levels, followed by a sharp drop at the time of plagioclase nucleation (for example, Hollister et al., 1971). The models reproduce the sharp reversal in plagioclase content of the melt (Figure 15).

In the models, plagioclase crystallizes mainly at very high ΔT, and some reverse zoning occurs: these results correlate with the skeletal morphology of the natural plagioclase crystals and observed reverse zoning (Crawford, 1973).

Of the natural and synthetic basalts, the most important feature *not* reproduced by the numerical models is the bimodal grain-size distribution of pyroxene; pyroxene in the groundmass of both the natural rocks and the laboratory equivalents is about the same size as plagioclase. Lofgren et al. (1974) suggested that a change in slope of the liquidus surface at the pyroxene-plagioclase cotectic could cause a second increase in super-cooling of pyroxene and a new episode of nucleation. However, super-cooling of pyroxene is not noticeably affected when the plagioclase liquidus is reached, or even when plagioclase is growing rapidly. The precise mechanism of production of a bimodal grain size distribution of a mineral during single-stage cooling thus remains obscure. Lofgren (this volume) summarizes the conditions under which such textures have been found to occur. It should be noted that this problem is independent of the question of the complexity of the cooling history of the natural pyroxene-phyric basalts (e.g. Grove and Walker, 1977); a bimodal grain-size distribution of pyroxene is obtained in all the dynamic crystallization experiments as well as the natural rocks. It was suggested by Dowty et al. (1974) that the groundmass was heterogeneously nucleated, and the extremely high nucleation rate necessary in the models to account for the fine grain size of plagioclase is certainly consistent with this. It is not clear, however, why little or no heterogeneous nucleation of pyroxene occurred during the long period of crystallization of the pyroxene phenocrysts, and why the initial appearance of plagioclase is so long delayed if its nucleation is totally heterogeneous. It is perhaps possible that cavitation is involved, there being ample evidence of very rapid and chaotic crystallization in the groundmass stage (see, for example, Boyd and Smith, 1971).

When rates deduced from the constant-temperature-drop models are applied to the constant-heat-loss model, the difference in size between phenocrysts and groundmass is decreased. This exacerbates a problem, pointed out by Grove and Walker (1977), that this difference tends to be too small in the laboratory experiments as compared to the natural rocks. The experiments of Grove and Walker were done in iron capsules and they obtained generally smaller phenocrysts than Lofgren et al. (1974), but phe-nocrysts are too small even in the latter's experiments. The most plausible explanations for the discrepancy are either a complex cooling history of the natural rocks (Grove and Walker, 1977), or excess heterogeneous nucleation in the experimental charges (Donaldson et al., 1975). Further experiments and modeling seem to be required to ascertain the degree to which experimental results can be applied *quantitatively* to these and other natural rocks.

CONCLUSIONS

Application of crystal growth and nucleation theory, combined with laboratory experiments, has resulted in a number of recent advances in the understanding of igneous crystallization. However, the use of theory has so far been almost entirely qualitative. Before more quantitative results can be obtained, some improvements in theory and the acquisition of many critical data are required.

For quantitative prediction of nucleation rates, much more must be learned about heterogeneous nucleation. Recognizing heterogeneous nucleation, either in natural igneous rocks or in experimental charges, is usually a problem of interpreting textures. Developing reliable criteria for distinguishing modes of nucleation will be very important in relating the observed textures to the cooling histories of igneous rocks.

The most notable shortcoming of crystal-growth theory seems to be the inability of the classical expressions for the surface-nucleation mechanism to describe growth rates at small ΔT. Clearing up this problem will be very important in dealing with plutonic rocks, because surface nucleation seems to be a common mechanism in minerals, as shown by morphology and measured growth rates as a function of ΔT.

To model the kinetics of a system, the equilibrium phase relations, both stable and metastable, must usually be known. At present, it is not possible to represent analytically the equilibrium phase relations in complex igneous systems with accuracy, even if the measurable thermodynamic data are available, because models of the entropy of mixing are unsatisfactory. This in turn can be attributed to the lack of knowledge of the actual atomic structures in silicate melts. Theories of crystal-growth and nucleation based on quantum-mechanical approaches, which may be necessary for truly quantitative predictions, will also require knowledge of melt structures.

The parameters that chiefly determine the dependence of growth and nucleation rates on ΔT—that is, the surface energy and the enthalpies of fusion—are also poorly known, as are the enthalpies of mixing of end-member igneous components: The latter enthalpies are required for modeling both equilibrium relations and growth rates.

Finally, despite the attention paid by petrologists to textures, there are essentially no quantitative data on grain size distribution with which to assess such phenomena as the prevalence in nature of heterogeneous versus homogeneous nucleation, or to test theory or models.

We can expect advances in many of these areas in the near future. The present situation with respect to crystallization kinetics is in many ways analogous to that in silicate phase equilibria when Bowen began his work at the beginning of this century. The most basic aspects of the thermodynamic theory of solid-liquid equilibria were known, and some possibly analogous systems, such as alloys and salt-water systems, had been investigated in detail. It remained however, to provide the all-important experimental data on silicates themselves. The acquisition of data on heats of melting, growth rates, surface energies, and the like, now appears to be a formidable task, which may take as much time and labor as the acquisition of the phase equilibria; but the ultimate benefits could be comparable.

ACKNOWLEDGMENTS

I thank Frank Spera for many helpful discussions. This work is supported in part by NSF Grant EAR76-22543.

REFERENCES

Albarede, F. and Bottinga, Y., 1972. Kinetic disequilibrium in trace-element partition between phenocrysts and host lava. *Geochim Cosmochim. Acta 36,* 141–156.

Anderson, A. T., 1967. Possible consequences of composition gradients in basalt glass adjacent to olivine phenocrysts. *Trans. Am. Geophys. Union. 48,* 227–228.

Baker, J. C. and Cahn, J. W., 1971. Thermodynamics of solidification. In *Solidification.* Am. Soc. for Metals, Metals Park, Ohio.

Barin, I. and Knacke O., 1973. *Thermochemical Properties of Inorganic Substances.* Springer, Berlin.

Barin, I., Knacke, O., and Kubaschewski, O., 1977. *Thermochemical Properties of Inorganic Substances: Supplement.* Springer, Berlin.

Becke, F., 1911. Uber die Ausbildung der Zwillingskristalle, *Fortschr. Mineral. Kristallogr. Petrographie 1,* 68–85.

Bennema, P. and Gilmer, G. H., 1973. Kinetics of crystal growth, In *Crystal Growth: An Introduction.* P. Hartman, (ed.), North-Holland, Amsterdam.

Benson, S. W., 1960. *The Foundations of Chemical Kinetics.* McGraw-Hill, New York.

Berner, R. A., 1975. Role of magnesium in the crystal growth of calcite and aragonite from seawater. *Geochim. Cosmochim. Acta. 39,* 489–504.

Bottinga, Y., Kudo, A., and Weill, D., 1966. Some observations on oscillatory zoning and crystallization of magmatic plagioclase, *Am. Mineral. 51,* 792–806.

Bowen, N. L., 1913. The melting phenomena of the plagioclase feldspars. *Am. J. Sci. 35,* 577–599.

Boyd, F. R. Jr. and Smith, D., 1971. Compositional zoning in pyroxenes from lunar rock 12021, Oceanus Procellarum. *J. Petrol. 12*, 439–464.

Brice, J. C., 1973. *The Growth of Crystals from Liquids.* North-Holland, Amsterdam.

Buckley, H. E., 1951. *Crystal Growth.* John Wiley and Sons, New York.

Burnham, C. W. and Davis, N. F., 1974. The role of H_2O in silicate melts: II. Thermodynamic and phase relations in the system $NaAlSi_3O_8$–H_2O to 10 kilobars, 700° to 1,100°C. *Am. J. Sci. 274*, 902–940.

Burton, J. A., Prim, R. C., and Slichter, W. P., 1953. The distribution of solute in crystals grown from the melt. *J. Chem. Phys. 21*, 1987–1989.

Burton, W. K., Cabrera, N. and Frank, F. C., 1951. The growth of crystals and the equilibrium structure of their surfaces. *Phil. Trans. Roy. Soc. 243*, 299–358.

Cahn, J. W., 1960. Theory of crystal growth and interface motion in crystalline materials. *Acta. Met. 8*, 554–562.

Cahn, J. W., 1967. On the morphological stability of growing crystals. In *Crystal Growth*, H. S. Peiser, (ed.), Pergamon Press, Oxford.

Cahn, J. W., Hillig, W. B., and Sears, G. W., 1964. The molecular mechanism of solidification. *Acta. Met. 12*, 1421–1439.

Calvert, P. D. and Uhlmann, D. R., 1972. Surface nucleation growth theory for the large and small crystal cases and the significance of transient nucleation. *J. Crystal Growth 12*, 291–296.

Carmichael, I. S. E., Turner, F. J., and Verhoogen, J., 1974. *Igneous Petrology.* McGraw-Hill, New York.

Chakraverty, B. K., 1973. Heterogeneous nucleation and condensation on substrates. In *Crystal Growth: An Introduction*, P. Hartman, (ed.), North-Holland, Amsterdam.

Chalmers, B., 1964. *The Principles of Solidification.* Wiley, New York.

Christian, J. W., 1965. *The Theory of Transformations in Metals and Alloys.* Pergamon Press, Oxford.

CNRS—Centre National de la Recherche Scientifique, 1965. *Adsorption et Croissance Cristalline.* CNRS, Paris.

Crawford, M. L., 1973. Crystallization of plagioclase in mare basalts. *Proc. 4th Lunar Sci. Conf.*, 705–717.

de Groot, S. R. and Mazur, P., 1962. *Non-Equilibrium Thermodynamics.* North-Holland, Amsterdam.

DeLuca, J. P., Eagen, R. J., and Bergeron, C. G., 1969. Crystallization of $PbO \cdot 2B_2O_3$ from its supercooled melt. *J. Am. Ceram. Soc. 52*, 322–326.

Denbigh, K., 1968. *The Thermodynamics of the Steady State.* Wiley, London.

Donaldson, C. H., 1975. Calculated diffusion coefficients and the growth rate of olivine in a basalt magma. *Lithos 8*, 163–174.

Donaldson, C. H., 1976. An experimental investigation of olivine morphology. *Contrib. Mineral. Petrol. 57*, 187–213.

Donaldson, C. H., Usselman, T. M., Williams, R. J., and Lofgren, G. E., 1975. Experimental modelling of the cooling history of Apollo 12 olivine basalts. *Proc. 6th Lunar Sci. Conf.*, 843–870.

Donnay, J. D. H. and Donnay, G., 1961. "Assemblage liaisons" et structure cristalline. *C. R. Acad. Sci. Paris 252*, 908–909.

Donnay, J. D. H. and Harker, D., 1937. A new law of crystal morphology extending the law of Bravais. *Am. Mineral. 22*, 446–467.

Dowty, E., 1976a. Crystal structure and crystal growth: I. The influence of structure on morphology. *Am. Mineral. 61*, 460–469.

Dowty, E., 1976b. Crystal structure and crystal growth: II. Sector zoning in minerals. *Am. Mineral. 61*, 460–469.

Dowty, E., 1977. Relative mobility of cations and anions in minerals [abstract] *Geol. Soc. Amer. Abstr. Prog. 9*, 955–956.

Dowty, E., 1978. The importance of adsorption in igneous partitioning of trace elements. *Geochim. Cosmochim Acta 41*, 1643–1646.

Dowty, E., Keil, K., and Prinz, M., 1974. Lunar pyroxene phyric basalts: crystallization under supercooled conditions. *J. Petrol. 15*, 419–453.

Dunning, W. J., 1969. General and theoretical introduction. In *Nucleation*. A. C. Zettlemoyer, (ed.), Marcel Dekker Inc., New York.

Eitel, W., 1954. *The Physical Chemistry of the Silicates*. University of Chicago Press.

Fenn, P. M., 1977. The nucleation and growth of alkali feldspars from hydrous melts. *Can. Mineral. 15*, 135–161.

Fleet, M. E., 1975. The growth habits of olivine—a structural interpretation. *Canadian Mineral. 13*, 293–297.

Flemings, M. C., 1974. *Solidification Processing*. McGraw-Hill, New York.

Folk, R. L., 1974. Natural history of crystalline calcium carbonate. Effect of magnesium content and salinity. *J. Sed. Petrol. 44*, 40–53.

Frank, F. C., 1949. The influence of dislocations on crystal growth. *Discuss. Faraday Soc. 5*, 48–54.

Frondel, C., 1978. Characters of quartz fibers. *Am. Mineral. 63*, 17–27.

Gebhardt, M., 1973. Epitaxy. In *Crystal Growth: An Introduction*. P. Hartman, (ed.), North-Holland, Amsterdam.

Gibb, F. G. F., 1974. Supercooling and the crystallization of plagioclase from a basaltic magma. *Mineral. Mag. 39*, 641–653.

Gilmer, G. H. and Bennema, P., 1972. Computer simulation of crystal surface structure and growth kinetics. *J. Crystal Growth 13/14*, 148–153.

Glicksman, M. E. and Schaeffer, R. J., 1967. Investigation of solid/liquid interface temperatures via isenthalpic solidification. *J. Crystal Growth 1*, 297–310.

Grey, N. H., 1970. Crystal growth and nucleation in two large diabase dikes. *Can. J. Earth Sci. 7*, 366–375.

Grove, T. L. and Walker, D., 1977. Cooling histories of Apollo 15 quartz normative basalts. *Proc. 8th Lunar Sci. Conf.* 1501–1520.

Hardy, S. C. and Coriell S. R., 1969. Morphological stability of cylindrical ice crystals. *J. Crystal Growth 5*, 329–337.

Harloff, C., 1927. Zonal structure in plagioclases. *Leidsche Geol. Med. 2*, 99–114.

Hartman P., 1973a. The equilibrium form in a phase of small dimensions. In *Crystal Growth: An Introduction*. P. Hartman, (ed.), North-Holland, Amsterdam.

Hartman, P., 1973b. Structure and morphology. In *Crystal Growth: An Introduction*, P. Hartman, (ed.), North-Holland, Amsterdam.

Hartman, P. and Perdok, W. G., 1955. On the relations between structure and morphology of crystals. *Acta. Cristallogr. 8*, 49–52; 521–524; 525–529.

Herring, C., 1951. Some theorems on the free energies of crystal surfaces. *Phys. Rev. 82*, 87–93.

Hillig, W. B., 1966. A derivation of classical two-dimensional nucleation kinetics and associated growth laws. *Acta. Met. 14*, 1868–1869.

Hollister, L. S. and Gancarz, A. J., 1971. Compositional sector zoning in clinopyroxene from the Narce area, Italy. *Am. Mineral. 56*, 959–979.

Hollister, L. S., Trzcienski, W. E., Jr., Hargraves, R. B., and Kulick, C. G., 1971. Petrogenetic significance of pyroxenes in two Apollo 12 samples. *Proc. 2nd Lunar Sci. Conf.* 529–557.

Hopper, R. W. and Uhlmann D. R., 1973. Temperature distributions during crystallization at constant velocity. *J. Crystal Growth 19*, 177–186.

Hunt, J. D. and Jackson, K. A., 1966. Nucleation of solid in an undercooled liquid by cavitation. *J. Appl Phys. 37*, 254–257.

Jackson, K. A., 1958a. Interface structure. In *Growth and Perfection of Crystals*. R. M. Doremus, B. W. Roberts, and D. Turnbull, (eds.), John Wiley and Sons, New York.

Jackson, K. A., 1958b. Kinetics of alloy solidification. *Can. J. Phys. 36*, 683–691.

Jackson, K. A., 1967. Current concepts in crystal growth from the melt. In *Progress in Solid State Chemistry*, Vol. 4. Pergamon, Oxford.

Jackson, K. A., 1969. On the theory of crystal growth; the fundamental rate equation. *J. Crystal Growth 5*, 13–18.

Jackson, K. A., Uhlmann, D. R., and Hunt, J. D., 1967. On the nature of crystal growth from the melt. *J. Crystal Growth 1*, 1–36.

JANAF, 1971. *Thermochemical Tables, 2nd Edition*. NSRDS-NBS37, U.S. Government Printing Office, Washington D.C.

JANAF, 1974. *Thermochemical Tables, 1974 Supplement*. J. Phys. Chem. Ref. Data 3, 311–480.

JANAF, 1975. *Thermochemical Tables, 1975 Supplement*. J. Phys. Chem. Ref. Data 4, 1–175.

Keith, H. D. and Padden, F. J., 1963. A phenomenological theory of spherulitic crystallization. *J. Appl. Phys. 34*, 2409–2421.

Kendall, M. G. and Moran, P. A. P., 1963. *Geometrical Probability*. Charles Griffin and Company, London.

Kirkpatrick, R. J., 1974. The kinetics of crystal growth in the system $CaMgSi_2O_6$ $CaAl_2SiO_6$. *Am. J. Sci. 273*, 215–242.

Kirkpatrick, R. J., 1975. Crystal growth from the melt: a review. *Am. Mineralogist 60*, 798–814.

Kirkpatrick, R. J., 1976. Towards a kinetic model for the crystallization of magma bodies. *J. Geophys. Res. 81*, 2565–2571.

Kirkpatrick, R. J., Gilpin, R. R., and Hays, J. F., 1976. Kinetics of crystal growth from silicate melts: anorthite and diopside. *J. Geophys. Res. 81*, 5715–5720.

Kirkpatrick, R. J., 1977. Nucleation and growth of plagioclase, Makaopuhi and Alae lava lakes, Kilauea Volcano, Hawaii. *Geol. Soc. Am. Bull. 88*, 78–84.

Kirkpatrick, R. J., Kuo, L-C., and Melchior, J., 1978. Rates of nucleation and growth from programmed cooling experiments: $An_{50}Ab_{50}$ (abstr.) *Geol. Soc. Am. Abstr. Prog. 10*, 435.

Kitamura, M. and Sunagawa, I., 1977. The effective distribution of solute and the anisotropy of a crystal. *Internat. Conf. Crystal Growth, Collected Abstracts 5*, 91.

Kittl, E., 1912. Experimentelle Untersuchungen über Kristallizationsgeschwindigkeit und Kristallisationsvermögen von Silikaten. *Z. Anorg. Chem. 77*, 335–364.

Klein, L. and Uhlmann, D. R., 1974. Crystallization behavior of anorthite, *J. Geophys. Res. 79*, 4869–4874.

Kotler, G. R. and Tiller, W. A., 1968. Stability of the needle crystal. *J. Crystal Growth 2*, 287–307.

Kotze, I. A. and Kuhlmann-Wilsdorf, D., 1966. A theory of the interfacial energy between a crystal and the melt. *Appl. Phys. Lett. 9*, 96–98.

Kraus, G., 1962. Gefüge, Kristallgrössen und Genese des Kristallgranites I in Vorderen Bayerischen Wald. *N. Jb. Miner., Abh. 97*, 357–434.

Kröger, F. A., 1973. *The Chemistry of Imperfect Crystals.* Vol. 1. 2nd edition. North-Holland, Amsterdam.

Krumbein, W. C. and Pettijohn, F. J., 1938. *Manual of Sedimentary Petrology.* New York.

Kubaschewski, O., Evans, E. L., and Alcock, C. B., 1967. *Metallurgical Thermochemistry*, 4th edition. Pergamon, Oxford.

Laidler, K. J., 1965. *Chemical Kinetics*, 2nd edition. McGraw-Hill, New York.

Laird, J. A. and Bergeron, C. G., 1968. Interface temperature of a crystallizing melt. *J. Am. Ceram. Soc. 51*, 60–61.

Laudise, P. A., 1970. *The Growth of Single Crystals.* Prentice-Hall, Englewood Cliffs, N. J.

Leonteva, A. A., 1948. Issledovaniye lineinoi skovosti kristallizatsii v sisteme albit-anortit-diopsid. *Zh. Fis. Khim 22*, 1205–1213.

Leonteva, A. A., 1949. Vichisleniya lineinoi skovosti kristallizatsii plagioklazov. *Dokl. Akad. Nauk SSSR 68*, 145–147.

Lofgren, G., 1974. An experimental study of plagioclase morphology: isothermal crystallization. *Am. J. Sci. 273*, 243–273.

Lofgren, G., Donaldson, C. H., Williams, R. J., Mullins, O., and Usselman, T. M., 1974. Experimentally reproduced textures and mineral chemistry of Apollo 15 quartz-normative basalts. *Proc. 5th Lunar Sci. Conf.* 549–567.

Loomis, T. P., 1980. Isothermal crystallization of plagioclase in the system *An-Ab-water* (preprint).

Lothe, J. and Pound, G. M., 1969. Statistical mechanics of nucleation. In *Nucleation.* A. C. Zettlemoyer, (ed.), Marcel Dekker, New York.

Matthews, J. W., 1975. *Epitaxial Growth.* Academic Press, New York.

McLachlan, D., Jr., 1974. Some factors in the growth of crystals. Part I. Extension of the Donnay-Harker Law. *Bull. Univ. Utah, Eng. Exptl. Sta. 57*, 1–12. Reprinted in *Crystal Form and Structure* (1977) C. J. Schneer (ed.), Dowden, Hutchinson and Ross, Stroudsburg, Penn.

Meiling, G. S. and Uhlmann, D. R., 1967. Crystallization and melting kinetics of sodium disilicate. *Phys. Chem. Glasses 8*, 62–68.

Murase, T. and McBirney, A. R., 1973. Properties of some common igneous rocks and their melts at high temperatures. *Geol. Soc. Am. Bull. 84*, 3563–3592.

Nagel, S. R. and Bergeron, C. G., 1972. Crystallization of SrB_4O_7 from its melt. In *Advances in Nucleation and Crystallization in Glasses.* L. L. Hench and S. W. Freiman, (eds.), American Ceramic Soc. Spec. Publ. 5.

Nakamura, Y., 1973. Origin of sector-zoning of igneous clinopyroxenes. *Am. Mineral. 58*, 986–990.

Ohara, M. and Reid, R. C., 1973. *Modeling Crystal Growth Rates from Solution.* Prentice Hall, Englewood Cliffs, N.J.

Pannetier, G. and Souchay, P., 1967. *Chemical Kinetics.* Elsevier, Amsterdam.

Prigogine, I., 1961. *Thermodynamics of Irreversible Processes.* Interscience, New York, 1961.

Ree, T. and Eyring, H., 1958. The relaxation theory of transport phenomena. In *Rheology,* Vol. 2. F. R. Eirich, (ed.), Academic Press, New York.

Robie, R. A., Hemingway, B. S., and Fischer, J. R., 1978. *Thermodynamic Properties of Minerals and Related Substances at 298.15 K and 1 Bar (10^5 Pascals) Pressure and at Higher Temperatures.* U.S. Geol. Survey Bull. 1452.

Scherer, G., 1974. Crystal Growth in Binary Silicate Glasses. Ph.D. Thesis, Mass. Inst. Tech; quoted by Kirkpatrick (1975).

Schneer, C. H., 1970. Morphological basis for the reticular hypothesis. *Am. Mineral. 55,* 1466–1488.

Schneer, C. H., 1978. The morphological complements of crystal structures. *Canadian Mineral. 16,* 465–470.

Schumoff-Deleano, V. and Dittler, E., 1911. Einige Versuche zur Bestimmung des Kristallisationsvermögens von Mineralien. *Neues Jahrb. Min. Geol. Paleon. Centralblatt.* 753–757.

Sekerka, R. F., 1973. Morphological Stability. In *Crystal Growth: An Introduction.* P. Hartman, (ed.), North-Holland, Amsterdam.

Shaw, H. R., 1965. Comments on viscosity, crystal settling and convection in granitic magmas. *Am. J. Sci. 263,* 120–152.

Sipling, P. J. and Yund, R. A., 1974. Kinetics of Al/Si disordering in alkali feldspar. In *Geochemical Transport and Kinetics.* A. W. Hofmann, B. J. Giletti, H. S. Yoder, Jr., and R. A. Yund, (eds.), Carnegie Inst. Wash. Publication 634.

Smith, J. V., 1974. *Feldspar Minerals,* Vol. 2. Springer, New York.

Strickland-Constable R. F., 1968. *Kinetics and Mechanism of Crystallization from the Fluid Phase and of the Condensation and Evaporation of Liquids,* Academic Press, New York.

Swanson, S. E., 1977. Relation of nucleation and crystal-growth rate to the development of granitic textures. *Am. Mineral. 62,* 966–978.

Tammann, G., 1898. Uber die Abhangigkeit der Zahl der Kerne. *Z. Phys. Chem. 25,* 441–479.

Tammann, G. and Mehl, R. F., 1925. *The States of Aggregation,* D. Van Nostrand, New York.

Temkin, D. E., 1966. Interface structures. In *Crystallization Processes,* Consultants Bureau, New York.

Temkin, M., 1945. Mixtures of fused salts as ionic solutions. *Acta Phys-chim. USSR 20,* 411–420.

'T Hart, J., 1978. The structural morphology of olivine: I. A qualitative derivation. *Can. Mineral. 16,* 175–186.

Tiller, W. A., 1963. Principles of solidification. In *The Art and Science of Growing Crystals,* J. J. Gilman, (ed.), John Wiley, New York.

Tiller, W. A., 1977. On the cross-pollination of crystallization ideas between metallurgy and geology. *Phys. Chem. Minerals 2,* 125–151.

Turnbull, D., 1948. Transient nucleation. *Trans. AIME 175,* 774–783.

Turnbull, D., 1950. Formation of crystal nuclei in liquid metals. *J. Appl. Phys. 21,* 1022–1028.

Turnbull, D., 1952. Kinetics of solidification of supercooled liquid mercury droplets. *J. Chem. Phys. 20*, 411–424.

Turnbull, D., 1956. Phase Changes. *Solid State Phys. 3*, 225–306.

Turnbull, D. and Cohen, M. H., 1960. Crystallization kinetics and glass formation. In *Modern Aspects of the Vitreous State*. S. D. Mackenzie, (ed.), Butterworths, London.

Turnbull, D. and Fischer, J. C., 1949. Rate of nucleation in condensed systems. *J. Chem. Phys. 17*, 71–73.

Uhlmann, D. R., 1972. Crystal growth in glass forming systems—a review. In *Advances in Nucleation and crystallization in Glasses*. L. L. Hench and S. W. Freiman, (eds.), Am. Ceram. Soc. Spec. Pub. 5.

Van Hook, A., 1961. *Crystallization: Theory and Practice*. Reinhold, New York.

Vergano, P. J. and Uhlmann, D. R., 1970a. Crystallization kinetics of germanium dioxide: The effect of stoichiometry on kinetics. *Phys. Chem. Glasses 11*, 39–45.

Vergano, P. J. and Uhlmann, D. R., 1970b. Melting kinetics of germanium dioxide. *Phys. Chem. Glasses 11*, 39–45.

Wagstaff, F. E., 1969. Crystallization and melting kinetics of cristobalite. *J. Am. Ceram. Soc. 52*, 650–654.

Walker, D., Kirkpatrick, R. J., Longhi, J., and Hays, J. F., 1976. Crystallization history and origin of lunar picritic basalt sample 12002: Phase-equilibria and cooling-rate studies. *Geol. Soc. Amer. Bull. 87*, 646–656.

Walton, A. G., 1969. Nucleation in liquids and solution. In *Nucleation*. A. C. Zettlemoyer, (ed.), Marcel Dekker, New York.

Winkler, H. G. F., 1947. Kristallgrösse und Abkühlung. *Heidelberger Beitr. Mineral. Petrol. 1*, 87–104.

Winkler, H. G. F., 1948. Zusammenhang zwischen Kristalogrösse und Salbandabstand bei magmatischen Gang-Intrusionen. *Heidelberger Beitr. Mineral. Petrol. 1*, 251–268.

Winkler, H. G. F., 1949. Crystallization of basaltic magma as recorded by variation of crystal size in dikes. *Mineral. Mag. 28*, 557–574.

Wood, B. J. and Fraser, D. G., 1978. *Elementary Thermodynamics for Geologists*. Oxford University Press, Oxford.

Woodruff, D. P., 1973. *The Solid-Liquid Interface*. Cambridge University Press.

Wulff, G., 1901. Zur Frage der Geschwindigkeit des Wachstums und der Auflösung der Kristallflächen. *Z. Kristallogr. 34*, 449–530.

Chapter 11

EXPERIMENTAL STUDIES ON THE DYNAMIC CRYSTALLIZATION OF SILICATE MELTS

GARY LOFGREN

Geology Branch, SN6, NASA Johnson Space Center, Houston

INTRODUCTION

Experimental crystallization of melts with cooling histories simulating those in nature (dynamic crystallization) is both a very old and a very new approach to studying the crystallization features of rocks and minerals. James Hall (1805) sealed powdered basalt from Arthur's Seat in Edinburgh in an iron pipe and placed it in a blacksmith's forge. He cooled the lava at different rates by either cooling in air, in hot coals, or quenching in water. Visual examination of the products convinced Hall that basalts crystallized from a silicate liquid and were not deposited from solution on the ocean floor as was believed by some at the time. Impressed by Hall's results, Watt (1804) melted "several hundred-weight" of basalt in a reverberatory furnace and let it flow out into a puddle. He covered the miniature lava flow with hot coals for eight days. The flow was 3.5 × 2.5 feet in plan and 4 inches thick at one end and 18 inches at the other. He observed a variety of textures similar to natural basalts and as a result strongly supported the work of Hall.

More systematic studies of the cooling and crystallization of basalt were conducted and reviewed by Foùque and Michel-Lévy (1882). They were hampered by the lack of thermocouples (yet to be invented) and could not relate their observations to accurate temperatures. They were successful, however, in reproducing textures that resembled those observed in natural basalts. Two striking watercolors (reproduced here as line drawings from their 1882 book, Figure 1) attest to their success. The attempts of Hall and Fouqué and Michel-Lévy were pioneering, but a thorough knowledge of the equilibrium phase relations of rock and mineral systems and a means to accurately measure temperature were needed before a basic understanding could be achieved. From the turn of the century, the study of phase equilibria in silicate systems (for which Bowen is famous) has progressed considerably, and these studies provide a foundation of data upon which to base dynamic crystallization studies.

Renewed interest in dynamic crystallization experiments can be attributed primarily to research in the semi-conductor industry aimed at understanding the details of the growth of ideal single crystals. These efforts not only provided a theoretical and empirical basis for understanding crystal growth, but also provided the technology from which the necessary new experimental apparatus were developed.

Contemporary studies are of two types. The more systematic approach involves the study of one-or-more-component mineral systems in which the detailed growth properties of a given mineral can be most easily isolated and understood. Another and perhaps more impatient approach is based on the belief that textures can be best interpreted by crystallizing whole-rock compositions. Both approaches have produced new and useful insight, with the studies on mineral systems in general providing a basis for interpreting the more complex experiments on rock systems. The first recent study of rock texture was an attempt to duplicate pegmatite (Jahns and Burnham, 1958). The current phase began with studies on mineral systems (e.g., Lofgren, 1974a,b; Fenn, 1974; Kirkpatrick, 1974) followed by studies on lunar basalts (e.g., Lofgren et al., 1974, 1975a; Donaldson et al., 1975a; Usselman et al., 1975; and Walker et al., 1976). Studies have also been made on terrestrial basaltic and granitic systems and their analogues (e.g., Lofgren and Donaldson, 1975a,b; Long, 1976; Swanson, 1977). Studies on lunar basalts have been reviewed, in part, by Donaldson et al. (1976) and studies on the feldspar and granitic systems have briefly by Luth (1976).

In this chapter I will review the above studies plus more recent studies and some as yet unpublished data for the insight they provide into the origin of rock textures. Crystal morphology, chemical zoning, and mineral intergrowths will be considered as major features of rock textures. I will discuss the philosophy behind dynamic crystallization studies, outline the techniques, and summarize some of the early results. Studies in this area are increasing rapidly and it is hoped that this chapter will provide a background for appreciation of future developments.

ARTIFICIAL OLIVINE BASALT

ARTIFICIAL LEUCITE BASALT

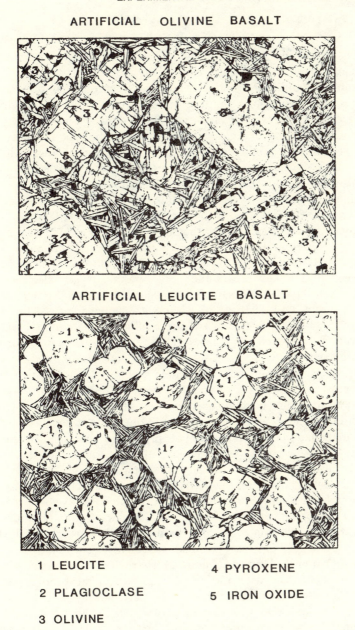

1 LEUCITE   4 PYROXENE

2 PLAGIOCLASE   5 IRON OXIDE

3 OLIVINE

Figure 1. Line drawing reproduced from Fouqué and Michel-Lévy (1882) showing their remarkable success at reproducing basaltic textures in olivine and leucite basalts.

TECHNIQUES

The basic technique for dynamic crystallization experiments differs significantly from crystallization experiments in which the object is to determine the equilibrium phase relations. To reproduce natural cooling histories, the starting material, whether it be powdered rock or synthetic analogues, is homogenized at or near the liquidus temperature of the system. The cooling and subsequent crystallization of the liquid are either continuous, usually at a linear cooling rate, or discontinuous, usually in isothermal steps. In these experiments, crystals are usually zoned and equilibrium obviously not obtained.

Dynamic crystallization studies utilize most of the standard experimental apparatus described in Ulmer (1971) or Edgar (1973). There are, however, a few new techniques that deserve mention. One of the most significant developments is the one-atmosphere, gas-mixing furnace. Williams and Mullins (1976) have improved the control and measurement of oxygen fugacity by use of the solid ceramic oxygen electrode first used by Sato (1971). The sample is suspended near the electrode and both the sample and electrode are exposed to the flowing gas. The oxygen fugacity can be adjusted to any desired value by changing the ratio of the oxidizing and reducing gases. The development of this furnance is particularly timely because it can be used to study volcanic rocks containing transition metals of variable oxidation states, which is one of the most fertile areas for dynamic crystallization studies.

Sample containment problems are severe with these new studies. The main difficulty continues to be in maintaining a constant amount of iron in the sample; to this is added the problem of controlling nucleation sites. In the dynamic studies it is generally preferable that the sample container itself not trigger the nucleation of phases during crystallization. With this in mind, Donaldson et al. (1975b) modified techniques described by Ribould and Muan (1962), Presnall and Brenner (1974), and G. McKay (personal comm.) in which a sample was suspended from a platinum wire in a gas-mixing furnance. By this modified technique, relatively large samples (up to 150–200 mg) can be suspended in a Pt-wire loop and form a bead 4–7 mm in diameter (Figure 2) from which thin sections can be easily prepared. The wire and exterior surface of the bead apparently have minimal effect on the nucleation properties of the liquid. Iron loss from the melt to the Pt-wire is still a problem and to that is added a new problem, the volatilization of sodium. The magnitude of both problems varies with bulk composition of the sample, oxygen fugacity, and temperature. The iron-loss problem can be minimized by pre-saturating the surface of the Pt-wire with iron but there is no obvious way to eliminate the sodium loss problem. The loss of sodium is less severe if a well-crys-

Pt-LOOP TECHNIQUE

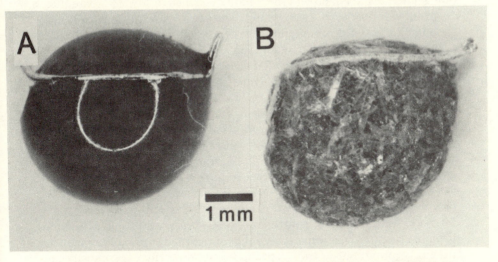

MELTED

CRYSTALLIZED

Figure 2. Glassy and crystallized beads on platinum wire loops as they appear upon removal
from the furnace. (A) Quenched from liquid state (the semicircle emanating downward from the platinum wire is a reflection of ring light used for illumination).
(B) Cooled from liquidus to subsolidus temperature.

tallized, coarse-grained sample, rather than a glassy sample, is used
as the starting material. Presumably the sodium in the feldspar is not
as easily vaporized as sodium in the glass.

For lunar basalts that are iron-saturated on the liquidus, the use
of iron crucibles sealed in evacuated silica tubes (Walker et al., 1976;
Grove and Walker, 1977) solve both the sodium and iron problems,
but under these conditions, the kinetics of nucleation appear to be different from those observed with the platinum-loop technique. The
iron-crucible technique is obviously restricted to lunar samples or
other materials (e.g., basaltic achondrites, Stolper, 1977) that crystallize
in a range of oxygen fugacity several orders of magnitude lower than
that of terrestrial volcanic rocks. In the platinum and gold capsules
used in high pressure experiments (Fenn, 1977a; Lofgren, 1974a), nucleation does not appear to be influenced by the capsule walls, but iron loss
to platinum is still a problem.

Several kinds of cooling histories have been devised: these attempt
either to duplicate natural histories or to form a basis for systematic
study of kinetic crystallization phenomena. The simplest cooling history

is the isothermal drop (Figure 3), in which the temperature of the melted or partially melted starting material is rapidly dropped to the crystallization isotherm, and either held there until quenched or lowered to another crystallization isotherm. In these experiments a known degree of supercooling can be induced initially; the degree of supercooling (ΔT) may change, of course, through evolution of latent heat and/or through constitutional changes as the charge crystallizes. The temperature increments can be of any size, depending on the goal of the experiment and the capabilities of the furnace. This cooling history does not, in general, attempt to duplicate natural circumstances, but is used to study nucleation and crystallization phenomena in relation to a known ΔT.

Continuous cooling histories, of which linear cooling (Figure 3) is the simplest, more closely resemble natural cooling histories, but the possibilities are infinite. Such studies require a means to cool a furnace continuously at known linear or non-linear rates that are repeatable. This has been accomplished by mechanical devices that in effect turn the potentiometer on a controller. Solid state devices have become available recently which require no moving parts. Linear cooling curves are easily programmed and reproduced so different experiments can

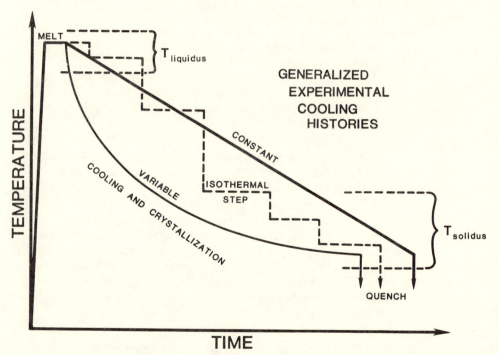

Figure 3. Schematic representation of commonly used experimental cooling histories.

be readily compared. They are also realistic because the magma cooling models of Jaeger (1968) show that the cooling is nearly linear through much of the crystallization interval. More complex histories can be programmed, in which isothermal plateaus or increases in temperature can be included. It should be remembered, however, that in cooling experiments, as opposed to isothermal experiments, ΔT is never known accurately, because as the melt crystallizes the residual melt may change composition so that its liquidus temperature is is constantly changing. This constitutional effect is augmented at fast cooling rates by difficulty in dissipation of the latent heat of crystallization. Both processes tend to lower ΔT.

KINETIC GROWTH PROPERTIES OF MINERALS

The kinetic growth properties of minerals, that is, their nucleation and growth mechanisms and rates, their shapes and zoning patterns, are important clues to unraveling their history, and that of the rock in which they occur. Theoretical and experimental aspects of nucleation and growth have been considered by Dowty in the previous chapter, but it is worth discussing here those aspects of nucleation that most directly affect the shape and zoning of minerals, mineral intergrowths, and the textural relations. Crystal shape is a valuable first-order indicator of the relative degree of supercooling of crystallization or cooling rate. Chemical zoning in minerals is less well studied and consequently less well understood, but is potentially a powerful tool for unraveling a mineral's history. Mineral intergrowths are also useful because they provide evidence of sequential crystallization.

NUCLEATION

Nucleation is generally defined as either homogeneous or heterogeneous: homogeneous nucleation occurs spontaneously in a melt, while heterogeneous nucleation occurs on some foreign substance. It is very difficult to prove whether or not a given nucleation event was homogeneous, because a foreign particle that acts as a nucleation site may be only hundreds of Angstroms in diameter or smaller and not readily visible. Heterogeneous nucleation can be obvious when the resulting crystals radiate from a capsule or crucible wall or when the distribution of nucleation centers is irregular. If the foreign particles are evenly distributed, however, heterogeneous and homogeneous nucleation can be nearly impossible to distinguish. It is usually most desirable in experimental studies to have nucleation take place within the charge and not on the sample container. Nucleation occurring *within* the melt homogeneously, or heterogeneously if triggered by a foreign particle or preexisting crystal in the melt, has been dubbed internal nucleation by Fenn (1977a). This term is appropriate because it does not attempt to resolve whether the nucleation event was homogeneous or heterogeneous, but clearly distinguishes it from nucleation on a container wall. The term external nucleation could then be used to describe nucleation on a foreign substrate.

Nucleation rate is very difficult and time-consuming to measure in silicate melts. Nucleation density, which is certainly related to rate, is easier to measure, to visualize, and usually is what is measured experimentally. The nucleation density is measured by assuming that each crystal represents a former nucleus. Spherulites are counted as one nucleus even though they contain many crystals. The nucleation rate or density is affected by many parameters, including temperature, pressure, melt composition, volatile content, and oxygen fugacity, all of which affect the viscosity of the melt. Increasing viscosity in melts of different composition tends to decrease nucleation potential for a given degree of supercooling.

The incubation period for nucleation (Turnbull, 1948) is another important aspect of nucleation that has received recent attention. The incubation period is the time required for nucleation to occur once a melt is lowered to a crystallization temperature, and, as can be easily imagined, it could be an especially important variable in cooling experiments. The effects of incubation time on crystallization have been discussed by Lofgren (1974a), Gibb (1974), Walker et al. (1976), Fenn (1977a), and Swanson (1977). In all these studies the incubation time for feldspar has been shown to be as much as a week or more and to vary considerably from run to run. For example, in identical charges run side by side in the same experiment, feldspar nucleated in one charge and not in the other in run times up to one week (Fenn, 1977a; Lofgren, 1974a). The incubation time would be expected to be a probabilistic process and does appear to mimic the nucleation rate curve; it is greatest at very low and very high ΔT values.

The nucleation of plagioclase from a Columbia River basalt collected at Picture Gorge has been studied by Gibb (1974). He cooled and crystallized a melt of this basalt at some isotherm, or cooled at some rate to a specified subliquidus temperature (Figure 4). In the time-temperature space above the upper dashed line (Figure 4) there was no nucleation; between the dashed lines, nucleation appeared to be random and did not always occur after a given amount of time. Below the lower dashed line nucleation always occurred. Gibb concluded by a variety of arguments that the rate of cooling to the isotherm, mechanical vibration, and melting time above the liquidus cannot be the cause of the random nucleation.

Heterogeneous nucleation on capsule walls of varying smoothness is a possible but unlikely cause. In an attempt to determine whether the presence of a crystalline phase could affect this plagioclase-nucleation behavior, Gibb added olivine to the starting material, so that olivine became the liquidus phase. The presence of olivine, however, did not alter the nucleation behavior of the plagioclase. Likewise, Gibb found that superheating the liquid 50 to 100°C for periods up to 20 hours did not suppress nucleation below the lower dashed line. This random nucleation of plagioclase is an important property to be considered in any dynamic crystallization study that includes plagioclase.

Based on theoretical considerations and the data represented schematically in Figure 4, Gibb suggested that for linear cooling rates the nucleation temperature of plagioclase should be lower at higher cooling rates. He further suggested that plagioclase could reverse its order of appearance with another phase because of its reluctance to nucleate. Walker et al. (1976) observed just such a phase reversal in cooling-rate studies on a lunar picritic basalt. They found that at the higher cooling rates, not only was plagioclase nucleation suppressed significantly compared to other phases, but that, contrary to equilibrium phase relations, it appeared after, not before, ilmenite. In cooling experiments on an oceanic tholeiite, Lofgren and Donaldson (1975a) found, contrary to Gibb's results, that

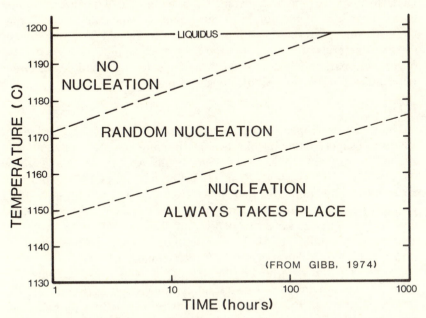

Figure 4. Nucleation behavior of plagioclase from a Columbia Plateau basalt in which plagioclase is the liquidus phase (from Gibb, 1974).

40–50°C superheat (amount by which melting temperature exceeds the liquidus) could suppress plagioclase nucleation to a greater degree than runs with only 5–10°C superheat.

A systematic study of the olivine incubation period in basaltic liquids has been completed by Donaldson (1979). He found (1) that the incubation time decreases as the degree of supercooling imposed increases, (2) that for a given degree of supercooling the incubation time increases as the degree of superheat increases, and (3) that the ΔT of nucleation increases as the cooling rate increases, that is, the temperature at which the olivine nucleates decreases with increasing cooling rate. As with the isothermal crystallization experiments, superheating the melt before cooling increases the incubation time. Increasing the amount of olivine in the liquid, however, lowers the incubation time for the same degree of superheat and cooling rate. These results are valid only for olivine in the basaltic liquid studied, but the behavior seems physically reasonable, and it is tempting to generalize the results to olivine in other liquids and even other minerals in any liquid.

Crystal Shape

The shapes of crystals, whether spherulitic, dendritic, skeletal, or equant, facted forms, have long been assumed to be a function of the rate of cooling or the degree of supercooling of the liquid in which the crystal grew. Only recently have these assumptions been confirmed by direct experiment for minerals of geologic interest. There is a significant body of data on the feldspars, olivine, and pyroxene collected from studies of either pure mineral systems or whole rock compositions.

Feldspar

The shape developed by a plagioclase crystal in response to a known amount of supercooling or a known cooling rate has been systematically studied by Lofgren (1973, 1974a, 1976) in the $Ab–An–H_2O$ and in the $Ab–An–Or–H_2O$ systems. Fenn (1977a,b) has studied the shapes of alkali feldspar as a function of ΔT in the $Ab–Or–H_2O$ and $Ab–SiO_2–H_2O$ systems. Feldspar in the $Ab–An–Or–Q–H_2O$ system has been studied by Swanson (1977), Lofgren (1976), and Long (1976). The shapes of plagioclase crystallizing from basaltic liquids have been studied by Lofgren (1974a), Lofgren et al. (1974), Usselman et al. (1975).

Lofgren (1974a) determined the shapes of plagioclase crystals that grew in response to a known amount of supercooling (Figure 5). The isothermal drop technique was used with initial melting periods from 6 to 25 hours and crystallization periods from 1 to 7 days. The plagioclase shape changes from tabular crystals, thin perpendicular to (010), to elongate skeletal crystals, to dendritic crystals, to spherulitic crystals

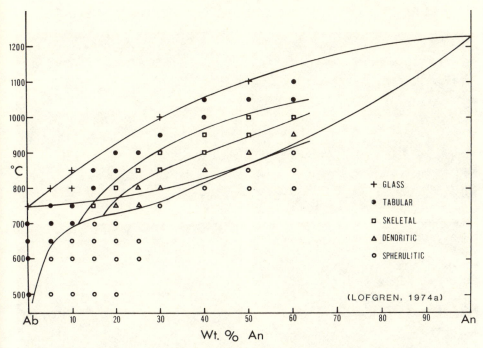

Figure 5. Plagioclase shape as a function of the degree of supercooling and composition in the Ab–An–H$_2$O system (from Lofgren, 1974a).

(Figures 5 and 6A–D), with increasing degree of supercooling. This progression in shape is found in bulk compositions of An$_{20}$ to An$_{70}$. In bulk compositions less than An$_{15}$ the shape changes directly from from tabular to radiate or spherulitic (Figure 5). The first spherulites are comprised of widely spaced, tabular fibers (Figure 6E). With increased supercooling the spherulites become more closed and finer (Figure 6F).

The crystal shapes observed by Lofgren (1974a) for An$_0$ to An$_{15}$ are similar to those observed by Fenn (1977a,b) for the alkali feldspar system. Fenn found that tabular crystals grew at the lowest ΔT's, generally less than 40°C, and that thereafter spherulitic forms grew up to the largest ΔT's. The spherulites at low ΔT have a coarse, open structure with large (generally tabular), widely spaced fibers. With increasing ΔT, the fibers become finer and are more closely spaced (see discussion of spherulite growth theory of Keith and Padden (1963) in Lofgren (1971b, 1974a). Fenn observed a dramatic change in the growth properties of the feldspars at or near the glass-transition temperature (i.e., crystallization was taking place from a glass, not a supercooled liquid). The growth rate was so suppressed that the charges retained large amounts of glass. The

EFFECTS OF SUPERCOOLING ON FELDSPAR SHAPE

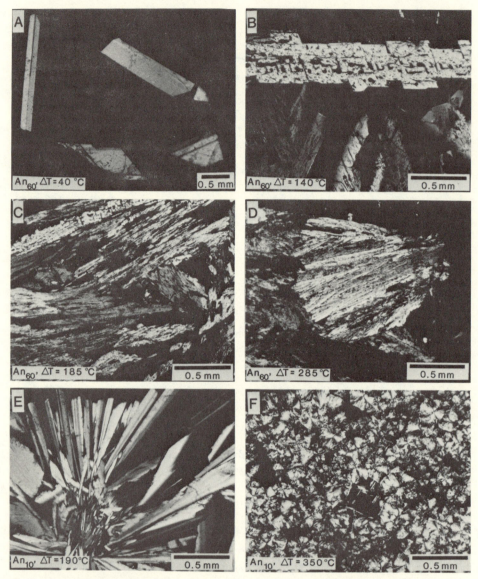

Figure 6. Photomicrographs of plagioclase showing a few of the typical feldspar shapes that are summarized in Figure 5. (A) Tabular. (B) Skeletal. (C) Dendritic. (D) Fan-spherulitic (closed, fine). (E) Fan-spherulitic (open, coarse). (F) Spherulitic (closed, fine).

spherulites grown at these large supercoolings appear to have coarser fibers with faceted tips; in reality they are probably very fine, closely packed fibers that have begun to recrystallize into tabular crystal fibers.

Fenn did not observe any effect that could be correlated with the amount of water present. Some of the water-undersaturated charges showed a greater range of shapes; this, however, may reflect growth under isothermal conditions at various ΔT's caused by the increasing water content that results from increasing fraction crystallized. Lofgren (1974a) observed a similar effect in the plagioclase feldspars which he noted was more pronounced in the undersaturated melts.

Fenn determined that the elongate growth direction in the experimentally grown alkali feldspars is parallel to the a crystallographic direction and that, in general, the growth rates along the crystallographic axes followed the trend $a > c > b$. This preferred growth direction is also found in natural crystals and other experimental studies. The only exception is that of anorthite grown by Klein and Uhlmann (1974), which grew elongate parallel to the c crystallographic direction; this is a natural growth direction for anorthite. The anorthite grown by Klein and Uhlmann (1974) from a pure anorthite melt was fibrous, but the tips of the fibers were faceted, suggesting a large entropy of fusion relative to the gas constant ($\Delta S_f > 4R$). Kirkpatrick et al. (1976) also observed faceted growth of anorthite even at large undercoolings.

Swanson (1977) found that in synthetic granite and granodiorite the plagioclase and alkali feldspar crystals have similar growth forms; that is, elongate parallel to the a crystallographic axes, which he attributes to their similar crystal structures. In the granodiorite he observed a progression of plagioclase crystal shapes with degree of supercooling that were difficult to divide by clear boundaries: "with an increase in undercooling, large single crystals give way to an open array of coarse skeletal to dendritic crystals, and finally a more compact radial mass of coarse crystals." The granite crystallized only spherulities, even at small values of ΔT. These results are in accord with both those of Lofgren (1974a) and Fenn (1977a,b). Long (1976), using a similar but more Orrich granitic melt, produced tabular, skeletal alkali feldspar at ΔT values of 20–50°C.

Cooling experiments have been conducted by Lofgren (1973) with the same plagioclase materials as those used in the isothermal drop experiments. The results are generally similar; that is, a progression from tabular to acicular to skeletal to spherulitic crystals, but there are some important distinctions (Figure 7). Dendritic forms are less common even in the An_{30}–An_{60} range of compositions. Crystals tend to be acicular (Figures 7B, C) at intermediate cooling rates and grade into fan spherulitic forms (Figure 7D) at the higher cooling rates. The acicular crystals are

EFFECTS OF COOLING RATE ON FELDSPAR SHAPE

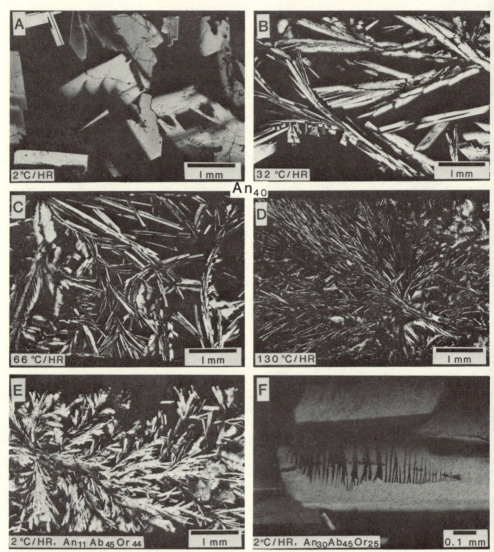

Figure 7. Photomicrographs of feldspar that were grown at different cooling rates. A–D show the effect of increased cooling rate on shapes of the plagioclase crystals that grow from an An_{40}–H_2O melt. (E) A much more An-poor, Or-rich melt cooled at the same 2°C/hr as the An_{40} melt shown in (A), demonstrates the effect of melt composition. (F) A tabular plagioclase with a skeletal interior; the external faces are planar, but the internal faces have become dendritic.

quite often arcuate (Figures 7B, C). The effect of composition is readily seen by comparing Figure 7A and 7E, both cooled at 2°C/hr; the first is from an An_{40} melt, the second is from an $An_{11}Ab_{45}Or_{44}$ melt. The $<Ab_{15}$ compositions crystallized only tabular crystals and spherulites. The effect of H_2O was not observed to be an important variable except that it may alter the value of ΔT at which a given shape is stable.

In one set of experiments the runs were serially quenched at intervals of 50°C. This allowed examination of the progressive growth of the crystals. The crystals quite often are initially skeletal, but, with continued growth, fill in to become tabular (Figure 8). The compositional effect on crystal shape observed in the isothermal drop experiments is still present. An example of infilling of a crystal by breakdown of an internal planar interface to a dendritic interface (see discussion below) is shown in Figure 7F.

Isothermal drop runs in which there is more than one crystallization plateau give an indication of the effect of ΔT on the transition from a planar to a dendritic interface. In the plagioclase system, Lofgren (1974b) reported the effect of an isothermal drop from one crystallization plateau to another in an An_{50} melt. For a 50°C drop the interface remains planar (Figure 9A). If the drop is 100°C, blocky or tabular projections occur on corners and on fast growing faces such as (001) (Figure 9B). For a 250° drop, the projections are fibrous and radiate and there may even be independent nucleation of spherulites (Figure 9C). For the An_{60} composition, spherulitic haloes often enclose preexisting tabular crystals and grow when the run is quenched; growth must take place in a few minutes (Figure 9D).

ALL CRYSTALS GROWN AT A COOLING RATE OF 2°/hr

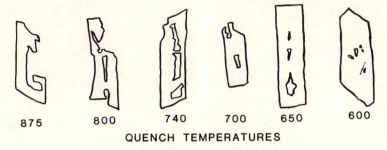

| 875 | 800 | 740 | 700 | 650 | 600 |

QUENCH TEMPERATURES

$Ab_{45}An_{37}Or_{18}$ (10 wt.% WATER)

Figure 8. Schematic representation of the observed sequence of plagioclase growth forms. The melts were quenched at the temperatures indicated. Skeletal crystals typical of the 875°C quench run must fill in when growth is allowed to proceed to lower temperatures as shown in the 600°C quench run.

FELDSPAR INTERFACE MORPHOLOGY

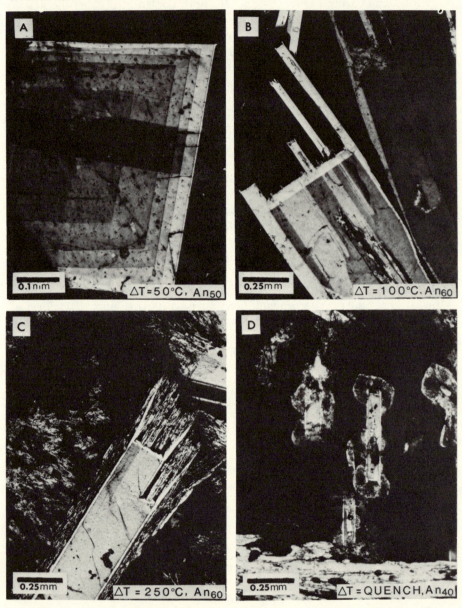

Figure 9. Photomicrographs showing the effect of changing the degree of supercooling (ΔT) during crystallization on the interface of an already growing crystal. The composition of melt and the magnitude of the change in ΔT are noted on the figure.

Plagioclase crystallized from an ocean-ridge basalt by the isothermal-drop technique (Lofgren, 1974a) showed a progression of plagioclase crystal shapes with increasing degree of supercooling similar to those observed in the pure mineral system. The shape changed from equant/tabular to acicular and skeletal to spherulitic. Complex dendritic forms were not observed even though the composition is about An_{50}, where complex dendrites are observed in the pure mineral system.

In the cooling studies on terrestrial and lunar basalts (see references in basalt section), the only important general conclusion is that, in the more complex basaltic systems, plagioclase does not display the variety of shapes evident in studies on the pure mineral systems. The crystals are generally tabular to tabular with a central cavity (belt-buckle plagioclase) to acicular, and may be skeletal. Spherulitic forms are not common, but acicular plagioclase fibers are often intergrown with pyroxene fan spherulites. Swallow-tail plagioclase is a common plagioclase shape in rapidly cooled basalts (Bryan, 1972; Downes, 1973).

The relationship between a crystal's shape and the conditions under which it grew was discussed for isothermal growth by Lofgren (1974a), but can be applied to cooling experiments as well. He summarized the treatment of spherulite growth by Keith and Padden (1963) and applied it to the varieties of plagioclase shapes observed in the experiments. Kirkpatrick (1975) treated this model more extensively and applied it more generally. The two critical parameters are the diffusion rate, D, of the rate-controlling component in the melt, and the growth rate, G, of the crystal. For growth at large degrees of supercooling, G will greatly exceed D and the ratio D/G will be on the order of 10^{-4} cm. With such a large disparity of D and G, a boundary layer enriched in the components rejected by the crystal during growth will enclose the crystal, creating a constitutionally supercooled zone. Random purturbations on the order of 10^{-4} cm occurring on a planar interface will encounter this supercooled zone and grow rapidly. The result will be the breakdown of the planar interface into a dendritic or spherulitic shape. As the ratio of D/G becomes larger, the size of the purturbations must be larger in order to be stable and grow: thus the relationship between the degree of supercooling and the size and spacing of dendritic or spherulitic fibers. If D/G becomes very large and ΔT is small, purturbations would have to be on the order of millimeters to be stable. Since these purturbations are rarely larger than a few microns, the interface remains planar and the crystal will assume its "equilibrium" shape.

Olivine

A study of the morphology of olivine as a function of crystallization conditions has been completed by Donaldson (1976). Descriptions of

Table 1. Ten Categories of Olivine Morphology (from Donaldson, 1976)

Polyhedral Olivine

    Crystals have well-formed faces (euhedral outline) and are generally equant or tabular ($a \simeq c \gg b$).

Granular Olivine

    Granular crystals are equant, anhedral and subspherical in shape. They may contain round or oval inclusions of glass or groundmass material.

Hopper Olivine ("Porphyritic Olivine" of Donaldson, 1974)

    Crystals are subequant skeletons which retain, with varying degrees of perfection, the gross outline of the common dome and prismatic forms of olivine (Figure 10A). They show a combination of planar faces, smooth curvilinear lobes and re-entrants (hopper shape), and may approach external completeness or be externally highly skeletal (Figure 10A).

Chain Olivine

    These crystals are elongate skeletons composed of linked units (Figure 10B; cf., Bryan, 1972, Figure 1B and C). They grow in random or parallel orientation. In sections parallel to either (001) or (100), the units are H-shaped, some with "tongues" connecting the bar of the H to the adjoining units (Figure 10B). The crystals may be bladed with $c > a \gg b$, or they may be thinly tabular with $a \simeq c \gg b$. Acicular varieties, elongate along a or c, are uncommon.

Lattice Olivine

    This morphology is dominated by a fine-grained dendritic mesh or lattice (Figure 10B, horizontal crystal). The dendrite fibres lie in the (010) plane and grown approximately along the poles to (101) and ($\bar{1}$01). When rotated on the Universal Stage until [101] is horizontal, the lattice morphology changes to chain shape; however, not all chain olivines have dendritic structure in the plane {010}. Crystals vary from tabular ($a \simeq c > b$) to bladed ($c > a > b$) in habit. In the former, the loci of the fibre terminations trace the directions a and c and straight extinction results. In bladed crystals the loci of dendrite terminations define the directions c and [101] and inclined extinction up to 40° results.

Plate Olivine

    These composite crystals consist of two or more parallel plates. The plates are either thinly tabular parallel to (010) or bladed ($c > a \gg b$), and both types are stacked like a pack of cards along the b crystallographic axis. Individual plates may be complete, or indented, or internally skeletal, but usually have well-developed faces at their terminations.

Branching Olivine

    Three types of this category are distinguished:
(a) Non-Crystallographic Branching Olivine. In this dendrite morphology, secondary bladed branches are inclined at 30° or less to the growth direction of the primary, bladed crystal (Donaldson, 1974, Figure 5). These branches have no fixed orientation with respect to the main crystal, other than sharing the b crystallographic axis. Hence, they are distinguished as "non-crystallographic" branches (cf. Keith and Padden, 1963).
(b) Crystallographic Branching Olivine. Since the primary crystal and the secondary branches of crystals in this category are in optical, and hence crystallographic, continuity, the morphology is distinguished as crystallographic branching (Figure 10C). Crystals are bladed ($c > a \gg b$) and the branches are inclined at approximately 60° to the c axis, growing along directions approximately perpendicular to (101) and ($\bar{1}$01).

Table 1 (*continued*)

Branching Olivine

    (c) Linked Parallel-Growth Olivine. These cellular crystals may be overall equant, bladed, or tabular in shape. They are composed of many parallel units, or "rods," elongate parallel to *a* or *c*. Adjacent "rods" are connected by "buds" or branches parallel to *a*, *b* and *c* (cf. Donaldson, 1976, Figure 5). Individual "rods" may be flattened parallel to (010) and can be regarded as substructure of an internally-skeletal plate crystal. The number of "rods" in a single crystal ranges from tens to thousands.

Radiate Olivine

    Olivine spherulites, up to .5 mm in diameter, have been found at the margins of basaltic pillows in Taiwan (Liou, 1974) and have been grown from the system forsterite-silica during rapid cooling (Blander et al., 1976). All other radiate forms of olivine known to the writer are fans of two to six blades ($c > a \gg b$) subtending 15° or less (cf. Lewis, 1972). The *a* axes of blades in a single fan are parallel, whereas the *c* axes diverge from the nucleus. Like plate olivine, the individual blades may have internal-skeletal or dendrite structure.

Feather Olivine

    Perpendicular to *a* and *c*, the shape resembles the veins of a leaf or feather (Lofgren et al., 1974, Figures 2A and 3A; Figure 10D). Growth initially involves a crystal, thinly tabular parallel to (010), which develops small, regularly spaced, spherical bumps on (010) and (0$\overline{1}$0), from which grow secondary fibers within the (100) and (001) planes. From these two sets of fibers grow tertiary fibers along *a* and *c* respectively. This gives rise to a complex mesh of regularly interconnected fibres, seen best in sections perpendicular to *b*. The loci of fibre terminations trace crystal faces of the forms {010}, {021} and {110}.

Swallow-Tail Olivine

    From an otherwise euhedral crystal, four gently arcing fibers extend from the crystal edges formed by intersection of faces of the forms {010} and {100}, or {010} and {110} (Donaldson, 1976, Figure 7A and B; Bryan, 1972, Figure 1D). Space between fibers may be partially filled by secondary fibers on the four primaries.

---

olivine shape as a function of cooling rate by Walker et al. (1976) and Bianco and Taylor (1977) are in accord with the results of Donaldson. The crystallization temperatures of forsteritic olivines are too high for conventional gas-mixing furnaces; for this reason, Donaldson studied olivine as a phase in several natural and synthetic rock compositions ranging from periodites to basalts. He performed both isothermal and cooling rate crystallization experiments at one atmosphere and at higher pressures with varying water content.

    Donaldson found that olivine has an apparently limitless variety of shapes and that it is necessary to examine the crystals on a universal stage to determine the shape in three dimensions. Using this technique he defined ten categories of olivine morphology (Table 1) that change systematically with increased cooling rate and with increased degree of supercooling (Table 2) from polyhedral or granular to hopper (Figure

Table 2a. Summary of olivine shapes in anhydrous, 1 atm cooling-rate crystallization experiments. Each entry indicates an experiment at that cooling rate (from Donaldson, 1976).

Cooling Rate °C/hr	Olivine Eucrite (E)	Apollo 12 Basalt	Apollo 11 Basalt[1]	Apollo 11 Basalt[2]	Apollo 15 Basalt	Chain 43 Basalt[3]	Chain 43 Basalt[4]	Leg 34 Basalt[5]	Leg 34 Basalt[6]
1	C + L	PO			No Olivine	No Olivine	No Olivine		
2	C + L	H	H + LPG		No Olivine	No Olivine	No Olivine	C + L	PO
5	C	H	H + LPG		H	No Olivine	No Olivine	C + L	PO
10	C	H	H + LPG		H	No Olivine	No Olivine	L + F	PO
20	C + L	L + LPG	H + C	H + C	H	No Olivine	No Olivine	No Olivine	ST
50	C + L	H	C	C	H	C	C	No Olivine	
100	C + L	C	C	C	C	L	C + L		
200	F	C + L	C	C	C + L	Glass Only	Glass Only		
300	F	C + L			L + F	Glass Only			
400	F	C + L			L + F	Glass Only			
500	F	F			F				

PO = polyhedral; H = hopper; LPG = linked parallel-growth branching; C = chain; L = lattice; G = granular; F = feather; ST = swallow-tail; PL = plate; B = branching; no oviline = charge is crystalline but lacks olivine; glass only = charge contains 100% glass; space = no experiment.

[1] .3 log f_{O_2} units less than iron-wüstite buffer
[2] 1.4 log f_{O_2} units less than iron-wüstite buffer
[3] Log f_{O_2} at melting is −7
[4] Log f_{O_2} at melting is −12
[5] Cooled from above liquidus
[6] Sample partially melted at 20°C below liquidus

Table 2b. Summary of olivine shapes in anhydrous, 1 atm isothermal crystallization experiments

Supercooling (ΔT)°C	Olivine Eucrite (E)	Apollo 12 Basalt	Apollo 11 Basalt[1]	Apollo 11 Basalt[2]	Apollo 15 Basalt	Chain 43 Basalt	Leg 34 Basalt
10	H	H	H		H		
20	H	H		H	H	H	C
30	H	H			H	H	No Olivine
40	H + LPG	H + LPG	C		H	H	No Olivine
50	H + LPG	H	C	C	H	H	No Olivine
75	LPG	C		C	C	H	No Olivine
100	L	C	No Olivine	No Olivine	C + L	C + L	No Olivine
150	L	C + L	No Olivine		C + L	L	
200	F	L	No Olivine		F	No Olivine	
250	F	C + L				No Olivine	

[1] .3 log f_{O_2} units less than iron-wüstite buffer
[2] .4 log f_{O_2} units less than iron-wüstite buffer

EFFECTS OF COOLING RATE ON OLIVINE SHAPE

Figure 10. Photomicrographs of typical olivine shapes that grow in cooling experiments on olivine-rich, basaltic melts. (A) Hopper. (B) Chain (vertical) and lattice (horizontal). (C) Branching. (D) Feather.

10A) to branching (Figure 10C) to randomly orientated chain (Figure 10B) to parallel-growth chain to chain plus lattice (Figure 10B) to plate or feather (Figure 10D). This pattern was consistent for all the melts investigated. In the isothermal-drop experiments polyhedral crystals were not observed, but Donaldson attributed this to the experimental difficulty of making runs at very small (1–2°C) degrees of supercooling. Many of the olivine shapes observed by Donaldson are very similar to those discussed by Drever and Johnston (1957) in natural rocks.

Donaldson also evaluates the effects of different properties of the melt on the shape of the olivine crystals in terms of the ratio D/G (Fleet, 1975a). Both D and G vary in a nonlinear manner with ΔT: in general, a decrease in the ratio of D/G will change the crystal growth form from, for example, polyhedral to skeletal or dendritic. Donaldson speculated that nucleation density may affect the growth rate and thus the ratio D/G. If nucleation density is high, growth will be slower because of competition for material, thus D/G is increased and polyhedral crystals might grow instead of skeletons.

The normative olivine content and viscosity of the melt can affect the crystal shape for a given cooling rate, but do not appear to affect the degree of supercooling in isothermal experiments at which a given shape is stable. For example, lattice crystals grow at a cooling rate of 3°C/hr in an olivine eucrite melt (40% normative olivine), but at 300°C/hr in an Apollo 12 olivine basalt (9% normative olvine). An ocean floor tholeiite with the same 9% normative olivine, but more viscous than the Apollo 12 basalt, contains chain crystals when cooled at 2°C/hr. The higher viscosity lowers both D and D/G thus the chain crystals are stable at slow cooling rates.

Donaldson found that the presence of water in the melt is not a major factor, but did notice some differences between wet and dry-melt crystallization. Hopper crystals often have rounded rather than planar interior faces in hydrous melts, and granular olivines are present. The water increases D and creates differences in the D/G of different crystallographic directions: these differences are sufficient to cause the faceted and nonfaceted growth. There is a greater variety of crystal forms in water-containing melts: this is attributed to a variation in the nucleation incubation period for crystals with different shapes. Variation in the incubation period can be caused by differences in the ease of nucleation in melts of differing water contents, and thus nucleation occurs over a greater range of ΔT values.

Pyroxene

The change in shape of pyroxene crystals in response to the variation in growth conditions has been studied almost exclusively in lunar basaltic systems. There has been no detailed experimental documentation of the crystal forms as a function of cooling rate like that done by Donaldson (1976) for olivine. The studies of Lofgren et al. (1974, see Figure 20, this chapter), Donaldson et al. (1975a), Usselman et al. (1975), Walker et al. (1976) and Bianco and Taylor (1977) all describe pyroxene shapes as a function of cooling rate in the range 1–2,000°C/hr. The sequence of pyroxene shapes is quite similar even though the compositions of the basalts studied varied considerably. A pyroxene of a given shape, however, did not grow at the same cooling rate in the different basalts and thus a given shape is not characteristic of a single cooling rate.

Some of the dominant crystal shapes are shown in Figure 11. At the slowest rates studied (<2°C/hr), the pyroxene is generally euhedral, nearly equant, and often slightly skeletal or blocky (Figure 11A, see upper right corner). With increasing cooling rate (2–10°C/hr) the pyroxenes become elongate and increasingly skeletal. Compare the progression from the large, central crystal in Figure 11A to the more elongate and skeletal crystals in Figure 11B and then to the even more acicular crystals in Figure 11C, which resemble chain olivines. At higher cooling rates (>10°C/hr), the pyroxenes become dendritic and then spherulitic. The first spherulites are coarse and open, so that individual fibers are readily visible and resemble curved spokes radiating from a central hub (Figure 11D). At the most rapid cooling rates (generally >50°C/hr) the pyroxenes are finely ornamented feathery or fan-spherulitic arrays of acicular fibers often complexly interwoven. All these pyroxene forms are represented in "spinifex-textured rock" described by Fleet (1975b).

ZONING

The dynamic crystallization of minerals provides the means to study chemical zoning in crystals grown under controlled conditions. Natural pyroxene and feldspar crystals show extensive variations in zoning patterns and many of these zoning features have been duplicated in laboratory grown crystals. The published literature on experimentally-grown, zoned crystals is sparse and this section will contain some unpublished results.

Feldspar

Various kinds of feldspar zoning in synthetic crystals have been discussed by Lofgren (1973, 1974a,b) and Lofgren and Gooley (1977). The starting materials were gels in either the pure plagioclase or ternary feldspar systems. Normal zoning (Lofgren, 1973; Lofgren and Gooley, 1977) such as is common in igneous feldspar, is produced in all experimentally grown feldspars that were formed while the melt cooled over a temperature

EFFECTS OF COOLING RATE ON PYROXENE SHAPE

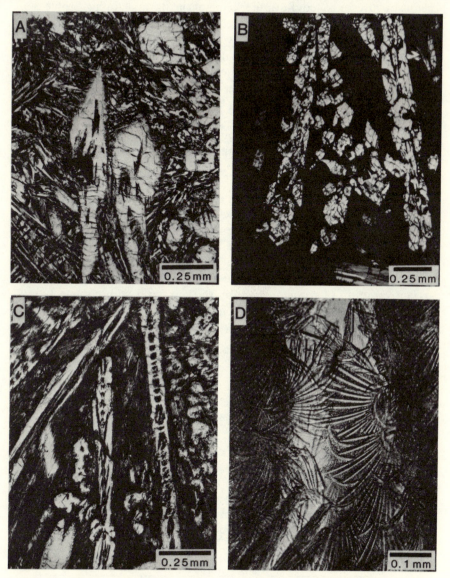

Figure 11.   Photomicrographs of typical pyroxene shapes in cooling experiments on basaltic melts. (A) Prismatic, skeletal shown in orthogonal orientations. (B) Elongate, skeletal. (C) Elongate, chain pigeonite (resembles chain olivine, but typically larger). (D) Spherulitic, wagon-wheel spoke variety typical of augite.

interval. Crystals grown at constant temperature were either homogeneous or reverse zoned. Both the shape of the normal zoning profile and the composition range vary with cooling rate. The shape of the normal zoning profile can be expressed as a function of the An content with distance from the core of the crystal. At slow rates ($\sim 2°C/hr$ for the system studied) the inner third to half of the crystal shows little zoning and the outer portion is strongly zoned (Figure 12). This zoning profile is similar to the Rayleigh fractionation zoning curves discussed by Klusman (1972). At more rapid cooling rates the zoning profile becomes increasingly linear (Figure 12). For a given composition of feldspar melt the core composition has a lower An content the higher the cooling rate (Figure 12). The rim compositions shown in Figure 12 appear to depend on quench temperature, not cooling rate. This suggests that, once the crystals start growing, the partitioning of major elements is primarily a function of temperature. Ultimately, since these are closed systems, the final rim compositions must reflect the differences in core composition. In general, the faster the cooling rate, the more restricted the zoning interval. If the crystal has time to react with the melt during growth, presumably at very slow cooling rates, a more restricted zoning interval would again result.

The variation of core composition of the crystal with cooling rate is probably a function of the nucleation temperature, although this has not

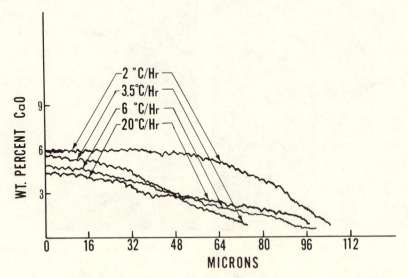

Figure 12, Normal zoning profiles in plagioclase crystals grown from an $An_{15}–H_2O$ melt at the indicated cooling rates. Crystals grown at the faster cooling rates have more linear profiles and a more restricted range of composition (see text for details).

been rigorously proven. The nucleation temperature is a function of the incubation time, which decreases with increasing cooling rate. This relationship is complicated, however, because the time needed to reach a given temperature decreases with increasing cooling rate. If the cooling rate is sufficiently rapid, the time needed to reach a given temperature can be less than the incubation time. Thus the core composition will depend on the interplay of incubation time and cooling rate. The core composition of the feldspar that forms is a function of the nucleation temperature: with increased cooling rate the core is more albitic in the case of either plagioclase or alkali feldspar.

Reverse zoning is not a common mode of zoning in natural igneous feldspar, but has been observed (e.g., Eichelberger et al., 1973). In many cases this zoning is clearly attributable to temperature fluctuations or mixing of magmas or other such phenomena. Reverse zoning, however, has been observed in experimentally grown feldspars and can be attributed to the kinetics of growth (Lofgren, 1974b); it is possible that this mechanism operating in laboratory grown crystals is responsible for some natural, reverse-zoned feldspars. All that is required is a rapidly induced ΔT that, for example, could accompany a volcanic eruption, movement of magma in a volcanic plumbing system, or the rapid release of volatiles in a water saturated magma.

Hopper and Uhlmann (1974) discuss possible models for solute (Ab-component) redistribution during isothermal growth in ceramic material. Their analysis resulted in two possibilities that they could not decide between on a theoretical basis. One model, based on local equilibrium at the interface during growth, predicts a progressive decrease in solute (increase in An content) during growth. The second model, based on a maximum free energy change resulting in growth, predicts a progressive increase in solute during growth. The reverse zoning observed in the plagioclase crystals supports the model based on local equilibrium.

The model for the reverse zoning presented by Lofgren (1974b) is consistent with the Hopper and Uhlman model for local equilibrium. It presumes that supercooling greater than 50–75°C will cause nucleation within the binary feldspar loop (Figure 13). As the crystals grow, the bulk liquid composition moves towards the liquidus at L and ΔT_1 decreases. The liquid at the interface with which the crystal is in local equilibrium, however, is initially enriched in the Ab component relative to the bulk liquid, and becomes progressively more calcium-rich as growth proceeds. This decrease in ΔT_1 reduces the driving force for crystallization and the initial disparity in diffusion rate versus growth rate (D/G) that resulted in a sodium-rich boundary layer being created is reduced. As the diffusion rate becomes more commensurate with the growth rate, the partition coefficient between liquid and crystal will approach equilibrium, and the crystal moves toward the equilibrium composition (S in Figure 13). The process can occur by repeatedly inducing another large ΔT (e.g., ΔT_2), but reverse zoning need not always accompany these changes in ΔT: this lack of zoning may reflect the random nucleation process with respect to the composition of the nuclei. If the stable nuclei happen to be close to the equilibrium composition there would be no change in crystal composition required. The intriguing feature of this reverse zoning is that it is produced isothermally and does not require reheating of the magma, an unlikely mechanism in volcanic rocks suffering heat loss during emplacement.

Sector zoning has also been observed in experimentally grown plagioclase feldspar (Lofgren, 1973) and in alkali feldspar (Long, 1976, 1978). It is most readily observed in those crystals in which the plane of the thin-section intersects the core of the crystal and is nearly parallel to a crystallographic axis. Because these requirements are not often fulfilled, sector zoning is rarely apparent in a thin section, and only portions of a sector boundary are usually visible. Some of the crystals are apparently not sector-zoned, but it is difficult to determine the percentage of each because of the orientation problem.

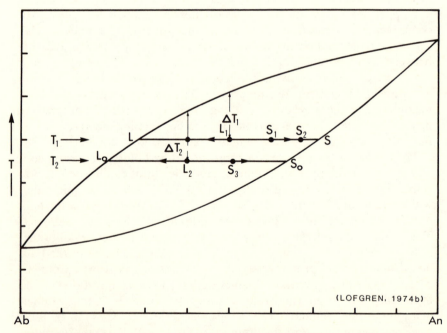

Figure 13. Schematic representation of sequence of cooling events that could result in reversely zoned crystals (see text for detailed explanation).

Sector-zoned crystals observed in cooling experiments are also normally zoned. They occur as tabular crystals (Figure 14A) in runs of intermediate and low cooling rates (1–50°C/hr), and as small crystals that often have one skeletal sector (Figure 14C) at higher cooling rates. The numerical values of the cooling rate, of course, apply only to these experiments. Sector-zoned plagioclase has also been observed in cooling experiments on lunar basalts (Lofgren, unpublished data).

In normally zoned crystals, the compositional break at the sector boundary (Figure 14B) can be observed as an offset in the zone of extinction as the microscope stage is rotated under crossed nicols and the extinction sweeps outward. This offset is largest at high cooling rates. Crystals grown at the slowest laboratory cooling rates (1–2°C/hr) often display sector boundaries with no detectable offset of the extinction band because there is very little compositional difference between sectors.

The experimentally-grown, sector-zoned plagioclase crystals resemble, and are zoned in the same sense as, natural occurrences reported recently by Bryan (1972). Bryan reported sector-zoned crystals that had either normally zoned or homogeneous sectors. The latter must have grown isothermally or nearly so, or in a chemically open system. In both the synthetic and natural crystals, the (001) sector is the most sodic and (010) the most calcic. The origin of sector zoning is currently in dispute (cf. Dowty, 1976; Kitamura and Sunagawa, 1977) and will not be discussed here.

Oscillatory zoning in plagioclase has long been a topic of discussion (cf. papers by Vance, 1962; Bottinga et al., 1966; Sibley et al., 1976); it is not clear whether it results from effects external to the liquid in which the crystal is growing (for example, periodic eruption leading to loss of water pressure), or whether it is controlled by the growth kinetics of individual crystals in a liquid. Bottinga et al., 1966 and Sibley et al., 1976 give more recent variations on the original diffusion-supersaturation model of Harloff (1927) that support the kinetic growth model. They all can explain a single zone's chemical variation, but do not provide a concrete mechanism capable of producing multiple oscillations.

Variations in pressure and temperature of the magma caused by eruption or movement can unquestionably cause compositional zones in plagioclase, as was demonstrated experimentally by Lofgren (1974b). Such a mechanism for oscillatory zoning has generally been dismissed (Vance, 1962) because of the rapidity with which the variations must occur to produce the extremely fine-scale oscillatory zoning commonly observed.

Plagioclase crystals that display oscillatory zoning have been produced in controlled laboratory cooling experiments (Lofgren, unpublished data). The crystals are few in number in a given charge, are best developed in

FELDSPAR ZONING

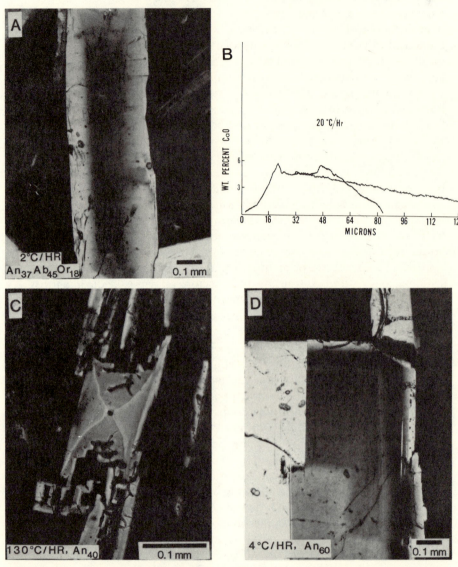

Figure 14. Photomicrographs of experimentally-grown, zoned feldspars. (A) Sector zoned crystal with planar crystal faces. (B) Microprobe traverses for CaO on a crystal that is sector-zoned in a manner very similar to that shown in (A); the traverses are perpendicular. (C) Sector-zoned crystal with planar crystal faces in one pair of sectors and dendritic interfaces in the other. (D) Oscillatory zoning in a plagioclase crystal grown at a linear cooling rate of 4°C/hr.

the An_{40-60} bulk compositional range, and show only very fine zones with very small compositional variation (Figure 14D). The zones are superimposed on a normal-zoning trend and are observable within 1–3 degrees of rotation to extinction, suggesting that the oscillations are less than one weight percent An. This resembles closely the oscillatory zoning observed by Gutmann (1977) in plagioclase from a basalt. The chemical variations are below the analytical capability of the electron microprobe. The most reasonable explanation of these experimentally-grown, oscillatory-zoned crystals is kinetic growth control.

Oscillatory-zoned crystals have been found in so few experiments that it has been difficult to explore the growth mechanism in detail. This rarity may suggest that some critical parameters are still unidentified in the experiments.

Zoning in the ternary feldspar system is not significantly different from that observed in plagioclase except for the mantling relationships of the alkali and plagioclase feldspars and the reverse zoning in sanidine. Figure 15A shows a typical example for a feldspar of $An_{19}Ab_{45}Or_{36}$. Plagioclase nucleated first near $An_{35}Or_{10}$ and zoned continuously to $An_{22}Or_{20}$, where a sanidine overgrowth nucleated at An_4Or_{75} and zoned continuously to An_3Or_{50}. This experiment was cooled 2°C/hr from 1,050°C to 770°C and about 20–30 percent glass remained. In a more calcic example, $An_{36}Ab_{45}Or_{19}$ (Figure 15B) was cooled in a series of isothermal drops of 50°C; the plagioclase core is near $An_{80}Or_{.7}$. Each zone is reverse-zoned, but the average composition of successive shows a normal zoning trend. The last plagioclase to grow has about the same An content (22) as the previous example but is less Or-rich (8–10). The first sanidine is $Or_{65}An_5$; it is normally zoned initially but becomes reverse zoned about half way through the zone. The zoning relations involving sanidine have not been explored as a function of cooling rate; all the current experiments have been done at 2°C/hr.

Pyroxene

Most published studies of pyroxene zoning in experimentally grown crystals are on lunar basalts; however, in these the pyroxenes that crystallized display a complex array of zoning trends that cover much of the pyroxene quadrilateral (Figure 16). The details of the zoning can be found in Lofgren et al. (1974); Donaldson et al. (1975a); Usselman et al. (1975); Walker et al. (1976); Usselman and Lofgren (1975); Grove and Bence (1977); and Bianco and Taylor (1977). The experiments have all involved cooling from a liquid to some partially or completely crystalline state. Pyroxene is either the first, second, or third phase to crystallize and the initial composition can range from pigeonite to augite.

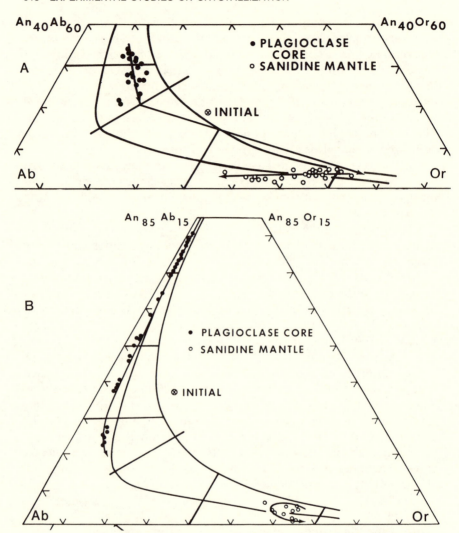

Figure 15. Zoning profiles in feldspar crystals grown from ternary feldspar melts $+H_2O$. (A) Profile in a plagioclase crystal with a sanidine overgrowth that grew from an $An_{19}Ab_{45}Or_{36}$ melt at a cooling rate of $2°C/hr$. (B) Profile in a plagioclase grown by the isothermal drop technique; each individual zone is reverse-zoned, but the overall sense of the zoning is normal. The sanidine zone is unusual in that the initial zoning trend is normal, but about half way the trend is reversed.

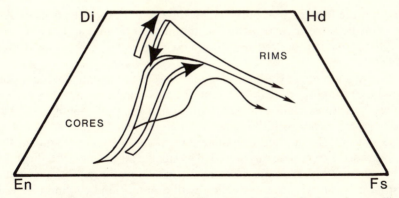

Figure 16. Generalized pyroxene zoning trends summarized from the dynamic crystallization studies of the lunar basalts.

There is a complex relation between major, minor, and trace-element chemistry and cooling rate. The major-element chemistry expressed by the end members En, Fs, and Wo was found to vary with cooling rate in all studies except those of Grove and Bence (1977). In general, pyroxenes grown at the slowest cooling rates tend to be initially more En-rich and Fs-poor and are zoned to more Fs-rich compositions. They may have either a low, medium, or high initial Wo content and are usually markedly zoned in this component, either to higher or lower values. Pyroxenes grown from a given basalt as a function of cooling rate will usually have parallel or overlapping Wo zoning trends that are systematically more Fs-rich and Wo-rich with increasing cooling rate. The extent of the range of zoning of the Wo component tends to be more restricted at higher cooling rates, but this behavior was not seen by Donaldson et al. (1975a). This increase in Fs content with increased cooling rate could be caused, at least in part, by iron loss to the platinum sample container. In a recent study (Lofgren, unpublished data) in which there was no iron loss, the same major element zoning trends were observed in pyroxene as a function of cooling rate.

Grove and Bence (1977) did not find major element zoning to be affected to a significant degree by cooling rate. They did observe a slight decrease in Mg content at the highest cooling rates. They point out that in the iron-capsule technique, which they employed, numerous Fe-metal particles are present in the melt at the initiation of cooling and provide nucleation sites for pyroxene; as a result, regardless of cooling rate (up to 150°C/hr) the nucleation temperature of pyroxene could not be suppressed. Because the first-generation pyroxene crystals all nucleated at approximately the same temperature, their core compositions are similar. The second generation or matrix pyroxene crystals are quite variable in composition. This phenomenon is apparently related to the presence of Fe-metal in the melt and not the form of the container, because recent experiments using Fe-wire in the same manner as the platinum-loop technique (Lofgren, unpublished data) produced results similar to those of Grove and Bence (1977).

Grove and Bence (1977) make a good case that this unique nucleation condition is appropriate for the lunar basalts they studied; these show a range of textures that obviously result from a range of cooling rates, but all have pyroxenes with the same core composition. Walker et al. (1976) also used the Fe-capsule technique, but in the basalt they studied the appearance of pyroxene was preceded by significant olivine crystallization. This presumably used most of the Fe-metal nucleation sites, and the suppression of pyroxene nucleation with a more rapid cooling rate was observed. The differences between the results of Grove and Bence (1977) and those of the other workers demonstrate the importance of nucleation and suggest that cooling rate must be considered in conjunction with the various other nucleation possibilities in order to interpret zoning trends in minerals. This nucleation effect will be treated in more detail in the section on textures.

All workers found similar trends for the incorporation of trace elements as a function of cooling rate. In general, TiO_2 and Al_2O_3 are preferentially incorporated into the pyroxene at high cooling rates. The Ti:Al ratio usually remains constant throughout the crystal (Figure 17A). At slower rates, the range of Ti and Al zoning is far more extensive. They are present in lesser amounts initially because growth is slower and closer to equilibrium. The Ti:Al ratio is dependent on the appearance of plagioclase, which depletes the melt in Al and enriches it in Ti. At faster rates, even if plagioclase does crystallize, its effect is minimal and the ratio does not change. At slower rates where plagioclase and pyroxene coprecipitate over a significant interval, the Ti:Al ratio can change from 1:4 to 1:1 (Grove and Bence, 1977). There has to be significant plagioclase crystallization before this effect is observed (Lofgren et al., 1974).

Grove and Bence (1977) have made the most systematic study of the nature of the substitutions of Al_2O_3 and TiO_2 in low Ca clinopyroxene as a function of cooling rate. Their data is summarized schematically in Figure 17B. Octahedral aluminum (Al^{VI}) is present in early-formed pyroxene in the rapidly cooled experiments, but absent in the more slowly cooled experiments except when plagioclase does not nucleate. The bulk partition coefficients for augite show the same rate dependence as the low Ca clinopyroxene except that the amounts of Cr, Ti, and Al are greater in the augite.

Sector zoning has been detected in experimentally grown pyroxenes in quartz-normative lunar basalts (Lofgren et al., 1974) and in a terrestrial

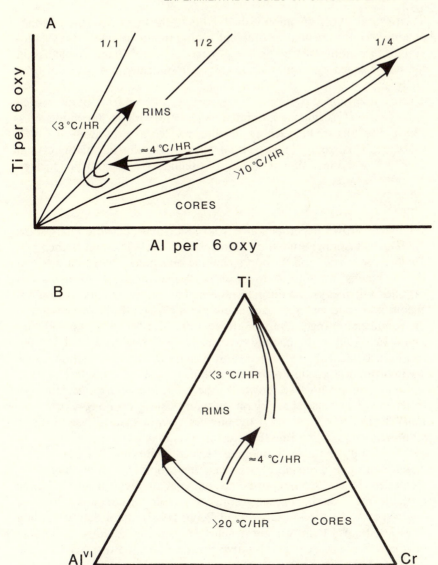

Figure 17. Generalized minor element zoning trends in pyroxene summarized from Lofgren et al. (1974) and Grove and Bence (1977).

leucite basalt (Lofgren, unpublished data). In the lunar case, the presence of zoning is documented by different zoning trends in different crystallo-graphic directions and is difficult to observe optically. In the leucite basalt, the sector zoned pyroxenes are readily seen optically, but have not been analyzed.

Oscillatory-zoned pyroxenes have been crystallized from titanium-rich lunar basalts (Usselman and Lofgren, 1975, 1976) with linear cooling rates. The oscillations are between augite and subcalcic augite with little En/Fs variation. This zoning must be controlled by the kinetics of the crystal growth in a manner similar to that of oscillatory zoning in plagioclase.

Olivine

Olivines grown experimentally from lunar mare basaltic melts show only normal zoning. Studies by Donaldson et al. (1975a) and Bianco and Taylor (1977) using the Pt-loop technique on similar olivine basalts had similar results, in contrast to those of Walker et al. (1976) using Fe-capsules. Olivines grown during isothermal drop experiments are homogeneous and appear to approach equilibrium closely: the K_D values are in the equilibrium range, .30 to .33 (Roeder and Emslie, 1970). The olivines that crystallized in the cooling experiments of Donaldson et al. (1975a) and Bianco and Taylor (1977) are zoned, and the core composition shows a significant decrease in Fo content with increased cooling rate. Contrary to zoning in plagioclase, however, the compositional range tends to increase with increased cooling rate in spite of the lower Fo core composition. The rim compositions of the olivines are systematically less Fo rich with increased cooling rate (Figure 18). This may have something to do with the ability of olivine to maintain surface equilibrium during crystallization and/or suppression of pyroxene nucleation at high cooling rates. Donaldson et al. (1975a) reported no zoning in an olivine from a 2.7°C/hr cooling rate experiment, suggesting that equilibrium was achieved. Lofgren et al. (1974) and Grove and Bence (1977) found that at cooling rates less than 3°C/hr, olivine which forms on the liquidus of the less olivine-rich lunar, quartz-normative basalts, had time to react with the liquid to form pyroxene as the equilibrium relations suggest before the run was quenched.

Walker et al. (1976) did not observe any decrease in the Fo content of the cores of the olivine as a function of cooling rate for a constant bulk composition. They studied a lunar picrite similar to the lunar olivine basalts discussed above, but the difference can be attributed to the use of the Fe-capsule technique. Presumably the Fe-blebs in the melt provide nucleation sites for the olivine as discussed by Lofgren (1975) and Grove and Bence (1977).

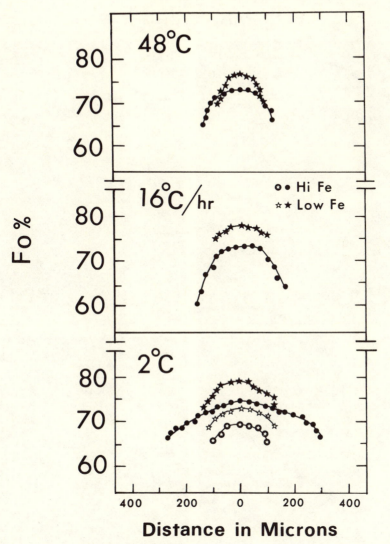

Figure 18. Normal zoning profiles in olivine grown from a lunar olivine basalt (from Bianco and Taylor, 1977).

The incorporation of minor elements into olivine as a function of cooling rate was reported by Donaldson et al. (1975a). He found that the amount of Ca, Ti, and Cr in the olivine increased as a direct function of the cooling rate under which the olivine grew. In the isothermal drop experiments Ca and Ti increased with increased ΔT of crystallization, but Cr remained constant.

FELDSPAR INTERGROWTHS

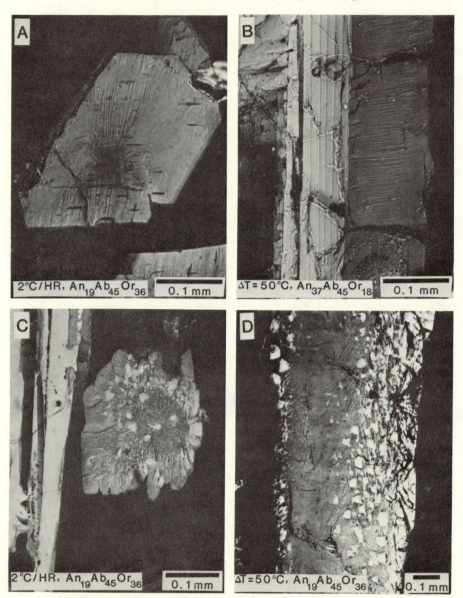

Figure 19. Feldspar intergrowths that developed by growth directly from the melt. (A) A fine lamellar intergrowth that resembles microperthite with the intergrowths in sectorial relationship. (B) Fine-lamellae in the outer zone of a crystal grown with 50°C isothermal drops. (C) Patchy intergrowth in a crystal grown from the same liquid and at the same cooling rate as the crystal shown in (A), but at a lower temperature. (D) Patchy intergrowths in a step grown crystal; all the patches on the right side of the crystal are in optical continuity.

Intergrowths

In experiments in the An–Ab–Or–H$_2$O system, Lofgren and Gooley (1977) reported the growth directly from the melt of perthite-like, feldspar intergrowths. They described two types of intergrowths. The most common type is a fine lamellar intergrowth that resembles microperthite. In this case, the lamellae are on the order of 1-5 microns (Figures 19A, B), but some reach 10–20 microns in diameter. The lamellae occur in patches in a sanidine host and in some instances these patches have a sectorial distribution in the crystal (Figure 19A). The other type is a patchy intergrowth with patches ranging up to 20–50 microns (Figures 19C, D). The patchy intergrowths are more common in the Or-rich compositions, and can occur together with lamellar intergrowths in the same crystal.

Lofgren and Gooley (1977) presented evidence that the intergrowths grew directly from the melt and are not the products of exsolution, and these arguments have been amplified by Morse and Lofgren (1978). In the original discussion, Lofgren and Gooley interpreted data from the alkali-feldspar solvus to indicate that the melt did not enter the solvus region (Morse, 1970); Morse (1969) has data on a natural feldspar that suggest that the solvus rises so rapidly with An content of the feldspar that all the experimentally grown crystals were below the solvus throughout their growth. While this facilitates simultaneous growth, it also makes exsolution a possibility. It is the intergrowth textures and the spatial relations of zoning in the crystal that are probably the most convincing evidence for growth directly from the melt. For example, in any given crystal, both the host for the lamellar intergrowths and the intergrowth-free parts of the crystal are zoned over the same compositional range. To suppose that the lamellar intergrowths represent exsolution requires that one part of a crystal have a composition significantly different from the other, such that when the first part exsolves, the composition of the host changes to become exactly the same as the unexsolved part elsewhere in the crystal. Lofgren and Gooley (1977) suggest that the lamellar intergrowth may result from the breakdown of a planar interface, but the patchy intergrowth would require independent nucleation and growth and the mechanism is not well understood.

THE EXPERIMENTAL DUPLICATION OF ROCK TEXTURES

The nucleation and growth of crystals in rock melts are more complex than in mineral melts because there are many more components and phases. There are many kinetically controlled growth features in rocks, such as mineral zoning, shape, composition, and intergrowth, that are important to unraveling the history of a rock. These features have been discussed already in the preceeding section on minerals, but the results in pure mineral systems are different when compared to more complex rock melts. The study of the rock textures concerns primarily the interaction of nucleation and growth of many phases during cooling.

The experimental duplication of rock textures by imposition of cooling histories resembling those that might occur in nature is an open-ended problem. The number of possible natural cooling histories is infinite within the spectrum ranging from a rapid quench to a slow cooling that takes millions of years. The slowest linear cooling rate that can be implemented in the laboratory is about .1°C/hr. At that rate it would take 70–80 days to cool through the crystallization interval of an average basalt. That is quite long for laboratory experiments, but still short geologically. There are, however, a large number of rocks that cool through the crystallization interval within that time frame. The center of lava flows roughly 3 to 4 meters thick would cool at about that rate, as would the outer portions of thicker flows. A small and unknown fraction of plutonic rocks may fall within that range.

As discussed earlier, linear cooling curves are the easiest to use experimentally and are representative of the cooling of natural lavas through the crystallization interval in most parts of a lava flow. Linear cooling histories have been used almost exclusively in the work that will be summarized here. The size of the experimental charge is an important aspect of the duplication of textures of rocks, in that the charge must be significantly larger than the largest crystals. Nucleation must not be dominated by the container, and should preferably occur uniformly throughout the charge. The wire-loop technique has proven successful on both counts for samples up to 150 mg, depending on the viscosity and surface tension of the melt.

In an attempt to replicate a pegmatitic texture, Jahns and Burnham (1958) used a 1 cm diameter gold tube sealed at both ends to enclose several grams of material in an internally heated pressure vessel. Gold and platinum capsules can hold large quantities of material for high pressure experiments, subject to the Fe-loss problems stated earlier. Smaller Pt-capsules have been used successfully in Fe-free systems for 20 to 150 mg samples, in which crystals up to several mm have been grown.

BASALTIC VOLCANIC ROCKS

Early dynamic crystallization studies have been concentrated on basalts for a variety of reasons, not the least of which is their widespread occurrence and frequency of eruption. As pointed out in the introduction, Hall's experiments conducted in the 1790's in support of Hutton's magmatic

theory for the origin of basalt are not only the first dynamic crystallization studies, but the first systematic experimental petrologic studies of any
kind.

Effect of Cooling Rate on Texture

Quench textures, in the broadest sense, contain skeletal or dendritic or
spherulitic crystals, or some combination of these, often in a glassy matrix.
Such textures are presumed to result from quenching the liquid. The true
quench product for most natural rock magmas is a glass. The broad range
of textures between a glass and coarse-grained, plutonic rock textures
result from a spectrum of cooling rates or histories and should not be
considered quench textures.

In the first basalt studies, a completely melted basalt composition was
cooled and a particular rock texture was related to a known cooling rate
(Lofgren et al., 1974; Donaldson et al., 1975a; Usselman et al., 1975;
Walker et al., 1976). The slowest laboratory rates ($< 1°C/hr$) in general
produce textures in which some of the minerals have an equilibrium or
near-equilibrium shape, and others are skeletal or acicular. The progression to more rapid cooling rates produces textures with crystals that
are increasingly skeletal, dendritic, and spherulitic (Figure 20). A given
range of cooling rates will, in general, produce a unique texture. This
relationship, however, is true only for the basalt in question, e.g., the same
range of cooling rates may result in an entirely different texture in a basalt
of a different bulk composition.

More recent studies (Lofgren and Donaldson, 1975a; Usselman and
Lofgren, 1976; Lofgren, 1977; Bianco and Taylor, 1977) have shown that
the precooling melt history greatly affects the subsequent nucleation rate
and density for a given cooling rate. Usselman and Lofgren (1976) also
demonstrated the importance of oxygen fugacity, especially where small
changes in oxygen fugacity greatly affect the phase relations. Therefore,
while cooling is an obvious factor in controlling texture, the interaction
of cooling with the nucleation kinetics, oxygen fugacity, precooling melt
history, and volatiles is important.

Porphyritic Texture

In the light of geologic and petrographic evidence, Hollister et al. (1971)
and Dowty et al. (1974) suggested that porphyritic texture can develop as a
single-stage cooling event; a suggestion made very early by Bowen (1914,
p. 257). Lofgren et al. (1974) confirmed by direct experiment that porphyritic textures could develop from basaltic liquids cooled at constant,
linear rates. The results have been substantiated by subsequent studies on

COOLED FROM COMPLETE LIQUID

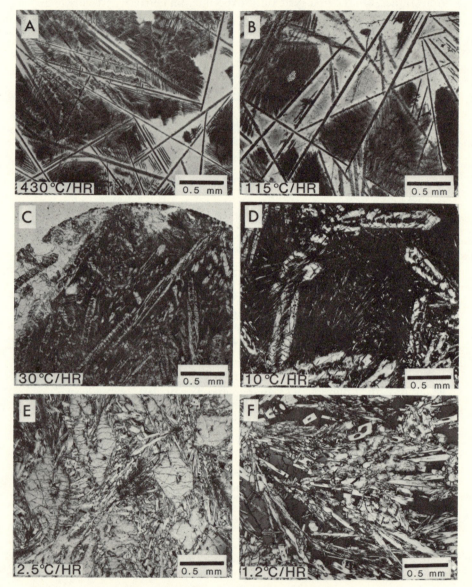

Figure 20. Photomicrographs that show the effect of cooling rate on the textures of a lunar quartz-normative, pyroxene-phyric basalt (from Lofgren et al., 1974).

other lunar basaltic compositions (Donaldson et al., 1975a; Usselman et al., 1975; Walker et al., 1976; Lofgren, 1977; Bianco and Taylor, 1977; Grove and Bence, 1977). In fact, the porphyritic texture is one of the most common experimentally produced textures in melts cooled from a totally liquid state.

The development of porphyritic texture with a single-stage cooling history can be understood by examining the phase relations of basalts and taking into account the effects of supercooling on nucleation and growth (Lofgren et al., 1975b). To produce two generations of a phase, it is necessary to have two distinct episodes of nucleation. For this to occur, the liquidus or near-liquidus phase must be modally dominant and have a significant temperature interval of crystallization exclusive of the other modally significant phases. This case is represented schematically in Figure 21, using the lunar quartz normative basalts as an example (Lofgren et al., 1974). As the liquid cools, the pyroxene nucleates at A and grows, without competition from other crystallizing phases, to phenocryst size.

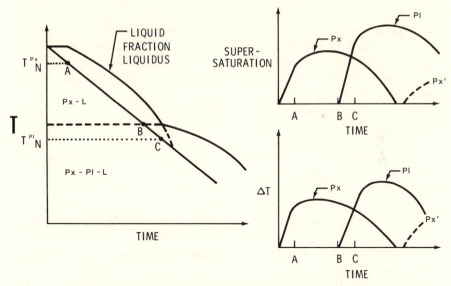

Figure 21. Schematic representation of the conditions that foster the development of porphyritic texture in a melt cooled at a constant, linear rate. Pyroxene is the liquidus phase followed by plagioclase. The pyroxene begin to crystallize at (A), at which point the liquid begins to fractionate. The fractionation is represented by the change in the liquidus temperature of the liquid fraction that remains at any given temperature. Plagioclase should nucleate at (B), but if it does not nucleate until (C) the liquid will be highly supersaturated with plagioclase: this will result in numerous nuclei resulting in small plagioclase crystals enclosing pyroxene. The variation of supersation of pyroxene and plagioclase and ΔT with time are shown to the right in the figure.

If plagioclase exhibits its natural reluctance to nucleate at its equilibrium appearance temperature B, the melt will rapidly become supersaturated because of the flattening of the liquidus that usually accompanies the appearance of a second phase (cf., Wyllie, 1963). When plagioclase finally precipitates at C, so many nuclei form that growth will be restricted, and the crystals will be much smaller than the preexisting pyroxene. The early formed pyroxene crystals continue to grow because pyroxene is still a major constituent in the liquid, but a new generation of smaller pyroxene crystals may form as the melt rapidly becomes supersaturated with pyroxene component as a result of the massive precipitation of plagioclase.

Porphyritic textures need not have two generations of the same phase. If the phenocryst phase is olivine, it tends not to nucleate as a matrix phase because it is usually not present in sufficient quantity. The matrix would most likely be pyroxene or plagioclase or both, but a fine-grained matrix could form as a result of essentially the same increase in supersaturation of these phases described above (Donaldson et al., 1975a, 1976).

Dynamic crystallization of a near-multiply saturated ocean tholeiite (Lofgren and Donaldson, 1975a) did not produce a porphyritic texture with linear cooling, but a wide variety of other textures depending on which phase nucleated first: olivine, pyroxene, or plagioclase. This demonstrates that a porphyritic texture will not develop with linear cooling unless there is a significant temperature interval during which the liquidus phase crystallizes alone.

Nucleation and Texture

It has been concluded earlier in this chapter that the temperature (and thus the ΔT) of crystallization is the controlling factor in determining the shape and composition of the crystal that begins to grow. Because of the variation of incubation time with cooling rate, nucleation does not take place at the same temperature in different experiments conducted on the same liquid cooled at a given rate. In general, the nucleation temperature is depressed as the cooling rate increases. Nucleation, however, is a random, statistical process, and even for the same cooling rate, nucleation may not occur at the same temperature or rate. Thus the shapes of the crystals may vary somewhat for different runs at the same cooling rate because the nucleation temperature (or ΔT) is different, or the sizes of the crystals may vary because the density of nuclei is different.

Because nucleation is a random process, it is subject to changes in the physical state of the liquid (see discussion of Donaldson's unpublished data earlier in the chapter). We can conclude from Donaldson's data that, for a given cooling rate, the incubation time for the liquidus phase is increased for a liquid that has been superheated. This will result in that phase having a more intricate shape than otherwise. If a liquid is only partially melted at subliquidus temperature before cooling, the opposite effect is observed, and the apparent incubation period is either decreased or eliminated if crystals are already present.

The partial melting or cooling from sub-liquidus temperatures is more interesting geologically because it most certainly occurs more often than the crystallization of superheated magmas. The difficulty experienced in getting plagioclase to nucleate from an ocean thoeliite prompted Lofgren and Donaldson (1975a) to melt only partially the crystalline starting material (a finely-ground, well-crystallized basalt) before cooling. The liquidus temperature of the basalt is $1{,}184 \pm 4°C$ and it was partially melted between $1{,}165-1{,}170°C$. The resulting liquid-crystal mixture contained $\sim 5\%$ plagioclase crystals and $<1\%$ olivine as widely dispersed crystals at the initiation of cooling. The resulting texture was dramatically different from those in charges cooled from complete liquids. When this basalt was cooled at $\sim 2°C/hr$ from a modestly superheated state ($1{,}220°C$), plagioclase did not nucleate and the result was an olivine-pyroxene vitrophyre (Figure 22A). When cooled at $\sim 2°C/hr$ from a slightly superheated liquid ($1{,}190°C$), the plagioclase nucleated, but the texture depended on the time of melting (compare Figures 22B and C). Apparently the longer melt time eliminated more potential nucleation sites in the melt, so that on cooling nucleation did not occur as readily and took place at a lower temperature, resulting in plagioclase and pyroxene crystals that are spherulitic in G-61 (24 hours melt time, Figure 22B) as opposed to acicular or coarser-grained in G-57 ($4\text{-}\frac{1}{2}$ hours melt time, Figure 22C). The three textural varieties above contrast markedly with the even-grained, subophitic texture of the partially melted charge (Figure 22D). The plagioclase is more evenly distributed and tabular and the pyroxene is much coarser grained. If the previous conclusions about the relation of crystal shape to cooling rate are valid, then the plagioclase and pyroxene nucleated and grew at a higher temperature in the partially melted liquid than in the completely melted liquids. This presumably occurred because plagioclase nuclei already existed at the initiation of cooling in partially melted liquid and nucleation was not required. The same technique has been used by Usselman and Lofgren (1976), Lofgren (1977), Bianco and Taylor (1977), Lofgren and Smith (1978), and Nabelek et al. (1978) on a variety of lunar materials with similar results. Bianco and Taylor (1977) observed that plagioclase did not need to be a residual phase of the partial melting to increase its facility to nucleate. Therefore, cooling the liquid from subliquidus temperatures that are above the appearance temperature of plagioclase increases the probability of plagioclase nucleating when the liquid is cooled.

COOLED AT 2°C/HR

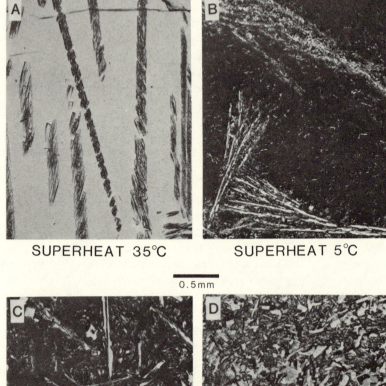

SUPERHEAT 35°C SUPERHEAT 5°C

0.5mm

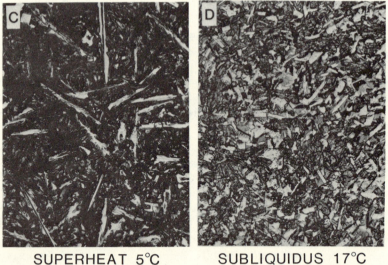

SUPERHEAT 5°C SUBLIQUIDUS 17°C

Figure 22. Photomicrographs that show the effects on the resulting textures of heating the liquid above its melting temperature (superheat) before cooling or cooling from a subliquidus temperature where crystals are present.

Bianco and Taylor (1977) have speculated that the increased plagioclase nucleation from partial melts is related to the degree of polymerization in the melt (see Hess, this volume). They presume that, because a material is partially melted at a subliquidus temperature, the melt retains a feldspar structure and this fosters increased feldspar nucleation upon cooling. This general concept is not new, but does not appear to be supported by viscosity data on lunar rocks (Cukierman et al., 1973). They observed no sharp change in the melt viscosity with temperature that might be attributed to polymerization. These data do not eliminate polymerization as a possible cause, but it does not appear to be a major factor. Another possibility, but one that is difficult to prove, is the presence of submicroscopic impurities that act as nuclei. It is not clear, however, how small these impurities, which presumably are refractory minerals, can be before they would be considered polymerized melt species. The important fact is that the texture can vary dramatically as a function of the potential for nucleation in the melt, which is in turn a function of the precooling, superliquidus, or subliquidus melt history.

Given their presence, the density and distribution of the nuclei that result from a partial melt history can be equally important in determining the nature of the resulting texture. Lofgren (1977) showed that, by varying the density of nuclei—in this case $1-5$ μm feldspar crystals—the textures of the run product can vary from poikilitic to intersertal to subophitic (Figure 23) within a few millimeters. The variations in density of nuclei were produced by first crystallizing the synthetic glass, then partially melting it at about 25°C below the liquidus. There appears to be some convection in the charge during melting: a typical density and distribution pattern of crystals is shown in Figure 23A. The sample in this figure was quenched after six hours of melting. Because the variations in density and distribution of nuclei result primarily from convection in the charge, they would be expected to vary from charge to charge, even though the experimental conditions are identical. Furthermore, the crystal-free areas, because they have few if any nuclei, should ultimately have larger crystals than the parts of the charge with a high density of crystals. It is readily apparent that the completely crystallized charge (Figure 23B) had a slightly different distribution of crystals than that shown in Figure 23A.

The poikilitic textures are typical of the areas of the charge most densely choked with small plagioclase crystals (Figure 23C). Few if any plagioclase crystals nucleate and grow in these parts of the charge on cooling, but olivine nucleates and poikilitically encloses the numerous plagioclase crystals. As the density of plagioclase decreases, the plagioclase can grow larger and the textures vary from intersertal to subophitic (Figure 23D). In crystal-free areas were plagioclase must nucleate, the texture could even be porphyritic if the conditions discussed in the previous section prevail.

COOLED AT 2°C/HR
FROM PARTIALLY MELTED LIQUID

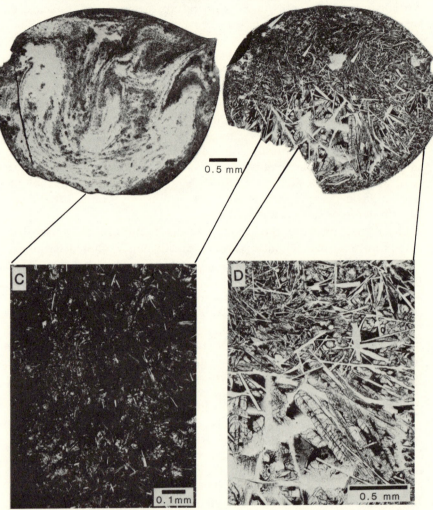

Figure 23. Photomicrographs of the textural variations that can result from the irregular distribution of crystals at the initiation of cooling. (A, upper left) The distribution of crystals. (B) The crystallized charge. (C) Poikilitc texture enlarged from a portion of (B). (D) Intersertal to subophitic texture enlarged from a portion of (B).

Intersertal and Ophitic Texture

The results of experiments just described can be applied to the origin of intersertal and ophitic textures. More experiments are needed, however, before the details of how nucleation and crystallization contribute to the formation of the textures will be fully understood. It has long been speculated that nucleation and growth rates relative to the degree of supercooling play a role (e.g., see Carmichael et al., 1974, p. 24) in the origin of ophitic textures. The experimental studies of Lofgren and Donaldson (1975a) and Lofgren (1977) demonstrate the importance of nucleation, especially as it is manifested in the nucleation density. They found that basalts which have either porphyritic or uniformly fine-grained, spherulitic textures when cooled from a complete liquid, will have intersertal to ophitic textures when cooled at the same rate from a liquid that already contains some plagioclase.

In the porphyritic case, a few plagioclase crystals nucleate, grow large, and the matrix pyroxene and plagioclase develop from a massive nucleation event (Figure 24A). Intersertal and ophitic textures can develop from the same liquids at the same cooling rate if the density of plagioclase is higher and uniform (Lofgren 1978). The transition from intersertal to subophitic to ophitic textures appears to depend on the plagioclase nucleation density and the timing of the pyroxene nucleation. With a high relative density of plagioclase crystals, less growth is necessary before the crystals can impinge upon one another and form a network in which pyroxene would necessarily be interstitial and the texture intersertal (Figure 24B). With relatively fewer plagioclase nuclei, the crystals would grow larger; because they would, of necessity, be farther apart, it would take longer for them to impinge. The timing or kinetics of pyroxene nucleation and growth rate would then become important in determining whether the texture by truly ophitic or subophitic. The ophitic textures developed only locally in the plagioclase enriched liquid (Figure 24D), in close juxtaposition with intersertal textures.

Ophitic and subophitic textures developed more commonly in two experimentally studied basalt compositions that have more nearly cotectic phase relations. In the ocean thoelite already discussed, a subophitic texture developed in a subliquidus cooling run, as opposed to the spherulitic in the liquidus cooling runs (Figure 22). In material of another composition resembling a basaltic komatiite (Arndt et al., 1977), subophitic textures (Figure 24D) develop when the material is cooled from subliquidus temperatures, as opposed to the radiate textures (Figure 24C) in the liquidus experiments. At higher magnification, plagioclase-pyroxene granophyric intergrowths (Figure 24E) are apparent in runs cooled from complete liquid and resemble those described by Arndt et al. (1977) and Schulz (1977) in basaltic komatiites. The subliquidus run produced some

COMPLETE LIQUID   LIQUID + CRYSTALS

Figure 24. Photomicrographs of crystallized charges that compare textures which developed upon cooling a complete liquid (on the left) or the same liquid plus crystals (on the right) (A) and (B). Feldspathic basalt crystallized at 2°C/hr forms a porphyritic texture when cooled from a complete liquid (A) and forms an intersertal texture when the liquid contains some crystals at the initiation of cooling (B). (C–F.) Basaltic composition that resembles basaltic komatiite which when cooled at 2°C/hr from a complete liquid has a radiate texture (C) with granophyric integrowths of plagioclase and pyroxene (E). When that same liquid contains crystals at the initiation of cooling, a subophitic texture results (D). The subophitic relationship shown at higher magnification in (F).

of the coarsest pyroxenes (Figure 24F) yet observed experimentally and come the closest to natural ophitic pyroxenes.

At this early stage in the experimental studies it seems safe to say that the ophitic textures are best developed experimentally in cotectic (or nearly) liquids when cooled from subliquidus temperatures.

The Importance of Oxygen Fugacity

The oxygen fugacity is another important variable in dynamic crystallization experiments on basalts. It is necessary to reproduce the proper redox state of the magma to obtain the equilibrium phase relations and chemistry of the phases. An iron-rich metal phase is present in many lunar basalts, indicating that the redox conditions are near those for iron saturation. In some of the experiments performed by Lofgren et al., (1974), the redox conditions were slightly lower than desired during crystallization. When this occurred, significant amounts of Fe-metal precipitated, raising the liquidus of the remaining liquid and thus increasing the degree of supercooling. The result was a massive precipitation of very fine-grained spherulitic-crystals that are nearly opaque to transmitted light (Figure 25A). This combination of pyroxene phenocrysts and nearly opaque matrix closely resembles a few of the Apollo 15 quartz-normative basalts (Figure 25B). There is the appearance of a sudden change in cooling rate to produce the large phenocrysts in a fine-grained, spherulitic matrix which could be equally attributed to a sudden change in the f_{O_2}. The phase relations can change as a function of the oxygen fugacity even within the range of appropriate values. For the high-titanium lunar basalts the order of phase appearance changes within less than one order of magnitude f_{O_2} variation from olivine, armalcolite, pyroxene to olivine, pyroxene, armalcolite (Usselman and Lofgren, 1976). Cooling runs in each of these f_{O_2} regimes produce distinctly different textures. If armalcolite crystallizes before pyroxene, it forms large dendrites with widely spaced parallel plates (Figure 25C) that dominate the texture and pyroxene crystallizes between the plates. When pyroxene crystallizes first it forms euhedral, elongate crystals in a matrix of randomly oriented armalcolite blades and feldspar (Figure 25D).

ACIDIC VOLCANIC ROCKS

The relative reluctance of phases to nucleate in basaltic rocks is amplified in the more acidic volcanic rocks and consequently they are not well studied experimentally. The liquids become more viscous from basalts to rhyolites, and feldspar and quartz become the only modally significant phases. The frequent occurrence of glassy acidic volcanic rocks is vivid

EFFECTS OF OXYGEN FUGACITY

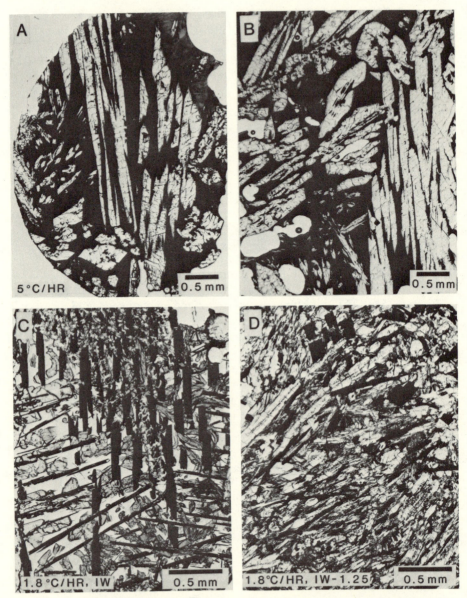

Figure 25. Photomicrographs of charges in which the textures were affected by the oxygen fugacity at which they were crystallized. (A). Experimentally crystallized lunar quartz normative basalt showing porphyritic texture with nearly cryptocrystalline matrix. (B.) Lunar basalt 15486 which has essentially the same texture. (C.) Lunar basalt 74275 crystallized at an oxygen fugacity where armalcolite appears before pyroxene. (D.) Same as (C), but crystallized at lower oxygen fugacity so that pyroxene appears before armalcolite.

evidence of the reluctance of feldspar and quartz to nucleate and grow during eruption and flow. Most of the feldspar and quartz crystals present, often in a glassy matrix, can be attributed to growth prior to eruption. The model for the development of porphyritic texture with linear cooling described for basaltic rocks, however, could also explain some porphyritic textures in felsic rocks.

Attempts to crystallize liquids of rhyolitic composition by either isothermal drops (Swanson, 1977; Lofgren, 1976) or cooling experiments (Lofgren, 1976) have resulted in the growth of feldspar spherulites and quartz that is either isolated or poikilitically encloses the feldspar. Long (1976), working at small ΔT's of approximately 20–30°C, obtained isolated skeletal alkali feldspar crystals in a bulk composition that was more Or rich than that studied either by Swanson or Lofgren, but still a rhyolitic composition. The spherulitic shapes follow the same general sequence of shapes with ΔT discussed earlier in feldspar mineral systems. At low ΔT, the spherulites have broad tabular fibers or branches; at higher ΔT's or cooling rate the fibers are finer and more closely spaced.

For felsic bulk compositions approximating a granodiorite without the ferromagnesian phases, Swanson (1977) found a sequence of plagioclase shapes similar to those observed by Lofgren (1974a) in isothermal drop experiments. In cooling experiments (Lofgren, unpublished data) on the same granodioritic composition, skeletal tabular crystals grew at cooling rates of 1°C/hr; these changed to more ornate skeletal crystals and then to coarse, open spherulitic forms at higher cooling rates.

Devitrification (crystallization below the glass transition temperature) of natural rhyolite glass (Lofgren, 1971a) with a range of ΔT's similar to that employed by Swanson (1977) produced similar spherulitic textures. The nucleation density, however, is higher and nucleation does not occur throughout the charge simultaneously but along a front that moves inward from the outer surface of the heated glassy charge. At low ΔT's, the front is not sharp, and there are numerous isolated nuclei that result in open, coarse spherulites not unlike those produced by cooling a liquid. At higher ΔT's the nucleation front is sharp and the spherulites are very closely spaced and intergrown. The only distinctive criteria to distinguish devitrification from primary crystallization textures appear to be the presence of a sharp devitrification front suggesting devitrification inward from a free surface such as a fracture. If the ΔT of devitrification was low, or the temperature high, and the front is diffuse, the difference would be difficult to detect.

The crystallization of quartz in both the primary crystallization (Lofgren, 1976) and devitrification (Lofgren, 1971a) experiments is most often as poikilitic crystals enclosing the feldspar spherulites. This poikilitic texture is referred to as micropoikilitic quartz or snowflake texture

(Anderson, 1969; Green, 1970; Torske, 1975) and is a common texture in rhyolitic lava flows and welded tuffs. Torske (1975) suggests that snow-flake texture develops from glass that has undergone phase separation into alkali-rich and silica-rich domains, and makes a strong case for this being the origin of the microgranophyric textures that display an interconnected network of feldspar and quartz. However, phase separation cannot explain the development of snowflake texture in melts cooled directly from a liquid state, because the feldspar has crystallized before the phase separation temperature is reached. In the devitrification case, either phase separation or the degree of polymerization of the melt may affect the density of nucleation sites and control the size of spherulites and the poikilitic quartz crystals. Phase separation in the glass may greatly facilitate the development of microgranophyric snowflake texture, but it is doubtful that phase separation is the sole cause of snowflake texture.

Plutonic Rocks

Plutonic rocks crystallize at depth and may require millions of years to cool. This view is supported by thermal modeling of plutonic bodies (for example, Jaeger, 1968). Experiments by Lofgren (1974a), Fenn (1977a), Swanson (1977), and others have shown that the feldspars of the size and shape typical of plutonic rocks can be grown in days to a few weeks. This does not necessarily mean that plutonic feldspars grew in these short times, but merely that it is possible, if the conditions are right.

Even though feldspars that resemble plutonic feldspars have been grown in the laboratory, plutonic textures have not been duplicated. Feldspar is usually the liquidus phase, and its nucleation and growth within laboratory time appear to be feasible. The crystallization of the remainder of the liquid (usually enriched in SiO_2) as large crystals may well take more time than is practically available in the laboratory. Work on liquids that contain ferromagnesian minerals is limited (Naney, 1975; Lofgren, unpublished data). Experiments by Lofgren on a granodioritic liquid (cf., Swanson, 1977) to which diopside has been added showed that while the diopside nucleated and formed crystals with shapes and sizes similar to those found in plutonic rocks, the feldspar would not nucleate, even though, in the same liquid without the diopside added, it nucleated readily.

As with the volcanic rocks, then, nucleation could prove to be an im-portant factor. Fenn (1977a) and Swanson (1977) conclude from their studies on nucleation and growth rates in granitic systems that the marked differences in these rates could allow for modest growth rates at low nucleation densities and thus a coarse grained plutonic texture. Swanson concludes, from the evidence of the nucleation and growth rate data he determined, that porphyritic textures should be more common in granodiorites. The porphyritic texture could form by the same model as presented for volcanic rocks in a continuous cooling event. Porphyritic textures are not common in granodiorites, however, and Swanson suggests that either the early-formed plagioclase phenocrysts are separated by differentiation processes, or they are masked by later crystallization of plagioclase. Swanson suggests another likely possibility. He points out that granodioritic magmas (and for that matter most plutonic magmas) may already contain plagioclase crystals when cooling starts. Because many nucleation sites exist, none of the plagioclase crystals reach phenocryst dimensions being prevented by lack of space. An irregular distribution of preexisting crystals (as discussed earlier for the basalts), caused possibly by irregularities of flow during intrusion, could result in local variation in textures.

The possibility that plutonic magmas may commonly have crystals when cooling is initiated, or at least Angstrom-sized nuclei or prenuclei, is an important factor in evaluating the effect of nucleation and growth-rate data on texture. As just suggested in the previous paragraph, many of the qualitative statements made earlier in the chapter for volcanic rocks could apply equally to plutonic rocks. The lack of textural variation right up to the contact of a plutonic rock may reflect the presence of nuclei in the magma at the time of intrusion. A nucleation event would not be required after cooling had begun, and, if the distribution of nuclei is uniform, the resulting texture would be homogeneous throughout.

Comb Layering

Comb layering in plutonic rocks was first described as a result of crystallization from a supercooled magma by Taubeneck and Poldervaart (1960). Lofgren and Donaldson (1975b) confirmed by cooling experiments on feldspar systems that the curved, branching crystal shapes typical of feldspars in comb-layered rocks grow only in a more supercooled or supersaturated environment than do the tabular feldspars found in the rock that encloses the comb-layered material (compare Figures 7B and C with 7A). They were unable, however, to place quantitative limits on the degree of supercooling or supersaturation necessary, because of the limited range of liquid compositions studied.

Lofgren and Donaldson (1975b) also proposed a mechanism for the origin of the highly differentiated, alternating layers in comb-layered rocks. They suggest it may be related to the development of a boundary layer, and thus constitutional supercooling, at the crystal-melt interface during steady state growth. Under this condition, latent heat is extracted through the growing crystals, rather than dissipated into the melt, and a positive temperature gradient exists in the melt adjacent to the crystal

(Figure 26). If the crystal growth rate is more rapid than the diffusion away from the moving interface of components rejected by the growing crystal ("impurity"), a gradational boundary layer enriched in these components forms. This boundary layer has a lower liquidus temperature than the bulk melt and thus a zone of constitutionally supercooled melt exists ahead of the interface, as shown by the hachured area (Figure 26). The maximum supercooling, ΔT_{cs}, will occur at some small distance ahead of the interface, depending on such parameters as the growth rate, diffusion rate, cooling rate and the specific heats of the melt and crystals. In magmas, where the proportion of "impurity" is large, its buildup will supersaturate the boundary layer with a different phase, which nucleates in the constitutionally supercooled zone, and then becomes the directionally growing phase. The growth direction is controlled by the temperature gradient, which is perpendicular to the layer. The elongate mineral(s) of the previous layer are terminated by the nucleation and growth of the new phase. The thickness of a layer dominated by one or more elongate minerals will vary with how rapidly the boundary layer

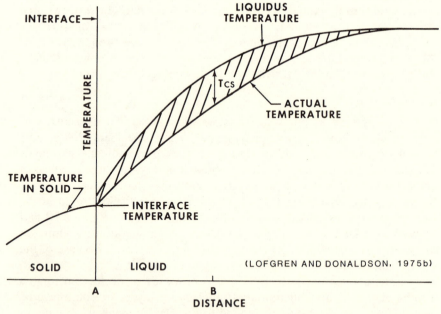

Figure 26. Schematic representation of constitutional supercooling in advance of a growing crystal interface where the heat is extracted through the crystal. This supercooling exists because low temperature components rejected by the growing crystal do not diffuse away as fast as the crystal interface is advancing (from Lofgren and Donaldson, 1975b).

builds up and how long nucleation of the new elongate crystal phase is delayed.

Donaldson (1977) reproduced the textural change from 'cumulate' to comb-layered harrisite in dynamic-crystallization experiments on olivine-eucritic and peridoditic varieties of harrisite. The liquids were cooled continuously at linear rates between 14 and 30°C/hr; these rates are probably higher than those that would prevail in natural magmas. In a single charge the entire sequence of olivine crystal shapes is present from the granular or polyhedral cumulate olivines at the bottom to the overlying, more skeletal, hopper olivines, to the larger branching olivine that comprises the comb-layered part of the sequence. Donaldson describes the sequence of events as follows. The first olivine nuclei are abundant and growth is slow, so many equant crystals form and settle to the bottom. Stoke's Law calculations suggest that the crystals will settle in minutes. These crystals form a cumulate layer at the bottom of the charge containing 50–80% olivine. With further cooling, the nucleation rate must decrease markedly and the growth rate must increase in order to grow the hopper olivines. As cooling continues and the degree of supercooling increases further, the nucleation rate must become vanishingly small. The growth rate, however, must be high so that those olivines whose a axis is approximately perpendicular to the liquid-solid interface will grow most rapidly as branching crystals. This relative behavior of decreasing nucleation and increasing growth rate with increasing ΔT is compatible with the experimental observation that the peak in the nucleation rate occurs at a lower ΔT value than the growth rate maximum. This explanation is, however, empirical, and not based on direct experimental verification of the nucleation and growth-rate curves versus ΔT for the system studied.

SUMMARY

The nucleation behavior, whether it be the incubation time, the rate, the density, or the distribution, appears to be the most important variable in determining the growth kinetics of mineral and rock textures. The growth rate of the crystals, although a significant variable, does not seem to be as important in producing various mineral features or rock textures. The nucleation behavior can be affected by a multitude of variables, most of which are interdependent. Foremost among them are the viscosity, which depends on the melt composition and its volatile content, temperature, degree of supercooling, and oxygen fugacity.

The ease with which a given mineral nucleates is reflected most accurately in the incubation time. It is possible to define a sequence of ease of nucleation for the common rock-forming minerals based on differences in incubation time determined or deduced from available experimental data. Olivine and oxide minerals such as the spinels, ilmenite, and armalcolite nucleate most readily, followed by pyroxene, with a distinct resistance to nucleation. The feldspars and quartz nucleate with the greatest difficulty. This sequence is subject to considerably more refinement, though the basic principle of correlation of ease of nucleation with simplicity of crystal structure is unlikely to change. Feldspar shows the greatest range of nucleation behavior. When on the liquidus of a rock melt, it will nucleate much more easily than if it is the second or third phase to appear according to the equilibrium phase relations. It has been shown experimentally, however, that plagioclase may not nucleate even though it is the liquidus phase, resulting in a petrogenetically misleading pyroxene-olivine vitrophyre (Lofgren and Donaldson, 1975a).

The shape that a crystal assumes during growth is most directly controlled by the ratio of the diffusion rate to the growth rate, D/G. This ratio is related to the degree of supercooling existing during growth. For isothermal drop experiments it is possible to relate the variation in crystal shape to the initial value of ΔT. In cooling experiments, the initial ΔT is not known accurately and may vary under identical cooling conditions because nucleation is a statistical process. Crystal shape can be related to the cooling rate, however, much as with ΔT. At high cooling rates, the ratio of D/G will be small, and spherulitic and dendritic forms will be stable. At slower cooling rates, growth and diffusion rates will be more nearly equal and planar crystal faces are stable, resulting in crystals with shapes that more closely approximate their equilibrium habit.

The temperature or ΔT at which a crystal nucleates will dictate the composition of the crystal, or the core composition if the crystal is zoned. In cooling experiments the core composition of zoned crystals will be a function of cooling rate if nucleation is delayed below its equilibrium appearance temperature. The composition will be determined by the ΔT of nucleation, which is a complex function of the nucleation incubation time and cooling rate.

The effect of nucleation on texture is dramatic. In volcanic rocks where cooling rate is often greater than $.1°C/hr$, the nucleation incubation period and the cooling rate interact to provide a variety of effects. An attempt is made in Figure 27 to represent schematically the interplay of nucleation and cooling. The data that support the sequence of events portrayed in Figure 27 are from the studies of basalts. These studies are far from complete; further studies will most assuredly add more detail to such a diagram.

The most important distinction is whether crystals are present in the liquid at the initiation of cooling; that is, whether the liquid is cooled

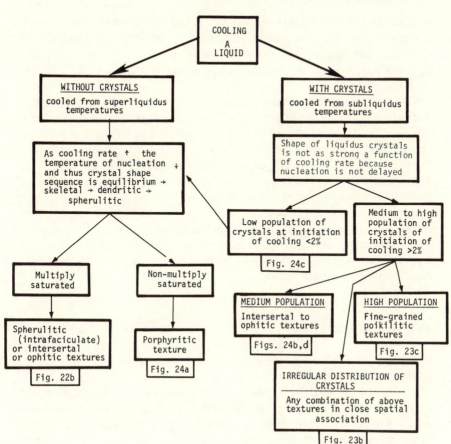

Figure 27. A flow diagram summarizing the complex interaction of cooling rate and nucleation in basaltic liquids either with or without crystals present in the liquid at the initiation of cooling.

from a superliquidus or subliquidus temperature. If there are no crystals present in the liquid, nucleation must occur before growth can proceed, and the shapes of the crystals are a strong function of the cooling rate. If crystals are present, nucleation of the liquidus phase is not necessary and growth will proceed as soon as cooling begins. Consequently the shapes of the liquidus crystals are not as strongly a function of cooling rate as when nucleation must precede growth. Porphyritic and spherulitic or radiate (intrafasiculate) textures are characteristic of cooling experiments on crystal-free liquids, although intersertal or ophitic textures will most likely be found in multiply saturated liquids if the cooling rate is very slow (generally slower than rates attainable in the laboratory).

Intersertal and ophitic textures are typical of liquids containing crystals that are evenly distributed and of medium population (5–15% by volume). Liquids with higher populations usually have fine-grained poikilitic textures and liquids with significantly lower populations behave like crystal-free liquids.

The material in this chapter must, in the broad sense, be considered preliminary. This is the beginning of a new line of study; the experiments are recent and the time to reflect has been minimal. In this area of dynamic crystallization studies, imagination is the key to deducing and replicating the sometimes complicated paths by which natural rocks may form. To watch a lava lake form a crust that is then torn into large rafts by renewed upwelling of magma, to watch these rafts buckle and sink from their own weight into a churning cauldron and then wonder if they are remelted either entirely or in part, and to know that this is only the final episode in the journey of that magma to the surface, makes us realize that we can only hope to have a chance to unravel some of the complexities of the rock-forming process.

ACKNOWLEDGMENTS

It would not have been possible to write this chapter without the influence of W. F. Weeks, who introduced me to the sciences of crystal growth and ceramics. Out of that broadening of my background came the desire to experimentally reproduce crystal growth and textural features of natural rocks. Much of the experimental work described herein was completed at the Johnson Space Center Experimental Petrology Laboratory, where the development and maintenance are ably carried out by O. Mullins, Jr. and D. P. Smith. Without their continued and often inspired help much of the experimental work upon which this chapter is based would not have been possible. The chapter has benefited greatly from the comments of R. H. McCallister, C. H. Donaldson, P. M. Fenn, R. B. Hargraves, and D. Walker.

REFERENCES

Anderson, J. E., Jr., 1969. Development of snowflake texture in a welded tuff, Davis Mountains, Texas. *Geol. Soc. Amer. Bull*, *80*, 2075–2080.

Arndt, N. T., A. J. Naldrett, and D. R. Pyke, 1977. Komatiitic and iron-rich tholeiitic lavas of Munro Township, northeast Ontario. *J. Petrol.*, *18*, 319–369.

Bianco, A. S., and L. A. Taylor, 1977. Applications of dynamic crystallization studies: lunar olivine normative basalts. *Proc. 8th Lunar Sci. Conf.*, 1593–1610.

Blander, M., H. M. Planner, K. Keil, L. S. Nelson, and N. L. Richardson, 1976. The origin of chondrules: experimental investigation of metastable liquids in the system $Mg_2SiO_4-SiO_2$. *Geochim. Cosmochim. Acta, 40*, 889–896.

Bottinga, Y., A. Kudo, and D. Weill, 1966. Some observations on oscillatory zoning and crystallization of magmatic plagioclase. *Amer. Mineral., 51*, 792–806.

Bowen, N. L., 1914. The Ternary system: Diopside-forsterite-silica. *Amer. J. Sci.,* suppl. 2, 1, 559–574.

Bryan, W. B., 1972. Morphology of quench crystals in submarine basalts. *J. Geophys. Res., 77*, 5812–5819.

Carmichael, I. S. E., F. J. Turner, and J. Verhoogen, 1974. *Igneous Petrology,* McGraw-Hill, New York.

Cukierman, M., L. Klein, G. Scherer, R. W. Hopper, and D. R. Uhlmann, 1973. Viscous flow and crystallization behavior of selected lunar compositions. *Proc. 4th Lunar Sci. Conf.,* 2685–2696.

Donaldson, C. H., 1974. Olivine crystal types in harrisitic rocks of the Rhum pluton and Archean spinifex rocks. *Geol. Soc. Am. Bull., 85*, 1721–1726.

Donaldson, C. H., 1976. An experimental investigation of olivine morphology. *Contrib. Mineral. Petrol., 57*, 187–213.

Donaldson, C. H., 1977. Laboratory duplication of comb layering in the Rhum pluton. *Mineral. Mag., 41*, 323–336.

Donaldson, C. H., 1979. An experimental investigation of the delay in nucleation of olivine in mafic magas. *Contr. Mineral. Petrol.,* v. 69, 21–32.

Donaldson, C. H., T. M. Usselman, R. J. Williams, and G. E. Lofgren, 1975a. Experimental modeling of the cooling history of Apollo 12 olivine basalts. *Proc. 6th Lunar Sci. Conf.,* 843–870.

Donaldson, C. H., R. J. Williams, and G. E. Lofgren, 1975b. A sample holding technique for study of crystal growth in silicate melts. *Am. Mineral., 60*, 324–326.

Donaldson, C. H., H. I. Drever, and R. Johnston, 1976. Supercooling on the lunar surface: a review of analogue information. *Phil. Trans. R. Soc. Lond. A., 285*, 207–217.

Downes, M. J., 1973. Some experimental studies on the 1971 lavas from Etna. *Phil. Trans. R. Soc. Lond. A., 274*, 55–62.

Dowty, E., 1976. Crystal structure and crystal growth: II. Sector zoning in minerals. *Amer. Mineral, 61*, 460–469.

Dowty, E., K. Keil, and M. Prinz, 1974. Lunar pyroxene-phyric basalts: crystallization under supercooled conditions. *J. Petrol., 15*, 419–453.

Drever, H. I., and R. Johnston, 1957. Crystal growth of forsteritic olivine in magmas and melts. *Royal Soc. Edin. Trans., 63*, 289–315.

Edgar, A. D., 1973. *Experimental Petrology,* Clarendon Press, Oxford.

Eichelberger, J., T. R. McGetchin, and D. Francis, 1973. Mode of emplacement of a basalt flows at Pacaya, Guatemala. *EOS Trans. AGU, 54*, 511.

Fenn, P. M., 1974. Nucleation and growth of alkali feldspars from a melt, in *The Feldspars,* W. S. Mackenzie and J. Zussman, (eds.), Manchester University Press, Manchester, 360–361.

Fenn, P. M., 1977a. The nucleation and growth of alkali feldspars from hydrous melts. *Canadian Mineral., 15*, 135–161.

Fenn, P. M., 1977b. The nucleation and growth of albite from the melt in the presence of excess SiO_2 (abstract). *Geol. Soc. Amer. Abs. with Programs, 9,* 973–974.

Fleet, M. E., 1975a. The growth habits of olivine—a structural interpretation. *Canadian Mineral., 13,* 293–297.

Fleet, M. E., 1975b. Growth habits of clinopyroxene. *Canadian Mineral., 13,* 336–341.

Fouqué, F., and A. Michel-Lévy, 1882. *Synthese des Minéraux et des Roches,* Paris.

Gibb, F. G. F., 1974. Supercooling and the crystallization of plagioclase from a basaltic magma. *Mineral. Mag., 39,* 641–653.

Green, J. C., 1970. Snowflake texture not diagnostic of devitrified ash-flow tuffs: discussion. *Geol. Soc. Amer. Bull., 81,* 2527–2528.

Grove, T. L., and A. E. Bence, 1977. Experimental study of pyroxene-liquid inter-action in quartz-normative basalt 15597. *Proc. 8th Lunar Sci. Conf.,* 1549–1580.

Grove, T. L., and D. Walker, 1977. Cooling histories of Apollo 15 quartz normative basalts. *Proc. 8th Lunar Sci. Conf.,* 1501–1520.

Gutmann, J. T., 1977. Textures and genesis of phenocrysts and megacrysts in basaltic lavas from the Pinacate volcanic field. *Amer. J. Sci., 277,* 833–861.

Hall, J., 1805. Experiments on whinstone and lava, 1798. *Trans. R. Soc. Edin., 5,* 43–76.

Harloff, C., 1927. Zonal structure in plagioclase. *Leidsche Geol. Medeleel., 2,* 99–114.

Hollister, L. S., W. E. Trzcienski, Jr., R. B. Hargraves, and C. G. Kulick, 1971. Petrogenetic significance of pyroxenes in two Apollo 12 samples. *Proc. 2nd. Lunar Sci. Conf.,* 529–557.

Hopper, R. W., and D. R. Uhlmann, 1974. Solute redistribution during crystal-lization at constant velocity and constant temperature. *J. Crystal Growth, 21,* 203–213.

Jaeger, J. C., 1968. Cooling and solidification of igneous rocks, in *Basalts: The Poldervaart Treatise on Rocks of Basaltic Composition,* H. H. Hess and A. Poldervaart, (eds.), 2, pp. 503–536, Wiley-Interscience, New York.

Jahns, R. H., and C. W. Burnham, 1958. Experimental studies of pegmatite genesis: Melting and crystallization of granite and pegmatite (abstract). *Geol. Soc. Amer. Bull., 69,* 1592.

Keith, H. D., and F. J. Padden, 1963. A phenomenological theory of spherulitic crystallization. *J. Appl. Phys., 34,* 2409–2421

Kirkpatrick, R. J., 1974. Kinetics of crystal growth in the system $CaMgSi_2O_6$–$CaAl_2SiO_6$. *Amer. J. Sci., 274,* 215–242.

Kirkpatrick, R. J., 1975. Crystal growth from the melt: a review. *Amer. Mineral., 60,* 798–814.

Kirkpatrick, R. J., R. R. Gilpin, and J. F. Hays, 1976. Kinetics of crystal growth from silicate melts: Anorthite and diopside. *J. Geophys. Res., 81,* 5715–5720.

Kitamura, M., and I. Sunagawa, 1977. The effective distribution of solute and the anisotropy of a crystal. *Internat. Conf. Crystal Growth, Collected Abstracts, 5,* 91.

Klein, L., and D. R. Uhlmann, 1974. Crystallization behavior of anorthite. *J. Geophys. Res., 79,* 4869–4874.

Klusman, R. W., 1972. Calcium fractionation in zoned plagioclase from the Tobacco Root batholith, southwestern Montana. *Chem. Geol., 9,* 45–56.

Lewis, J. D., 1972. "Spinifex texture" in a slag, as evidence for its origin in rocks, *Western Australia Geol. Surv. Ann. Report*, 60–68.

Liou, J. G., 1974. Mineralogy and chemistry of glassy basalts, Coast Range Ophiolites, Taiwan, *Geol. Soc. Am. Bull.*, *85*, 1–10.

Lofgren, G. E., 1971a. Experimentally produced devitrification textures in natural rhyolite glass. *Geol. Soc. Amer. Bull.*, *82*, 111–124.

Lofgren, G. E., 1971b. Spherulitic textures in glassy and crystalline rocks. *J. Geophys. Res.*, *76*, 5635–5648.

Lofgren, G. E., 1973. Experimental crystallization of synthetic plagioclase at prescribed cooling rates (abstract). *EOS Trans. AGU*, *54*, 482.

Lofgren, G. E., 1974a. An experimental study of platioclase crystal morphology: isothermal crystallization. *Amer. J. Sci.*, *274*, 243–273.

Lofgren, G. E., 1974b. Temperature induced zoning in synthetic plagioclase feldspar, in *The Feldspars*, W. S. Mackenzie and J. Zussman, (eds.), Manchester University Press, Manchester, 362–375.

Lofgren, G. E., 1975. Dynamic crystallization experiments on mare basalts, in *Origins of Mare Basalts*, Lunar Science Institute, Houston, 99–103.

Lofgren, G. E., 1976. Nucleation and growth of feldspar in dynamic crystallization experiments (abstract). *Geol. Soc. Amer. Abs.* with *Programs*, *7*, 982.

Lofgren, G. E., 1977. Dynamic crystallization experiments bearing on the origin of textures in impact generated liquids. *Proc. 8th Lunar Sci. Conf.*, 2079–2095.

Lofgren, G. E., 1978. An experimental study of intersertal and subophitic texture (abstract). *EOS Trans. AGU*, *59*, 396.

Lofgren, G. E., and C. H. Donaldson, 1975a. Phase relations and nonequilibrium crystallization of ocean ridge tholeiite from the Nazca plate (abstract). *EOS Trans. AGU*, *56*, 468.

Lofgren, G. E., and C. H. Donaldson, 1975b. Curved, branching crystals and differentiation in comb-layered rocks. *Contrib. Mineral. Petrol.*, *49*, 309–319.

Lofgren, G. E., and R. Gooley, 1977. Simultaneous crystallization of feldspar intergrowths from the melt. *Amer. Mineral.*, *62*, 217–228.

Lofgren, G. E., D. Smith, and R. W. Brown, 1978. Dynamic melting and crystallization studies on a lunar soil. *Proc. 9th Lunar Sci. Conf.*, 959–975.

Lofgren, G. E., C. H. Donaldson, R. J. Williams, O. Mullins, Jr., and T. M. Usselman, 1974. Experimentally reproduced textures and mineral chemistry of Apollo 15 quartz normative basalts. *Proc. 5th Lunar Sci. Conf.*, 549–567.

Lofgren, G. E., C. H. Donaldson, and T. M. Usselman, 1975a. Geology, petrology, and crystallization of Apollo 15 quartz normative basalts. *Proc. 6th Lunar Sci. Conf.*, 79–99.

Lofgren, G. E., R. J. Williams, C. H. Donaldson, and T. M. Usselman, 1975b. An experimental investigation of porphyritic texture (abstract). *Geol. Soc. Amer. Abs. with Programs*, *7*, 1173–1174.

Long, P. E., 1976. Precambrian granitic rocks of the Dixon-Peñasco area, northern New Mexico: a study in contrasts. Unpublished Ph.D. thesis, Stanford University.

Long, P. E., 1978. Experimental determination of partition coefficients for Rb, Sr, and Ba between alkali feldspar and silicate liquid. *Geochim. Cosmochim. Acta*, *42*, 833–846.

Luth, W. C., 1976. Granitic rocks, in *The Evolution of the Crystalline Rocks*, D. K. Bailey and R. Macdonald, (eds.), Academic Press, New York. 333–417.

Morse, S. A., 1969. Ternary feldspars, *Carnegie Inst. Wash. Year Book*, *67*, 124–126.

Morse, S. A., 1970. Alkali feldspars with water at 5 kb pressure. *J. Petrol.*, *11*, 221–251.

Morse, S. A., and G. E. Lofgren, 1978. Simultaneous crystallization of feldspar intergrowths from the melt, a discussion. *Amer. Mineral.*, *63*, 419–421.

Nabelek, P. I., L. A. Taylor, and G. E. Lofgren, 1978. Nucleation and growth of plagioclase and development of textures in a high-alumina basaltic melt. *Proc. 9th Lunar Sci. Conf.*, 725–741.

Naney, M. T., 1975. Nucleation and growth of biotite and clinopyroxene from haplogranitic magmas (abstract). *EOS Trans. AGU*, *56*, 1075.

Presnall, D. C., and N. L. Brenner, 1974. A method for studying iron-silicate liquids under reducing conditions with negligible iron loss. *Geochim. Cosmochim. Acta*, *38*, 1785–1788.

Ribould, R. V., and A. Muan, 1962. Phase equilibria in a part of the system "FeO"-MnO-SiO_2, *Trans. Metall. Soc. AIME*, *224*, 27–33.

Roeder, P. L., and R. F. Emslie, 1970. Olivine-liquid equilibrium. *Contrib. Mineral. Petrol.*, *29*, 275–289.

Sato, M., 1971. Electrochemical measurements and control of oxygen fugacity and other gaseous fugacities with solid electrolyte sensors, in *Research Techniques for High Pressure and Temperature*, G. C. Ulmer, (ed.), Springer-Verlag, New York, 43–99.

Shulz, K. J., 1977. The petrology and geochemistry of Archean volcanics, western Vermilion district, northeastern Minnesota. Unpublished Ph.D. thesis, University of Minnesota.

Sibley, D. F., T. A. Vogel, B. M. Walker, and G. Byerly, 1976. The origin of oscillatory zoning in plagioclase: a diffusion and growth controlled model. *Amer. J. Sci.*, *276*, 275–284.

Stolper, E., 1977. Experimental petrology of eucritic meteorites. *Geochim. Cosmochim. Acta*, *41*, 587–611.

Swanson, S. E., 1977. Relation of nucleation and crystal-growth rate to the development of granitic textures. *Amer. Mineral.*, *62*, 966–978.

Taubeneck, W. H., and A. Poldervaart, 1960. Geology of the Elkhorn Mountains, Northeastern Oregon: Part 2. Willow Lake intrusion. *Geol. Soc. Amer. Bull.*, *71*, 1295–1322.

Torske, T., 1975. Snowflake texture in Ordovician rhyolite from Stord. Hordaland. *Norges Geol. Unders.*, *319*, 17–27.

Turnbull, D., 1948. Transient nucleation, *Trans. AIME*, *175*, 774–783.

Ulmer, G. C., (ed.), 1971. *Research Techniques for High Pressure and High Temperature*, Springer-Verlag, New York.

Usselman, T. M., and G. E. Lofgren, 1975. Crystallization of Mare Basalts: Pyroxene Zoning (abstract). *EOS Trans. AGU*, *56*, 471.

Usselman, T. M., and G. E. Lofgren, 1976. The phase relations, textures, and mineral chemistries of high-titanium mare basalts as a function of oxygen fugacity and cooling rate. *Proc. 7th Lunar Sci. Conf.*, 1345–1363.

Usselman, T. M., G. E. Lofgren, C. H. Donaldson, R. J. Williams. 1975. Experimentally reproduced textures and mineral chemistries of high-titanium mare basalts. *Proc. 6th Lunar Sci. Conf.*, 997–1020.

Vance, J. A., 1962. Zoning in igneous plagioclase: normal and ocillatory zoning. *Amer. J. Sci.*, *260*, 746–760.

Walker, D., R. J. Kirkpatrick, J. Longhi, and J. F. Hays, 1976. Crystallization history of lunar picrite basalt sample 12002: phase equilibria and cooling-rate studies. *Geol. Soc. Amer. Bull.*, *87*, 646–656.

Watt, G., 1804. Observations on basalt, and on the transition from the vitreous to the stony texture, which occurs in the gradual refrigeration of melted basalt; with some geological remarks. *Phil. Trans. R. Soc. Lond.*, *5*, 279–314.

Williams, R. J. and O. Mullins, Jr., 1976. *A system using solid ceramic oxygen electrolyte cells to measure oxygen fugacities in gas-mixing systems.* NASA TMX-58167, Houston.

Wyllie, P. J., 1963. *Effects of the changes in slope occurring on liquidus and solidus paths in the system diopside-anorthite-albite.* Mineral. Soc. Amer. Sp. Paper 1, 204–212.

AUTHOR INDEX

The indexes were compiled by the editor with the kind help of B.W. Bathurst, S.C. Bergman, D.P. Bhattacharyya, M.S. Chyi, M.C. Donnelly, B.J. Douglas, A.M. Hargraves, C.A. Lawson, W.B. Maze, J.A. Miller, B.W. Murck, V.B. Sisson. They were typed ready for reproduction on a Vydec 1400 word processer by Sally Shaginaw.

SUBJECT INDEX

LIBRARY OF CONGRESS CATALOGING IN PUBLICATION DATA
Main entry under title:

Physics of magmatic processes.

Includes indexes.
1. Magmatism.   I.   Hargraves, Robert B.
QE461.P59   552'.3   80-7525
ISBN 0-691-08259-6
ISBN 0-691-08261-8 (pbk.)